A-PLUS NOTES
For
SAT MATH

One Book Covers All the SAT Math You'll Ever Need
(SAT / PSAT and Subject Test Levels 1 & 2)

Aligned To the College Board Official Guidance

By
Rong Yang

A-Plus Notes Learning Center

Library of Congress Control Card Number: 2008904794
Washington, DC 20540-4320.

ISBN–13: 978-09654352-6-0
ISBN–10: 09654352-6-1

A-Plus Notes Learning Center
A publisher
Redondo Beach, CA 90278, USA
Tel: (310) 542-0029
Fax: (310) 542-6837

A-Plus Notes Book Company
Distributor in Asia
Taipei, Taiwan
Tel: (02) 2631–5976
Fax: (02) 2631–6345

WHERE TO BUY:

Available at Barnes & Noble and local educational bookstores across the United States, such as Teacher Supply, School Supply,····, etc.

HOW TO ORDER:

Quantity discounts are available from the publisher. Please include information on your letterhead concerning the number of books you wish to purchase. See last page for the order form.

Revised and Updated, 2nd Printing

PREFACE

I am a math teacher in Los Angeles. In order to provide a clear and easy way for my students to learn math in my classes, I designed **notes** which simplify and outline the concepts and skills in course textbooks. The notes focus on the correct and easy steps in the problem-solving process. Each topic is organized in one or two pages of explanation for easy understanding. I wrote this book following the same pattern of problem-solving in math books used by schools across Asia. This pattern has been proven to be excellent for student-learning.

The SAT program and tests are designed by the College Board. The College Board is an association of colleges whose aim is to assist the admissions or placement of new students in colleges. Educational Test Service (ETS) develops and administers the SAT tests for the College Board. The tests questions are written by the experts and reviewed by the SAT Committee.

This book focuses on the **Math** portion of SAT Tests (SAT/PSAT and Subject Tests). The content is simple and easy to read like a math dictionary for the students to study and review all the math topics and skills necessary to ace the SAT Math Tests. This book will help you to be familiar with different types of questions on the tests and facilitate the learning process. If you are not familiar with a specific topic in math, you can easily find it and learn it in this book.

Many SAT test-prep books emphasize a variety of test-taking strategies. I believe the best strategy for a student's success is to practice , practice, and more practice. Since SAT Math Tests are not tied to specific textbooks or teaching methods, the more practice tests a student does, the better the students can ace the test on the real test day. The practice tests in this book can help you find out which areas you need to work on. The content of this book covers all the math you need to review for the SAT Math tests.

The goal of this book is to give the students the tools to master the SAT Math tests with confidence and optimism. If you are a math professional and have comments or suggestions regarding the content of this book, you can fax your comments to me at (310) 542-6837.

Rong Yang
Los Angeles, California

HOW TO USE THIS BOOK

1. Questions are sometimes repeated in actual SAT Math Tests. Therefore, you should study and get used to all of the test questions in this book.
2. Read the additional up-to-date sample questions at the end of each chapter.
3. If you want more practice tests, other SAT practice books are available in the libraries or bookstores. Use this book as your SAT Math dictionary when you need help.
4. If there are any math areas or topics you cannot find in this book or you need to learn how to use the graphing calculator, check out A-Plus Notes Series:
   "A-Plus Notes for Algebra, A Graphing Calculator Approach"

HOW TO STUDY THE PRACTICE TESTS IN THIS BOOK

1. Work on each practice test, check your answers, and count your score using the answer key at the end of each test.
2. There are many overlap in math areas between SAT / PSAT and Subject Tests. work on the 10 practice tests in this book from test #1 up to your grade level.
3. In the practice tests in this book, similar questions of different cases are often put together to study them simultaneously, (See page 350, 360, 388, and 392)
4. If your answer to a question is incorrect, find out how to reach the right answer by reading the related areas and topics in this book. You can find the areas and topics on Pages 7 and 8, Table of Contents on Pages 9~12, and Index on Page 465.
5. If you can not find how to reach the right solutions to any questions in the practice tests, you should ask for help from your classmates or math teachers. Or, write down your questions and fax them to the author for a free consultation.
   Fax number: Mr. Yang (310) 542-6837.

ATTENTION: MATH TEACHERS AND STUDENTS

This book will make the teachers' job and the students' learning process easier if the students have it on hand in class.

It makes a teacher's job easier because the teacher need not " write notes " on the board in the class. It makes a student's learning process easier because the student needs not spend time to " take notes " in the class. This book is a complete set of lecture "notes" for the students. They only need to read it.

Most schools in the United States take back the textbook from the student after the course is completed. The student has no material on hand for future review. This book is simple and easy like a math dictionary for the student to review when needed in the future.

This book is strongly recommended for math teachers and students. You may use it as a reference book together with the textbook, or use it as the textbook in your classes.

You may visit this book and readers' comments, or write your comments at
www.amazon.com or www.barnes & noble.com

The Publisher

UNDERSTANDING THE SAT / PSAT MATH TESTS

1. The SAT / PSAT are **reasoning tests** including three parts (critical reading, writing, and math). SAT is a 3-hour and 45-minute test to complete all three parts. PSAT is a practice test for SAT.

2. Most colleges use SAT reasoning tests as one of the factors to assist the admissions and placements of new students.

3. SAT reasoning tests are given seven times a year around the world. You can search for information and SAT Registration Booklet about the tests at www.collegeboard.com.

4. In the **SAT MATH** section, you have 70 minutes to complete 54 questions. Each multiple-choice question has five choices. The three parts of the math section include:
   a. 20 multiple-choice questions (25 minutes)
   b. 8 multiple-choice questions and 10 grid-in questions (25 minutes)
   c. 16 multiple-choice questions (20 minutes)

5. The use of calculators is allowed. The answers in the test are rounded. Do not round any calculation at any intermediate step before you have the final answer. In the grid-in questions, do not grid "0" before the decimal point. Grid ".5" or ".50" for 0.50.

6. You get one point for each right answer. You lose one-fourth (0.25) of a point for each wrong answer on multiple-choice questions. Nothing is subtracted for wrong answers on grid-in questions. No points are added or subtracted for unanswered questions. If your final score includes a decimal, the score is rounded to the nearest integer. The point you obtained is your **raw score** which ranges from −11 to 54.

7. **Raw scores** are converted to **scaled scores** which provide the information of your math skills on which colleges can compare your level background with those of students who have taken difference tests at difference dates. The final scores range from 200 to 800.

8. Taking the test, you may answer only the easy questions your are sure of, and skip the harder questions you can't answer. Guessing could lower your score unless you can eliminate one or two choices that you know are wrong.

9. For the simple questions when you understand what the question is asking, it may be easier to find the answer by "testing each choice" or "eliminating choices". If you finish before time is called, you should check your work or try a few of the harder questions you skipped.

10. Taking the PSAT/NMSQT (Preliminary SAT/National Merit Scholarship Qualifying Test) is the best practice for real SAT reasoning test. It is a 2-hour and 10-minute test including critical reading, writing, and math. You have 50 minutes to complete 38 math questions. There are 20 multiple-choice questions for first 25 minutes, 8 multiple-choice questions and 10 grid-in questions for second 25 minutes. The math scores range from 20 to 80. The test can qualify you to apply for scholarships.

UNDERSTANDING THE SAT MATH SUBJECT TESTS
(Levels 1 & 2)

1. The SAT **Subject Tests** measure how well you know a particular subject area which you will need in college, such as Biology, Physics, Chemistry, ···. You can take up to three Subject Tests on a single test day.

2. The SAT Math Subject Tests include Level 1 and Level 2.

3. Most colleges use Math Level 1 to assist the admissions or placements of new students. The Math level 2 test is needed if you are interested in a college of science or technology.

4. The SAT Math Subject Tests are given about seven times a year around the world. You can search for information about which Subject Test is needed for colleges that interest you and SAT Registration Booklet about the tests at www.collegeboard.com.

5. A graphing calculator is allowed on both Math Subject Tests. Read the book "A-Plus Notes for Algebra". It covers many graphing calculator instructions.

6. The answers in the test are rounded. Do not round any calculation at any intermediate step before you have the final answer.

7. Each Math Subject Test has 50 multiple-choice questions in which you must choose the **BEST** of the five choices given. You have one hour to complete it. You get one point for each right answer. You lose one-fourth (0.25) of a point for each wrong answer.
   No points are added or subtracted for unanswered questions. If your final score includes a decimal, the score is rounded to the nearest integer.
   The point you obtained is your raw score which ranges from -13 to 50 .

8. **Raw scores** are converted to **scaled scores** which provide the information of your math skills on which colleges can compare your **level** background with those of students who have taken different tests at different dates. The final scores range from 200 to 800.

9. Taking the test, you may answer only the easy questions you are sure of, and skip the harder questions you can't answer. Guessing could lower your score unless you can eliminate one or two choices that you know are wrong.

10. For the simple questions when you understand what the question is asking, it may be easier to find the answer by "testing each choice" or "eliminating choices". If you finish before time is called, you should check your work or try a few of the harder questions you skipped.

Areas and Topics Covered

(SAT / PSAT Math Tests)

The SAT / PSAT Math Tests cover the following areas:

1. Number and Operations **2.** Algebra and Functions
3. Geometry and Measurements **4.** Data Analysis, Statistics, and Probability

The SAT/PSAT Math Tests cover the following areas and topics. You are not expected to have studied every topics on the tests.

Areas	SAT/ PSAT Topics	Pages
Number and Operations 15%	Integers and Number Lines, Word Problems	13~15
	Number Theory (Factors, Multiples, Remainders, Prime Numbers)	17~20
	Ratios, Proportions, Percents, Fractions, Rational Numbers	21~27
	Powers and Exponents, Square Roots, Radicals	28~32
	Sequences, Sets, Logical Reasoning, Counting Principal	34, 36, 37, 38
Algebra and Functions 40%	Operations on Algebraic Expressions	61
	Solving Equations and Direct Translation of Word Problems	62~68
	Linear Equations, Functions, Graphs, New Definitions, Domain & Range	69~75, 148~149
	Direct Variations	76
	Inequalities & Absolute Values	77~84
	Quadratic Equations, Functions, Graphs	85~88
	Translations (Transformations) of Graphs	97, 100
	Rational Equations and Inequalities	111~117
	Inverse Variations	118
	Radical (Exponential) Equations, Functions, Graphs	119~121
	Systems of linear Equations and Inequalities	122~125
Geometry and Measurements 35%	Terms, Notations, Points, Lines, Angles	151~158
	Triangles (Equilateral, Isosceles, Right, Triangle-Inequality)	159~162
	Pythagorean Theorem	163~164
	Special Right Triangles (30-60-90 and 45-45-90)	165~166
	Similar and Congruent Triangles	167~168
	Polygons, Quadrilaterals, Rectangles, Squares	169~172
	Circles (Radius, Diameter, Circumference, Arc, Area, Tangent Line)	173~176
	Areas and Perimeters (Squares, Rectangles, Triangles, Parallelograms, Polygons)	177~180
	Coordinate Geometry (Slopes, Parallel and Perpendicular lines, Distance and Midpoint)	71~75 181~182
	Symmetry and Transformations (Translation, Rotation, Reflection)	191~194
	Solid Geometry (Surface Area and Volume of Cylinders, Prisms, Cones, Pyramids, Spheres)	195~208
Data Analysis, Statistics, and Probability 20%	Mean (Average), Median, Mode, Range	281~282
	Graphs and Plots (Line, Bar, Pictograph, Circle or Pie, scatterplot)	285~258
	Probability (Independent and Dependent Events, Geometric Figures)	291~296

Source: College Board Official Guidance of PSAT & SAT Math Reasoning Tests

Areas and Topics Covered
(SAT Math Subject Test Levels 1 / 2)

The SAT Math Subject Test Levels 1 / 2 cover the following areas:

1. Number and Operations
2. Algebra and Functions (Add Pre-Calculus for Level 2)
3. Geometry and Measurements
4. Trigonometry
5. Data Analysis, Statistics, and Probability

The SAT Math Subject Tests cover the following areas and topics. You are not expected to have studied every topics on the tests.

Areas	Levels 1 & 2 Topics	Level 2 Only
Number and Operations 15%	Ratio, Proportion & Percent, Number Theory Imaginary & Complex Numbers, Exponents Logarithms, Logical Reasoning, Counting Principal Permutation, Combination, Matrix, Determinant Arithmetic and Geometric Sequences	Arithmetic Series Geometric Series Infinite Geometric Series Vectors
Algebra and Functions 40%	Variable and Expressions, Equations, Inequalities Representation and Modeling of Word Problems Linear Functions, Polynomial Functions Rational Functions, Exponential Functions	Logarithmic Functions Piecewise-Defined Function Recursive Functions for sequences Parametric Equations Pre-Calculus of Elementary Functions (Limits, Continuity, Derivatives)
Geometry and Measurements 35%	Plane (Euclidean) Geometry & Measurements Coordinate (Analytic) Geometry (Line, Parabola, Circle, Symmetry, and Transformations) Solid Geometry in Three-Dimensions (Surface Area and Volume of Cylinders, Prisms, Cones, Pyramids, Spheres)	Coordinate (Analytic) Geometry (Ellipses, Hyperbolas) Polar Coordinates Coordinate (Solid Analytic) Geometry in Three-Dimensions
Trigonometry 10%	Degree Measure of Angles Six Trigonometric Functions in a Right Triangle Solving a Right Triangle Trigonometric Identities	Radian Measure of Angles Law of Sines Law of Cosines Double-Angle Formulas Periodic (Trigonometric) Functions Trigonometric Equations Inverse Trigonometric Functions
Data Analysis, Statistics, and Probability 10%	Mean, Median, Mode, Range, Interquartile Range Graphs and Plots Probability Least-Squares Regression (Linear)	Standard Deviation Least-Squares Regression (Quadratic and Exponential)

Source: College Board Official Guidance of SAT Math Subject Tests

Table of Contents

Review the Topics and Related Skills

(SAT Math Subject Test Levels 1 / 2)

There is some overlap between both tests.
You are not expected to have studied every topic on the tests.

Chapter 1: NUMBER and OPERATIONS

Levels 1 & 2

Level 2 Only

Table of Contents

Chapter 2: ALGEBRA and FUNCTIONS

Levels 1 and 2

Level 2 Only

Table of Contents

Chapter 3: GEOMETRY and MEASUREMENTS

Chapter 4: TRIGONOMETRY

Table of Contents

Chapter 5: DATA ANALYSIS, STATISTICS, and PROBABILITY

NUMBER and Operations

1-1 Real Numbers and Number Line

In mathematics, the following numbers are called **real numbers**:

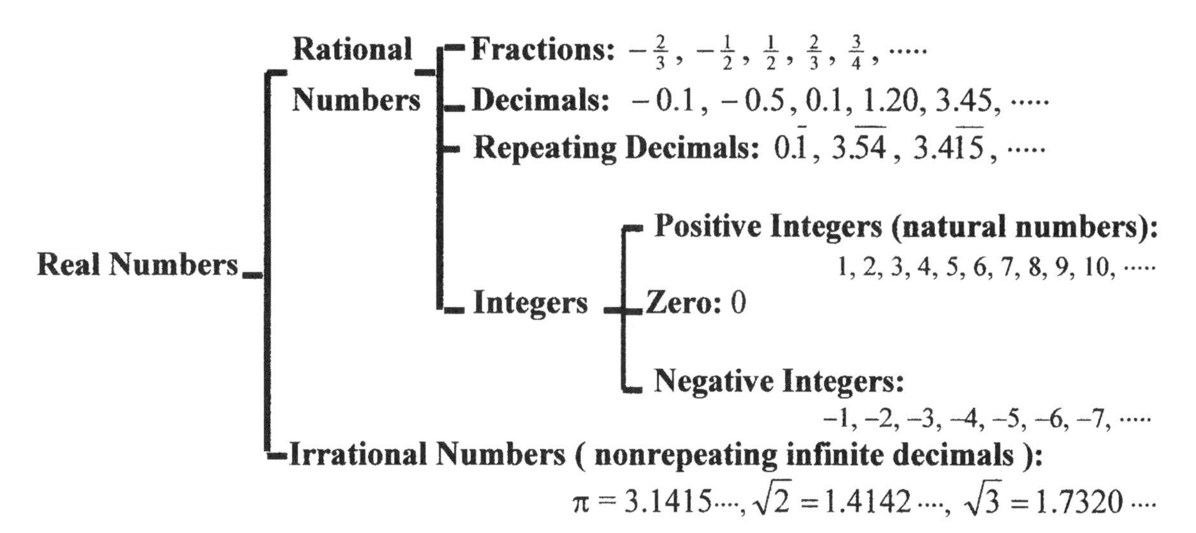

Definition of Real Numbers: They are numbers which can be located on the number line.

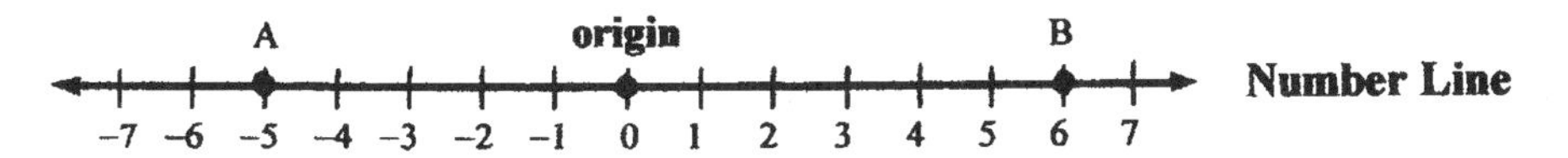

Real numbers include **rational numbers** and **irrational numbers**.

Rational Numbers: They are numbers which can be written as the ratio of two integers.

$0.1 = \frac{1}{10}$, $0.5 = \frac{5}{10} = \frac{1}{2}$, $1.2 = \frac{12}{10} = \frac{6}{5}$, $0.\overline{1} = \frac{1}{9}$, $0.\overline{54} = \frac{54}{99} = \frac{6}{11}$.

Every rational number can be expressed as either a terminating decimal or as a repeating decimal.

Irrational Numbers: They are numbers which cannot be written as the ratio of two integers. They are nonrepeating, and nonterminating (infinite) decimals.

Integers: They are positive integers, negative integers, and zero. Zero is an integer, neither positive nor negative.

Positive Integers: They are numbers which we use to count objects. (1, 2, 3, 4, 5, 6, ·····) They are also called **natural numbers** or **counting numbers**.

Whole Numbers: They are 0 and positive integers. (0, 1, 2, 3, 4, 5, 6, ·····)

Prime Number: It is a whole number other than 0 and 1 which is divisible only by 1 and itself. (2, 3, 5, 7, 11, 13, 17, 19, 23, 29, 31, 37, 41, ·····)

Composite Number: It is a whole number other than 0 and 1 which is not a prime number. (4, 6, 8, 9, 10, 12, 14, 15, 16, 18, 20, 21, 22, 24, ·····)

1-2 Order of Operations

To simplify an expression with numbers, we use the following **order of operations**:

1) Do all operations within the grouping symbols first, start with the innermost grouping symbols.
2) Do all operations with exponents.
3) Do all multiplications and divisions in order from left to right.
4) Do all additions and subtractions in order from left to right.

In mathematics, we have agreed the above order of operations to ensure that there is exactly one correct answer.

It is important that we always multiply and divide before adding and subtracting.

Grouping Symbols: A grouping symbol is used to enclose an expression that should be simplified first. We first simplify the expression in the innermost grouping symbol, then work toward the outmost grouping symbol.
Grouping symbols are also called **inclusion symbols**.

Parentheses ( ), Brackets [ ], Braces { },
Fraction Bar ——

Examples

1. $1+2\cdot3=1+6=7$ **2.** $(1+2)\cdot3=3\cdot3=9$ **3.** $2\cdot3-4=6-4=2$

4. $4-6\div2=4-3=1$ **5.** $(4-6)\div2=-2\div2=-1$ **6.** $2(3-4)=2(-1)=-2$

7. $3+2^3=3+8=11$ **8.** $(3+2)^3=5^3=125$ **9.** $3-3^3=3-27=-24$

10. $2^3\cdot3^2=8\cdot9=72$ **11.** $2^3-3^2=8-9=-1$ **12.** $2^3+3^2=8+9=17$

13. $1^5+2\cdot3^4=1+2\cdot81=1+162=163$ **14.** $12-35\div7+2=12-5+2=7+2=9$

15. $7-\frac{72-30\div5}{3}+2=7-\frac{72-6}{3}+2=7-\frac{66}{3}+2=7-22+2=-15+2=-13$

16. $2+3\{4-[6-2(3-1)]\div2\}$
$=2+3\{4-[6-2(2)]\div2\}$
$=2+3\{4-[6-4]\div2\}$
$=2+3\{4-2\div2\}$
$=2+3(4-1)$
$=2+3(3)=2+9=11$

17. $3-2\{5-4^2\cdot[2-3(1-3)]\}$
$=3-2\{5-16\cdot[2-3(-2)]\}$
$=3-2\{5-16\cdot[2+6]\}$
$=3-2\{5-16\cdot8\}$
$=3-2(5-128)$
$=3-2(-123)=3+246=249$

18. Joe finished 15 assignments and 3 tests (which he received 2 A's and 1 B) in math class. Each assignment earns him 2 points. Each test he can earn 10 points for an A and 7 points for a B. What is his total point ?
Solution:

$$15\times2+10\times2+7\times1=30+20+7=57 \text{ points}$$

1-3 Scale Drawings and Scale Factors

Scale Drawing: It is a reduced or enlarged figure that is similar to the actual object.
Scale Factor: It is a ratio of the measurement of a scale drawing to the actual measurement.

If two geometric figures are similar (the same shape), then the ratios of all corresponding dimensions (length, width, height, radius, circumference) are equal. These ratios in lowest terms are called the **scale factors**.
If the scale factor of two similar figures is $a : b$ (or $\frac{a}{b}$) , we have the following formulas:

The ratio of the **surface areas**: $\frac{A_1}{A_2} = \left(\frac{a}{b}\right)^2$; The ratio of the **volumes**: $\frac{V_1}{V_2} = \left(\frac{a}{b}\right)^3$

Examples

1. The scale of the drawing on a map is 1 inch = 10 inches. What is the actual length of a line if the length on the map is 8 inches ?
Solution:

Actual length $= 8 \times 10 = 80$ inches

2. The scale of the drawing on a map is $\frac{1}{2}$ inch = 10 inches. What is the actual length of a line if the length on the map is $5\frac{1}{2}$ inches ?
Solution:

The scale is 1 in. = 20 in.. The actual length $= 5\frac{1}{2} \times 20 = 110$ in.

3. The measurement on the scale drawing is 4 inches. The actual measurement is 5 yards. What is the scale factor ? (Hint: 1 yard = 36 inches)
Solution:

Scale factor $= \frac{4}{5 \times 36} = \frac{4}{180} = \frac{1}{45}$

4. The following two rectangular solids are similar,

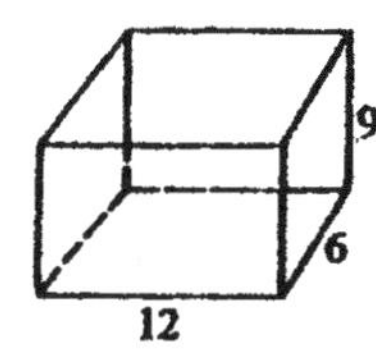

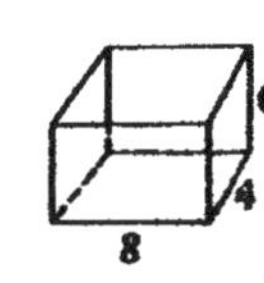

a) what is the ratio of their corresponding sides ?
b) what is the ratio of their surface areas ?
c) what is the ratio of their volumes ?

Solution: a) $\frac{a}{b} = \frac{6}{4} = \frac{9}{6} = \frac{12}{8} = \frac{3}{2}$ It is the **scale factor**.

b) $\frac{A_1}{A_2} = \frac{2(12 \cdot 6 + 12 \cdot 9 + 9 \cdot 6)}{2(8 \cdot 4 + 8 \cdot 6 + 6 \cdot 4)} = \frac{2(234)}{2(104)} = \frac{468}{208} = \frac{9}{4}$ **OR**: $\frac{A_1}{A_2} = \left(\frac{3}{2}\right)^2 = \frac{9}{4}$

c) $\frac{V_1}{V_2} = \frac{12 \cdot 6 \cdot 9}{8 \cdot 4 \cdot 6} = \frac{648}{192} = \frac{27}{8}$ **OR**: $\frac{V_1}{V_2} = \left(\frac{3}{2}\right)^3 = \frac{27}{8}$

1-4 Imaginary Numbers and Complex Numbers

To find the solutions of the equation $x^2 - 1 = 0$, we have the solutions $x = \pm\sqrt{1} = \pm 1$. However, there is no real number solution that satisfies the equation $x^2 + 1 = 0$. The solutions of the equation $x^2 + 1 = 0$ is $x = \pm\sqrt{-1}$. The radical $\sqrt{-1}$ is not a real number. There is no real number x whose square is -1. Therefore, we introduce a new mathematical symbol using the **imaginary unit**, denoted by i.

Definition of i: $i = \sqrt{-1}$ and $i^2 = -1$, where i is called "**the square root of** -1".

There is no real number square root of any negative number, such as:

$\sqrt{-1}$, $\sqrt{-2}$, $\sqrt{-4}$, $\sqrt{-9}$, $\sqrt{-10}$, ...→ $\sqrt[n]{b}$ if n is even and $b < 0$.

Examples: **1.** $\sqrt{-2} = \sqrt{2(-1)} = \sqrt{2} \cdot \sqrt{-1} = \sqrt{2}\, i$. **2.** $\sqrt{-4} = \sqrt{4(-1)} = \sqrt{4} \cdot \sqrt{-1} = 2i$.
3. $\sqrt{-9} = \sqrt{9(-1)} = \sqrt{9} \cdot \sqrt{-1} = 3i$. **4.** $\sqrt[3]{-8} = \sqrt[3]{-2 \cdot -2 \cdot -2} = -2$.

$i^2 = -1$, $\quad i^3 = i^2 \cdot i = -i$, $\quad i^4 = i^2 \cdot i^2 = (-1)(-1) = 1$, $\quad i^7 = i^4 \cdot i^3 = 1 \cdot -i = -i$

i **and** $-i$ **are reciprocals:** $\frac{1}{i} = \frac{1}{i} \cdot \frac{i}{i} = \frac{i}{i^2} = \frac{i}{-1} = -i$, $\quad \frac{1}{-i} = \frac{1}{-i} \cdot \frac{i}{i} = \frac{i}{-i^2} = \frac{i}{-(-1)} = i$.

Simplify each imaginary number, we must eliminate negative by i.

$\sqrt{-2} \cdot \sqrt{-18} = \sqrt{2}\, i \cdot \sqrt{18}\, i = \sqrt{36}\, i^2 = 6(-1) = -6$ is correct.

$\sqrt{-2} \cdot \sqrt{-18} = \sqrt{(-2)(-18)} = \sqrt{36} = 6$ is incorrect.

$\sqrt{a} \cdot \sqrt{b} = \sqrt{ab}$ does not hold if a and b are negative.

Complex Numbers: Numbers in the form $a + b\,i$ or $a - b\,i$, such as:

$1 + 2i$, $1 - 2i$, $1 + \sqrt{2}\, i$, $1 - \sqrt{2}\, i$, $5 - 7\sqrt{3}\, i$, $5\sqrt{2} + 7\sqrt{3}\, i$,.......

Conjugates: Numbers in the form $a \pm b\,i$, such as: $1 \pm 2i$, $5\sqrt{2} \pm 7\sqrt{3}\, i$ ....

The conjugate of $z = 2 - 5i$ is denoted by $\bar{z} = 2 + 5i$.

The product of two conjugate complex numbers is always a nonnegative real number (no imaginary part). $z\bar{z} = a^2 + b^2$

$$(a + b\,i)(a - b\,i) = a^2 - (b\,i)^2 = a^2 - b^2 i^2 = a^2 - b^2(-1) = a^2 + b^2$$

Examples

1. $i^{30} = (i^2)^{15} = (-1)^{15} = -1$ $\qquad$ **2.** $(2 + 3i) - (4 - 5i) = 2 + 3i - 4 + 5i = -2 + 8i$

3. $(2 + 3i)(2 - 3i) = 2^2 - (3i)^2 = 4 - 9i^2 = 4 - 9(-1) = 4 + 9 = 13$

4. $\frac{2+3i}{2-3i} = \frac{2+3i}{2-3i} \cdot \frac{2+3i}{2+3i} = \frac{4+12i+9i^2}{4-9i^2} = \frac{4+12i+9(-1)}{4-9(-1)} = \frac{-5+12i}{13} = -\frac{5}{13} + \frac{12}{13} i$

1-5 Number Theory

Number Theory is the study of the properties of whole numbers.
In this section, we discuss a few properties of whole numbers. We will discuss the other properties of whole numbers in the other sections in this chapter..
Whole numbers: They are 0 and positive integers (0, 1, 2, 3, 4, 5, 6, 7, 8, and so on).
Divisible: A whole number is said to be divisible by another number if the remainder is 0.
Even numbers: They are numbers which are divisible by 2 (such as 2, 4, 6, 8, 10, ·····).
Odd numbers: They are numbers which are not divisible by 2 (such as 1, 3, 5, 7, 9, ·····).
0 is considered to be an even number, neither positive nor negative.

Factor: The whole numbers we multiply are called the factors of their product.
$2 \times 6 = 12$ 2 and 6 are factors of 12.
The factors of 12 are 1, 2, 3, 4, 6, and 12.
To find the factors of a number, we divide that number by 1, 2, 3, ·····. If the reminder is 0, the divisor of a number is a factor of that number.
1 is the smallest factor of every whole number. The only factor of 1 is 1.
A number divided by 0 is undefined. Therefore, 0 is not a factor of any number.
0 divided by any nonzero integer is 0. Therefore, every nonzero integers is a factor of 0.

Multiple: A multiple of a number is the product of that number and any whole number.
To find the multiples of a number, we multiply that number by 0, 1, 2, 3, ·····.
The multiples of 3 are 0, 3, 6, 9, 12, 15, 18, 21, ·····.
The first five multiples of 6 are 0, 6, 12, 18, and 24.
0 is a multiple of any nonzero integer.

Conventionally, we apply factors and multiples on whole numbers (no negative) only.

A whole number is divisible (the remainder is 0) by each of its factors.
To find the factors of a number, sometimes it is helpful to test the **divisibility** by inspecting the patterns of the digits without actually dividing.

Divisibility Tests: A whole number is:
divisible by **2** : if its last digit is 0, 2, 4, 6, or 8.
divisible by **3** : if the sum of its digits is divisible by 3.
divisible by **4** : if its last two digits are divisible by 4.
divisible by **5** : if its last digit is 0 or 5.
divisible by **6** : if it is divisible by both 2 and 3.
divisible by **8** : if its last three digits are divisible by 8.
divisible by **9** : if the sum of its digits is divisible by 9.
divisible by **10** : if its last digit is 0.
divisible by **11** : if the difference between the sum of the odd digits and the sum of the even digits is 0 or 11.
25894 is divisible by 11 because $(2+8+4)-(5+9)=0$.
86801 is divisible by 11 because $(8+8+1)-(6+0)=11$.

1-6 Prime Factorizations

Prime Number: A whole number other than 0 and 1 which is divisible only by 1 and itself.
Here are the prime numbers: 2, 3, 5, 7, 11, 13, 17, 19, 23, 29, 31, 37,·····.
A prime number has exactly two factors, 1 and itself.

0 is divisible by every nonzero integer (0 has many factors). Therefore, 0 is not a prime number.
1 is divisible only by itself (1 has only one factor, itself). Therefore, 1 is not a prime number.

All even numbers greater than 2 are not prime numbers. 2 is the only even prime number.
Therefore, to identify whether or not an odd number is a prime number, we test the divisibility by the prime numbers 2, 3, 5, 7, 11, 13, 17, 19,·····.
121 is divisible by 11. Therefore, 121 is not a prime number.

Composite Number: A whole number other than 0 and 1 which is not a prime number.
Here are the composite numbers: 4, 6, 8, 9, 10, 12, 14, 15, 16, ·····.
A composite number has more than two factors.
The numbers 0 and 1 are neither prime nor composite.

Prime Factorization: It is to write a composite number as a product of prime numbers. We can factor a number into prime factors by using either of the following two methods:
Shortcut Division and **Factor Tree**

In prime factorization, we divide the number by a prime number (from small to large) until the bottom row is a prime number. Then, we write the prime factors in order from least to greatest.

Example: Find the prime factorization of 24.
Solution:
Method 1: **Shortcut Division** Method 2: **Factor Tree**

$$\begin{array}{r|l} 2 & 24 \\ \hline 2 & 12 \\ \hline 2 & 6 \\ \hline & 3 \end{array}$$

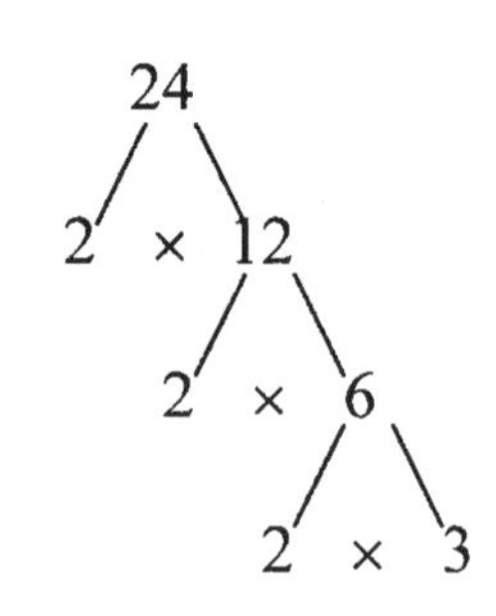

Answer: $24 = 2 \times 2 \times 2 \times 3 = 2^3 \times 3$.

Exponents are often used to express the prime factorization.

1-7 Greatest Common Factor & Least Common Multiple

Greatest Common Factor (GCF): It is the greatest number that is a factor of two or more given numbers.

Factors of 12: 1, 2, 3, 4, **6**, 12 ; Factors of 18: 1, 2, 3, **6**, 9, 18

The common factors of 12 and 18 are 1, 2, 3, **6**. The GCF of 12 and 18 is **6**.

To find the GCF of two or more given numbers, we write the prime factorizations of the given numbers, and find the **lowest power** of each factor that occurs in **all** of the factorizations. The product of these lowest powers is the GCF.

Examples

1. Find the GCF of 12 and 18.

Solution:

$$12 = 2^2 \times 3 \; ; \; 18 = 2 \times 3^2$$
$$\therefore \text{ GCF} = 2 \times 3 = 6$$

2. Find the GCF of 54 and 90.

Solution:

$$54 = 2 \times 3^3 \; ; \; 90 = 2 \times 3^2 \times 5$$
$$\therefore \text{ GCF} = 2 \times 3^2 = 18$$

Least Common Multiple (LCM): It is the smallest number that is a multiple of two or more given numbers.

Multiples of 12: 0, 12, 24, **36**, 48, 60, **72**, ···· , Multiples of 18: 0,18, **36**, 54, **72**, ····

The common multiples of 12 and 18 are **36**, **72**, ····. The LCM of 12 and 18 is **36**.

To find the LCM of two or more given numbers, we write the prime factorizations of the given numbers, and find the **highest power** of each factor that occurs in **any** of the factorizations. The product of these highest powers is the LCM.

Examples

3: Find the LCM of 12 and 18.

Solution:

$$12 = 2^2 \times 3, \quad 18 = 2 \times 3^2$$
$$\therefore \text{ LCM} = 2^2 \times 3^2 = 36$$

4: Find the LCM of 54 and 90.

Solution:

$$54 = 2 \times 3^3, \quad 90 = 2 \times 3^2 \times 5$$
$$\therefore \text{ LCM} = 2 \times 3^3 \times 5 = 270$$

5: Find the GCF and the LCM of $24x^2y^2$ and $84x^3y$.

Solution: $24x^2y^2 = 2^3 \cdot 3 \cdot x^2 \cdot y^2, \quad 84x^3y = 2^2 \cdot 3 \cdot 7 \cdot x^3 \cdot y$

$\therefore \text{GCF} = 2^2 \cdot 3 \cdot x^2 \cdot y = 12x^2y.$, $\text{LCM} = 2^3 \cdot 3 \cdot 7 \cdot x^3 \cdot y^2 = 168x^3y^2$.

Relationship between GCF and LCM

Let (a, b) represents the GCF, $[a, b]$ represents the LCM of two numbers a and b.
The product of two numbers a and b is equal to the product of their GCF and LCM.

Formula: $a \times b = (a, b) \times [a, b]$

Examples

6. If the product of two numbers is 216 and their GCF is 6, what is their LCM?

Solution: $216 = 6 \times \text{LCM} \quad \therefore \text{LCM} = 36$

7. If the product of two numbers is 4860 and their LCM is 270,What is their GCF?

Solution: $4860 = 270 \times \text{GCF} \quad \therefore \text{GCF} = 18$

The Shortcut Method

1. To find the GCF of two or more given numbers, we choose a prime number (from small to large) that can divide into **all** of the given numbers. Continue to choose the divisors until no divisor can be used. The product of all the divisors is the GCF. Or, find the largest number that can divide into all of the given numbers.

2. To find the LCM of two or more given numbers, we choose a prime number (from small to large) that can divide into **any two or more** of the given numbers. Write the quotients and the remaining numbers on the next row. Continue to choose the divisors until no divisor can be used. The product of all the divisors and the numbers on the bottom row is the LCM.

3. To find GCF and LCM of two very large numbers, we use the **Rollabout Divisions** (read page 54).

Examples

8: Find the GCF and the LCM of 54 and 90.

Solution:

$$\begin{array}{r|rr} 2 & 54 & 90 \\ \hline 3 & 27 & 45 \\ \hline 3 & 9 & 15 \\ \hline & 3 & 5 \end{array} \qquad \textbf{Or: } \begin{array}{r|rr} 18 & 54 & 90 \\ \hline & 3 & 5 \end{array}$$

$\therefore$ GCF $= 2\times3\times3=18$

LCM $= 2\times3\times3\times3\times5=270$.

9: Find the GCF and the LCM of 12, 30 and 45.

Solution:

$$\begin{array}{r|rrrl} 3 & 12 & 30 & 45 & \\ \hline 2 & 4 & 10 & 15 & \\ \hline 5 & 2 & 5 & \mathbf{15} & \leftarrow 15 \text{ is the remaining number.} \\ \hline & \mathbf{2} & 1 & 3 & \leftarrow 2 \text{ is the remaining number.} \end{array}$$

$\therefore$ GCF $= 3$. (3 is the only common divisor.)

LCM $= 3\times2\times5\times2\times1\times3=180$.

10: Find the GCF and the LCM of 15, 21 and 35.

Solution:

$$\begin{array}{r|rrrl} 3 & 15 & 21 & 35 & \leftarrow \text{no common divisor except 1.} \\ \hline 5 & 5 & 7 & \mathbf{35} & \leftarrow 35 \text{ is the remaining number.} \\ \hline 7 & 1 & \mathbf{7} & 7 & \leftarrow 7 \text{ is the remaining number.} \\ \hline & \mathbf{1} & 1 & 1 & \leftarrow 1 \text{ is the remaining number.} \end{array}$$

$\therefore$ GCF $= 1$. (There is no common divisor except 1.)

LCM $= 3\times5\times7\times1\times1\times1=105$.

11: If a and b are integers and $a<b$, GCF = 18, LCM = 270, find a and b.

Solution: (See Example 8)

$270\div18=15$, 15 is the product of the numbers on the bottom row.

$15=1\times15$ or $15=3\times5$

$a=1\times18=18$ and $b=15\times18=270$

Or: $a=3\times18=54$ and $b=5\times18=90$

1-8 Ratio, Proportion, and Percents

A **ratio** is used to compare two numbers. We write a ratio as a fraction in **lowest terms.** If an improper fraction represents a ratio, we do not change it to a mixed number. Note that the lowest terms of $\frac{18}{10}$ is $\frac{9}{5}$, the simplest form of $\frac{18}{10}$ is $1\frac{4}{5}$.

If there are 15 boys and 25 girls in a class, we can compare the number of boys to the number of girls by writing a ratio (fraction). A comparison of 15 boys to 25 girls is $\frac{15}{25}$. It is read as " 15 to 25 ". A ratio can be written in three ways: 15 to 25, 15:25, or $\frac{15}{25}$.

Just as with fractions, a ratio can be written in lowest terms: $\frac{15}{25} = \frac{3}{5}$. We say that the ratio of the number of the boys to the number of the girls is " 15 to 25 ", or " 3 to 5 ".
A ratio $\frac{15}{25}$ can be also represented to the statement " 15 out of 25 ". It is equivalent to the statement " 3 out of 5 ".
If a ratio is used to compare two measurements, we must use the measurements in the same unit (like measures).

A **rate** is a ratio of two unlike measures. A rate is usually simplified to a **per unit form**, or called a **unit rate**. The rate of 120 miles to 8 gallon is $\frac{120}{8} = 15$ miles per gallon.
To find a rate in a problem, it always means to find the unit rate.

Examples

1. Write the ratio " 6 to 9 " as a fraction in lowest terms. Solution: $\frac{6}{9} = \frac{2}{3}$
2. Write the ratio $\frac{1.8}{1.2}$ in lowest terms. Solution: $\frac{1.8}{1.2} = \frac{18}{12} = \frac{3}{2}$.
3. You drive 240 miles in 4 hrs. At that rate, how many miles could you drive in 7 hrs.?
Solution:

$$\text{Unit rate (average speed)} = \frac{240}{4} = 60 \text{ mph (miles per hour)}$$

$$\therefore \text{ distance} = 60 \times 7 = 420 \text{ miles in 7 hrs.}$$

A **proportion** is formed by two equivalent ratios. In a proportion, we say that the given two ratios are equal or **proportional.**

$\frac{1}{2} = \frac{2}{4}$ It is a proportion.

If two ratios are not equal, we say that the two given ratios are **not proportional.**
$\frac{1}{2}$ and $\frac{3}{4}$ are not equal (not proportional). We can not form them as a proportion.

$\frac{1}{2} \neq \frac{3}{4}$ It is not a proportion.

To determine if two given ratios are equal (proportional), we use the following statement:

Property of Proportions: In a proportion, the cross products are equal.

If $\frac{a}{b} = \frac{c}{d}$, then $ad = bc$. (b and d are nonzero numbers.)

Examples

1. Are the ratios $\frac{5}{8}$ and $\frac{15}{24}$ proportional ? Solution: $5 \times 24 = 8 \times 15$ (proportional)

2. Are the ratios $\frac{5}{6}$ and $\frac{7}{8}$ proportional ? Solution: $5 \times 8 \neq 6 \times 7$ (not proportional)

3. Solve $\frac{n}{24} = \frac{5}{8}$. Solution: $n \times 8 = 24 \times 5$, $8n = 120$, $n = \frac{120}{8} = 15$

4. Two pounds of cheese cost $7. How much do five pounds cost ?
   Solution:
   $$\frac{7}{2} = \frac{x}{5}, \quad 7 \times 5 = 2 \times x, \quad 35 = 2x, \quad x = \frac{35}{2} = \$17.50$$

A **percent** is a ratio of a number to 100. The symbol for percent is " %".
Percent means: **1. per hundred (or hundredths)**
2. out of 100

20% means " 20 out of 100 ". Therefore, $20\% = \frac{20}{100}$

Therefore, we have: 20% of 100 is 20. We multiply: $100 \times 20\% = 100 \times \frac{20}{100} = 20$
20% of 200 is 40. $200 \times 20\% = 200 \times \frac{20}{100} = 40$

$\frac{20}{100} = \frac{2}{10} = \frac{1}{5}$ " 20 out of 100 " is equivalent to " 2 out of 10 " or " 1 out of 5 ".
Therefore, $\frac{20}{100}$, $\frac{2}{10}$, and $\frac{1}{5}$ are all equivalent to 20%.

Examples:

1. Find 20% of 300. Solution: $300 \times 20\% = 300 \times \frac{20}{100} = 60$

2. What number is 5% 0f 40 ? Solution: $n = 40 \times 5\% = 40 \times \frac{5}{100} = 2$

3. 60 out of 300 students are boys. What percent of the students is boy ?
   Solution:
   $$p = \frac{60}{300} = \frac{20}{100} = 20\%$$

4. Write $\frac{1}{5}$ as a percent.
   Solution:
   $\frac{1}{5} = \frac{n}{100}$
   $5n = 100$
   $n = \frac{100}{5} = 20$
   $\therefore \frac{1}{5} = 20\%$

5. What percent is 60 out of 300 ?
   Solution:
   $\frac{60}{300} = \frac{n}{100}$
   $300n = 6{,}000$
   $n = \frac{6{,}000}{300} = 20$
   $\therefore \frac{60}{300} = 20\%$

1-9 Decimals, Fractions, and Percents

The following examples show us a general relationship between decimal, fraction, and percent. $45\% = \frac{45}{100} = 0.45$, $120\% = \frac{120}{100} = 1.20$, $25.4\% = \frac{25.4}{100} = 0.254$

Decimal and Percent:

1. To change a percent to a decimal, we move the decimal point 2 places to the left and remove the % sign.

1. 20% = 0.2 ; 2% = 0.02 ; 0.2% = 0.002 ; 0.02% = 0.0002.
2. 25% = 0.25 ; 2.5% = 0.025 ; 0.25% = 0.0025.
3. 125% = 1.25 ; 225% = 2.25 ; 925% = 9.25.
4. 25.7% = 0.257 ; 125.7% = 1.257 ; 1325.7% = 13.257.

2. To change a decimal to a percent, we move the decimal point 2 places to the right and add the % sign.

1. 0.2 = 20% ; 0.02 = 2% ; 0.002 = 0.2% ; 0.0002 = 0.02%.
2. 0.25 = 25% ; 0.025 = 2.5% ; 0.0025 = 0.25%.
3. 1.25 = 125% ; 2.25 = 225% ; 9.25 = 925%.
4. 0.257 = 25.7% ; 1.257 = 125.7% ; 13.257 = 1325.7%

Examples

1. What number is 5% of 40 ?
Solution:

$$\begin{aligned} n &= 40 \times 5\% \\ &= 40 \times 0.05 \\ &= 2 \end{aligned}$$

2. Find 80% of 50.
Solution:

$$\begin{aligned} &50 \times 80\% \\ &= 50 \times 0.8 \\ &= 40 \end{aligned}$$

3. What percent is 15% of 80% ?
Solution:

$$\begin{aligned} &80\% \times 15\% \\ &= 0.80 \times 0.15 \\ &= 0.12 = 12\% \end{aligned}$$

4. Find the sales tax on $50 if the tax rate is 8.25 %.
Solution:

$$50 \times 8.25\% = 50 \times 0.0825 = \$4.13$$

5. 30% of teens do not consider themselves to be risk for sexuality-related diseases and 4% of this group accounts for new sexuality-related disease cases each year. How many new sexuality-related disease cases are there each year at a high school with 3,000 students ?
Solution:

$$30\% = 0.30,\ 4\% = 0.04$$
$$3{,}000 \times 0.30 \times 0.04 = 36 \text{ cases}$$

Fraction and Percent

1. **To change a fraction to a percent, we first change it to a decimal and then to a percent.**

Examples

1. $\frac{1}{2} = 0.5 = 50\%$, $\frac{1}{5} = 0.2 = 20\%$, $\frac{2}{5} = 0.4 = 40\%$, $\frac{3}{4} = 0.75 = 75\%$

$\frac{3}{8} = 0.375 = 37.5\%$, $\frac{5}{8} = 0.625 = 62.5\%$

2. $\frac{1}{3} = 0.333\cdots = 33.\bar{3}\ \% = 33.3\ \%$ (It is the nearest tenth of a percent)

$\frac{1}{3} = 0.33\frac{1}{3} = 33\frac{1}{3}\ \%$ (It is an exact percent)

3. $\frac{3}{7} = 0.42\overline{857142} = 0.42\frac{857142}{999999} = 0.42\frac{857142 \div 142857}{999999 \div 142857} = 0.42\frac{6}{7} = 42\frac{6}{7}\%$

(Hint: To find the GCF = 142857, read "Rollabout Divisions" on Page 54.)

2. **To change a percent to a fraction, we change it to an equivalent fraction with a denominator of 100.**

Examples

1. $50\ \% = \frac{50}{100} = \frac{1}{2}$, $20\ \% = \frac{20}{100} = \frac{1}{5}$, $40\ \% = \frac{40}{100} = \frac{2}{5}$, $125\ \% = \frac{125}{100} = 1\frac{1}{4}$

$12.5\ \% = \frac{12.5}{100} = \frac{125}{1000} = \frac{1}{8}$, $62.5\ \% = \frac{62.5}{100} = \frac{625}{1000} = \frac{5}{8}$

2. $\frac{1}{2}\% = \frac{1}{2} \times \frac{1}{100} = \frac{1}{200}$, $2\frac{1}{2} = \frac{5}{2}\% = \frac{5}{2} \times \frac{1}{100} = \frac{5}{200} = \frac{1}{40}$

$33\frac{1}{3}\% = \frac{100}{3}\% = \frac{100}{3} \times \frac{1}{100} = \frac{1}{3}$

Examples

1. Find $5\frac{1}{3}\%$ of 33. Solution: $33 \times 5\frac{1}{3}\% = 33 \times \frac{16}{3}\% = 33 \times \frac{16}{3} \times \frac{1}{100} = \frac{528}{300} = 1.76$
2. What percent is 12 of 30 ? Solution: $\frac{12}{30} = 0.4 = 40\%$
3. What percent is 2.25 out of 15 ? Solution: $\frac{2.25}{15} = 0.15 = 15\%$
4. What percent of 40 is 5 ? Solution: $p = \frac{5}{40} = 0.125 = 12.5\%$
5. 40 is 5% of what number ? Solution: $n \times 5\% = 40$, $n = \frac{40}{5\%} = \frac{40}{0.05} = 800$

1-10 Percent and Applications

A) Sales, Price, and Tax

1. Joe spent \$50 and sales tax \$4 to buy a pair of shoes. What percent is the sales tax ?

Solution: $50 \times p = 4$, $p = \frac{4}{50} = \frac{2}{25} = 0.08 = 8\%$

2. Joe paid \$4 sales tax for a pair of shoes he bought. The sales-tax rate is 8%. What is the price of the shoes ?

Solution: Let p = the price, $p \times 8\% = 4$, $p \times 0.08 = 4$, $p = \frac{4}{0.08} = \$50$

B) Percent of Increase or Decrease

1. What is the percent of increase from 4 to 6 ?

Solution: percent of increase $= \frac{6-4}{4} = \frac{1}{2} = 0.5 = 50\%$

2. What is the percent of decrease from 6 to 4 ?

Solution: percent of decrease $= \frac{6-4}{6} = \frac{2}{6} = \frac{1}{3} = 0.33\frac{1}{3} = 33\frac{1}{3}\%$

3. The bill of electricity was \$60 last month. It decreased to \$45 this month. Find the percent of decrease.

Solution:

$$\text{percent of decrease} = \frac{60-45}{60} = \frac{15}{60} = \frac{1}{4} = 0.25 = 25\%$$

4. The population of a city increased from 125,000 to 180,000 in 10 years. Find the percent of increase.

Solution:

$$\text{percent of increase} = \frac{180{,}000 - 125{,}000}{125{,}000} = \frac{55{,}000}{125{,}000} = \frac{11}{25} = 0.44 = 44\%.$$

C) Discount and Commission

A discount is a decrease in the price of a product. A discount can be expressed as a percent of the original price.

A commission is a payment to the salesman for selling products. A commission can be expressed as a percent of the total sales.

1. A suit that is regularly \$45.90 is on sale for 30% discount. What is the sale price of the suit ?

Solution:

the discount: $\$45.90 \times 30\% = 45.90 \times 0.30 = \$13.77$

the sale price: $\$45.90 - \$13.77 = \$32.13$. Ans.

2. Mary sold sport products worth \$32,000 this week. She receives 3% commission on her sales. How much is her commission this week ?

Solution:

commission: $\$32{,}000 \times 3\% = 32{,}000 \times 0.03 = \$960$

D) Simple Interest

Interest is the amount of money charged for the use of borrowed money. Interest is also the the amount of money earned for the money deposited in a bank.
The interest rate is usually expressed as a percent of the principal over a period of time.
If the interest rate is computed on a yearly basis, it is called a yearly rate or an annual rate.
When interest is computed on the principal only (no interest on the interest previously earned), it is called **simple interest**. The formula to compute simple interest is:

$$\text{Interest} = \text{principal} \times \text{rate} \times \text{time}$$
$$\mathbf{I = p\,r\,t}$$

1. Find the simple interest you pay if you borrow \$500 for 3 years at a yearly rate 7%.
Solution:

$$I = p\,r\,t, \quad I = \$500 \times 7\% \times 3 = \$500 \times 0.07 \times 3 = \$105$$

2. You have a saving account with \$500. The annual interest rate is 3.5%. How much simple interest will your account earn in 3 months ? How much do you have in your account after 3 months ?
Solution:

$$I = p\,r\,t, \quad \text{Interest} = 500 \times 3.5\% \times \frac{3}{12} = 500 \times 0.035 \times \frac{1}{4} = \$4.38$$
$$\text{Total} = \$500 + \$4.38 = \$504.38$$

E) Compound Interest

Compound interest is computed on "the principal plus the interest previously earned".
The principal increases when the interest is compounded. The compound interest is usually given in daily, monthly, quarterly, semiannually, or yearly.
The formula to compute the compound amount (principal plus compound interest) is:

$$A = p(1+r)^n, \quad r: \text{annual interest rate} \quad n: \text{number of years}$$

1. You have a savings account with \$500. The annual interest rate 7%, compounded yearly. How much do you have in your account after 3 years ?
Solution:

$$A = p(1+r)^n = 500(1+0.07)^3 = 500(1.07)^3 = 500(1.225043) = \$612.52$$

2. \$500 is deposited in a bank at 6%, compounded monthly. Find the new principal after 3 months ?
Solution:

$$\text{monthly rate } r = 0.06 \div 12 = 0.005$$
$$A = p(1+r)^n = 500(1+0.005)^3 = 500(1.005)^3$$
$$= 500(1.01508) = \$507.54$$

1-11 Averages

The average of a group of numbers is the sum of the numbers divided by the number of the the numbers. It is also called **arithmetic mean**. The average of the numbers 21, 32, 40, 50, and 62 is:

$$\text{Average} = \frac{21+32+40+50+62}{5} = \frac{205}{5} = 41$$

A **weighted average** is the average of two or more groups in which there are more members in one group than there are in another.

Examples

1. The average of five numbers is 28. Four of the five numbers are 15, 13, 42, and 33. What is the fifth number ?

Solution:

The sum of the five numbers $= 28 \times 5 = 140$

The fifth number $= 140 - (15+13+42+33) = 140 - 103 = 37$

2. Roger took six math tests. His average was 78. The average of his first four tests was 72. What is the average of his fifth and sixth tests ?

Solution:

The sum of the six tests $= 78 \times 6 = 468$

The sum of the first four tests $= 72 \times 4 = 288$

The average of the last two tests $= \frac{468-288}{2} = \frac{180}{2} = 90$

3. The average of the first 3 numbers is 12. The average of the next 9 numbers is 25. What is the overall average ?

Solution:

$$\text{Overall average} = \frac{(3 \times 12)+(25 \times 9)}{12} = \frac{261}{12} = 21.75$$

4. The height of the first building is 360 feet. The height of the second building is 426 feet. The average height of the three buildings is 410 feet. What is the height of the third building ?

Solution:

The height of the third building

$= 410 \times 3 - (360+426) = 1230 - 786 = 444$ feet

5. If the average of 7, 13, 15, 32, and *n* is 28, what is the value of *n* ?

Solution:

$$7+13+15+32+n = 28 \times 5, \quad 67+n = 140, \quad n = 73$$

6. Find the average of 3^{15}, 3^{20}, and 3^{35}.

Solution:

$$\frac{3^{15}+3^{20}+3^{35}}{3} = \frac{3^{15}}{3} + \frac{3^{20}}{3} + \frac{3^{35}}{3} = 3^{14}+3^{19}+3^{34}$$ (Read page 28)

7. A group of 20 students has an average SAT Math score of 620. Another group of 30 students has an average of 560. What is the average score of the entire 50 students ?

Solution:

$$\text{Average (weighted)} = \frac{620 \times 20 + 560 \times 30}{50} = \frac{29200}{50} = 584$$

1-12 Powers and Exponents

Exponent (power): In the expression 5^3, 3 is called the **exponent** (or **power**) of 5.
5 is called the **base.** 5^3 is the exponential (or power) form of $5 \times 5 \times 5$.
The exponent 3 in the expression 5^3 means that there are three **repeated factors** of 5.
Exponential form or **power form** a^n is a short way to write a **multiplication (product) of *n* repeated factors of** a: $a^n = a \cdot a \cdot a \cdots\cdot a \to n$ repeated factors of a

$5^1 \to$ read " five to the first power".
$5^2 \to$ read " five to the second power " or " five squared ".
$5^3 \to$ read " five to the third power " or " five cubed ".
$5^n \to$ read " five to the nth power ".

Any number can be written as " **the number to the first power** ".
$1 = 1^1$, $2 = 2^1$, $(-3) = (-3)^1$, $4 = 4^1$, $-5 = (-5)^1$, $99 = 99^1$, $100 = 100^1$, ...

Examples: Write each product using exponents.
$2 \cdot 2 = 2^2$, $2 \cdot 2 \cdot 2 = 2^3$, $2 \cdot 2 \cdot 2 \cdot 2 = 2^4$, $-2 \cdot -2 \cdot -2 \cdot -2 \cdot -2 = (-2)^5$
$3 \cdot 3 = 3^2$, $3 \cdot 3 \cdot 3 = 3^3$, 7 squared $= 7^2$, 7 cubed $= 7^3$

The expression $3a^2$ means that a is squared, but not 3. $3a^2 = 3 \cdot a \cdot a$
The expression $(3a)^2$ means that 3 and a are both squared. $(3a)^2 = 3a \cdot 3a = 9a^2$

Notice that: $-3^2 \neq (-3)^2$. $-3^2 = -3 \cdot 3 = -9$, $(-3)^2 = (-3) \cdot (-3) = 9$
$-2^4 = -2 \cdot 2 \cdot 2 \cdot 2 = -16$, $(-2)^4 = (-2) \cdot (-2) \cdot (-2) \cdot (-2) = 16$

$(-a)^n$ is positive if n is an even number. $(-1)^2 = 1$, $(-1)^{50} = 1$, $(-1)^{100} = 1$
$(-a)^n$ is negative if n is an odd number. $(-1)^3 = -1$, $(-1)^{49} = -1$, $(-1)^{99} = -1$
$(-2)^{19} = -2^{19}$, $(-2)^{20} = 2^{20}$

General rules of exponents (powers): If a is any nonzero real number, then:

1. $a^m \cdot a^n = a^{m+n}$, **2.** $\dfrac{a^m}{a^n} = a^{m-n}$, **3.** $a^0 = 1$ if $a \neq 0$

Rules: 1. To multiply two powers of the same base, we add the powers.
2. To divide two powers of the same base, we subtract the powers.
3. If a is any nonzero number, then $a^0 = 1$, $1^0 = 1$, $2^0 = 1$, $3^0 = 1, \cdots, 100^0 = 1$.
0^0 is undefined (it means that the power "0" does not apply to 0 itself).

Examples:

1. $2^{30} \cdot 2^{10} = 2^{30+10} = 2^{40}$, **2.** $\dfrac{2^{30}}{2^{10}} = 2^{30-10} = 2^{20}$, **3.** $\dfrac{2^{10}}{2^{10}} = 1$, or $\dfrac{2^{10}}{2^{10}} = 2^{10-10} = 2^0 = 1$

1-13 Scientific Notations

To write very large numbers or very small numbers, we use the **scientific notations.**

The distance from the Sun to the Earth = 93 million miles = 9.3×10^7 miles = 1.5×10^{11} meters

The distance from the Earth to the Moon = 250,000 miles = 2.5×10^5 miles = 4×10^8 meters

The mass of one atom of hydrogen = 1.674×10^{-24} gram

Scientific Notation: **(A number)** = $m\times10^n$, $1\le m<10$

m is a number greater or equal to 1, but less than 10. n is an integer.

The following powers of 10 are useful in scientific notations.

$10=10^1$, $100=10^2$, $1{,}000=10^3$, $10{,}000=10^4$, $100{,}000=10^5$,.......

$0.1=10^{-1}$, $0.01=10^{-2}$, $0.001=10^{-3}$, $0.0001=10^{-4}$, $0.00001=10^{-5}$,.......

To write a number in scientific notation, we move the decimal point to get a number that is at least 1 but less than 10. Count the number of decimal places it moves and use it as the power of 10.

$123{,}000{,}000=1.23\times10^8$; $0.0000000123=1.23\times10^{-8}$

The Milky Way is the home galaxy of our solar system and contains about 200 billion stars ($200\times10^9=2\times10^{11}$ stars).

A **light-year** is the distance light travels in one year in a vacuum. A light-year is about 5.9 trillion miles (5.9×10^{12} miles).

The nearest star to our Sun, Alpha Centauri, is 4.25 light-years (2.51×10^{13} miles) away.

The diameter of our solar system is 1.18×10^{11} miles. The size of the universe is estimated at least 2 million light-years (2×10^6 light-years $=1.18\times10^{19}$ miles) in diameter.

Examples (Write in scientific notations.)

1. $100=1\times100=1\times10^2$

2. $5{,}000=5\times1{,}000=5\times10^3$.

3. $0.02=2\times0.01=2\times10^{-2}$

4. $0.005=5\times0.001=5\times10^{-3}$.

5. $12{,}000=1.2\times10{,}000=1.2\times10^4$.

6. $0.00012=1.2\times0.0001=1.2\times10^{-4}$.

7. $235=2.35\times100=2.35\times10^2$.

8. $0.235=2.35\times0.1=2.35\times10^{-1}$

9. $2{,}350{,}000=2.35\times10^6$.

10. $0.00000235=2.35\times10^{-6}$

11. $2\times10^5\times6\times10^7=12\times10^{12}=1.2\times10\times10^{12}=1.2\times10^{13}$

12. $2.35\times10^2\times5.2\times10^4=(2.35)(5.2)\times10^6=12.22\times10^6=1.222\times10\times10^6=1.222\times10^7$.

13. An atom of oxygen has a mass of 2.658×10^{-23} gram. Write it in decimal numeral.

Solution: $2.658\times10^{-23}=0.00000000000000000000002658$ gram

14. Light travels at the rate of 186,000 miles per second. It takes 8 minutes and 20 seconds for the light from the Sun to reach the Earth. Find the distance from the Sun to the Earth in scientific notation.

Solution: $d=186{,}000\times(8\times60+20)=93{,}000{,}000=9.3\times10^7$ miles

1-14 Laws of Exponents

Laws of Exponents (Rules of Powers)

If a and b are real numbers, m and n are positive integers, $a \neq 0$, $b \neq 0$, then:

1) $a^0 = 1$ 2) $a^m \cdot a^n = a^{m+n}$ 3) $\dfrac{a^m}{a^n} = a^{m-n}$ 4) $\dfrac{a^m}{a^n} = \dfrac{1}{a^{n-m}}$

5) $(a^m)^n = a^{mn}$ 6) $(ab)^m = a^m b^m$ 7) $a^{-m} = \dfrac{1}{a^m}$ 8) $\dfrac{1}{a^{-m}} = a^m$

9) $\left(\dfrac{b}{a}\right)^m = \dfrac{b^m}{a^m}$ 10) $\left(\dfrac{b}{a}\right)^{-m} = \left(\dfrac{a}{b}\right)^m$ 11) $a^{\frac{1}{m}} = \sqrt[m]{a}$ 12) $a^{\frac{n}{m}} = (\sqrt[m]{a})^n$

Examples

1. $x^7 \cdot x^3 = x^{7+3} = x^{10}$

2. $\dfrac{x^7}{x^3} = x^{7-3} = x^4$, $\dfrac{x^7}{x^3} = \dfrac{1}{x^{3-7}} = \dfrac{1}{x^{-4}} = x^4$

3. $\dfrac{x^3}{x^7} = \dfrac{1}{x^{7-3}} = \dfrac{1}{x^4}$, $\dfrac{x^3}{x^7} = x^{3-7} = x^{-4} = \dfrac{1}{x^4}$

4. $\dfrac{x^3}{x^3} = 1$, $\dfrac{x^3}{x^3} = x^{3-3} = x^0 = 1$

5. $a^{\frac{1}{2}} = \sqrt{a}$, $a^{\frac{1}{3}} = \sqrt[3]{a}$; $a^{\frac{1}{5}} = \sqrt[5]{a}$

6. $1^0 = 1$, $2^0 = 1$, $10^0 = 1$, $x^0 = 1$ if $x \neq 0$

7. $(a^3)^{-2} = a^{-6} = \dfrac{1}{a^6}$

8. $\dfrac{ac^5}{a^3c^2} = a^{1-3}c^{5-2} = a^{-2}c^3 = \dfrac{c^3}{a^2}$

9. $\dfrac{x^2}{y^2}\left(\dfrac{y^3}{x^2}\right)^4 = \dfrac{x^2}{y^2} \cdot \dfrac{y^{12}}{x^8} = \dfrac{y^{10}}{x^6}$

10. $25^{\frac{1}{2}} = \sqrt{25} = 5$, $32^{\frac{3}{5}} = (\sqrt[5]{32})^3 = 2^3 = 8$

11. $25^{-\frac{1}{2}} = \dfrac{1}{25^{\frac{1}{2}}} = \dfrac{1}{\sqrt{25}} = \dfrac{1}{5}$

12. $32^{-\frac{3}{5}} = \dfrac{1}{32^{\frac{3}{5}}} = \dfrac{1}{(\sqrt[5]{32})^3} = \dfrac{1}{2^3} = \dfrac{1}{8}$

13. $9^{2-\pi} \cdot 3^{2+2\pi} = (3^2)^{2-\pi} \cdot 3^{2+2\pi} = 3^{4-2\pi} \cdot 3^{2+2\pi} = 3^{4-2\pi+2+2\pi} = 3^6 = 726$

1-15 Square Roots and Radicals

Because $5^2 = 25$, we say that the square root of 25 is 5. We write $\sqrt{25} = 5$.
Because $5^3 = 125$, we say that the cubed root of 125 is 5. We write $\sqrt[3]{125} = 5$.
Because $2^4 = 16$, we say that the 4th root of 16 is 2. We write $\sqrt[4]{16} = 2$.

The symbol $\sqrt[n]{b}$ is called a **radical**. We read $\sqrt[n]{b}$ as " **the** n **th root of** b ". b is called the **radicand**, n is called the **index**.
Index "2" of the square root is usually omitted from the radical. $\sqrt{b}$ means $\sqrt[2]{b}$.

To find the square root of a given number, we find a number that, when squared, equals the given number. However, $\sqrt[n]{b}$ means only the **principal root** of b.

$\sqrt{4} = \sqrt{2 \cdot 2} = \sqrt{2^2} = 2$
$\sqrt{4} = \sqrt{(-2)^2} = -2$ is incorrect.
$\sqrt{(-2)^2} = \sqrt{4} = 2$
$\sqrt{0.01} = \sqrt{(0.1)(0.1)} = \sqrt{(0.1)^2} = 0.1$

$\sqrt{9} = \sqrt{3 \cdot 3} = \sqrt{3^2} = 3$
$\sqrt{9} = \sqrt{(-3)^2} = -3$ is incorrect.
$\sqrt{(-3)^2} = \sqrt{9} = 3$
$\sqrt{1.44} = \sqrt{(1.2)(1.2)} = \sqrt{(1.2)^2} = 1.2$

To find the n th root of a given number, we follow the same idea as the above examples, and so on.

$\sqrt[3]{8} = \sqrt[3]{2 \cdot 2 \cdot 2} = \sqrt[3]{2^3} = 2$; $\sqrt[4]{16} = \sqrt[4]{2 \cdot 2 \cdot 2 \cdot 2} = \sqrt[4]{2^4} = 2$
$\sqrt[3]{-8} = \sqrt[3]{-2 \cdot -2 \cdot -2} = \sqrt[3]{(-2)^3} = -2$; $\sqrt[4]{(-2)^4} = \sqrt[4]{16} = 2$, **not** -2

Rules of Radicals

For all nonnegative real numbers a and b:

1) $\sqrt{a^2} = a$ **2)** $\sqrt{ab} = \sqrt{a} \cdot \sqrt{b}$ **3)** $\sqrt{\frac{b}{a}} = \frac{\sqrt{b}}{\sqrt{a}}$, where $a \neq 0$

4) $\sqrt[n]{a} = a^{\frac{1}{n}}$ **5)** $\sqrt[n]{a^m} = a^{\frac{m}{n}}$

For all real numbers a, b, and x:

6) $\sqrt[n]{a^n} = |a|$ **if** n is even. **7)** $\sqrt{x^2} = |x|$
$\sqrt[n]{a^n} = a$ **if** n is odd. **8)** $\sqrt{x^3} = x\sqrt{x}$

1-16 Operations on Radicals

To simplify radicals, we use the following steps:

1. Simplify each radical in simplest form.
2. Combine (add or subtract) with like radicands.
3. Use the **Rules of Radicals** to multiply or divide two radicals.
4. Multiply binomials by using the distributive property or formulas:
   $(a+b)^2 = a^2 + 2ab + b^2$, $(a-b)^2 = a^2 - 2ab + b^2$, $(a+b)(a-b) = a^2 - b^2$
5. Rationalize the denominator. (No radicals are in the denominator.)

Examples (Assume that all variables are nonnegative real numbers.)

1. $4\sqrt{2} - \sqrt{2} = 3\sqrt{2}$, **2.** $6\sqrt{5} - 4\sqrt{5} = 2\sqrt{5}$ **3.** $3\sqrt{7} - 2\sqrt{11} + 5\sqrt{7} = 8\sqrt{7} - 2\sqrt{11}$

4. $5\sqrt{3} + \sqrt{12} = 5\sqrt{3} + 2\sqrt{3} = 7\sqrt{3}$ **5.** $\sqrt{24} - \sqrt{54} = 2\sqrt{6} - 3\sqrt{6} = -\sqrt{6}$

6. $3\sqrt{18} + 2\sqrt{8} - 6\sqrt{50} = 3\cdot 3\sqrt{2} + 2\cdot 2\sqrt{2} - 6\cdot 5\sqrt{2} = 9\sqrt{2} + 4\sqrt{2} - 30\sqrt{2} = -17\sqrt{2}$

7. $5\sqrt{7} + 6\sqrt{3} - \sqrt{7} + 2\sqrt{3} = (5\sqrt{7} - \sqrt{7}) + (6\sqrt{3} + 2\sqrt{3}) = 4\sqrt{7} + 8\sqrt{3}$

8. $\sqrt{2}\cdot\sqrt{8} = \sqrt{2\cdot 8} = \sqrt{16} = 4$ **9.** $\sqrt[3]{2}\cdot\sqrt[3]{16} = \sqrt[3]{2\cdot 16} = \sqrt[3]{32} = 2\sqrt[3]{4}$

10. $2\sqrt{5}\cdot 6\sqrt{12} = 2\cdot 6\cdot\sqrt{5\cdot 12} = 12\sqrt{60} = 12\cdot 2\sqrt{15} = 24\sqrt{15}$

11. $\sqrt{3}(\sqrt{4} + \sqrt{6}) = \sqrt{3}\cdot\sqrt{4} + \sqrt{3}\cdot\sqrt{6} = \sqrt{12} + \sqrt{18} = 2\sqrt{3} + 3\sqrt{2}$

12. $(4+\sqrt{2})(5-\sqrt{2}) = 20 - 4\sqrt{2} + 5\sqrt{2} - 2 = 18 + \sqrt{2}$

13. $(\sqrt{2}+\sqrt{3})^2 = (\sqrt{2})^2 + 2\cdot\sqrt{2}\cdot\sqrt{3} + (\sqrt{3})^2 = 2 + 2\sqrt{6} + 3 = 5 + 2\sqrt{6}$

14. $(\sqrt{2}-\sqrt{3})^2 = (\sqrt{2})^2 - 2\cdot\sqrt{2}\cdot\sqrt{3} + (\sqrt{3})^2 = 2 - 2\sqrt{6} + 3 = 5 - 2\sqrt{6}$

15. $(\sqrt{2}+\sqrt{3})(\sqrt{2}-\sqrt{3}) = (\sqrt{2})^2 - (\sqrt{3})^2 = 2 - 3 = -1$

16. $\sqrt{\frac{5}{3}}\cdot\sqrt{\frac{2}{5}} = \sqrt{\frac{5}{3}\cdot\frac{2}{5}} = \frac{\sqrt{2}}{\sqrt{3}}\cdot\frac{\sqrt{3}}{\sqrt{3}} = \frac{\sqrt{6}}{3}$ **17.** $\frac{\sqrt{15}}{\sqrt{6}} = \sqrt{\frac{15}{6}} = \sqrt{\frac{5}{2}} = \frac{\sqrt{5}}{\sqrt{2}}\cdot\frac{\sqrt{2}}{\sqrt{2}} = \frac{\sqrt{10}}{2}$

18. $\sqrt[3]{\frac{5}{3}}\cdot\sqrt[3]{\frac{2}{5}} = \sqrt[3]{\frac{5}{3}\cdot\frac{2}{5}} = \frac{\sqrt[3]{2}}{\sqrt[3]{3}}\cdot\frac{\sqrt[3]{9}}{\sqrt[3]{9}} = \frac{\sqrt[3]{18}}{\sqrt[3]{27}} = \frac{\sqrt[3]{18}}{3}$

1-17 Laws of Logarithms

In the expression $2^3 = 8$, 3 is the power (exponent) of 2.
In order to express the statement " 2 to what power gives 8 ", we use the form of **logarithm**.

$\log_2 8 = 3$ is the equivalent form of $2^3 = 8$. Read " The logarithm of 8 to the base 2 is 3. "
We may simply say that $\log_2 8 = 3$ represents " 2 to the third power gives 8 ".
$\log_{10} 100 = 2$ is the equivalent form of $10^2 = 100$. When we write a logarithm with the base 10, the subscript 10 is omitted. $\log_{10} 100 = \log 100 = 2$
Logarithms with base 10 are called **Common Logarithms**.

Laws of Logarithms (Rules of Logarithms)

For all positive real numbers *a, p, q* with $a \neq 1$, and *n* is any real number:

1) $\log_a 1 = 0$ **2)** $\log_a a = 1$ **3)** $\log_a a^n = n$ **4)** $a^{\log_a n} = n$

5) $\log_a pq = \log_a p + \log_a q$ **6)** $\log_a \frac{p}{q} = \log_a p - \log_a q$ **7)** $\log_a p^n = n \log_a p$

8) Formula for Changing Base: $\log_a n = \dfrac{\log_b n}{\log_b a}$ where $b \neq 1$

In practice, we use $b = 10$. Therefore, $\log_a n = \dfrac{\log_{10} n}{\log_{10} a}$

9) Formula for Antilogarithms: If $\log x = a$, then $x = anti\log a$.

Also: $anti\log_a p = a^p$ and $anti\log p = 10^p$

$anti\log_a (p+q) = a^{p+q} = a^p \cdot a^q = (anti\log_a p) \cdot (anti\log_a q)$

Examples: Given $\log_{10} 2 = 0.301$, $\log_{10} 3 = 0.477$, $\log_{10} 5 = 0.699$, we can find:
(Hint: We may omit the subscript 10.)

1. $\log_{10} 20 = \log_{10}(4 \times 5) = \log_{10} 4 + \log_{10} 5 = \log_{10} 2^2 + \log_{10} 5 = 2\log_{10} 2 + \log_{10} 5$
$= 2(0.301) + 0.699 = 1.301$

2. $\log_{10} 0.2 = \log_{10} \frac{2}{10} = \log_{10} 2 - \log_{10} 10 = 0.301 - 1 = -0.699$

3. $\log_{10} 81 = \log_{10} 3^4 = 4\log_{10} 3 = 4(0.477) = 1.908$ **4.** $\log_{0.1} 3 = \dfrac{\log_{10} 3}{\log_{10} 0.1} = \dfrac{0.477}{-1} = -0.477$

Examples:

1. $\log 1 = 0$, $\log 10000 = 4$, $\log 0.1 = -1$, $\log 0.0001 = -4$

2. $\log_2 8 = \log_2 2^3 = 3$ **3.** $5^{\log_5 9} = 9$ **4.** $4^{\log_4 \pi} = \pi$

5. Expand: $\log_a (x^2 \sqrt{x+1}) = \log_a x^2 + \log_a (x+1)^{1/2} = 2\log_a x + \frac{1}{2}\log_a (x+1)$

6. Condense:

$4\log_a x + \frac{1}{2}\log_a (x-1) - 2\log_a (x+2) = \log_a x^4 + \log_a \sqrt{x-1} - \log_a (x+2)^2 = \log_a \dfrac{x^4\sqrt{x-1}}{(x+2)^2}$

7. $anti\log 4 = 10^4 = 10{,}000$ **8.** $anti\log 0.9445 = 10^{0.9445} = 8.8$

9. $anti\log(-1.8763) = 10^{-1.8763} = 1/10^{1.8763} = \frac{1}{75.2142} = 0.0133$

1-18 Sequences and Number Patterns

Patterns appear in nature and our daily life. Studying sequences and number patterns helps us to identify, understand patterns and make predictions.

Sequence: A sequence is a consecutive nature numbers in a certain order.
Each number is called a term of the sequence.

To find the pattern of a sequence, we find the **differences** between any two successive terms.

Examples

1: Find the next five terms of the sequence 2, 4, 6, 8, 10, ·····.
Solution: The terms increase by 2. The differences are constant.
The next five terms are 12, 14, 16, 18, and 20.

2: Find the next five terms of the sequence 106, 95, 84, 73, 62, ····.
Solution: The terms decrease by 11. The differences are constant.
The next five terms are 51, 40, 29, 18, and 7.

3: Find the next five terms of the sequence 1, 3, 6, 10, 15, ·····.
Solution: The differences are not constant.
The next five terms are continued by adding 6, 7, 8, 9, and 10.
The next five terms are 21, 28, 36, 45, and 55.

1 , 3 , 6 , 10 , 15 , **21** , **28** , **36** , **45** , **55**
2 3 4 5 **6** **7** **8** **9** **10** (differences)

4: Find the next five terms of the sequence 2, 6, 12, 20, 30, ·····.
Solution: The differences are not constant.
The next five terms are continued by adding 12, 14, 16, 18, and 20.
The next five terms are 42, 56, 72, 90, and 110.

2 , 6 , 12 , 20 , 30 , **42** , **56** , **72** , **90** , **110**
4 6 8 10 **12** **14** **16** **18** **20** (differences)

5: Each of the five teams plays each of the other teams just once. How many games will there be ? How many games will there be for 8 teams ?
Solution:

Number of teams	1	2	3	4	5	6	7	8
Number of games	0	1	3	6	10	15	21	28

There will be 10 games for 5 teams.
There will be 28 games for 8 teams.

1-19 Arithmetic and Geometric Sequences

Arithmetic Sequence: A sequence which has common different between any two successive terms.

Arithmetic Mean (Average): A single term between two terms of an arithmetic sequences. If p, m, q is an arithmetic sequence, then $m = \frac{p+q}{2}$.

Geometric Sequence: A sequence in which the ratios of consecutive terms are the same.

Geometric Mean: If p, m, q is a geometric sequence, then $m = \pm\sqrt{pq}$.

Formulas

1. The general or *n*th term of an arithmetic sequence a_1, $a_1 + d$, $a_1 + 2d$, $a_1 + 3d$, ..., a_n is given by: $a_n = a_1 + (n-1)\,d$

Where a_1 :1st term, n: position of a_n, d: common difference.

2. The general or *n*th term of a geometric sequence a_1, $a_1 r$, $a_1 r^2$, $a_1 r^3$, ..., a_n is given by: $a_n = a_1 r^{n-1}$

Where a_1: 1st term, n: position of a_n, r: common ratio

Examples

1. Find the 10th term of 25, 33, 41, ..., a_{10}.

Solution:

$$d = 33 - 25 = 8\ ,\quad a_{10} = a_1 + (n-1)d = 25 + (10-1)8 = 97$$

2. Find the arithmetic mean between 17 and 53.

Solution:

$$m = \frac{p+q}{2} = \frac{17+53}{2} = 35.$$ Hint: the sequence is 17, **35**, 53, 71,.....

3. Find the 10th term of the geometric sequence 2, 6, 18, ...

Solution: The common ratio $r = \frac{6}{2} = 3$, $a_{10} = a_1 r^{10-1} = 2 \cdot 3^9 = 39{,}366$

4. Find the 8th term of the geometric sequence 2, $\frac{2}{3}$, $\frac{2}{9}$, $\frac{2}{27}$, ...

Solution: The common ratio $r = \frac{2/3}{2} = \frac{2}{3} \cdot \frac{1}{2} = \frac{1}{3}$, $a_8 = a_1 r^{8-1} = 2 \cdot \left(\frac{1}{3}\right)^7 = \frac{2}{3^7} = \frac{2}{2187}$

5. Find the geometric mean of –4 and –9.

Solution

$$m = \pm\sqrt{pq} = \pm\sqrt{-4 \cdot -9} = \pm 6$$

(Hint: The sequence is –4, **6**, –9, ... or –4, **–6**, –9, ...)

1-20 Sets and Venn Diagrams

Set: Any particular group of elements or members. We use braces, { }, to name a set.

U: The **universal** set. It is the set consisting of all the elements.

$A \cap B$: The **intersection** of two sets. The set consists of the elements that are common to both sets A and B.

$A \cup B$: The **union** of two sets. The set consists of all members of both sets A and B.

ϕ : The **empty** set, or null set. The set consists of no member.
The set {0} contains exactly one member "0". The set {0} is not an empty set.

$\in$: It indicates " **it is a member of** ". $\notin$: It indicates " **it is not a member of** ".

$A \subset B$ or $A \subseteq B$: If every member in set A is also a member of set B, then A is a **subset** of B, B is a **superset** of A. Every set is the subset of itself. It is agreed that the empty set is also a subset of every set.

If A is a subset of B and $A \neq B$, we write $A \subset B$, then A is a **proper subset** of B. If A is a subset of B and $A = B$, we write $A \subseteq B$.

$A \not\subset B$: Set A is **not a subset** of set B.

A' or A^c: The **complement** of set A. A' is the set consisting of all the elements not in A.

$A \cup A' = U$; $A \cap A' = \phi$

Venn Diagram: A diagram used to indicate relations between sets.

1. $n(A \cup B) = n(A) + n(B) - n(A \cap B)$
2. $n(A \cup B \cup C) = n(A) + n(B) + n(C) - n(A \cap B) - n(B \cap C) - n(A \cap C) + n(A \cap B \cap C)$

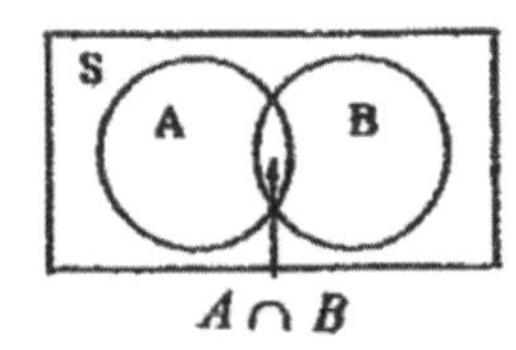

$A \cap B$

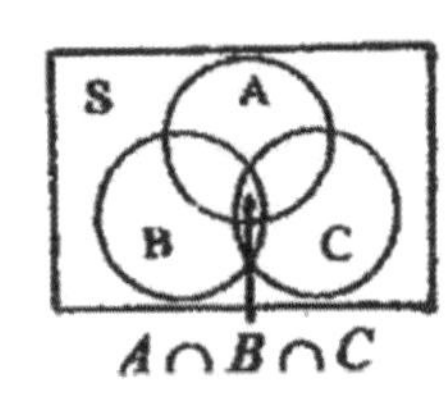

$A \cap B \cap C$

Examples

1. If $A = \{1, 3, 5, 7\}$, $B = \{2, 4, 6, 8\}$, $U = \{1, 2, 3, 4, 5, 6, 7, 8\}$, we have:

$3 \in \{1, 3, 5, 7\}$, $6 \notin \{1, 3, 5, 7\}$, $A \neq B$, $\{1, 3, 5, 7\} = \{1, 5, 7, 3\}$

$A \subset U$, $B \subset U$, $A \not\subset B$, $B \subset B$, $\phi \subset A$, $\phi \subset B$, $A' = B$, $A \cup A' = U$

$A \cup B = U$, $A \cap B = \phi$, $A \cap U = \{1, 3, 5, 7\}$, $U \cap B = \{2, 4, 6, 8\}$

2. In a high school class, there are 25 students who went to see movie A, 18 students went to see move B. These figures include 9 students went to see both movies. How many students are in this class ?

Solution:

$25 + 18 - 9 = 34$ students.

16 9 9
A B.

3. In a high school class, there are 17 students in the basketball team, 18 in the volley ball team, 20 in the baseball team. Of these, there are 9 students in both the basket ball and baseball teams, 6 in both volleyball and baseball teams, and 5 in both volleyball and basketball teams. These figures include 2 students who are in all three teams. How many students are in these class ?

Solution:

$17 + 18 + 20 - 5 - 6 - 9 + 2 = 37$ students.

Basketball Valleyball Baseball
5 3 9 7 2 4 7

1-21 Logical Reasoning

Logical Reasoning: It is a thinking process to prove a desired conclusion from given facts. Logical Reasoning includes analyzing relevant information, observing patterns, and estimating unknown data to reach a logical conclusion. The logical questions in mathematics may not cover any of the topics in mathematics. We use the following symbols to express a statement.

Suppose A is a statement, we use $\sim A$ to represent the negation of A. We read $\sim A$ as "not A" or " it is false that A.".
$A \rightarrow B$ is a conditional statement in the form " if A, then B.".
$B \rightarrow A$ is the **converse** of $A \rightarrow B$. It tells that " If B, then A."
$\sim A \rightarrow \sim B$ is the **inverse** of $A \rightarrow B$. It tells that " If not A, then not B.".
$\sim B \rightarrow \sim A$ is the **contraposition** of $A \rightarrow B$. It tells that " If not B, then not A.".
In logical reasoning, we have the following rule:

Rule on Contraposition: $A \rightarrow B$ is logical equivalent to $\sim B \rightarrow \sim A$.

The rule states that given a **If-then** true statement, then there is only one other **If-then** statement (the **contraposition**) is logically true.
The rule also indicates that " If A, then B " does not necessarily mean " If B, then A.", or " If not A, then not B." are true.

A **contradiction** is a false statement because it directly contradicts either the original statement of " If A, then B " with " If A, then $\sim B$ " or " the Rule on Contraposition ".

Examples

1. $A \rightarrow B$: If two angles are vertical angles, then they are congruent. **True**
$B \rightarrow A$: If two angles are congruent, then they are vertical angles. **False**
$A \rightarrow \sim B$: If two angles are vertical angles, then they are not congruent. **False**
$\sim B \rightarrow A$: If two angles are not congruent, then they are vertical angles. **False**
$\sim A \rightarrow B$: If two angles are not vertical angles, then they are congruent. **False**
$\sim A \rightarrow \sim B$: If two angles are not vertical angles, then they are not congruent. **False**
$\sim B \rightarrow \sim A$: If two angles are not congruent, then they are not vertical angles. **True**

2. Given the true statement " If you have measles, then you have a rash.". Write another If-then statement which must be true.
Solution:
The contraposition is " If you have no rash, then you have no measles.". Ans.

3. Given the true statement " If you have measles, then you have a rash. ". Write a If-then statement that cannot be true.
Solution:
The contradiction is " If you have measles, then you have no rash. ". Ans.
or " If you have no rash, then you have measles. ". Ans.

Most logical questions apply to Rule on Contraposition. However, it may not be easy to prove a statement false. To prove a statement false, we find a contradiction or find one exception. A **counterexample** is an exception that proves a statement false. A single counterexample is sufficient to prove a statement which is false.

Examples

4. Give a counterexample to show that the statement can be false: " $|x+y| = |x| + |y|$ ".

   Solution: If $x = 4$ and $y = -1$, then $|x+y| = |4+(-1)| = |3| = 3$,

   and $|x| + |y| = |4| + |-1| = 4 + 1 = 5$.

   We have $|x+y| \neq |x| + |y|$ if x and y are in opposite signs. The statement is false.

5. Give a contradiction to show that the given statement is false:
   " If two angles are congruent, then they are vertical angles."
   Solution:
   A contradiction: "If two angles are not vertical angles, then they are not congruent. "
   Hint: It is a contradiction because the contraposition of the given statement is false.
   A contraposition must be true if the original statement is true.

6. Give a counterexample to show that the statement is false"
   " If two angles are congruent, then they are vertical angles"
   Solution:
   If two angles are congruent, then they could be two of the three angles in a triangle.

Logical Argument: A logical argument has two given statements (premises) and one conclusion. There are five patterns of logical arguments. Arguments that use these patterns correctly will have a valid conclusion. Arguments that use these patterns incorrectly will have a invalid conclusion.

1. **Direct Argument**
   If p is true, then q is true
   P is true.
   Therefore, q is true.

2. **Indirect Argument**
   If p is true, then q is true
   q is not true.
   Therefore, p is not true.

3. **Chain Rule**
   If p is true, then q is true.
   If q is true, then r is true.
   Therefore, if p, then r.

4. **Or Rule**
   P is true or q is true.
   P is not true.
   Therefore, q is true.

5. **And Rule**
   p and q not both true.
   But q is true.
   Therefore, p is not true.

Examples

7. In the logical proof of the statement
   " If $x = 4$, then $\sqrt{x}$ is an integer. ",
   decide whether the following conclusion is true or false.

   $\sqrt{x}$ is an integer. Therefore, $x = 4$.

   Solution:

   The conclusion is false.

8. In an indirect proof of the statement
   " If $x = 4$, then $\sqrt{x}$ is an integer. ",
   what is the assumption used to arrive a valid conclusion ?

   Solution:

   $\sqrt{x}$ is not an integer.
   Therefore, $x \neq 4$.

1-22 Counting Principles

1) Principle of Multiplication

If we select one from m ways, then we select another one from n ways; then all the possible selections are:

$$m \times n \text{ different ways}$$

2) Principle of Addition

If we select one from m ways, then **mutually exclusive**, we select a another one from n ways; then all the possible selections are:

$$m + n \text{ different ways}$$

Examples

1. How many different ways can a 10-question true-false test be answered if each question must be answered ?

Solution: each question has two selections:

$2\times2\times2\times2\times2\times2\times2\times2\times2\times2 = 1{,}024$ ways Or: $2^{10} = 1{,}024$ ways

2. How many license plates of 6 symbols can be made using letters for the first 2 symbols and digits for the remaining symbols ?

Solution:

$26 \times 26 \times 10 \times 10 \times 10 \times 10 =$ 6,760,000 license plates

3. How many different selections are possible to assign 4-letter from 26 alphabets ?

Solution:

$26\times26\times26\times26 = 456{,}976$ codes

4. How many positive even integers less than 1,000 can be made using the digits 5, 6, 7, and 8 ?

Solution:

1-digit even integers: 2

2-digit even integers: $4 \times 2 = 8$

3-digit even integers: $4 \times 4 \times 2 = 32$

Total 42 even integers

5. There are 3 routes from City A to City B, and 4 routes from B to City D. Mutually exclusive, there are 5 routes from City A to City C, and 2 routes from City C to City D. How many different routes can be selected from City A to City D ?

Solution:

$3\times4 = 12$ routes

$5\times2 = 10$ routes

Total 22 routes

Examples (Read Chapter 1–23 for the following 2 examples)

6. How many short-cut paths (moving upward or to the right along the grid lines) can be selected from A to B ?

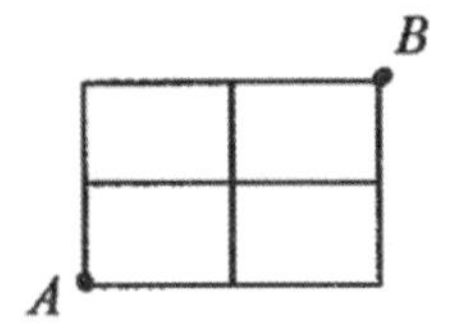

Solution:

Method 1:

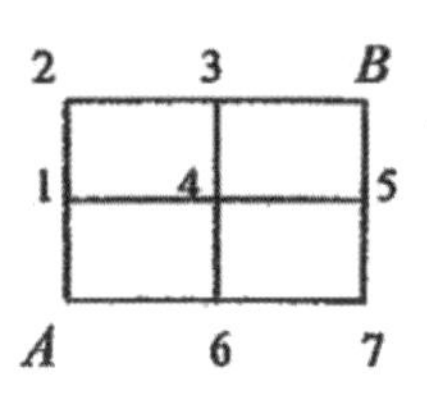

$A \to 1 \to 2 \to 3 \to B$
$A \to 1 \to 4 \to 3 \to B$
$A \to 1 \to 4 \to 5 \to B$
$A \to 6 \to 4 \to 3 \to B$
$A \to 6 \to 4 \to 5 \to B$
$A \to 6 \to 7 \to 5 \to B$

Ans: 6 paths

Method 2: Using factorial !: $\to\to\uparrow\uparrow$

$$\frac{4!}{2! \cdot 2!} = \frac{4 \cdot 3 \cdot 2 \cdot 1}{2 \cdot 1 \cdot 2 \cdot 1} = 6$$

Ans: 6 paths

7. How many short-cut paths (moving upward or to the right along the grid lines) can be select from A to D that do not include either B or C ?

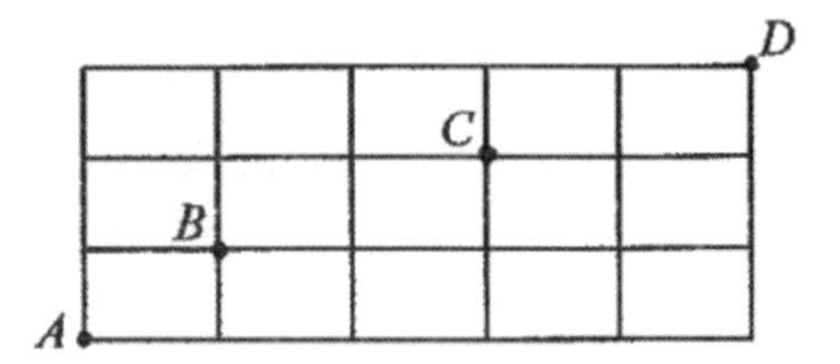

Solution:

Total paths of $A \to D$: $\to\to\to\to\to\uparrow\uparrow\uparrow$

$$\frac{8!}{5! \cdot 3!} = \frac{8 \cdot 7 \cdot 6 \cdot 5!}{5! \cdot 6} = 56$$

Find the paths that include B and C:

$$A \to B \to D \qquad \frac{2!}{1! \cdot 1!} \cdot \frac{6!}{4! \cdot 2!} = 30$$

$$A \to C \to D \qquad \frac{5!}{3! \cdot 2!} \cdot \frac{3!}{2! \cdot 1!} = 30$$

The paths of overlap within the above are:
$A \to B \to C \to D$

$$\frac{2!}{1! \cdot 1!} \cdot \frac{3!}{2! \cdot 1!} \cdot \frac{3!}{2! \cdot 1!} = 18$$

The paths that include B and C are :
$30 + 30 - 18 = 42$

The paths that does not include B and C are
$56 - 42 = 14$

Ans: 14 paths

In the shopping mall, a boy said to his father.
Boy: Why are you always holding mom's hand, dad ?
Father: If I don't hold her hand, she will keep on buying new clothes.

1-23 Permutations and Combinations

Permutation: It is an arrangement of elements without repetitions in a definite order.
Factorial Notation (*n*!): It is the product of consecutive integers beginning with *n*, decreasing and ending with 1. Read as " *n* factorial ".

The number of permutations of the 3 letters a, b, c taken 3 letters at a time:
abc, acb, bac, bca, cab, cba → total 6 permutations. $3! = 3 \cdot 2 \cdot 1 = 6$

The number of permutations of the 4 letters a, b, c, d taken 3 letters at a time:
abc, acb, bac, bca, cab, cba
bcd, bdc, cbd, cdb, dbc,dcb
cda, cad, dca, dac, acd, adc
dab, dba, adb, abd, bda, bad → total 24 permutations. $\frac{4!}{(4-3)!} = \frac{4 \cdot 3 \cdot 2 \cdot 1}{1} = 24$

The number of permutations of *n* different elements taken *n* elements at a time is :

$${}_nP_n = n \cdot (n-1) \cdot (n-2) \cdot (n-3) \cdots\cdots 3 \cdot 2 \cdot 1$$

Formula: ${}_nP_n = n!$ **Example:** ${}_{10}P_{10} = \frac{10!}{(10-10)!} = \frac{10!}{1!} = 10! = 3{,}628{,}800$

The number of permutations of *n* different elements taken *r* elements at a time is:

$${}_nP_r = n \cdot (n-1) \cdot (n-2) \cdot (n-3) \cdots\cdots (n-r+1)$$

Formula: ${}_nP_r = \frac{n!}{(n-r)!}$ **Example:** ${}_{18}P_2 = \frac{18!}{(18-2)!} = \frac{18 \cdot 17 \cdot \cancel{16!}}{\cancel{16!}} = 306$

The symbols $P(n, \mathrm{r})$ and P_r^n **are often used to denote** ${}_nP_r$**.**

Examples

1. Find the number of permutations of the letters a, b, c taken 3 letters at a time ?
   Solution:
   $${}_3P_3 = 3! = 3 \cdot 2 \cdot 1 = 6 \text{ permutations}$$

2. How many permutations can be made from the letters a, b, c, d taken 4 letters at a time ?
   Solution: ${}_4P_4 = 4! = 4 \cdot 3 \cdot 2 \cdot 1 = 24$ permutations

3. How many permutations can be made from the letters a, b, c, d taken 3 letters at a time ?
   Solution: ${}_4P_3 = \frac{4!}{(4-3)!} = \frac{4 \cdot 3 \cdot 2 \cdot 1}{1!} = \frac{24}{1} = 24$ permutations

4. How many arrangements can be selected on a shelf from 10 different books taken 5 books for each arrangements ?
   Solution:
   $${}_{10}P_5 = \frac{10!}{(10-5)!} = \frac{10 \cdot 9 \cdot 8 \cdot 7 \cdot 6 \cdot \cancel{5!}}{\cancel{5!}} = 30{,}240 \text{ arrangements}$$

Combinations: It is an arrangement of elements without repetitions and regardless of the order.

The number of combinations of the letters a, b, c taken 3 letters at a time:

abc → 1 combination.

The number of combinations of the letters a, b, c, d, take 3 letters at a time:

abc, bcd, cda, dab → 4 combinations.

Note that the only difference between a permutation and a combination is that we consider "abc, acb, bac, bca, cab, cba" ($3! = 6$) objects as one combination (regardless of the order). Therefore, we divide the formula of ${}_nP_r$ by $r!$ to obtain the formula of ${}_nC_r$.

The number of combinations of n different elements taken r elements at a time:

Formula: ${}_nC_n = 1$ **Example:** ${}_{10}C_{10} = \dfrac{10!}{(10-10)!\cdot 10!} = \dfrac{1}{0!} = \dfrac{1}{1} = 1$ (Hint: $0! = 1$)

The number of combinations of n different elements taken r elements at a time:

Formula: ${}_nC_r = \dfrac{n!}{(n-r)!\cdot r!}$; ${}_nC_r = {}_nC_{n-r}$

Examples: ${}_{18}C_2 = \dfrac{18!}{(18-2)!\cdot 2!} = \dfrac{18!}{16!\cdot 2} = \dfrac{18\cdot 17\cdot \cancel{16!}}{\cancel{16!}\cdot 2} = 153$

${}_{18}C_{16} = \dfrac{18!}{(18-16)!\cdot 16!} = \dfrac{18!}{2!\cdot 16!} = \dfrac{18\cdot 17\cdot \cancel{16!}}{2!\cdot \cancel{16!}} = 153$

The symbols $C(n, r)$, C_r^n and $\binom{n}{r}$ **are often used to denote** ${}_nC_r$.

Examples

1. Find the number of combinations of the letters a, b, c taken 3 letters at a time ?

   Solution: ${}_3C_3 = \dfrac{\cancel{3!}}{(3-3)!\cdot \cancel{3!}} = \dfrac{1}{0!} = \dfrac{1}{1} = 1$ combination

2. How many combinations can be made from the letters a, b, c, d taken 4 letters at a time ?

   Solution: ${}_4C_4 = 1$ combination

3. How many combinations can be made from the letters a, b, c, d taken 3 letters at a time ?

   Solution: ${}_4C_3 = \dfrac{4!}{(4-3)!\cdot 3!} = \dfrac{24}{1!\cdot 6} = 4$ combinations

4. There are 20 boys in a class. How many different committees of 3 boys can be formed ?

   Solution:

   ${}_{20}C_3 = \dfrac{20!}{(20-3)!\cdot 3!} = \dfrac{20\cdot 19\cdot 18\cdot \cancel{17!}}{\cancel{17!}\cdot 6} = 1{,}140$ committees

5. There are 20 boys and 15 girls in a class. A committee to be formed consisting of 3 boys and 3 girls. How many different committees are possible ?

   Solution:

   ${}_{20}C_3 \cdot {}_{15}C_3 = \dfrac{20!}{(20-3)!\cdot 3!} \cdot \dfrac{15!}{(15-3)!\cdot 3!} = \dfrac{20\cdot 19\cdot 18}{6} \cdot \dfrac{15\cdot 14\cdot 13}{6} = 1140\cdot 455 = 518{,}700$ committees

1-24 Matrices and Determinants

Matrix: It is a rectangular array of numbers enclosed by brackets.
Element: Each number in a matrix is called the element of the matrix.
Dimensions of a matrix: It is the number of rows and the number of columns in a matrix. The number of rows is given first. A 3×4 matrix is a matrix having 3 rows and 4 columns (read " 3 by 4 matrix ").
Row Matrix: It is a matrix having only one row.
Column Matrix: It is a matrix having only one column.
Square Matrix: It is a matrix having the same numbers in rows and columns.
Zero Matrix: It is a matrix having all elements to be zero.
Identity Matrix: It is a square matrix whose main diagonal has all elements by 1 and all other elements by 0.
Two matrices are equal if and only if they have the same dimensions and the same elements in all corresponding positions.

Matrix addition: We add the corresponding elements of the matrices being added.
Matrix Subtraction: We subtract the corresponding elements of the matrices being subtracted. Or, add the additive inverse of the second matrix.
Addition and subtraction of matrices of different dimensions are not allowed.

Matrix Multiplication: The product (multiplication) of two matrices is the matrix in which each element is the sum of the product of corresponding elements in row of the first matrix and in column of the second matrix.

$$A_{m\times n} \cdot B_{n\times p} = X_{m\times p}$$

The dimension of column in the first matrix must equal the dimension of row in the second matrix. Otherwise, they cannot be multiplied.

Examples

1. Let $A=\begin{bmatrix} 5 & -4 \\ -1 & 2 \end{bmatrix}$, $B=\begin{bmatrix} 3 & 6 \\ 2 & -5 \end{bmatrix}$

a. $A+B=\begin{bmatrix} 8 & 2 \\ 1 & -3 \end{bmatrix}$ **b.** $A-2B=A+(-2B)=\begin{bmatrix} 5 & -4 \\ -1 & 2 \end{bmatrix}+\begin{bmatrix} -6 & -12 \\ -4 & 10 \end{bmatrix}=\begin{bmatrix} -1 & -16 \\ -5 & 12 \end{bmatrix}$

2. $\begin{bmatrix} 2 & 3 \end{bmatrix}\cdot\begin{bmatrix} 5 \\ 7 \end{bmatrix}=\begin{bmatrix} 2\cdot 5+3\cdot 7 \end{bmatrix}=\begin{bmatrix} 31 \end{bmatrix}_{1\times 1}$

3. $\begin{bmatrix} 5 \\ 7 \end{bmatrix}_{2\times 1}\cdot\begin{bmatrix} 2 & 3 \end{bmatrix}_{1\times 2}=\begin{bmatrix} 5\cdot 2 & 5\cdot 3 \\ 7\cdot 2 & 7\cdot 3 \end{bmatrix}=\begin{bmatrix} 10 & 15 \\ 14 & 21 \end{bmatrix}_{2\times 2}$

4. $\begin{bmatrix} 3 & -1 \\ 2 & 0 \end{bmatrix}\cdot\begin{bmatrix} -1 \\ 2 \end{bmatrix}_{2\times 1}=\begin{bmatrix} 3(-1)+(-1)\cdot 2 \\ 2(-1)+0\cdot 2 \end{bmatrix}_{2\times 1}=\begin{bmatrix} -5 \\ -2 \end{bmatrix}_{2\times 1}$

5. $\begin{bmatrix} -1 \\ 2 \end{bmatrix}_{2\times 1}\cdot\begin{bmatrix} 3 & -1 \\ 2 & 0 \end{bmatrix}_{2\times 2}$ They cannot be Multiplied

Determinants

The form of a determinant is similar to a matrix. A matrix is formed with brackets enclosing the entries. A **determinant** is formed with vertical bars enclosing the entries. Each entry is called an **element** of the determinant.

The number of elements in any row or column is called the **order** of the determinant.

Determinant of a 2 × 2 matrix A is defined as follows:

$$\det A = \begin{vmatrix} a & b \\ c & d \end{vmatrix} = ad - bc$$

Determinant of a 3 × 3 matrix A is defined as follows:(Three ways)

1. $\det A = \begin{vmatrix} a_1 & b_1 & c_1 \\ a_2 & b_2 & c_2 \\ a_3 & b_3 & c_3 \end{vmatrix} = a_1b_2c_3 + a_2b_3c_1 + a_3b_1c_2 - a_3b_2c_1 - a_1b_3c_2 - a_2b_1c_3$

The products of elements
1. run down from left to right are positive.
2. run down from right to left are negative.

2. Expansion by its minors across any row

$$\det A = a_1\begin{vmatrix} b_2 & c_2 \\ b_3 & c_3 \end{vmatrix} - b_1\begin{vmatrix} a_2 & c_2 \\ a_3 & c_3 \end{vmatrix} + c_1\begin{vmatrix} a_2 & b_2 \\ a_3 & b_3 \end{vmatrix}$$

3. Copy the first two columns to the right side of the determinant

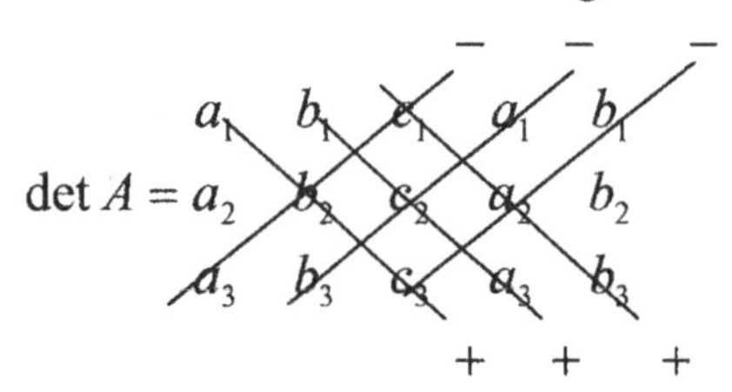

The above formulas are not valid for $n \times n$ if $n \geq 4$.

Examples (Evaluate each determinant)

1. $\begin{vmatrix} 1 & 5 \\ 4 & 8 \end{vmatrix} = 1 \cdot 8 - 5 \cdot 4 = 8 - 20 = -12$

2. $\begin{vmatrix} -3 & 6 \\ -2 & 4 \end{vmatrix} = -3 \cdot 4 - 6 \cdot (-2) = -12 + 12 = 0$

3. $\begin{vmatrix} 2 & 1 & -3 \\ 3 & 5 & -4 \\ -4 & 2 & 6 \end{vmatrix} = 60 + (-18) + 16 - 60 - (-16) - 18 = 60 - 18 + 16 - 60 + 16 - 18 = -4$

4. $\begin{vmatrix} \sqrt{2} & -\sqrt{3} \\ \sqrt{15} & \sqrt{6} \end{vmatrix} = \sqrt{2} \cdot \sqrt{6} - (-\sqrt{3} \cdot \sqrt{15}) = \sqrt{12} + \sqrt{45} = 2\sqrt{3} + 3\sqrt{5}$

5. $\begin{vmatrix} \log 5 & \log 2 \\ \log 2 & \log 5 \end{vmatrix} = (\log 5)^2 - (\log 2)^2 = (\log 5 + \log 2)(\log 5 - \log 2) = \log(5 \times 2) \cdot \log \frac{5}{2}$

$= \log 10 \cdot \log \frac{5}{2} = 1 \cdot \log \frac{5}{2} = \log \frac{5}{2}$

1-25 Arithmetic and Geometric Series

Arithmetic Series: It is the sum of an arithmetic sequence.
Geometric Series: It is the sum of a geometric sequence.

Formulas:

1. The sum of the first n terms of an **arithmetic series** is given by:

$$S_n = a_1 + a_2 + a_3 + \cdots + a_n = \frac{n}{2}(a_1 + a_n)$$

Where $a_1 \rightarrow 1^{st}$ term , $a_n \rightarrow$ last term , $n \rightarrow n$th terms

2. The sum of the first n terms of a **geometric series** is given by:

$$S_n = a_1 + a_1 r + a_1 r^2 + a_1 r^3 + \cdots + a_1 r^{n-1} = \frac{a_1(1-r^n)}{1-r} \ , \ r \neq 1$$

Where $a_1 \rightarrow$ 1st term , $n \rightarrow$ The nth terms , $r \rightarrow$ common ratio

(**Hint:** Find n from $a_n = a_1 r^{n-1}$ if n is not given.)

We use the above formula, the Index must begin at $k = 1$.
(see example 5 below)

Examples

1. Find the sum of the arithmetic sequence $1 + 2 + 3 + 4 + \cdots + 500$.
Solution: $a_1 = 1$, $a_{500} = 500$, $n = 500$

$$S_{500} = \frac{n}{2}(a_1 + a_{500}) = \frac{500}{2}(1 + 500) = 250(501) = 125{,}250$$

2. Evaluate the series $\sum_{k=1}^{30}(-5k + 10)$.

Solution: $5, 0, -5, -10, \cdots$

$a_1 = 5$, $d = -5$, $a_{30} = -5(30) + 10 = -140$

$S_{30} = \frac{30}{2}[5 + (-140)] = -2025$

3. Find the sum of the geometric sequence $1 + 3 + 9 + 27 + \cdots$ for the first 8 terms.
Solution:

$$a_1 = 1, n = 8, r = 3, \ S_8 = \frac{a_1(1-r^n)}{1-r} = \frac{1 \cdot (1-3^8)}{1-3} = \frac{1-6561}{-2} = \frac{-6560}{-2} = 3{,}280$$

4. $$\sum_{k=1}^{12}\left(\frac{1}{2}\right)^k = \frac{1}{2} + \frac{1}{4} + \frac{1}{8} + \cdots + \left(\frac{1}{2}\right)^{12} = \frac{\frac{1}{2}[1 - (\frac{1}{2})^{12}]}{1 - \frac{1}{2}} = 1 - \left(\frac{1}{2}\right)^{12}$$

5. $$\sum_{k=0}^{12}\left(\frac{1}{2}\right)^k = \left(\frac{1}{2}\right)^0 + \sum_{k=1}^{12}\left(\frac{1}{2}\right)^k = 1 + \frac{\frac{1}{2}[1 - (\frac{1}{2})^{12}]}{1 - \frac{1}{2}} = 1 + \left[1 - \left(\frac{1}{2}\right)^{12}\right] = 2 - \left(\frac{1}{2}\right)^{12}$$

1-26 Infinite Geometric Series

Infinite Geometric Series: It is a geometric series having its terms increases without limit (without bound).

$$a_1 + a_1 r + a_1 r^2 + a_1 r^3 + \cdots\cdots + a_1 r^{n-1} + \cdots\cdots \quad \text{as } n \to \infty .$$

Formula: The sum of a finite geometric series is given by:

$$S_n = \sum_{k=1}^{n} a_1 r^{k-1} = \frac{a_1(1-r^n)}{1-r}$$

Find the formula for the sum of an infinite geometric series.

1) If $|r| \geq 1$, then $|r^n|$ increases without limit as $n \to \infty$, the sum increases without limit as n increases without limit (the series has no sum).
2) If $|r| < 1$, then $|r^n|$ approaches 0 as $n \to \infty$, the sum is a finite number given by:

$$S_\infty = \sum_{k=1}^{\infty} a_1 r^{k-1} = a_1 + a_1 r^2 + a_1 r^3 + \cdots\cdots + a_1 r^{n-1} + \cdots\cdots = \frac{a_1}{1-r}, \quad |r| < 1$$

Examples

1. Find the sum of the infinite geometric series: $9 - 6 + 4 - \frac{8}{3} + \cdots\cdots$

   Solution:

   $$r = \tfrac{-6}{9} = -\tfrac{2}{3}, \quad |r| = \left|-\tfrac{2}{3}\right| = \tfrac{2}{3} < 1$$

   The series has a sum (a finite number).

   $$S_\infty = \frac{a_1}{1-r} = \frac{9}{1-(-\frac{2}{3})} = \frac{9}{5/3} = \frac{27}{5}$$

2. Find the sum of the infinite geometric series: $1 - \sqrt{2} + 2 - 2\sqrt{2} + 4 - 4\sqrt{2} + \cdots$

   Solution:

   $$r = \tfrac{-\sqrt{2}}{1} = -\sqrt{2}, \quad |r| = \left|-\sqrt{2}\right| = \sqrt{2} = 1.414 > 1$$

   The series has no sum. The sum increases without limit as n increases without limit.

3. Find the sum of the infinite geometric series: $1 + \frac{1}{2} + \frac{1}{4} + \frac{1}{8} + \frac{1}{16} + \cdots\cdots$

   Solution:

   $$r = \tfrac{1}{2} \div 1 = \tfrac{1}{2} < 1$$

   The series has a sum (a finite number).

   $$S_\infty = \frac{a_1}{1-r} = \frac{1}{1-\frac{1}{2}} = 2$$

1-27 Vectors

A quantity, such as a velocity or a force, that is given by both its magnitude and direction is represented by a **vector**.

A vector from point A to point B on a coordinate plane written by an arrow $\overrightarrow{AB}$.

We sometimes use a boldface letter to represent a vector. To write a vector, an arrow is placed over the letter.

A force of 10 units at a 40° angle.

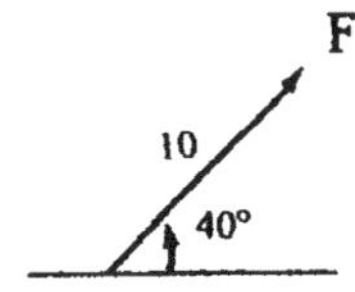

$\mathbf{v} = \overrightarrow{AB}$

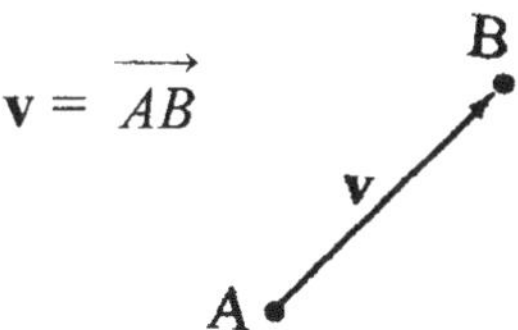

Point A is the **initial point** and Point B is the **terminal point**. The length of the arrow arrow indicates the **magnitude** (or **norm**) of the vector. The arrow-head of the arrow indicates the **direction** of the vector. The direction of the vector is the angle θ that the vector makes with the positive $x-axis$. The magnitude of a vector **v** is written as $||\mathbf{v}||$.

Zero Vector 0: A vector whose magnitude is zero. A zero vector has no direction.
Equivalent Vectors: They are vectors that have the same magnitude and direction.

$$\overrightarrow{AB} = \overrightarrow{CD}$$

The **Opposite** of a vector: – **v** is a vector with the same magnitude but the opposite direction as **v**.

Addition (Resultant) of two vectors: To add two vectors, we place the initial point of the second vector at the terminal point of the first vector , connect the initial point of the first vector to the terminal point of the second vector.

Subtraction (difference) of two vectors: To subtract two vectors, we add the opposite of the second vector. $\mathbf{u} - \mathbf{v} = \mathbf{u} + (-\mathbf{v})$, $\mathbf{v} - \mathbf{u} = \mathbf{v} + (-\mathbf{u})$

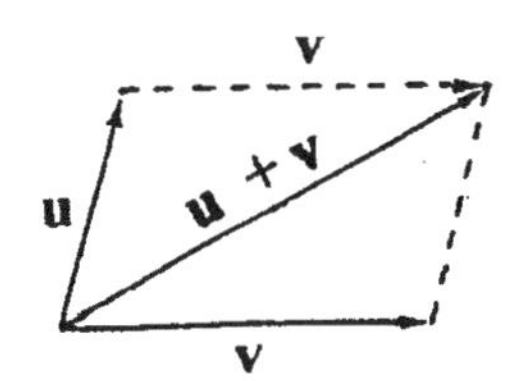

(Form a parallelogram)

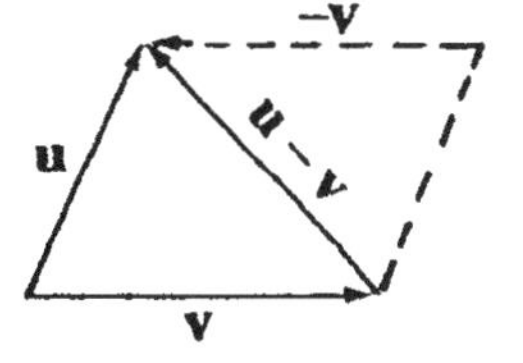

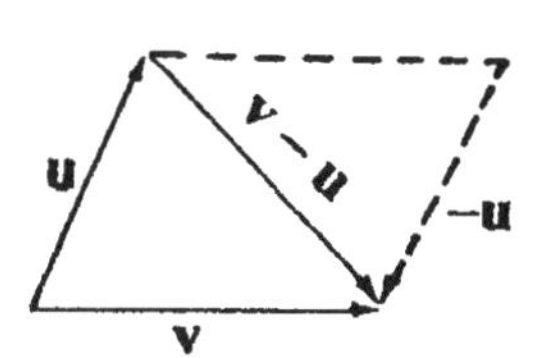

The addition of vectors is **commutative** and **associative**:

$$\mathbf{v} + \mathbf{u} = \mathbf{u} + \mathbf{v}, \quad \mathbf{u} + (\mathbf{v} + \mathbf{w}) = (\mathbf{u} + \mathbf{v}) + \mathbf{w}$$

In a coordinate plane, a vector **v** with initial point at the origin and terminal point at $p(a,\ b)$ can be represented by a (the $x-$component) and b (the $y-$component) in the form of a **position vector** :

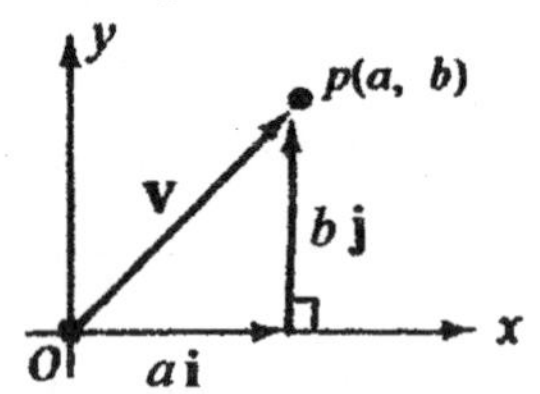

$$\mathbf{v} = \overrightarrow{op} = (a,\ b) \quad \text{or} \quad \mathbf{v} = a\,\mathbf{i} + b\,\mathbf{j} \quad \text{and} \quad \|\mathbf{v}\| = \sqrt{a^2 + b^2}$$

where **i** is the **unit vector** in the same direction as the positive $x-axis$, **j** is the **unit vector in** the same direction as the positive $y-axis$, $\mathbf{i} = (1, 0)$ and $\mathbf{j} = (0, 1)$.

To add two vectors, we add the corresponding components. To subtract two vectors, we subtract the corresponding components.

If any vector **v** whose initial point $p_1(x_1,\ y_1)$ is not at the origin, and the terminal point is at $p_2(x_2,\ y_2)$, then we have:

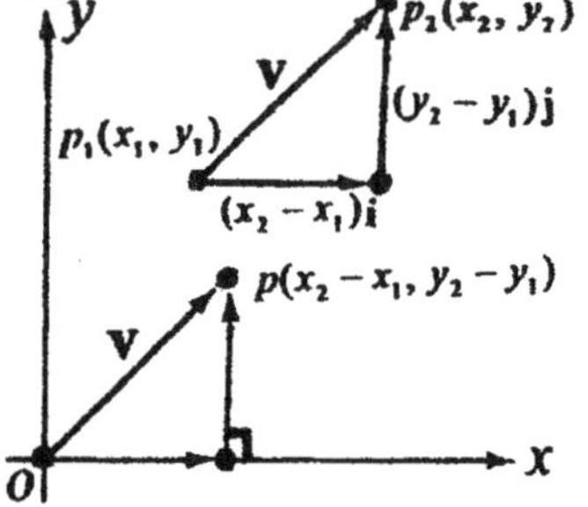

$\overrightarrow{p_1p_2} = \overrightarrow{op}$ (They have the same magnitude and direction.)

$$\therefore\ \mathbf{v} = \overrightarrow{p_1p_2} = (x_2 - x_1)\mathbf{i} + (y_2 - y_1)\mathbf{j}$$

A vector **u** whose magnitude $\|\mathbf{u}\| = 1$ is called a **unit vector**. To find the **unit vector** having the same direction as a given nonzero vector **v** , we use the following formula:

$$\textbf{unit vector} = \frac{\mathbf{v}}{\|\mathbf{v}\|}$$

Dot product (inner product) of two vectors

If $\mathbf{v} = a_1\mathbf{i} + b_1\mathbf{j}$ and $\mathbf{u} = a_2\mathbf{i} + b_2\mathbf{j}$ are two vectors, then the **dot product** $\mathbf{v} \cdot \mathbf{u} = a_1a_2 + b_1b_2$.

The dot product of two vectors are **commutative** and **distributive**:

$$\mathbf{v} \cdot \mathbf{u} = \mathbf{u} \cdot \mathbf{v} \quad ; \quad \mathbf{v} \cdot (\mathbf{u} + \mathbf{w}) = \mathbf{v} \cdot \mathbf{u} + \mathbf{v} \cdot \mathbf{w}$$

Two vectors **v** and **u** are **perpendicular** (or **orthogonal**) if and only if $\mathbf{v} \cdot \mathbf{u} = 0$.

Two vectors **v** and **u** are **parallel** if there is a nonzero scalar c and $\mathbf{v} = c\,\mathbf{u}$.

Angle between two vectors: If **v** and **u** are two nonzero vectors, the angle θ between **v** and **u** is given by the following formula:

$$\cos\theta = \frac{\mathbf{u} \cdot \mathbf{v}}{\|\mathbf{u}\| \cdot \|\mathbf{v}\|}$$

The vector $\mathbf{v} = 4\mathbf{i} - 2\mathbf{j}$ and $\mathbf{u} = 2\mathbf{i} + 4\mathbf{j}$ are perpendicular since $\mathbf{v} \cdot \mathbf{u} = (4)(2) + (-2)(4) = 0$. Or

$$\cos\theta = 0 \quad \therefore \theta = 90°.$$

The vector $\mathbf{v} = 4\mathbf{i} - 2\mathbf{j}$ and $\mathbf{u} = 2\mathbf{i} - \mathbf{j}$ are parallel since $\mathbf{v} = 2\,\mathbf{u}$. Or

$$\cos\theta = \frac{(4)(2) + (-2)(-1)}{\sqrt{4^2 + (-2)^2} \cdot \sqrt{2^2 + (-1)^2}} = \frac{10}{\sqrt{20} \cdot \sqrt{5}} = \frac{10}{\sqrt{100}} = 1 \quad \therefore \theta = 0°$$

Examples

1. If $\overrightarrow{AB} = (4,\ 1)$ and $\overrightarrow{BC} = (1,\ 2)$, find $\overrightarrow{AB} + \overrightarrow{BC}$.
   Solution:
$$\overrightarrow{AB} + \overrightarrow{BC} = (4,\ 1) + (1, 2) = (4 + 1, 1 + 2) = (5, 3)$$

2. If $\mathbf{v} = (-4, 2)$ and $\mathbf{u} = (3, 4)$, find the **resultant** and the **norm** (magnitude) of $\mathbf{v} + \mathbf{u}$. Express $\mathbf{v}$ in terms of $\mathbf{i}$ and $\mathbf{j}$.
   Solution:
$$\text{Resultant } \mathbf{v} + \mathbf{u} = (-4 + 3, 2 + 4) = (-1, 6)$$
$$\text{Norm } \|\mathbf{v}+\mathbf{u}\| = \sqrt{(-1)^2 + 6^2} = \sqrt{37}$$
$$\mathbf{v} = -4\mathbf{i} + 2\mathbf{j}$$

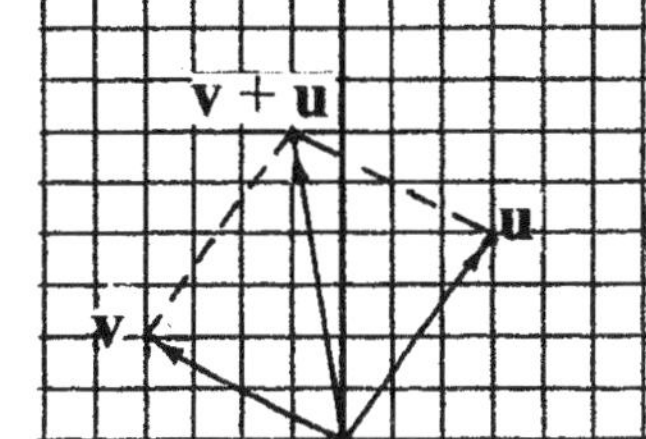

3. If $\mathbf{u} = (-4, 2)$ and $\mathbf{u} + \mathbf{v} = (-1, 6)$, find $\mathbf{v}$.
   Solution:
$$\mathbf{u} + \mathbf{v} = (-4 + v_1, 2 + v_2) = (-1, 6)$$
$$\text{We have } v_1 = 3,\ v_2 = 4 \quad \therefore\ \mathbf{v} = (3, 4)$$

4. Find the position vector of $\mathbf{v} = \overrightarrow{p_1 p_2}$ if its initial and terminal points are $p_1(5,\ 3)$ and $p_2(-1, -4)$.
   Solution: $\mathbf{v} = (-1 - 5)\mathbf{i} + (-4 - 3)\mathbf{j} = -6\mathbf{i} - 7\,\mathbf{j}$

5. If $\mathbf{v} = 3\mathbf{i} + 2\mathbf{j}$ and $\mathbf{u} = 5\mathbf{i} - 4\mathbf{j}$, find $\mathbf{v} + \mathbf{u}$, $\mathbf{v} - \mathbf{u}$, $\mathbf{u} - \mathbf{v}$, $4\mathbf{u}$, $3\mathbf{v} - 4\mathbf{u}$, and $\|\mathbf{v}\|$.
   Solution:
$$\mathbf{v} + \mathbf{u} = (3\mathbf{i} + 2\mathbf{j}) + (5\mathbf{i} - 4\mathbf{j}) = (3 + 5)\mathbf{i} + (2 - 4)\mathbf{j} = 8\mathbf{i} - 2\mathbf{j}$$
$$\mathbf{v} - \mathbf{u} = \mathbf{v} + (-\mathbf{u}) = (3\mathbf{i} + 2\mathbf{j}) + (-5\mathbf{i} + 4\mathbf{j}) = (3 - 5)\mathbf{i} + (2 + 4)\mathbf{j} = -2\mathbf{i} + 6\mathbf{j}$$
$$\mathbf{u} - \mathbf{v} = \mathbf{u} + (-\mathbf{v}) = (5\mathbf{i} - 4\mathbf{j}) + (-3\mathbf{i} - 2\mathbf{j}) = (5 - 3)\mathbf{i} + (-4 - 2)\mathbf{j} = 2\mathbf{i} - 6\mathbf{j}$$
$$4\mathbf{u} = 4(5\mathbf{i} - 4\mathbf{j}) = 20\mathbf{i} - 16\mathbf{j}$$
$$3\mathbf{v} - 4\mathbf{u} = (9\mathbf{i} + 6\mathbf{j}) + (-20\mathbf{i} + 16\mathbf{j}) = -11\mathbf{i} + 22\mathbf{j}$$
$$\|\mathbf{v}\| = \|3i + 2j\| = \sqrt{3^2 + 2^2} = \sqrt{13}$$

6. Find a **unit vector** having the same direction as $\mathbf{u} = 3\mathbf{i} + 4\mathbf{j}$.
   Solution:
$$\|\mathbf{u}\| = \|3i - 4j\| = \sqrt{3^2 + 4^2} = 5 \quad \therefore \textbf{ unit vector} = \frac{\mathbf{u}}{\|\mathbf{u}\|} = \frac{3i + 4j}{5} = \frac{3}{5}i + \frac{4}{5}j$$

Hint: Its magnitude (norm) $= \sqrt{(\frac{3}{5})^2 + (\frac{4}{5})^2} = \sqrt{\frac{9}{25} + \frac{16}{25}} = 1$

Examples

7. Express $\mathbf{v} = -4\mathbf{i} + 2\mathbf{j}$ in terms of $\mathbf{a} = (2, 3)$ and $\mathbf{b} = (4, 2)$.

Solution:

$$\mathbf{v} = (-4, 2)$$

$$\mathbf{v} = v_1\,\mathbf{a} + v_2\,\mathbf{b} = v_1(2, 3) + v_2(4, 2) = (2v_1, 3v_1) + (4v_2, 2v_2)$$

$$= (2v_1 + 4v_2, 3v_1 + 2v_2)$$

Solve $\begin{cases} 2v_1 + 4v_2 = -4 \\ 3v_1 + 2v_2 = 2 \end{cases}$ we have $v_1 = 2$, $v_2 = -2$ $\therefore \mathbf{v} = 2\mathbf{a} - 2\mathbf{b}$

Check : $\mathbf{v} = 2\mathbf{a} - 2\mathbf{b} = 2(2, 3) - 2(4, 2) = (4, 6) - (8, 4) = (-4, 2)$

8. Find the x – component and y – component of a vector $\mathbf{v}$, 10 units long at a direction of $60°$.

Solution:

$$x = 10\cos 60° = 10(0.5) = 5$$

$$y = 10\sin 60° = 10(0.866) = 8.66$$

$$\mathbf{v} = (5, 8.66)$$

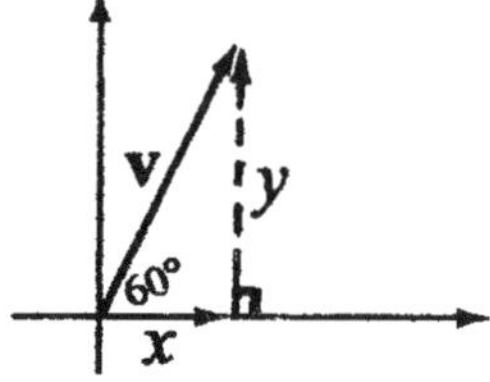

9. If $\mathbf{v} = 3\mathbf{i} + 4\mathbf{j}$ and $\mathbf{u} = 4\mathbf{i} + 3\mathbf{j}$, find the dot product $\mathbf{v} \cdot \mathbf{u}$ and the angle between $\mathbf{v}$ and $\mathbf{u}$.

Solution:

$$\mathbf{v} \cdot \mathbf{u} = 3(4) + 4(3) = 24$$

$$\cos\theta = \frac{\mathbf{v} \cdot \mathbf{u}}{\|\mathbf{v}\| \cdot \|\mathbf{u}\|} = \frac{24}{\sqrt{3^2 + 4^2} \cdot \sqrt{4^2 + 3^2}} = \frac{24}{25} = 0.96 \qquad \therefore \theta = 16.26°$$

10. An object is pulled by a force of 10 units to the south and a force of 6 units at a heading (clock-wise from due north) of $60°$. Find the magnitude and direction of the resultant.

Solution:

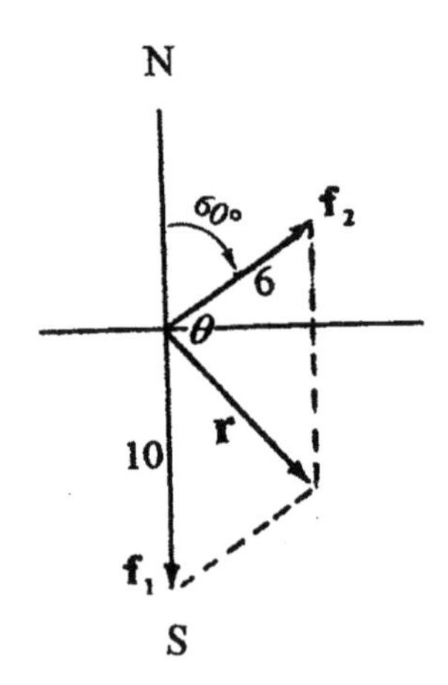

The force of 10 units to the south is $\mathbf{f}_1 = -10\,\mathbf{j}$

The force of 6 units at a heading of $60°$ is

$$\mathbf{f}_2 = (6\cos 30°)\,\mathbf{i} + (6\sin 30°)\,\mathbf{j} = 3\sqrt{3}\,\mathbf{i} + 3\,\mathbf{j}$$

The resultant $\mathbf{r} = (-10\,\mathbf{j}) + (3\sqrt{3}\,\mathbf{i} + 3\,\mathbf{j}) = 3\sqrt{3}\,\mathbf{i} - 7\,\mathbf{j}$

The magnitude of the resultant is

$$\|\mathbf{r}\| = \sqrt{(3\sqrt{3})^2 + (-7)^2} = \sqrt{76} = 8.72$$

The direction of the resultant is given by

$$\cos\theta = \frac{\mathbf{f}_2 \cdot \mathbf{r}}{\|\mathbf{f}_2\| \cdot \|\mathbf{r}\|} = \frac{(3\sqrt{3})(3\sqrt{3}) + 3(-7)}{(6)(8.72)} = \frac{6}{52.32} = 0.1147$$

$\therefore\ \theta = 83.4°$ (It is at a heading of $143.4°$.)

SAMPLE QUESTIONS

CHAPTER 1: NUMBER and OPERATIONS

1. If $a+b=15$, what is the value of $5a+5b$?
Solution:
$$5a+5b=5(a+b)=5(15)=75. \text{ Ans.}$$

2. If $3a+2b=c$, what is a in terms of b and c ?
Solution:
$$3a=c-2b,\quad a=\frac{c-2b}{3}. \text{ Ans.}$$

3. The price of a television went up 20% since last month. If last month's price was x, what is this month's price in terms of x ?
Solution:
$$x+0.2x=1.2x. \text{ Ans.}$$

4. The average price of gasoline was \$2.95. This year, the average price of gasoline is \$3.50. By what percent has the price of gasoline increased ?
Solution:
$$\frac{3.50-2.95}{2.95}\approx 0.186\approx 18.6\%. \text{ Ans.}$$

5. In a 5-student basketball team, the average height of 3 players is 176 *cm*. What does the average height in *cm* of the other 2 players have to be if the average height of the entire team equals 182 *cm* ?
Solution:
$$182\times 5=910\ cm$$
$$176\times 3=528\ cm$$
$$\frac{910-528}{2}=191\,cm. \text{ Ans.}$$

6. A car traveling at an average of 60 miles per hour made a trip in 6 hours. If it had traveled at an average of 50 miles per hour, how many minutes longer the trip would have taken ?
Solution:
$$60\times 6=360 \text{ miles}$$
$$360\div 50=7.2 \text{ hours}$$
$$7.2-6=1.2 \text{ hours}=72 \text{ minutes. Ans.}$$

7. Find the 30th terms of the arithmetic sequence
$$25, 33, 41, \cdots\cdots, a_{30}.$$
Solution:
$$d=33-25=8$$
$$a_{30}=a_1+(n-1)\cdot d$$
$$=25+(30-1)\cdot 8=257. \text{ Ans.}$$

8. Find the sum of the arithmetic sequence
$$25+33+41+\cdots\cdots+257$$
Solution:
$$a_n=a_1+(n-1)\cdot d,\quad d=33-25=8$$
$$257=25+(n-1)\cdot 8$$
$$232=8n-8,\quad \therefore n=30$$
$$S_n=\frac{n(a_1+a_n)}{2}=\frac{30(25+257)}{2}=4{,}230. \text{ Ans.}$$

9. Find a_{20} of the arithmetic sequence if
$$a_5=36,\quad a_{11}=48.$$
Solution:
$$d=\frac{48-36}{11-5}=2\quad \therefore a_1=28$$
$$a_{20}=a_1+(n-1)\cdot d$$
$$=28+(20-1)\cdot 2=66. \text{ Ans.}$$

10. The 3rd term of an arithmetic sequence is 32 and the 11st term is 48. Find the sum of the first 15 terms.
Solution: $d=\frac{48-32}{11-3}=2\quad \therefore a_1=28,\ a_{15}=56$
$$S_{15}=\frac{15(a_1+a_{15})}{2}=\frac{15(28+56)}{2}=630. \text{ Ans.}$$

SAMPLE QUESTIONS

CHAPTER 1: NUMBER and OPERATIONS

1. Write all the whole-number factors of 12.
Solution:
1, 2, 3, 4, 6, and 12. Ans

2. Write all the whole-number factors of 36.
Solution:
1, 2, 3, 4, 6, 9, 12, 18, and 36. Ans.

3. Write the first 5 multiples of 12.
Solution:
0, 12, 24, 36, 48. Ans.

4. Write the first 5 positive-integer multiples of 12.
Solution: 12, 24, 36, 48, and 60. Ans.

5. How many of the first 250 positive integers are multiples of 3 ?
Solution:
$1\times3,\ 2\times3,\ 3\times3, \cdots\cdots, 83\times3=249$
There are 83 multiples of 3. Ans.

6. How many of the first 250 positive integers are multiples of 4 ?
Solution:
$1\times4,\ 2\times4,\ 3\times4, \cdots, 62\times4=248$
There are 62 multiples of 4. Ans.

7. How many multiples of 6 are between 100 and 1,000 ?
Solution:
$1\times6,\ 2\times6,\ 3\times6, \cdots, 16\times6=96,$
$17\times6=102, \cdots, 166\times6=996$
$166-16=150$ of these integers are multiples of 6. Ans.

8. How many of the first 250 positive integers are multiples of 3 and 4 ?
Solution:
An integer is a multiple of both 3 and 4 if and only it is a multiple of 12.
$1\times12,\ 2\times12,\ 3\times12, \cdots\cdots, 20\times12=240$
There are 20 multiples of 12. Ans.

9. How many of the first 250 positive integers are multiples of 3 or 4 ?
Solution:
$1\times3,\ 2\times3,\ 3\times3,\ 4\times3, \cdots, 8\times3, \cdots, 83\times3=249$ There are 83 multiples of 3.
$1\times4,\ 2\times4,\ 3\times4, \cdots, 6\times4, \cdots\cdots, 62\times4=248$ There are 62 multiples of 4.
The following multiples of 12 are overlap within the multiples of 3 and 4.
$1\times12,\ 2\times12,\ 3\times12, \cdots\cdots, 20\times12=240$ There are 20 multiples of 12.
$83+62-20=125$ of these integers are multiples of 3 or 4 or both 3 and 4. Ans.

10. How many of the first 250 positive integers are multiples of neither 3 nor 4 ?
Solution: (See above example) $250-125=125$ of these integers are multiples of neither 3 nor 4.

11. How many of the first 250 positive integers are multiples of 6 or 15 ?
Solution:
$1\times6=6,\ 2\times6=12,\ 3\times6=18, \cdots\cdots\ 41\times6=246$ There are 41 multiples of 6.
$1\times15=15,\ 2\times15=30,\ 3\times15=45, \cdots, 16\times15=240$ There are 16 multiples of 15.
The following multiples of 30 are overlap within the multiples of 6 and 15.
$1\times30,\ 2\times30,\ 3\times30, \cdots, 8\times30=240$ There are 8 multiples of 30.
$41+15-8=48$ of these integers are multiples of 6 or 15 or both 6 and 15. Ans.
(Hint: $250-48=202$ of these integers are multiples of neither 6 nor 15.)

SAMPLE QUESTIONS

CHAPTER 1: NUMBER and OPERATIONS

1. If $\log_3 x = 15$, what is the value of x ?

Solution:

$\log_3 x = 5$

$x = 3^{15}$. Ans.

2. If $\log_3 x^2 = 15$, what is the value of $\log_3 x$?

Solution:

$\log_3 x^2 = 15$

$2\log_3 x = 15 \quad \therefore \log_3 x = \frac{15}{2}$. Ans.

3. If $\log_a x^3 = 15$, What is the value of $\log_a x$?

Solution:

$\log_a x^3 = 15$

$3\log_a x = 15$

$\log_a x = \frac{15}{3} = 5$. Ans.

4. If $\log_2 x^3 = 15$, what is the value of x ?

Solution:

$\log_2 x^3 = 15$

$3\log_2 x = 15, \quad \log_2 x = \frac{15}{3} = 5$

$\therefore x = 2^5 = 32$. Ans.

5. If $(1, k)$ is a solution to the system of inequalities: $y > x^2 + 2$

$y < x + 5$

What is the intervals of the values k ?

Solution:

Substitute 1 for x. We have:

$y > 3$ and $y < 8$

Substitute k for y. We have:

$3 < k < 8$. Ans.

6. If $-4 < a < 12$ and $-2 < b < 11$, what is the range of values of $a - b$?

Solution: $-4 - 11 < a - b < 12 - (-2)$

$-15 < a - b < 14$. Ans.

7. If $-4 < a < 12$ and $-2 < b < 11$, what is the range of values of ab ?

Solution: The greatest value is $12 \times 11 = 132$.

The least value is $-4 \times 11 = -44$.

$-44 < ab < 132$. Ans.

8. The graph of a function $y = f(x)$ is given. Graph the following functions using the techniques of transformations. **a.** $y = |f(x)|$ **b.** $y = f(|x|)$

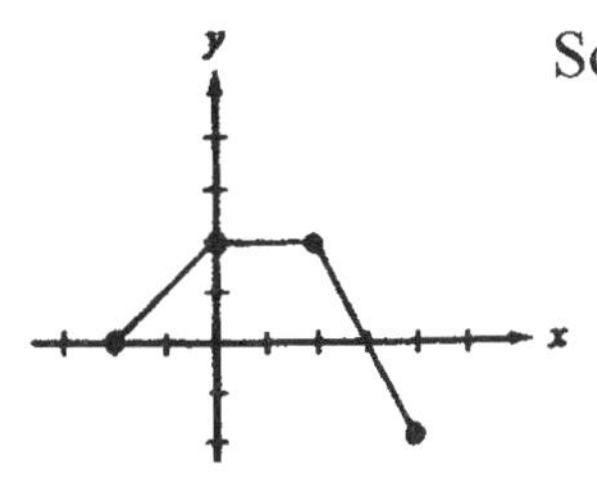

Solution: **a.** $y = |f(x)|$

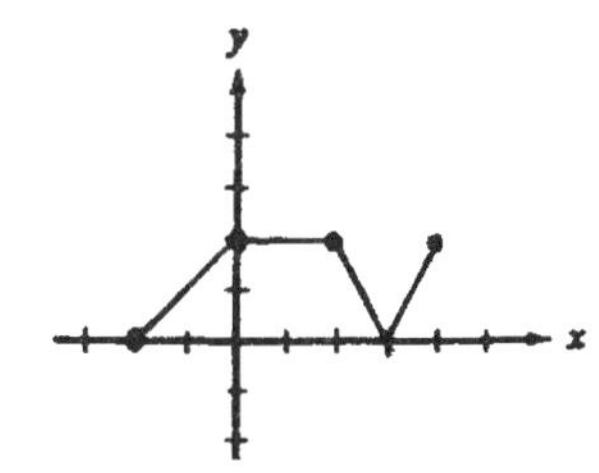

b. $y = f(|x|)$

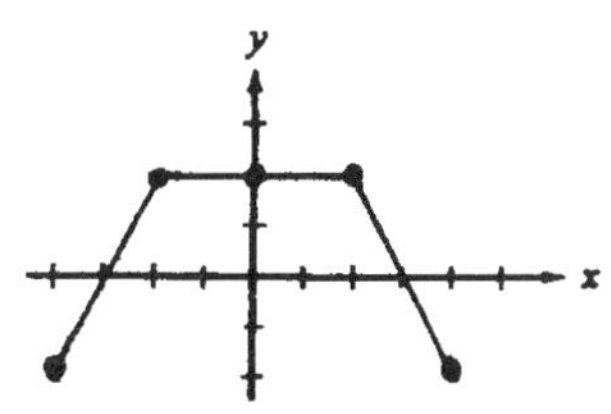

Hint: **1.** Graphing $y = |f(x)|$, we graph $y = f(x)$ and reflect the part that is below the x-axis through the *x-axis*.

2. Graphing $y = f(|x|)$, we graph $y = f(x)$ and replace the graph on the left of the *y-axis* by reflecting the part that is on the right of the *x-axis* about the *y-axis*.

3. Read " Graphing Techniques: Transformations " on page 99.

SAMPLE Question

Finding GCF and LCM by Rollabout Divisions

We have learning how to find the GCF and the LCM of two given numbers. However, it is not easy to find the GCF and the LCM if the numbers are very large.

To find the GCF and the LCM of two large numbers, we use the **Rollabout Divisions**.

The Rollabout Divisions was first invented by Euclid, a famous Greek mathematician who lived about 300 B.C.

Rollabout Divisions

Step 1: Divide the larger number by the smaller number to find the remainder.
Step 2: Divide the smaller number by the remainder in Step 1.
Step 3: Divide the previous remainder by the new remainder.
Step 4: Repeat the Step 3 until no remainder is found.
Step 5: The divisor of the last division is the GCF if the remainder is zero.
Step 6: Find LCM by the same method on page 50.

Example: Find the GCF and the LCM of 3,473 and 1,963.

Solution:

1.
$$1963\overline{)3473}\quad \text{quotient } 1;\quad 3473 - 1963 = 1510 \text{ remainder}$$

2.
$$1510\overline{)1963}\quad \text{quotient } 1;\quad 1963 - 1510 = 453 \text{ remainder}$$

3.
$$453\overline{)1510}\quad \text{quotient } 3;\quad 1510 - 1359 = 151 \text{ remainder}$$

4.
$$151\overline{)453}\quad \text{quotient } 3;\quad 453 - 453 = \mathbf{0}$$

We have: GCF $=151$ Ans.

$$151 \,\underline{|\, 3473 \quad 1963}$$
$$\quad 23 \quad\quad 13$$

$3473 \div 151 = 23$ and $1963 \div 151 = 13$

We have: LCM $= 151 \times 23 \times 13 = 45{,}149$ Ans.

SAMPLE QUESTIONS

CHAPTER 1: NUMBER and OPERATIONS

1. There are 9 students in the party. Each students shakes hands exactly once with each other. How many handshakes take place ?
Solution: (Apply sequence)

Students	1	2	3	4	5	6	7	8	9
Handshakes	0	1	3	6	10	15	21	28	36

Ans. 36 handshakes

2. The average (arithmetic mean) of *a*, *b*, c, and *d* is 12. The average of *a* and *b* is 14. What is the average of *c* and *d* ?
Solution:

$$\frac{4\times 12-14\times 2}{2}=\frac{48-28}{2}=\frac{20}{2}=10. \text{ Ans.}$$

3. The ratio of the numbers of red balls and white balls in a basket is 5 : 9. There are 280 balls in the basket. How many red balls in the basket ?
Solution:

$$280\times\frac{5}{9+5}=280\times\frac{5}{14}=100. \text{ Ans.}$$

4. ΔABC is a triangle and $<A:<B:<C=3:4:5$. What is the degree measure of $<C$?
Solution:

$$180^o\times\frac{5}{3+4+5}=180^o\times\frac{5}{12}=75^o. \text{ Ans.}$$

5. How many seat arrangement can be made for five students in a row if one of them does not want to take the first seat ?
Solution:

5	4	3	2	1	$\rightarrow$ $5\times4\times3\times2\times1=120$
×	4	3	2	1	$\rightarrow$ $4\times3\times2\times1=24$

$120-24=96$. Ans.

6. The average (arithmetic mean) of 4, 8, 12, 20, and *a* is 18, what is the value of *a* ?
Solution:

$$a=18\times5-(4+8+12+20)=90-44=46. \text{ Ans.}$$

SAMPLE QUESTIONS

CHAPTER 1: NUMBER and OPERATIONS

1. If a is 20% of 30 and b is 50% of 48, what is the value of $a \times b$?
   Solution:

   $$a \times b = 30 \times 0.2 + 48 \times 0.5 = 6 \times 24 = 144. \text{ Ans.}$$

2. If $a \otimes b$ be defined as the product of all integers larger than a and smaller than b. What is the value of $(4 \otimes 8) - (3 \otimes 7)$?
   Solution:

   $$a \otimes b = (5 \times 6 \times 7) - (4 \times 5 \times 6) = 210 - 120 = 90. \text{ Ans.}$$

3. The third term and each term thereafter of the following sequence is the average of the two preceding terms. What is the value of the first term in the sequence that is not an integer ?

   18, 14, ...

   Solution:

   18, 14, 16, 15, 15.5, ...

   Ans: 15.5

4. The population of males of a city increased 12% and females decreased 20% this year. The ratio of males to females at the end of this year was how many times the ratio at the beginning of this year ?
   Solution:

   The ratio of the population of male to female at the beginning of year is:

   $$\frac{M}{F}$$

   The ratio of the population of male to female at the end of year is:

   $$\frac{1.12M}{0.8F} = \frac{112}{80}\left(\frac{M}{F}\right) = \frac{7}{5}\left(\frac{M}{F}\right)$$

   Ans: $\frac{7}{5}$

5. What is the maximum number of 3-digit codes can be made from 0 to 9 if no 0 is used for the first digit in the 3-digit code ?
   Solution:

   $$9 \times 10 \times 10 = 900. \text{ Ans.}$$

SAMPLE QUESTIONS

CHAPTER 1: NUMBER and OPERATIONS

1. How many arrangements can be made from the letters *a*, *b*, *c*, *d*, and *e* in a row if letter *a* is not at either end ?
Solution:

5	4	3	2	1

$\rightarrow\ 5\times4\times3\times2\times1=120$

×	4	3	2	1

$\rightarrow\ 4\times3\times2\times1=24$

4	3	2	1	×

$\rightarrow\ 4\times3\times2\times1=24$

$120-24-24=72$. Ans.

2. A club consists of 30 boys and 25 girls. They want to select a girl as president, a boy as vice president, and a treasurer who may be either sex. How many choices are possible ?
Solution:

$30\times25\times53=39{,}750$ choices. Ans.

3. The ratio of sugar to water in a 230-gram sugar-water solution is 3 : 20. How many grams of sugar are there in the solution ?
Solution:

$230\times\frac{3}{23}=30$ grams. Ans.

4. What is the 10th term of the following sequence ?
5, 11, 18, 26, 35, ...

Solution:

5, 11, 18, 26, 35, 45, 56, 68, 81, **95**
+6 +7 +8 +9 +10 +11 +12 +13 +14

Ans: 95

5. What is the smallest number which is divided by 6, divided by 8, or divided by 11. All divisions have the same remainder of 5 ?
Solution:

$$\begin{array}{r|lll} 2 & 6 & 8 & 11 \\ \hline & 3 & 4 & 11 \end{array}$$

LCM $=2\times3\times4\times11=264$, $264+5=269$. Ans.

SAMPLE QUESTIONS

CHAPTER 1: NUMBER and OPERATIONS

1. In the school theater, there are 372 students at present and is $\frac{6}{17}$ full. What is the total number of students that the theater can hold when it is full ?
Solution:

$$372 \div \frac{6}{17} = 372 \times \frac{17}{6} = 1{,}054 \text{ students. Ans.}$$

2. If x is divided by 4, the remainder is 1. If x is divided by 5, the remainder is 2. If x is divided by 6, the remainder is 3. What is the smallest value of x ?
Solution:

$$\begin{array}{r|lll} 2 & 4 & 5 & 6 \\ \hline & 2 & 5 & 3 \end{array}$$

$$\text{LCM} = 2 \times 2 \times 5 \times 3 = 60$$

$$60 - 3 = 57. \text{ Ans.}$$

3. Scientists tag 100 salmons in a river. Later, they caught 150 salmons. Of these, 5 have tags. Estimate how many salmons are in the river ?
Solution:

$$100 \times \frac{150}{5} = 3{,}000 \text{ salmons. Ans.}$$

4. How many lines can be determined by nine noncollinear points ?
Solution:

Points	2	3	4	5	6	7	8	9
Lines	1	3	6	10	15	21	28	**36**

Ans: 36 lines

5. What is the last digit written of 2^{40} ?
Solution:

$$2^1 = 2, \quad 2^2 = 4, \quad 2^3 = 8, \quad 2^4 = 16$$
$$2^5 = 32, \quad 2^6 = 64, \quad 2^7 = 128, \quad 2^8 = 256$$
$$\vdots$$
$$2^{40} = \cdots\cdots 6$$

Ans: 6

SAMPLE QUESTIONS

CHAPTER 1: NUMBER and OPERATIONS

1. What is the remainder of $(2^{40} \div 5)$?
Solution:

The remainder of

$2^3 \div 5$ is 3, $2^4 \div 5$ is 1, $2^5 \div 5$ is 2, $2^6 \div 5$ is 4

$2^7 \div 5$ is 3, $2^8 \div 5$ is 1, $2^9 \div 5$ is 2, $2^{10} \div 5$ is 4

$\vdots$

$2^{40} \div 5$ is 1

Ans: 1

2. What is the sum of $1+2+3+\cdots\cdots+60$?
Solution:

$$Sum = (1+60)\times\frac{60}{2} = 1{,}830.$$ Ans.

3. What is the formula for the *n*th term of the arithmetic sequence
3, 8, 13, 18, 23, ····· ?
Solution:

$$a_n = a_1 + (n-1)d = 3 + (n-1)\cdot 5 = 5n - 2.$$ Ans.

4. What is the formula for the *n*th term of the geometric sequence
1, 2, 4, 8, 16, 32, ····· ?
Solution:

$$a_n = a_1 r^{n-1} = 1\cdot 2^{n-1} = 2^{n-1}.$$ Ans.

5. How many 4-digit passwords can be formed for an office security system, using digits from 1 to 9 if the digits can be used more than once in any password ?
Solution:

$9\times 9\times 9\times 9 = 6{,}561$ passwords. Ans.

6. How many 4-digits passwords can be formed for an office security system, using digits from 1 to 9 if the digits can not be used more than once in any password ?
Solution: $${}_9P_4 = \frac{9!}{(9-4)!} = \frac{9!}{5!} = \frac{9\cdot 8\cdot 7\cdot 6\cdot 5!}{5!} = 9\times 8\times 7\times 6 = 3{,}024.$$ Ans.

Notes

In the court room, three defendants are waiting for the judge to sentence them for the time they need to stay in jail for the crimes they have committed.

Judge: Do you have any good reason why you think I should sentence you lesser time in jail ?
Defendant A: I was your high school classmate.
Judge: 1 week.
Defendant B: I built your house.
Judge: 1 month.
Defendant C: I introduced a girl friend to you 20 years ago. She is your wife now.
Judge: 20 years.

CHAPTER TWO

ALGEBRA and FUNCTIONS

2-1 Variables and Expressions

Variable: It is a letter that can be used to represent one or more numbers.

If $x = 1$, then $2x = 2(1) = 2$. If $x = 2$, then $2x = 2(2) = 4$. x is a variable.

We use variables to form algebraic expressions and equations.

Numerical Expression: It is an expression that includes only numbers.

Algebraic Expression: It is an expression that combines numbers and variables by four operations (add, subtract, multiply or divide).

$xy + x + 2y - 3$ is an algebraic expression.

Algebraic Equation: It is a statement by placing an equal sign " = " between two expressions.

$2x + 5 = 7$ is an equation. $2x + 5 = 9$ is an equation.

Term: It is a single number or a product of numbers and variables in an expression.

A monomial: An expression with only one term: $5x$ is a monomial.

A polynomial: An expression formed by two or more terms: $xy + x + 2y - 3$

A polynomial or binomial: An expression formed by only two terms: $3x^2 + 6$

A polynomial or trinomial: An expression formed by three terms: $3x^2 - 5x - 2$

Constant: It is a number only. In the expression $x + 3$, 3 is a constant.

Coefficient: It is a number that is multiplied by a variable.

In the term $2y$, 2 is the coefficient. The coefficient of the term x is 1.

Degree of an expression: It is the greatest of the degrees of its terms after it has been simplified and its like terms are combined.

The degree of $2x + 1$ is 1, a linear form. The degree of $3x^2 + x - 1$ is 2, a quadratic form.

The degree of $4x^3 + 5x$ is 3, a cubic form. The degree of $x^4 - 1$ is 4, a quartic form.

The degree of $4a^2b + 2a^3b^4 - 2$ is 7, a polynomial form.

Evaluating an Expression: The process of replacing variables with numbers in an algebraic expression and finding its value.

Examples

1. Evaluate $2x + 15$ when $x = 4$.

Solution:

$2x + 15 = 2(4) + 15 = 23$

2. Evaluate $a^2 + b - 2$ when $a = 3$, $b = 1$.

Solution:

$a^2 + b - 2 = 3^2 + 1 - 2 = 9 + 1 - 2 = 8$

3. Evaluate $\frac{1}{2}x + 3$ if $x = 3$.

Solution:

$\frac{1}{2}x + 3 = \frac{1}{2}(3) + 3 = 4\frac{1}{2}$

4. Evaluate $\frac{1}{2}x - 3$ if $x = 3$.

Solution:

$\frac{1}{2}x - 3 = \frac{1}{2}(3) - 3 = -1\frac{1}{2}$

2-2 Solving Equations in One Variable

Equation: It is a statement formed by placing an equals sign " = " between two numerical or variable expressions.

Solution (root): It is any value of the variable that makes the equation true.

To write the solutions , we may say either "the solutions are 2 and 5" or "the solutions set is {2, 5}".

To find the solution of an equation, we transfer the given equation into a simpler and equivalent equation that has the same solution. The last statement of the given equation after we transfer it should isolate the variable to one side of the equation.

An easier way to simplify an equation is to transfer (isolate) all variables to one side of the equation and transfer (isolate) all numbers to the other side. Reverse (change) the signs (+, −) in the process. (See Example 5 and Example 6)

Examples

1. Solve $2x+6=4x$.

Solution:

$$2x+6=4x$$
$$2x+6-4x=\cancel{4x}-\cancel{4x}$$
$$-2x+6=0$$
$$-2x+\cancel{6}-\cancel{6}=-6$$
$$-2x=-6$$
$$\frac{\cancel{-2}x}{\cancel{-2}}=\frac{-6}{-2}$$
$$\therefore x=3$$

2. Solve $2x-5=7-6x$.

Solution:

$$2x-5=7-6x$$
$$2x-5+6x=7-\cancel{6x}+\cancel{6x}$$
$$8x-5=7$$
$$8x\cancel{-5}\cancel{+5}=7+5$$
$$8x=12$$
$$\frac{\cancel{8}x}{\cancel{8}}=\frac{12}{8}$$
$$\therefore x=1\tfrac{1}{2}$$

3. Solve $3(a-5+2a)=6$.

Solution:

$$3(a-5+2a)=6$$
$$3(3a-5)=6$$
$$9a-15=6$$
$$9a\cancel{-15}\cancel{+15}=6+15$$
$$9a=21$$
$$\frac{\cancel{9}a}{\cancel{9}}=\frac{21}{9}$$
$$\therefore a=2\tfrac{1}{3}$$

4. Solve $3(2y+3)=4y-7$.

Solution:

$$3(2y+3)=4y-7$$
$$6y+9=4y-7$$
$$6y+9-4y=\cancel{4y}-7-\cancel{4y}$$
$$2y+9=-7$$
$$2y+\cancel{9}-\cancel{9}=-7-9$$
$$2y=-16,\quad \frac{\cancel{2}y}{\cancel{2}}=\frac{-16}{2}$$
$$\therefore y=-8$$

5. Solve $2x+6=4x$.

Solution:

$$2x+6=4x$$
$$2x-4x=-6$$
$$-2x=-6\quad \therefore x=\tfrac{-6}{-2}=3$$

6. Solve $2x-5=7-6x$.

Solution:

$$2x-5=7-6x$$
$$2x+6x=7+5$$
$$8x=12\quad \therefore x=\tfrac{12}{8}=1\tfrac{1}{2}$$

2-3 Words Representations and Modeling

Mathematical Modeling: To write mathematical expressions, equations, or inequalities that represent real-life situations.

Before we learn how to solve word problems in algebra, we must learn how to translate or represent words into symbols containing variables.

A. Translate word phrases into variable expressions.

5 more than x	$\to x+5$	The sum of 5 and x	$\to 5+x$
5 less than x	$\to x-5$	The difference between 5 and x	$\to 5-x$
x less than 5	$\to 5-x$	The product of 5 and x	$\to 5x$
x minus 5	$\to x-5$	The quotient of 5 and x	$\to 5/x$
x is less than 5	$\to x<5$	x increased by 5	$\to x+5$
x is greater than 5	$\to x>5$	x decreased by 5	$\to x-5$
x is less than or equal to 5	$\to x\le 5$	x times 5	$\to 5x$
x is greater than or equal to 5	$\to x\ge 5$	x divided by 5	$\to x/5$

5 times the quantity x decreased by 3	$\to 5(x-3)$
5 times the difference of x and 3	$\to 5(x-3)$
5 times the sum of x and 3	$\to 5(x+3)$
3 more than 5 times x	$\to 5x+3$
3 increased by 5 times x	$\to 3+5x$
The difference between 5 times x and 3	$\to 5x-3$
3 decreased by 5 times x	$\to 3-5x$
3 less than 5 times x	$\to 5x-3$
3 less than half of x	$\to \frac{1}{2}x-3$
Half the difference between x and 3	$\to \frac{1}{2}(x-3)$
Twice the sum of x and 3	$\to 2(x+3)$
Twice x, increased by 3	$\to 2x+3$

Three consecutive integers (n is an integer)	$\to n,\ n+1,\ n+2$
Three consecutive even integers (n is an even integer)	$\to n,\ n+2,\ n+4$
Three consecutive odd integers (n is an odd integer)	$\to n,\ n+2,\ n+4$
Three consecutive positive multiples of 5 (n is a multiple of 5)	$\to n,\ n+5,\ n+10$

Examples: Translate each statement into a variable expression.

1. There are 15 fewer girls than boys in a class. How many girls are there if the number of boys is b ?
   Solution: $b-15$

2. You paid \$50 cash and made 10 equal monthly payments. What is the total cost of the product if the amount in dollars of each of the monthly payments is x ?
   Solution: $50+10x$

B. Write an Algebraic Model

1. Eight less than a number n is 15.
Translation:
$$n-8=15$$

2. Four more than a number n is 12.
Translation:
$$n+4=12$$

3. Five more than a number n is equal to 9.
Translation:
$$n+5=9$$

4. 5 times the quantity x decreased by 3 is 25.
Translation: $5(x-3)=25$

5. Twice the sum of a number x and 3 is 22.
Translation:
$$2(x+3)=22$$

6. A number n decreased by 12 is 30.
Translation:
$$n-12=30$$

7. The result of a number x multiplied by 5, decreased by 3 is 57.
Translation:
$$5x-3=57$$

8. The sum of three consecutive numbers is 48.
Translation:
$$n+(n+1)+(n+2)=48$$

9. Write an equation involving x if the perimeter of the figure is 24.

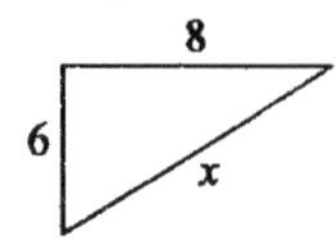

Solution: $x+8+6=24$
$x+14=24$

10. Write an equation involving x if the perimeter of the figure is 32.

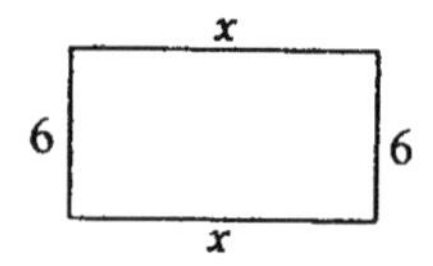

Solution: $x+x+6+6=32$
$2x+12=32$

11. One-third of a pizza sold for \$3.50. Write an equation that represents the cost c of the whole pizza.

Solution: $\frac{1}{3}c=3.50$

12. You paid \$50 cash and made 10 equal monthly payments. The total cost of the product was \$400. If the amount in dollars of each of the monthly payments is x, write an equation that represents the amount of each of the monthly payments.
Solution:
$$400=50+10x$$

13. John picks up two-thirds of the books in the shelf. There are 12 books left in the shelf. Write an equation that represents the number n of the books in the shelf to begin with.

Solution: $n-\frac{2}{3}n=12$

2-4 Solving Word Problems in One Variable

To solve a word problem using an equation in algebra, we use the following steps.

Steps for solving a word problem:

1. Read the problem carefully.
2. Choose a variable and assign it to represent the unknown (answer).
3. Write an equation based on the given fact in the problem.
4. Solve the equation and find the answer.
5. Check your answer with the problem (usually on a scratch paper).

Examples

1. Eight less than a number is 18. Find the number.

Solution:

Let $n =$ the number

The equation is

$$n - 8 = 18$$
$$n - \cancel{8} + \cancel{8} = 18 + 8$$
$$\therefore n = 26. \text{ Ans.}$$

Or:
$$n - 8 = 18$$
$$n = 18 + 8$$
$$n = 26. \text{ Ans.}$$

Check: $n - 8 = 18$

$26 - 8 = 18$ √

2. A number increased by 15 is 32. Find the number.

Solution:

Let $n =$ the number

The equation is

$$n + 15 = 32$$
$$n + \cancel{15} - \cancel{15} = 32 - 15$$
$$\therefore n = 17. \text{ Ans.}$$

Or:
$$n + 15 = 32$$
$$n = 32 - 15$$
$$n = 17. \text{ Ans.}$$

Check: $n + 15 = 32$

$17 + 15 = 32$ √

3. Find two consecutive integers whose sum is 37.

Solution:

Let $n = 1^{st}$ integer

$n + 1 = 2^{nd}$ integer

The equation is

$$n + (n + 1) = 37$$
$$2n + 1 = 37$$
$$2n = 37 - 1$$
$$2n = 36$$

$\therefore n = 18$ 1st integer

$n + 1 = 19$ 2nd integer. Ans.

Check: 18 +1 9 =37 √

Examples

4. Twice the sum of a number and 3 is 22. Find the number.
Solution:

Let $n =$ the number
The equation is

$$2(n+3) = 22$$
$$2n+6 = 22$$
$$2n = 22-6$$
$$2n = 16$$
$$\therefore n = 8. \text{ Ans.}$$

Check: $2(n+3) = 22$
$2(8+3) = 22$ √

5. Find three consecutive even integers whose sum is 120.
Solution:

Let $n = 1^{st}$ even integer
$n+2 = 2^{nd}$ even integer
$n+4 = 3^{rd}$ even integer
The equation is

$$n+(n+2)+(n+4) = 120$$
$$3n+6 = 120$$
$$3n = 114$$

$\therefore n = 38$ 1st integer
$n+2 = 40$ 2nd integer
$n+4 = 42$ 3rd integer

Ans: 38, 40, 42
Check: $38+40+42 = 120$ √

6. John has twice as much money as Carol. Together they have $36. How much money does each have ?
Solution: Let $c =$ Carol's money
$2c =$ John's money
The equation is

$$c+2c = 36$$
$$3c = 36$$

$\therefore c = 12 \rightarrow$ Carol's money.
$2c = 24 \rightarrow$ John's money. Ans.

Check: $12+24 = 36$ √

7. A number divided by 2 is equal to the number increased by 2. Find the number.
Solution: Let $n =$ the number

The equation is $\dfrac{n}{2} = n+2$

Multiply each side by 2: $n = 2n+4$, $n-2n = 4$, $-n = 4$

$\therefore n = -4$. Ans.

Check: $\frac{-4}{2} = -4+2$ √

Examples

8. John picked up two-thirds of the books in the shelf. There are 12 books left in the shelf. How many books were in the shelf to begin with ?

Solution:

Let $n =$ the number of books to begin with

There are $(1-\frac{2}{3})\ n$ books left in the shelf after John picked up $\frac{2}{3}$ of the books.

The equation is

$$(1-\tfrac{2}{3})n = 12$$

$$\tfrac{1}{3}n = 12$$

$$\therefore n = 36 \text{ books. Ans.}$$

Check: $\frac{2}{3}(36) = 24$

$36 - 24 = 12$ √

9. Roger picked up two-fifths of the books in the shelf. Maria picked up one-half of the remaining books. Jack picked up 6 books that were left. How many books were in the shelf to begin with ?

Solution:

Let $n =$ the number of books to begin with

There are $(1-\frac{2}{5})\ n$ remaining books in the shelf after Roger picket up $\frac{2}{5}$ of the books.

Then, Maria picked up $\frac{1}{2}(1-\frac{2}{5})n$ books after Roger picked up.

The equation is

$$(1-\tfrac{2}{5})n - \tfrac{1}{2}(1-\tfrac{2}{5})n = 6$$

$$\tfrac{3}{5}n - \tfrac{1}{2}\cdot\tfrac{3}{5}n = 6$$

$$\tfrac{3}{5}n - \tfrac{3}{10}n = 6$$

$$\tfrac{6n-3n}{10} = 6$$

$$\tfrac{3n}{10} = 6$$

$$3n = 60$$

$$n = 20 \text{ books. Ans.}$$

Check: $\frac{2}{5}(20) = 8$

$20 - 8 = 12$

$\frac{1}{2}(12) = 6$

$12 - 6 = 6$ √

10. There are one-dollar bills and five-dollar bills in the piggy bank. It has 60 bills with a total value of $228. How many ones and how many fives in the piggy bank ?

Solution:

Let $x =$ number of one-dollar bills

$60 - x =$ number of five-dollar bills

The equation is

$$1x + 5(60-x) = 228$$

$$x + 300 - 5x = 228$$

$$-4x = -72$$

$$\therefore x = 18 \rightarrow \text{one-dollar bills}$$

$$60 - x = 42 \rightarrow \text{five-dollar bills. Ans.}$$

Check: $18 + $5(42) = $228 √

Examples

11. The length of a rectangle is 14 *cm* longer than the width. The perimeter is 76 *cm*. Find its length and width.

Solution:

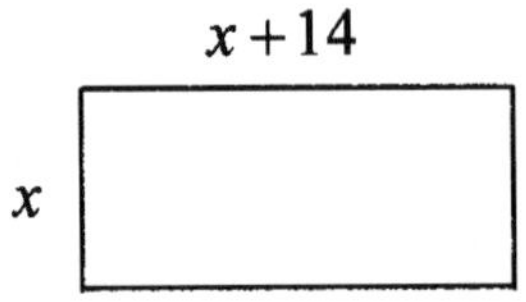

Let $x =$ the width

$x + 14 =$ the length

The equation is

$$2x + 2(x + 14) = 76$$
$$2x + 2x + 28 = 76$$
$$4x = 76 - 28$$
$$4x = 48$$
$$\therefore x = 12\ cm \rightarrow \text{the width.}$$
$$x + 14 = 26\ cm \rightarrow \text{the length.} \quad \text{Ans.}$$

Check: $(12 + 26) \times 2 = 76$ √

12. A suit is on sale for 20% discount. Roger paid \$8.82, including a 5% sales tax. Find the original price of the suit.

Solution:

Let $x =$ the original price

The sales price $= 80\% \cdot x = 0.80x$

The sales tax $= 0.80x(5\%) = 0.80x(0.05)$

The equation is

$$0.80x + 0.80x(0.05) = 8.82$$
$$0.80x + 0.04x = 8.82$$
$$0.84x = 8.82$$
$$\therefore x = \tfrac{8.82}{0.84} = \$10.50. \quad \text{Ans.}$$

Check: $\$10.50 \times 0.80 = \$8.40$

$\$\ 8.40 \times 0.05 = \$0.42$

$\$8.40 + \$0.42 = \$8.82$ √

13. A 16-gallon salt-water solution contains 25% pure salt. How much water should be added to produce the solution to 20% salt ?

Solution:

$16 \times 25\% = 16 \times 0.25 = 4$ gallons of pure salt in the original solution.

Let $x =$ gallons of water added

The equation is

$$\frac{4}{16 + x} = 20\%, \quad \frac{4}{16 + x} = 0.20$$
$$4 = 0.20(16 + x)$$
$$4 = 3.2 + 0.20x$$
$$0.8 = 0.20x$$
$$\therefore x = \tfrac{0.8}{0.20} = 4 \text{ gallons of water added.} \quad \text{Ans.}$$

Check: $\frac{4}{16+4} = \frac{4}{20} = 0.2 = 20\%$ √

2-5 Linear Equation and Functions

We have learned to solve equations that have only one variable.
In this section, we will learn to solve equations that have two variables of degree 1.
Here is an example in different forms:

$$2x - y = 3, \quad \text{or } 2x = y + 3, \quad \text{or } y = 2x - 3$$

The solutions to equations in one variable are numbers.
The solutions to equations in two variables are ordered pairs of numbers (x, y).

Example: Identify whether the ordered pair (0, –3) is a solution of $2x - y = 3$.
Solution: We substitute $x = 0$ and $y = -3$ in the equation:

$$2x - y = 2(0) - (-3) = 0 + 3 = 3 \ 4$$

Yes. It is a solution.

Therefore, $x = 0$ and $y = -3$ is a solution of $2x - y = 3$ and can be written as (0, –3).
The equation $2x - y = 3$ has infinitely many solutions. Here are some of its solutions:

(–2, –7), (–1, –5), (0, –3), (1, –1), (2, 1), ...

Each solution of the equation is a point (x, y) on the coordinate plane.
The graph of connecting all the solutions (points) of the equation $2x - y = 3$ forms a straight line. Therefore, we call the equation $2x - y = 3$ a **linear equation.**

We can write any linear equation into the form $ax + by = c$, where a, b, and c are real numbers with a and b not both zero.
The equation $ax + by = c$ is called the **standard form** of a linear equation. If a is negative, we change it to positive by multiplying each side of the equation by –1.

If we rewrite a linear equation by solving for y in terms of x, we have a **linear function.**
Example: Solve the equation $3x + 2y = 6$ for y in terms of x.
Solution: $3x + 2y = 6$, $2y = -3x + 6$
Divide each side by 2: $y = -\frac{3}{2}x + 3$. Ans.

In the equation $y = -\frac{3}{2}x + 3$, x is called the **independent variable**, and y is called the **dependent variable.** Each value of y depends on the value chosen for x.
To find the value of y in the equation, we could substitute any value (any number) chosen for x into the equation: $x = 2$, $y = -\frac{3}{2}(2) + 3 = 0$. The solution is (2, 0).
The equation $y = -\frac{3}{2}x + 3$ has infinitely many solutions. The x – values are called the **domain** of the function, and the y – values are called the **range** of the function.
In the equation $y = -\frac{3}{2}x + 3$, each value of x is assigned exactly one value of y. We say that "y **varies directly with** x", or "y **is a function of** x".
"y is a function of x" is written by **functional notation** $y = f(x)$.
Therefore, the equation $y = -\frac{3}{2}x + 3$ can be written by **functional notation**:

$$f(x) = -\tfrac{3}{2}x + 3 \quad ; \quad f(x) = -\tfrac{3}{2}x + 3 \text{ is equivalent to } y = -\tfrac{3}{2}x + 3.$$

Since two points determine a straight line, we need to find only two points to graph a linear equation.
The points where the line crosses the $x-axis$ (let $y=0$) and the $y-axis$ (let $x=0$) are the easiest two points to find. We call them the $x-$intercept and the $y-$intercept.

Example: Graph $2x-y=4$.

Solution:

Let $y=0$	Let $x=0$
$2x-0=4$	$2(0)-y=4$
$2x=4$	$-y=4$
$x=2$	$y=-4$
Point (2, 0)	Point (0, –4)
$x-$intercept = 2	$y-$intercept = –4

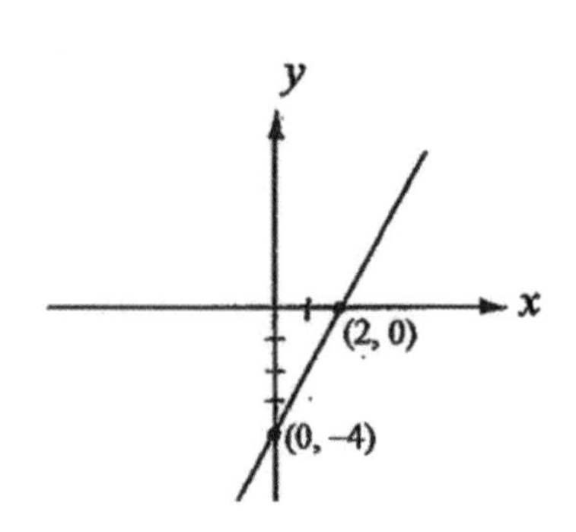

On the number line, the graph of an equation in one variable is a point.

On the coordinate plane, an equation in one variable is a linear equation. Its graph is a vertical line or a horizontal line.
A vertical line has no restriction on y. A horizontal line has no restriction on x.

Examples: **a.** Graph $x=4$ **b.** Graph $y=3$ **c.** Graph $x=-4$ **d.** Graph $y=-4$

Solution:

a.
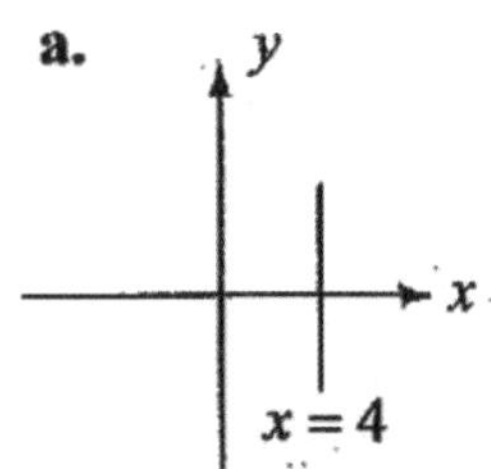

b.
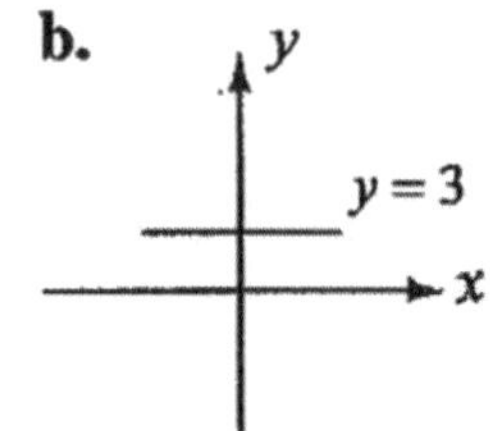

c.
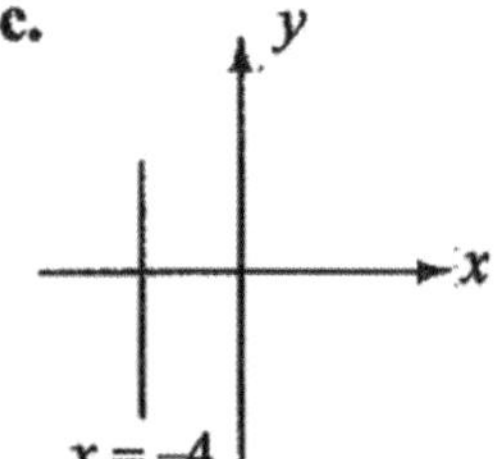

d.
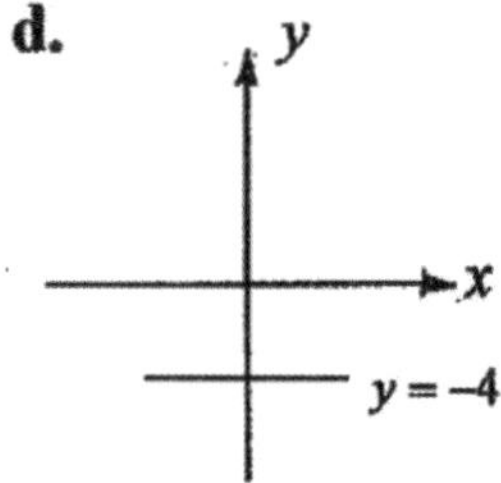

Using New Definitions (New Symbols, Defined Symbols)

A special symbol or sign (⊗, ⊕, ∗, Ω, #, !, ϒ, Δ, □, ……) is sometimes used to define a function for the particular case without using standard mathematical symbols.

Examples: 1. If $a^{\oplus}=a^2+3a-4$ is defined, then $3^{\oplus}=3^2+3(3)-4=14$.

2. If $a*b=(a+b)(a-b)$ is defined, then $5*3=(5+3)(5-3)=16$.

3. If $a\Delta b=a-2b$ and $a\square b=a+3b$ are defined, what is the value of x for

$$2\Delta(3x)=(4x)\square 5\ ?$$

Solution: $2-2(3x)=4x+3(5)$, $2-6x=4x+15$, $10x=-13$ $\therefore x=-1.3$

2-6 Slope of a Line

Slope is used to describe the **steepness** of a straight line. To find the slope of a line, we choose any two points on the line and form a right triangle which has the line as its hypotenuse. A line that rises more steeply has a greater slope. A line that runs more horizontally has a smaller slope. Therefore, the slope of a line is defined as the ratio of its vertical rise to horizontal run. The letter m is commonly used to represent the slope.

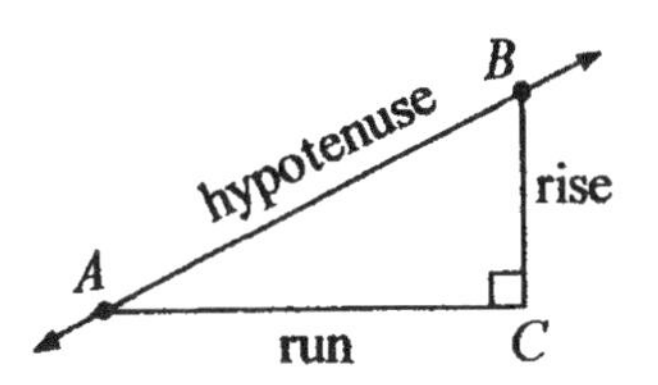

Definition of Slope:

$$\text{Slope of } \overleftrightarrow{AB} = \frac{rise}{run} = \frac{\overline{BC}}{\overline{AC}}$$

In the coordinate plane, the slope of a line between two points (x_1, y_1) and (x_2, y_2) is given in the following slope formula:

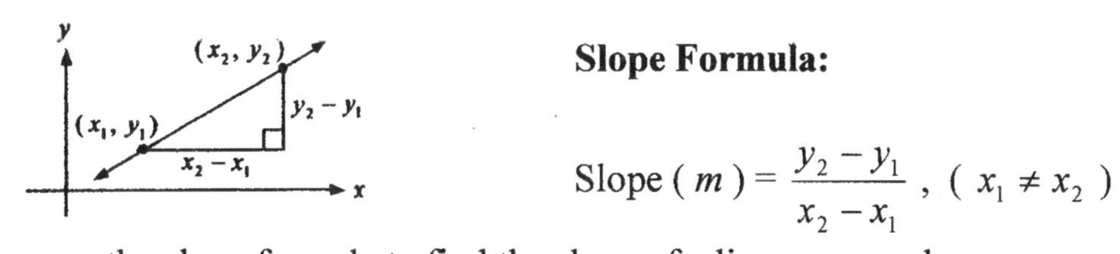

Slope Formula:

$$\text{Slope } (m) = \frac{y_2 - y_1}{x_2 - x_1}, \quad (x_1 \neq x_2)$$

When we use the slope formula to find the slope of a line, we can choose any point as the first point and the other as the second point. The result will be the same.

Examples

1. Find the slope of the line passing through the points (2, 3) and (6, 5).
   Solution:

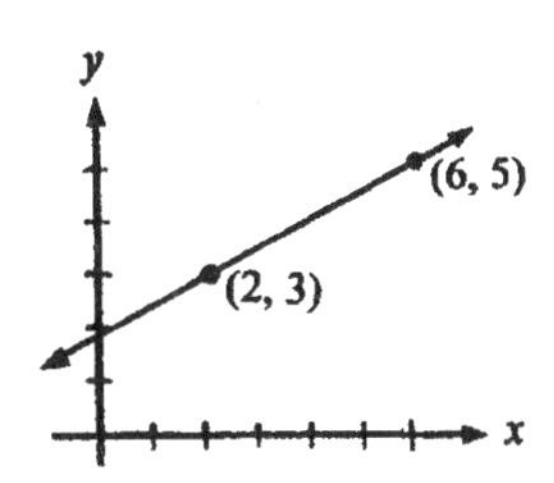

$$m = \frac{y_2 - y_1}{x_2 - x_1} = \frac{5-3}{6-2} = \frac{2}{4} = \frac{1}{2}. \text{ Ans.}$$

$$\text{Or: } m = \frac{y_2 - y_1}{x_2 - x_1} = \frac{3-5}{2-6} = \frac{-2}{-4} = \frac{1}{2}. \text{ Ans.}$$

2. Find the slope of the line containing the points (–2, 4) and (5, 2).
   Solution:

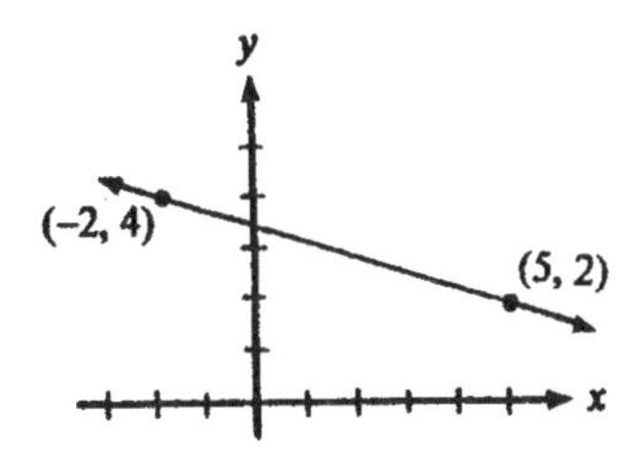

$$m = \frac{y_2 - y_1}{x_2 - x_1} = \frac{2-4}{5-(-2)} = \frac{-2}{7} = -\frac{2}{7}. \text{ Ans.}$$

$$\text{Or: } m = \frac{y_2 - y_1}{x_2 - x_1} = \frac{4-2}{-2-5} = \frac{2}{-7} = -\frac{2}{7}. \text{ Ans.}$$

Note that: 1. If a line slants up from left to right, the slope is positive.
2. If a line slants down from left to right, the slope is negative.

Examples

3. Find the slope of the line passing the points (–7, 4) and (3, 4).
Solution:

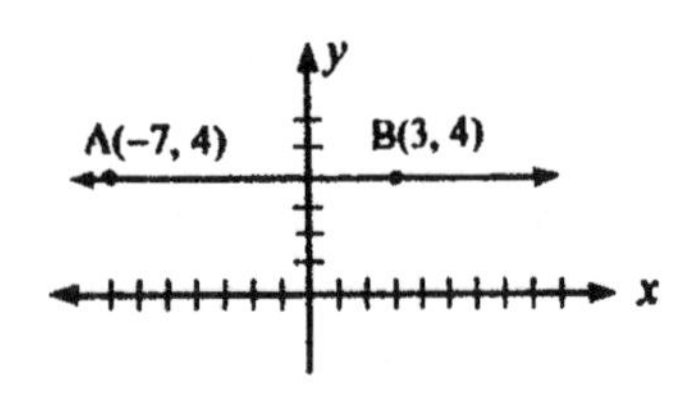

$$m = \frac{4-4}{3-(-7)} = \frac{0}{10} = 0$$

Ans: The slope is 0.

4. Find the slope of the line passing through the points (7, 5) and (7, –3).
Solution:

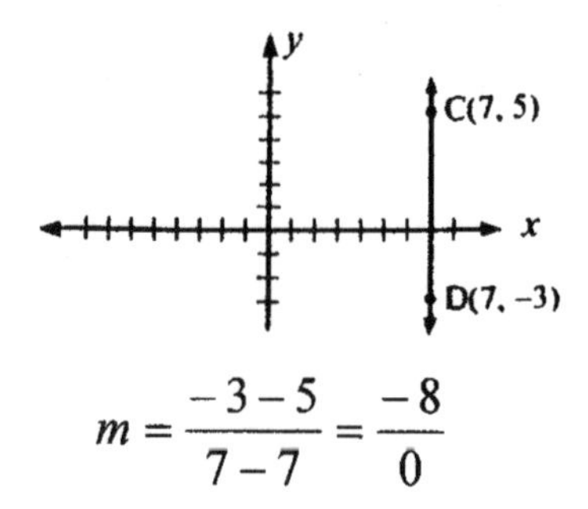

$$m = \frac{-3-5}{7-7} = \frac{-8}{0}$$

Ans: The slope is undefined.

Note that: 1) The slope of a horizontal line is 0.
2) The slope of a vertical line is undefined. It means that the slope of a vertical line does not exist.

Slope-Intercept Form of a Straight Line

We can choose any two points on a straight line and find its slope by the slope formula. However, an easier way to find the slope of a straight line is to rewrite the equation in the slope-intercept form:

$$y = mx + b$$

Then, the slope is m and $y-$ intercept is b.

5. Find the slope of the line $y + 2x = 3$.
Solution:

$$y + 2x = 3$$
$$y = -2x + 3$$
$\therefore m = -2$. Ans.

6. Find the slope of the line $y - 2x = 3$
Solution:

$$y - 2x = 3$$
$$y = 2x + 3$$
$\therefore m = 2$. Ans.

7. Find the slope and $y-$ intercept of the line whose equation is $2x + 3y - 6 = 0$.
Solution:

$$2x + 3y - 6 = 0$$
$$3y = -2x + 6$$
$$y = -\tfrac{2}{3}x + 2$$

Ans: slope $= -\frac{2}{3}$.
$y-$ intercept $= 2$.

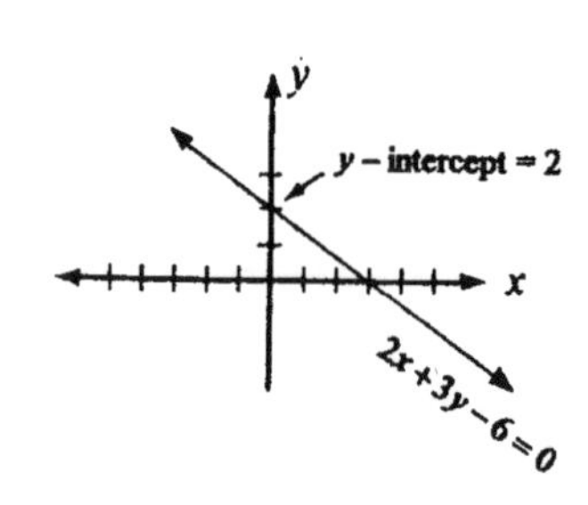

2-7 Finding the Equation of a Line

We have learned how to find the points, slope, and intercepts of a given line.
Now we will learn how to find the equation of a line under certain given information.
Such as, find the equation of a line:

1) having its slope and $y-$intercept.
2) having its slope and a point on the line.
3) having its two points on the line.

The following examples show how to find the equation of a line.
The equation of a line can be written in two different ways:

1) Slope-Intercept Form $y = mx + b$ **2) Standard Form** $ax + by = c$

Examples

1. Find the equation of the line having slope -5 and $y-$intercept 2.
Solution:
$m = -5,\ b = 2$
$y = mx + b$
$\therefore y = -5x + 2$. Ans. (slope-intercept form)
or: $5x + y = 2$. Ans. (standard form)

2. Find the equation of the line having slope $\frac{3}{4}$ and $y-$intercept $\frac{5}{2}$.
Solution:
$m = \frac{3}{4},\ b = \frac{5}{2}$
$y = mx + b$
$y = \frac{3}{4}x + \frac{5}{2}$. Ans. (slope-intercept form)
or: Multiply each side by 4:
$4y = 3x + 10$
$-3x + 4y = 10 \quad \therefore 3x - 4y = -10$. Ans. (standard form)

3. Find the equation of the line having slope 4 and passing through the point (2, –3).
Solution:
$m = 4$
$y = mx + b$
$y = 4x + b$
To find b, we substitute the point (2, –3) into the above equation.
$-3 = 4(2) + b$
$-11 = b$
$\therefore y = 4x - 11$. Ans. (slope-intercept form)
or: $4x - y = 11$. Ans. (standard form)

Examples

4. Find the equation of the line passing through the points (–2, 3) and (4, 6).
   Solution:

Find the slope: $m = \dfrac{y_2 - y_1}{x_2 - x_1} = \dfrac{6-3}{4-(-2)} = \dfrac{3}{6} = \dfrac{1}{2}$

$$y = mx + b$$
$$y = \tfrac{1}{2}x + b$$

To find b, substitute the point (–2, 3) into the above equation.

$$3 = \tfrac{1}{2}(-2) + b$$
$$3 = -1 + b$$
$$4 = b$$

$\therefore y = \tfrac{1}{2}x + 4$. Ans. (slope-intercept form)

or: $x - 2y = -8$. Ans. (standard form)

Point-Slope Form of a Line:

If m is the slope of a line and (x_1, y_1) is one of its points, we can write the equation of the line in point-slope form. To write the equation, we choose any other point (x, y) on the line.

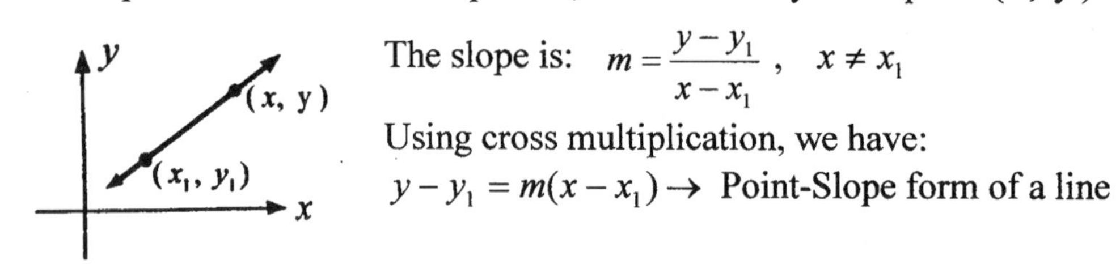

The slope is: $m = \dfrac{y - y_1}{x - x_1}$, $x \neq x_1$

Using cross multiplication, we have:

$y - y_1 = m(x - x_1) \rightarrow$ Point-Slope form of a line

Examples

5. Find the equation in point-slope form of a line having slope 4 and passing through the point (2, –3).
   Solution:

$$y - y_1 = m(x - x_1)$$
$$y - (-3) = 4(x - 2)$$

$\therefore y + 3 = 4(x - 2)$. Ans. (point-slope form)

6. Find the equation in point-slope form of a line passing through the points (–2, 3) and (4, 6). (Hint: There are two answers.)

Solution: Find the slope: $m = \dfrac{y_2 - y_1}{x_2 - x_1} = \dfrac{6-3}{4-(-2)} = \dfrac{3}{6} = \dfrac{1}{2}$

Substitute one of the points, say (–2, 3), and the slope into the equation.

$$y - y_1 = m(x - x_1)$$
$$y - 3 = \tfrac{1}{2}[x - (-2)]$$

$\therefore y - 3 = \tfrac{1}{2}(x + 2)$. Ans. (point-slope form)

If we choose (4, 6), the other answer is: $y - 6 = \tfrac{1}{2}(x - 4)$.

Both answers have the same standard form: $x - 2y = -8$.

2-8 Parallel and Perpendicular Lines

Now we will learn how to identify parallel lines and perpendicular lines by comparing their slopes.
Two different lines that have the same slope are **parallel**. Two different lines that intersect to form a right angle (90° angle) are **perpendicular**.

Rules:

1. **Two lines are parallel if and only if their slopes are equal:** $m_1 = m_2$
2. **Two lines are perpendicular if and only if the product of their slopes is** -1**:** (They are negative reciprocals of each other.)

$$m_1 \cdot m_2 = -1, \text{ or } m_1 = -\frac{1}{m_2}, \quad m_2 = -\frac{1}{m_1}$$

Examples

1. Find the slope of a line that is perpendicular to the line $8x - 4y = 3$.
   Solution:
   $8x - 4y = 3$
   Rewrite it in slope-intercept form: $-4y = -8x + 3$
   Divide each side by -4, we have $y = 2x - \frac{3}{4}$. slope $m = 2$
   The slope of the perpendicular line is the negative reciprocal of 2.
   It is $-\frac{1}{2}$. Ans.

2. Tell whether or not the following two lines are parallel, perpendicular, or neither.
   $3x - 4y = -5$ and $6x - 8y = 7$
   Solution:
   To find their slopes, we rewrite each equation in slope-intercept form.
   $3x - 4y = -5, \quad -4y = -3x - 5, \quad y = \frac{3}{4}x + \frac{5}{4} \quad \therefore m_1 = \frac{3}{4}$
   $6x - 8y = 7, \quad -8y = -6x + 7, \quad y = \frac{3}{4}x - \frac{7}{8} \quad \therefore m_2 = \frac{3}{4}$
   $m_1 = m_2$ Ans: They are parallel.

3. Tell whether or not the following two lines are parallel, perpendicular, or neither.
   $x - 2y = 4$ and $2x + y = 5$
   Solution:
   $x - 2y = 4, \quad -2y = -x + 4, \quad y = \frac{1}{2}x - 2 \quad \therefore m_1 = \frac{1}{2}$
   $2x + y = 5, \quad y = -2x + 5 \quad \therefore m_2 = -2$
   $m_1 \cdot m_2 = \frac{1}{2} \cdot (-2) = -1$ Ans: They are perpendicular.

4. Tell whether or not the following two lines are parallel, perpendicular, or neither.
   $x + 3y = 4$ and $6y = 4x - 5$
   Solution: $x + 3y = 4, \quad 3y = -x + 4, \quad y = -\frac{1}{3}x + \frac{4}{3} \quad \therefore m_1 = -\frac{1}{3}$
   $6y = 4x - 5, \quad y = \frac{2}{3}x - \frac{5}{6} \quad \therefore m_2 = \frac{2}{3}$
   $m_1 \neq m_2$ and $m_1 \cdot m_2 \neq -1$ Ans: Neither

2-9 Direct Variations

A linear equation (or function) with the form $y = kx$ is called an equation having a **direct variation** with x and y. k is a nonzero constant. k is called the **constant of variation** or **constant of proportionality**. In the equation $y = 3x$, 3 is a constant. We say that y varies directly as x. Therefore, $y = 3x$ is an example of a direct variation with x and y.
The graph of $y = 3x$ is a straight line with slope 3 and passes through the origin (0, 0).
If (x_1, y_1) and (x_2, y_2) are two ordered pairs of an equation having a direct variation defined by $y = kx$ and that neither x_1 nor x_2 is zero, we have: $y_1 = kx_1$ and $y_2 = kx_2$,

$k = \frac{y_1}{x_1}$ and $k = \frac{y_2}{x_2}$. Therefore: $\frac{y_1}{x_1} = \frac{y_2}{x_2}$. **The ratios are equal.**

If $y = kx$, we say that y varies directly as x, or y is directly proportional to x.
In direct variation, the variables that vary may involve powers (exponents).

Note that $y = 3x - 2$ does not represent a direct variation with x and y because the ratios of its ordered pairs $\frac{y_1}{x_1}$ and $\frac{y_2}{x_2}$ are not equal (not proportional).
However, the equation $y = 3x - 2$ shows x and $(y + 2)$ as a direct variation $(y + 2 = 3x)$.

Joint Variation: An equation having one variable varies directly as the product of two or more other variables.
For $y = kxz$, we say that y varies jointly with x and y.

Examples

1. If y varies directly as x, and if $y = 6$ when $x = 2$, find the constant of variation.
   Solution: Let $y = kx$, $6 = k \cdot 2$ $\therefore k = \frac{6}{2} = 3$. Ans.

2. If y varies directly as x, and if $y = 6$ when $x = 2$, find y when $x = 3$.
   Solution: Let $y = kx$, $6 = k \cdot 2$ $\therefore k = \frac{6}{2} = 3$, $y = 3x$ is the equation.
   When $x = 3$, $y = 3x = 3 \cdot 3 = 9$. Ans.

3. If (x_1, y_1) and (x_2, y_2) are ordered pairs of the same direct variation, find y_1.
   $x_1 = 3$, $y_1 = ?$, $x_2 = 12$, $y_2 = 8$
   Solution: The ratios are equal: $\frac{y_1}{x_1} = \frac{y_2}{x_2}$, $\frac{y_1}{3} = \frac{8}{12}$, $12y_1 = 24$ $\therefore y_1 = 2$. Ans.
   Or: $y_2 = kx_2$, $8 = k \cdot 12$, $k = \frac{8}{12} = \frac{2}{3}$ $\therefore y_1 = kx_1 = \frac{2}{3}x_1 = \frac{2}{3} \cdot 3 = 2$. Ans.

2-10 Solving Inequalities in One Variable

An **inequality** is formed by placing an inequality sign or symbol (>, <, ≤, ≥) between numerical or variable expressions.

$5 > 3$ is read " 5 is greater than 3."
$3 < 5$ is read " 3 is less than 5."
$x > -4$ is read " x is greater than –4."

An inequality containing a variable is called **an open sentence**.
The inequality $x > -4$ indicates that all numbers greater than –4 are solutions of x.

$x \geq -3$ is read "x is greater than or equal to –3."
$x \leq 2$ is read "x is less than or equal to 2."
$-3 < x < 3$ is read "x is greater than –3 and less than 3."
$-3 \leq x < 2$ is read "x is greater than or equal to –3 and less than 2."

We can graph an inequality on a number line.
On a number line, the number on the right is greater than the number on the left.
To graph an inequality on a number line, we use a **close circle** " • " to show "included", and use an **open circle** " O " to show "not included".

Properties for inequalities:

1. Adding or subtracting the same number to or from each side of an inequality does not change the direction (order) of its inequality sign.
2. Multiplying or dividing each side of an inequality by the same positive number does not change the direction (order) of its inequality sign.
3. Multiplying or dividing each side of an inequality by the same negative number reverses (changes) the direction (order) of its inequality sign.

Examples

1. Add 6 to each side of $5 > 2$. Then write a true inequality.
Solution:
$5 > 2$
$5 + 6 > 2 + 6$
$11 > 8$ (true)

2. Multiply each side of $5 > 2$ by -6. Then write a true inequality.
Solution:
$5 > 2$
$5(-6) < 2(-6)$
$-30 < -12$ (true)

To solve an inequality, we use similar methods which are used to solve equations. Simply transfer all variables to one side of the inequality (usually to the left side), and transfer the others to the right side. Reverses (changes) the signs (+, –) in the process. Reverses the direction (order) of the inequality sign if we **multiply** (or **divide**) each side by **the same negative number**.

$$-\tfrac{1}{2}x > 4$$
$$(-2) \cdot -\tfrac{1}{2}x < 4 \cdot (-2)$$
$$x < -8)$$

Examples

1. Solve $4x-1<11$.

Solution:

$$4x-1<11$$
$$4x<11+1$$
$$4x<12$$
$$\therefore x<3. \quad \text{Ans.}$$

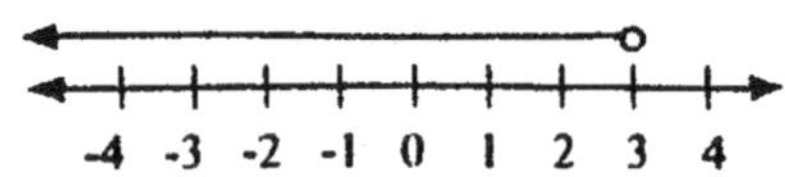

2. Solve $4x-1\geq 11$.

Solution:

$$4x-1\geq 11$$
$$4x\geq 11+1$$
$$4x\geq 12$$
$$\therefore x\geq 3. \quad \text{Ans.}$$

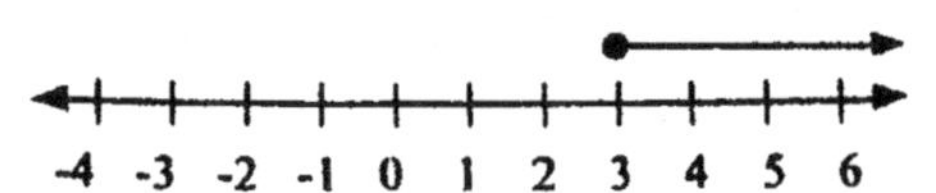

3. Solve $-4x-1<11$.

Solution:

$$-4x-1<11$$
$$-4x<12$$

Divide each side by –4:

$$\therefore x>-3. \quad \text{Ans.}$$

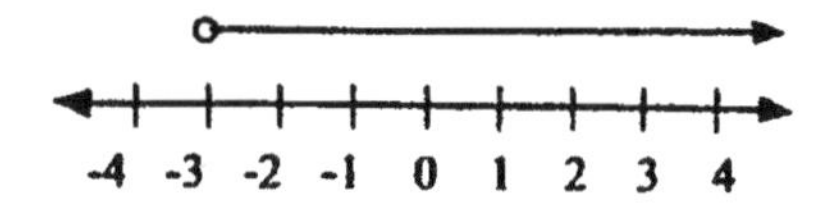

4. Solve $-\frac{1}{4}x-1\geq 11$.

Solution:

$$-\tfrac{1}{4}x-1\geq 11$$
$$-\tfrac{1}{4}x\geq 12$$

Multiply each side by –4:

$$\therefore x\leq -48. \quad \text{Ans.}$$

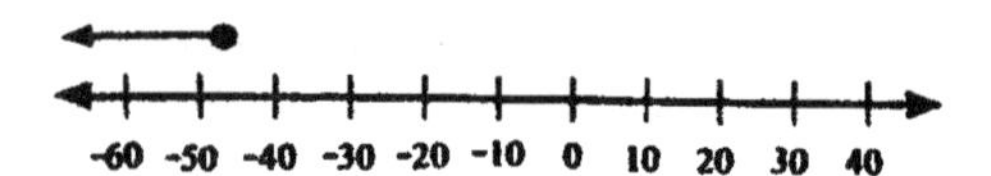

If the final statement is a "true statement", the inequality has solutions for all real numbers. If the final statement is a "false statement", the inequality has no solution.

Examples

5. Solve $3x<3(x+4)$.

Solution:

$$3x<3(x+4)$$
$$3x<3x+12$$
$$3x-3x<12$$
$$0<12 \text{ (true)}$$

Ans: The inequality is true for all real numbers.

6. Solve $3x>3(x+4)$.

Solution:

$$3x>3(x+4)$$
$$3x>3x+12$$
$$3x-3x>12$$
$$0>12 \text{ (false)}$$

Ans: The inequality has no solution. The solution set is ϕ (empty).

A kid said to his mother.
Kid: Did you wear contact lenses this morning ?
Mom: No. Why do you ask ?
Kid: I heard you yell to father this morning: You said "After 20 years of marriage, I have finally seen what kind of man you are"

2-11 Solving Combined Inequalities

A **combined inequality** (or **compound inequality**) is an inequality formed by joining two inequalities with "**and**" or "**or**".
There are two forms of combined inequalities, **conjunction** and **disjunction**.

1. **Conjunction of Inequality:** It is an inequality formed by two inequalities with the word "**and**". A conjunction is true when **both inequalities are true**.
   Example:
   $-5 < n$ and $n < 3$
   (or $-5 < n < 3$)

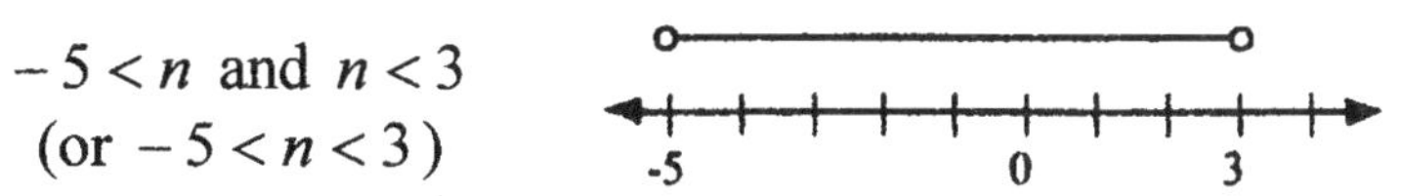

To solve a conjunction of inequality, we isolate the variable between the two inequality signs. Or, solve each part separately.

2. **Disjunction of Inequality:** It is an inequality forms by two inequalities with the word "**or**". A disjunction is true when **at least one of the inequalities is true.**
   Example:
   $n < -1$ or $n \geq 3$

To solve a disjunction of inequality, we solve each part separately.

Examples

1. Solve $2 < \frac{1}{3}x + 4 \leq 6$.
   Solution:

Method 1:
(Solve each part separately.)

$2 < \frac{1}{3}x + 4$ and $\frac{1}{3}x + 4 \leq 6$

$-2 < \frac{1}{3}x$	$\frac{1}{3}x \leq 2$
$-6 < x$	$x \leq 6$

$\therefore -6 < x \leq 6$. Ans.

Method 2:
(Solve between the inequality signs.)

$2 < \frac{1}{3}x + 4 \leq 6$

Subtract 4 to each expression:

$-2 < \frac{1}{3}x \leq 2$

Multiply each expression by 3:

$\therefore -6 < x \leq 6$. Ans.

2. Solve $1 - 2x < -3$ or $3x + 14 \leq 2 - x$.
   Solution:

Solve each part separately:

$1 - 2x < -3$ or $3x + 14 \leq 2 - x$

$-2x < -4$	$4x \leq -12$
$x > 2$	$x \leq -3$

Ans: $x \leq -3$ or $x > 2$

Examples

3. Solve $n-2 \le 2n \le 3n-1$.

Solution:

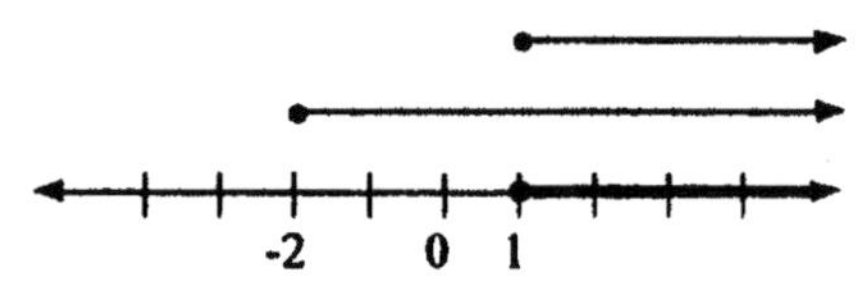

$$n-2 \le 2n \quad \text{and} \quad 2n \le 3n-1$$
$$-2 \le n \qquad\qquad 1 \le n$$

It is a conjunction "and" inequality. Both must be true.

Ans: $n \ge 1$.

4. Solve $3x-1 \le 2x$ or $3x+10 \ge -2x$.

Solution:

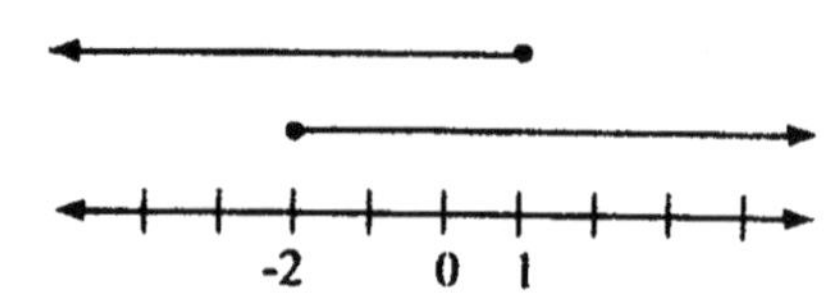

$$3x-1 \le 2x \quad \text{or} \quad 3x+10 \ge -2x$$
$$x \le 1 \qquad\qquad 5x \ge -10$$
$$x \ge -2$$

It is a disjunction "or" inequality. At least one is true.

Ans: All real numbers of x.

5. Solve $3a+5 < a+7$ and $3-a < 1$.

Solution:

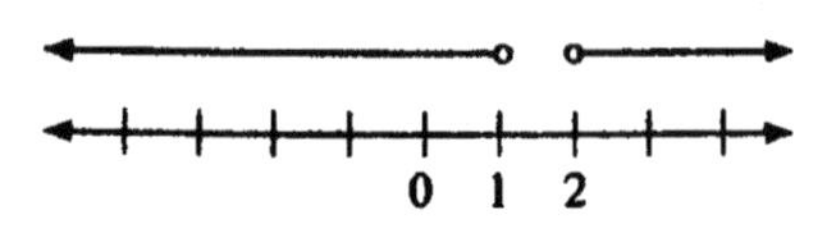

$$3a+5 < a+7 \quad \text{and} \quad 3-a < 1$$
$$2a < 2 \qquad\qquad -a < -2$$
$$a < 1 \qquad\qquad a > 2$$

It is a conjunction "and" inequality. Both must be true.

Ans: No solution. ϕ (The solution set is empty.)

6. Solve $x-1 \le 0$ and $x+2 \ge 0$.

Solution:

$$x-1 \le 0 \quad \text{and} \quad x+2 \ge 0$$
$$x \le 1 \qquad\qquad x \ge -2$$

It is a conjunction "and" inequality. Both must be true.

Ans: $-2 \le x \le 1$.

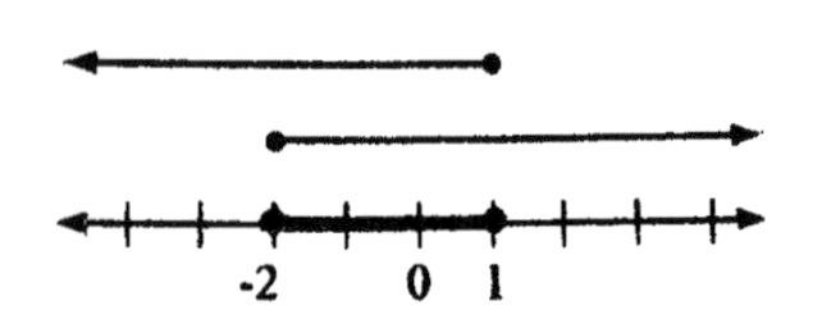

7. Solve $x-1 \le 0$ or $x+2 \ge 0$

Solution:

$$x-1 \le 0 \quad \text{or} \quad x+2 \ge 0$$
$$x \le 1 \qquad\qquad x \ge -2$$

It is a disjunction "or" inequality. At least one is true.

Ans: All real numbers of x.

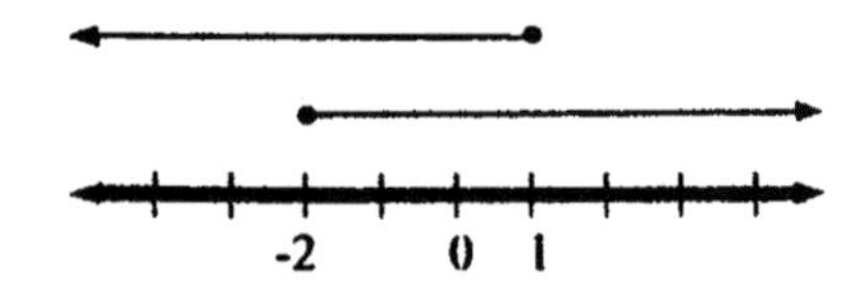

2-12 Solving Word Problems involving Inequalities

Steps for Solving a word problem involving inequality:

1. Read the problem.
2. Assign a variable to represent the unknown.
3. Determine which inequality sign ($<, >, \leq, \geq$) needed.
4. Write an inequality based on the given facts.
5. Solve the inequality to find the answer.
6. Check your answer (usually on a scratch paper).

Examples

1. You want to rent a car for a rental cost of $50 plus $30 per day, or a van for $80 plus $25 per day. For how many days at most is it cheaper to rent a car ?

   Soluion:

   Let d = days needed

   The inequality is:

   $50 + 30d \leq 80 + 25d$

   $5d \leq 30$

   $\therefore d \leq 6$

   Ans: At most 6 days (or less)

   Check: $50 + 30d = 50 + 30(6) = \$230$ √

   $80 + 25d = 80 + 25(6) = \$230$ √

   The rental costs are the same for 6 days. It is cheaper to rent the car for less than 6 days. It is cheaper to rent the van for more than 6 days.

2. Your scores on your 4 math tests were 71, 74, 75 and 86. What is the lowest score you need in your next test in order to have an average score of at least 80 ?

   Solution:

   Let x = score for next test

   The inequality is:

   $\frac{71 + 74 + 75 + 86 + x}{5} \geq 80$, $\frac{306 + x}{5} \geq 80$, $306 + x \geq 400$ $\therefore x \geq 94$

   Ans: 94 or more

3. John wants to rent a car and pay at most $250. The car rental costs $150 plus $0.25 a mile. How far can he travel ?

   Solution:

   Let x = mileage he can travel

   The inequality is:

   $150 + 0.25x \leq 250$

   $0.25x \leq 100$ $\therefore x \leq 400$ Ans: 400 miles or less.

4. The sum of two consecutive integers is less than 79. Find the integers with the greatest sum.

   Solution: Let $n = 1^{st}$ integer. $n + 1 = 2^{nd}$ integer.

   The inequality is:

   $n + (n + 1) < 79$, $2n < 78$, $\therefore n < 39$

   The greatest integer in n is 38. Ans. 38 and 39.

Notes

2-13 Operations and Graphs with Absolute Values

Absolute Value: The positive number of any real number a. It is written by $|a|$.

If $a > 0$, $|a| = a$ Example: $|5| = 5$

If $a = 0$, $|a| = |0| = 0$

If $a < 0$, $|a| = -a$ Example: $|-5| = -(-5) = 5$

The absolute value of a negative number is the opposite of the number.

Absolute value is commonly used as the distance between two points on a number line. The distance between two points is always a positive number. A distance can never be a negative number. Therefore, we use the absolute value to represent the distance.

A C B Number line
−4 0 3

$\overline{AB} = |3-(-4)| = |3+4| = |7| = 7$ Or: $\overline{AB} = |-4-3| = |-7| = 7$

The absolute value of a number is its distance from zero on a number line.

$\overline{BC} = |3-0| = |3| = 3$. $\overline{AC} = |-4-0| = |-4| = 4$.

The domain and range of the absolute-value function

$y = |x|$ is the most basic absolute-value function. Its domain is all real numbers of x. Its range is all real numbers of $y \geq 0$, or all non-negative real numbers of y.

The graph of the absolute-value function

To graph an absolute-value function, we can plot its ordered pairs. Or, we can plot its separated equations without absolute-value sign.

Examples

1. Find the domain and range of $y = |x-2|$.

   Solution: D = all real numbers. R = all non-negative real numbers.

2. Find the domain and range of $y = |x| - 2$.

   Solution: D = all real numbers. R = all real numbers of $y \geq -2$.

3. Find the domain and range of $y = -|x-2|$.

   Solution: D = all real numbers. R = all non-positive real numbers.

4. Graph $y = |x|$.

   Solution:

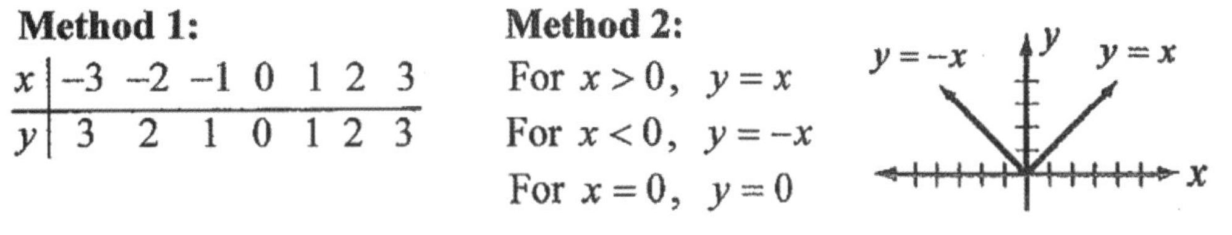

Method 1:

x	−3	−2	−1	0	1	2	3
y	3	2	1	0	1	2	3

Method 2:

For $x > 0$, $y = x$

For $x < 0$, $y = -x$

For $x = 0$, $y = 0$

2-14 Absolute-Value Equations and Inequalities

To solve equations and inequalities involving absolute values, we start with the following most basic patterns. Following these patterns, we can solve all inequalities easily.

Rules: On the number line:

$|x| = 5$ means " the distance from x to 0 is 5 ".

$|x - c| = 5$ means " the distance from x to c is 5 ".

$|x + c| = 5$ means " the distance from x to $-c$ is 5 ".

$|2x - c| = 5$ means " the distance from $(2x)$ to c is 5 ". (See Example 4)

These rules are also valid for most absolute-value inequalities ($<, \leq, >, \geq$).

1. Solve $|x| = 5$.

Solution: $|x| = 5$

Ans: $x = 5$ or -5 (or $x = \pm 5$)

2. Solve $|x| > 5$.

Solution: $|x| > 5$

Ans: $x < -5$ or $x > 5$

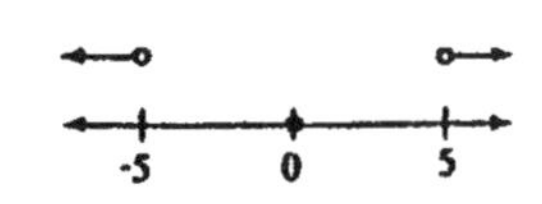

3. Solve $|x| < 5$.

Solution: $|x| < 5$

Ans: $-5 < x < 5$

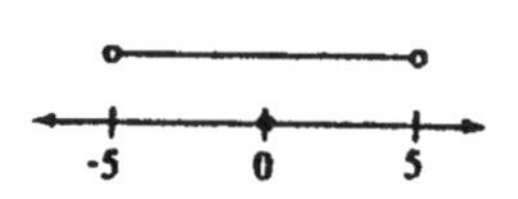

Examples

1. Solve $|x - 2| = 8$.

Solution: $x - 2 = \pm 8$

$x = \pm 8 + 2$

Ans: $x = 10$ or -6

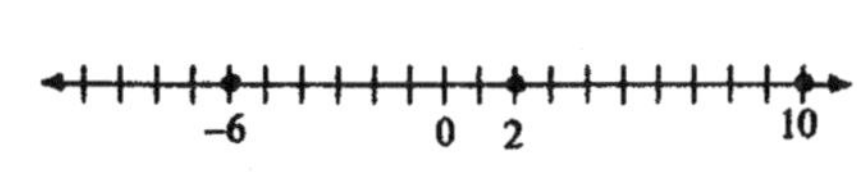

2. Solve $|x - 2| > 8$.

Solution: $x - 2 < -8$ or $x - 2 > 8$

$x < -6$ | $x > 10$

Ans: $x < -6$ or $x > 10$

3. Solve $|x + 2| < 8$.

Solution: $-8 < x + 2 < 8$

$-10 < x < 6$. Ans.

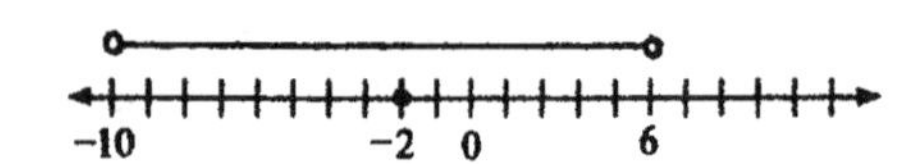

4. Solve $|2x - 7| = 5$.

Solution: $2x - 7 = \pm 5$

$2x = \pm 5 + 7$

$2x = 12$ or 2

Ans. $x = 6$ or 1

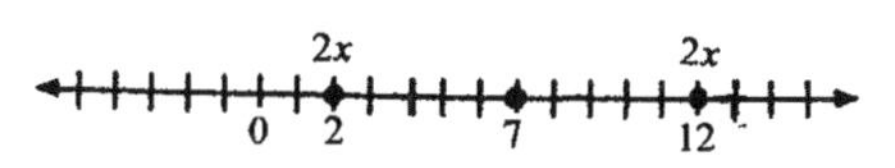

5. Solve $|3n + 5| - 3 \geq 7$.

Solution: $|3n + 5| \geq 10$

$3n + 5 \leq -10$ or $3n + 5 \geq 10$

$3n \leq -15$ | $3n \geq 5$

$\therefore n \leq -5$ or $n \geq \frac{5}{3}$. Ans.

6. Solve $|20 - 5y| \leq 25$.

Solution: $-25 \leq 20 - 5y \leq 25$

$-45 \leq -5y \leq 5$

$9 \geq y \geq -1$

Ans: $-1 \leq y \leq 9$

2-15 Solving Quadratic Equations

Steps for solving a quadratic equation:

1. Solve the equation in the form of perfect-square.
   If $x^2 = k$, then $x = \pm\sqrt{k}$.
2. Solve by factoring if it can be factored.
   If $ax^2 + bx + c = 0$ can be factored, then $(px + r)(qx + s) = 0$.
3. Solve by using the quadratic formula if factoring is not possible.
   If $ax^2 + bx + c = 0$, then $x = \dfrac{-b \pm \sqrt{b^2 - 4ac}}{2a}$ **(Quadratic Formula)**
4. Some equations are not quadratic equations, but they can be written in quadratic form by factoring, or by making a substitution. (see Example 3, 5)

Examples

1. Solve $2x^2 - 5x - 3 = 0$ by factoring.
Solution:

$$(2x+1)(x-3) = 0$$

$$\begin{array}{r|l} 2x+1=0 & x-3=0 \\ 2x=-1 & x=3 \\ x=-\frac{1}{2} & \end{array}$$

Ans: $x = -\frac{1}{2}$ or 3

2. Solve $2x^2 - 5x - 3 = 0$ by formula.
Solution: $a = 2$, $b = -5$, $c = -3$

$$x = \frac{-b \pm \sqrt{b^2 - 4ac}}{2a}$$
$$= \frac{-(-5) \pm \sqrt{(-5)^2 - 4 \cdot 2 \cdot (-3)}}{2(2)}$$
$$= \frac{5 \pm \sqrt{25 + 24}}{4} = \frac{5 \pm 7}{4}$$
$$= \frac{5+7}{4} \text{ or } \frac{5-7}{4} = 3 \text{ or } -\frac{1}{2}. \text{ Ans.}$$

3. Solve $x^4 - 2x^2 - 8 = 0$.
Solution: (By factoring)

$$(x^2 - 4)(x^2 + 2) = 0$$

$$\begin{array}{r|l} x^2 - 4 = 0 & x^2 + 2 = 0 \\ x^2 = 4 & x^2 = -2 \\ x = \pm 2 & x = \pm\sqrt{-2} = \pm\sqrt{2}\, i \end{array}$$

Ans: $x = \pm 2$ or $\pm\sqrt{2}\, i$

4. Solve $x^2 + 4x + 8 = 0$
Solution: $a = 1$, $b = 4$, $c = 8$

$$x = \frac{-b \pm \sqrt{b^2 - 4ac}}{2a}$$
$$= \frac{-4 \pm \sqrt{4^2 - 4 \cdot 1 \cdot 8}}{2(1)} = \frac{-4 \pm \sqrt{16 - 32}}{2}$$
$$= \frac{-4 \pm \sqrt{-16}}{2} = \frac{-4 \pm 4i}{2}$$
$$= -2 \pm 2i. \text{ Ans.}$$

5. Solve $2x - 2\sqrt{2x} - 8 = 0$.
Solution: (By making a substitution)

Let $u = \sqrt{2x}$, $u^2 = 2x$

$$u^2 - 2u - 8 = 0$$
$$(u-4)(u+2) = 0$$

$$\begin{array}{r|l} u - 4 = 0 & u + 2 = 0 \\ u = 4 & u = -2 \\ \sqrt{2x} = 4 & \sqrt{2x} = -2 \\ x = 8 & \text{no solution} \end{array}$$

Ans: $x = 8$

2-16 Quadratic Functions and Parabolas

Quadratic Equation: An equation of the form $ax^2 + bx + c = 0$ $(a \neq 0)$.
Quadratic Function: A quadratic equation of the form $y = ax^2 + bx + c$ $(a \neq 0)$.
The graph of a quadratic function $y = ax^2 + bx + c$ $(a \neq 0)$ on the coordinate plane is a **parabola**. The central line of a parabola is called **the axis of symmetry**, or simply the **axis**. The point where the parabola crosses its axis is the **vertex**.

General Form of a parabola: $y - k = a(x-h)^2$. The vertex is $(h,\ k)$.
Intercept Form of a parabola: $y = a(x-p)(x-q)$. The $x-$intercepts are p and q.
The $x-$coordinate of the vertex is the midpoint between the two $x-$intercepts.

If $a > 0$ (positive), the parabola opens upward. Vertex is a minimum point.
If $a < 0$ (negative), the parabola opens downward. Vertex is a maximum point.
$y = x^2$ is the simplest quadratic function. Its vertex is located on the origin (0, 0).

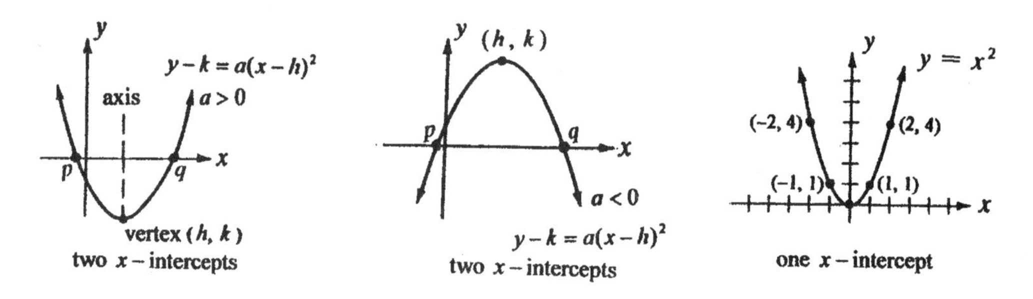

A parabola can have two, one, or no $x-$intercepts.
To find the $x-$intercepts of a parabola, we find the **zeros** (roots) of the related quadratic equation (let $y = 0$):

Solve $ax^2 + bx + c = 0$ by factoring, or by the formula: $x = \dfrac{-b \pm \sqrt{b^2 - 4ac}}{2a}$.

If $b^2 - 4ac > 0$, the equation has two roots. The parabola has two $x-$intercepts.
If $b^2 - 4ac = 0$, the equation has one root. The parabola has one $x-$intercept.
If $b^2 - 4ac < 0$, the equation has no real-number root. The parabola has no $x-$intercept.
The value of $b^2 - 4ac$ indicates the differences among the three cases. Therefore, we call the value of $b^2 - 4ac$ "**the discriminant**" of the quadratic equation.

Steps for graphing a quadratic function: $y = ax^2 + bx + c$ $(a \neq 0)$

1. Let $y = 0$. Find the zeros (roots). The roots are the $x-$intercepts.
2. Find the vertex $(h,\ k)$ by matching the general form $y - k = a(x-h)^2$.
3. If $a > 0$, the graph opens upward. If $a < 0$, the graph opens downward.

The graph of a parabola is symmetrical. One half of the graph on one side of the axis is the **mirror image** of the other half. The $x-$coordinate of the vertex is $-\frac{b}{2a}$.

Examples

1. Graph $y = 2x^2$.

   Solution:

   One x – intercept = 0

   Vertex (0, 0)

   $a = 2 > 0$

   Open upward

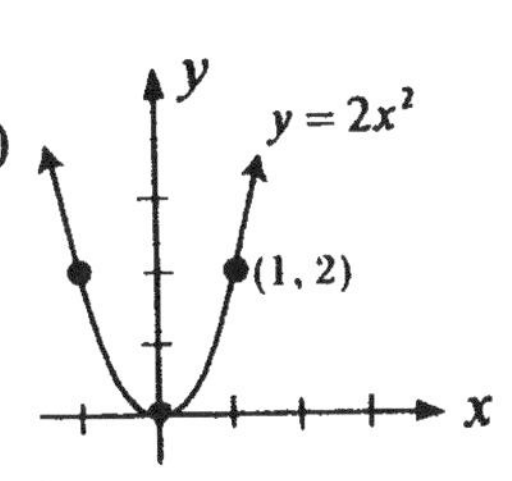

2. Graph $y = -2x^2$.

   Solution:

   One x – intercept = 0

   Vertex (0, 0)

   $a = -2 < 0$

   Open downward

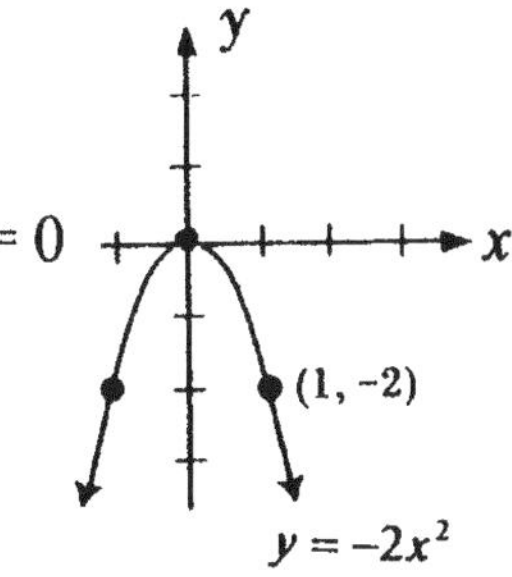

3. Graph $f(x) = x^2 - 2$.

   Solution:

   $x^2 - 2 = 0$, $x^2 = 2$

   $x = \pm\sqrt{2} \approx \pm 1.4$

   ∴ Two x – intercepts -1.4 and 1.4

   $y + 2 = x^2$, $y - (-2) = (x - 0)^2$

   ∴ Vertex (0, –2)

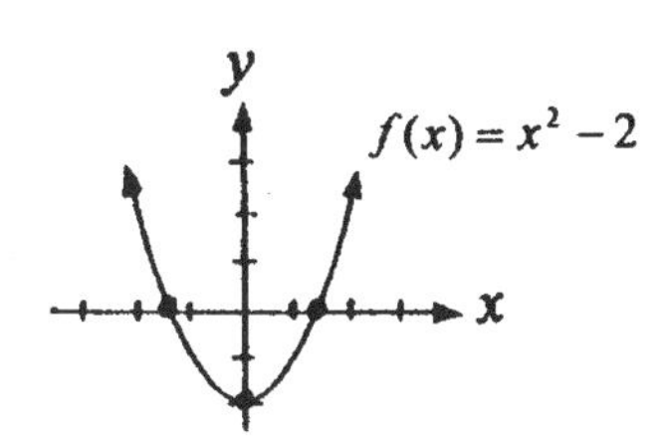

4. Graph $f(x) = x^2 - 4x$.

   Solution:

   $x^2 - 4x = 0$, $x(x - 4) = 0$

   $x = 0$ and 4

   ∴ Two x – intercepts = 0 and 4

   $y = (x^2 - 4x + 4) - 4$

   $y + 4 = x^2 - 4x + 4$

   $y - (-4) = (x - 2)^2$ ∴ Vertex (2, –4)

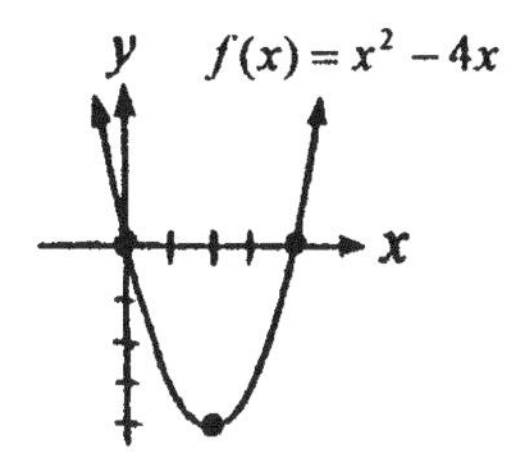

5. Graph $f(x) = x^2 - 2x + 1$.

   Solution:

   $x^2 - 2x + 1 = 0$, $(x - 1)^2 = 0$

   $x = 1$

   ∴ One x – intercept = 1

   $y = x^2 - 2x + 1$

   $y - 0 = (x - 1)^2$

   ∴ Vertex (1, 0)

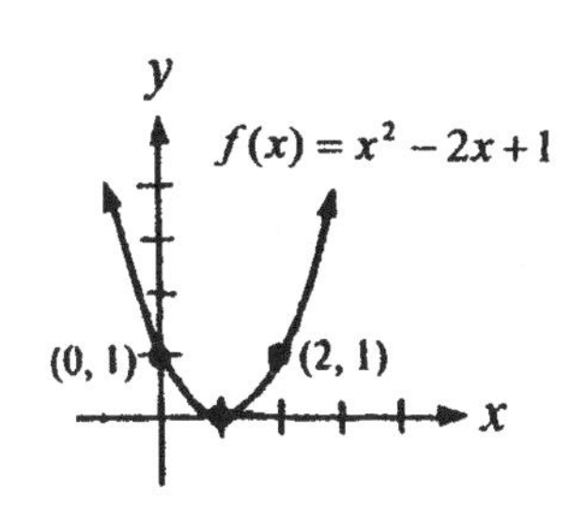

6. Graph $f(x) = (x - 3)^2 + 1$

   Solution:

   $(x - 3)^2 + 1 = 0$, $(x - 3)^2 = -1$

   No real-number root

   ∴ No x – intercept

   $y = (x - 3)^2 + 1$

   $y - 1 = (x - 3)^2$

   ∴ Vertex (3, 1)

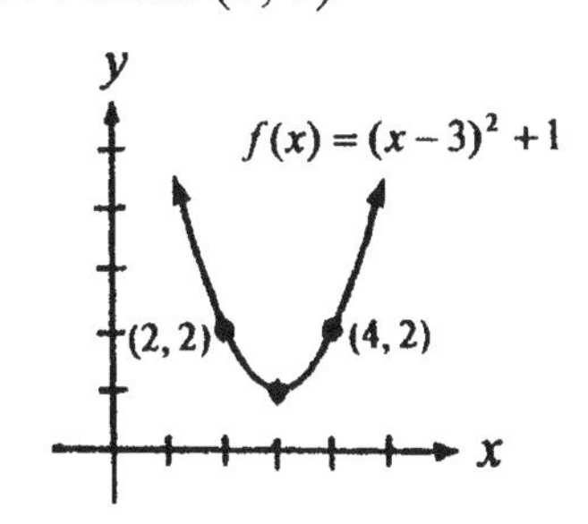

Graphing any Quadratic Function – Completing The Square

To graph any quadratic function $f(x) = ax^2 + bx + c\ (a \neq 0)$, we need to rewrite it to match the general form of parabola by the method of **completing the square.**

General Form of Parabola: $f(x) - k = a(x-h)^2$ or $y - k = a(x-h)^2$

1) Vertex is (h, k). **2)** If $a > 0$, open upward. **3)** If $a < 0$, open downward.

In science, we use the quadratic function to find the **extreme values** (maximum or minimum) of its application. There is a maximum value if it is open downward, a minimum value if it is open upward.

Rules of Completing the square for $y = ax^2 + bx + c$:

1. If $a = 1$, we add $\left(\frac{b}{2}\right)^2$ on right side to complete the square.

2. If $a \neq 1$, we rewrite the right side by $a(x^2 + \frac{b}{a}x + \frac{c}{a})$ so that the coefficient of x^2 is 1. Then follow Rule 1 to complete the square on the right side.

Examples

1. Graph $y = -x^2 - 2x + 1$.

Solution:

$$y = -x^2 - 2x + 1$$
$$y - 1 = -(x^2 + 2x)$$

Completing the square

$$y - 1 = -(x^2 + 2x + 1 - 1)$$
$$y - 1 = -(x^2 + 2x + 1) + 1$$
$$y - 1 = -(x+1)^2 + 1$$

$\therefore\ y - 2 = -(x+1)^2$ is the general form.

Vertex $(-1, 2)$, $a = -1 < 0$ open downward.

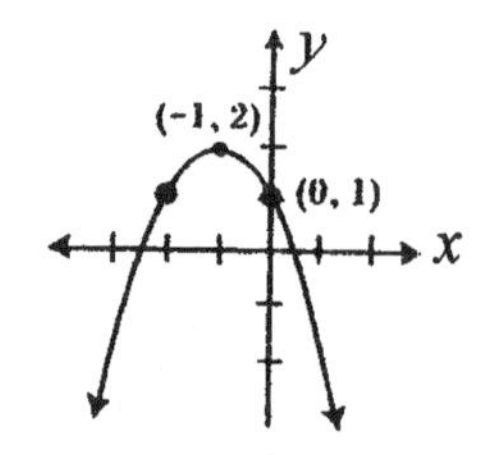

2. Graph $f(x) = 4x^2 + 12x + 10$.

Solution:

$$f(x) = 4x^2 + 12x + 10$$
$$f(x) - 10 = 4x^2 + 12x$$
$$f(x) - 10 = 4(x^2 + 3x)$$
$$f(x) - 10 = 4(x^2 + 3x + \tfrac{9}{4} - \tfrac{9}{4})$$
$$f(x) - 10 = 4(x^2 + 3x + \tfrac{9}{4}) - 9$$
$$f(x) - 1 = 4(x^2 + 3x + \tfrac{9}{4})$$

$\therefore\ f(x) - 1 = 4(x + \frac{3}{2})^2$ is the general form.

Vertex $(-\frac{3}{2}, 1)$, $a = 4 > 0$ open upward.

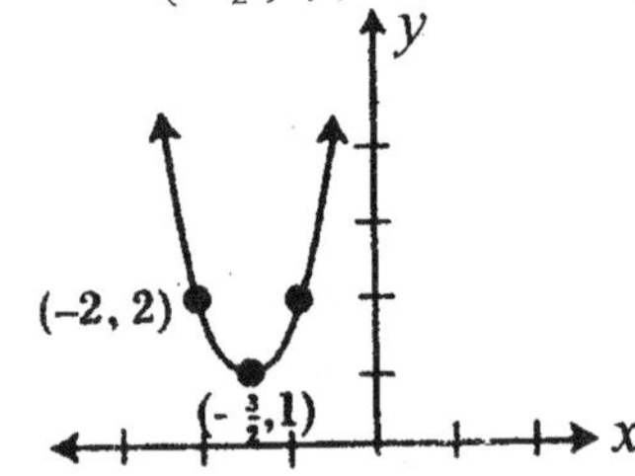

3. Find the extreme value of $f(x) = -2x^2 + 16x - 2$ and determine whether it is a maximum or a minimum.

Solution: $f(x) = -2x^2 + 16x - 2$

$$f(x) + 2 = -2(x^2 - 8x)$$
$$f(x) + 2 = -2(x^2 - 8x + 16 - 16)$$
$$f(x) + 2 = -2(x^2 - 8x + 16) + 32$$
$$f(x) - 30 = -2(x^2 - 8x + 16)$$

$f(x) - 30 = -2(x-4)^2$ is the general form.

Vertex $(4, 30)$, $a = -2 < 0$ open downward

There is a maximum value of $f(x)$ at $x = 4$

$f(x)_{max.} = 30$ at $x = 4$. Ans.

2-17 Solving Polynomial Equations by Factoring

Many polynomial equations can be solved by **factoring** and the **zero-product theory**.
Zero-Product Theory: For all real numbers a and b:

$a \cdot b = 0$ **if and only if** $a = 0$ or $b = 0$.

The zero-product theory tells us when an equation is written as a product of factors, we can find its solutions by setting each of the factors equal to 0 and solve for its roots.
Double Roots: If an equation has two identical factors, the equation is said to have a **double root** (see example 4).
The following formulas of special patterns are useful in the process to factor polynomials:

Formulas:

1. $a^2 + 2ab + b^2 = (a+b)^2$	**4.** $a^3 + b^3 = (a+b)(a^2 - ab + b^2)$
2. $a^2 - 2ab + b^2 = (a-b)^2$	**5.** $a^3 - b^3 = (a-b)(a^2 + ab + b^2)$
3. $a^2 - b^2 = (a+b)(a-b)$	

An x – intercept of a function $y = f(x)$ is a point $(x, 0)$ at which the graph crosses the $x-axis$ ($y = 0$). To find the x – intercepts of a function, we find the zeros of the function $y = f(x)$ by letting $f(x) = 0$ (see example 1).

Examples

1. Find the zeros of $f(x) = x^2 + x - 2$.

Solution: Let $f(x) = 0$, $x^2 + x - 2 = 0$

$(x+2)(x-1) = 0$

$x + 2 = 0 \mid x - 1 = 0$

$x = -2 \mid x = 1$

Ans: $x = -2$ and 1.

2. Solve $(2x-4)(x+5) = 0$.

Solution: $(2x-4)(x+5) = 0$

$2x - 4 = 0 \mid x + 5 = 0$

$2x = 4 \mid x = -5$

$x = 2$

Ans: $x = 2$ and -5.

3. Solve $2x^2 + 3x = 9$.

Solution: $2x^2 + 3x - 9 = 0$

$(2x-3)(x+3) = 0$

$2x - 3 = 0 \mid x + 3 = 0$

$2x = 3 \mid x = -3$

$x = 1\frac{1}{2}$ | Ans: $x = 1\frac{1}{2}$ and -3.

4. Solve $x^2 - 14x + 49 = 0$.

Solution: $x^2 - 14x + 49 = 0$

$(x-7)(x-7) = 0$

$x - 7 = 0 \mid x - 7 = 0$

$x = 7 \mid x = 7$

Ans: $x = 7$ (**a double root**).

5. Solve $2a^3 - 6a^2 - 20a = 0$.

Solution: $2a^3 - 6a^2 - 20a = 0$

$2a(a^2 - 3a - 10) = 0$

$2a(a-5)(a+2) = 0$

$2a = 0 \mid a - 5 = 0 \mid a + 2 = 0$

$a = 0 \mid a = 5 \mid a = -2$

Ans: $a = 0$, 5, and -2.

6. Solve $x^4 - 8x = 0$.

Solution: $x^4 - 8x = 0$, $x(x^3 - 8) = 0$

$x(x-2)(x^2 + 2x + 4) = 0$

$x = 0$, $x = 2$, $x^2 + 2x + 4 = 0$

$x = \frac{-b \pm \sqrt{b^2 - 4ac}}{2a} = \frac{-2 \pm \sqrt{2^2 - 4 \cdot 1 \cdot 4}}{2(1)} = -1 \pm \sqrt{3}\, i$

Ans: $x = 0$, 2, and $-1 \pm \sqrt{3}\, i$.

2-18 Synthetic Division

Dividing a polynomial in x by a binomial of the form $x-c$ is very much like dividing real numbers.
When we divide a polynomial by $x-c$, we must arrange the terms of the polynomial in order of decreasing degree of the variable. Keep space (using 0) on the missing terms in degrees of x.
The common process of dividing a polynomial by $x-c$ is shown in Example 1. In this section we use the easier method called **Synthetic Division** to get the same answer.

Synthetic Division (Horner's Algorithm):

Dividing a polynomial by $x-c$, we proceed only the coefficients of the terms of the polynomial.

1) If the divisor is $x+c$, rewrite it as $x-(-c)$.
2) If the divisor is $ax-b$, rewrite it as $a(x-\frac{b}{a})$, use the divisor $x-\frac{b}{a}$ to proceed the synthetic division. Then, multiply the result by $\frac{1}{a}$.

Examples

1. Divide $\dfrac{2x^3-8x^2-4x-6}{x-3}$.

Solution: (By common division)

$$\begin{array}{r} 2x^2-2x-10 \\ x-3\,\overline{)\,2x^3-8x^2-4x-6} \\ -)\ 2x^3-6x^2 \qquad\qquad \\ \hline -2x^2-4x-6 \\ -)\ -2x^2+6x \qquad \\ \hline -10x-6 \\ -)\ -10x+30 \\ \hline -36 \end{array}$$

(Remainder)

Ans: $\dfrac{2x^3-8x^2-4x-6}{x-3}=2x^2-2x-10-\dfrac{36}{x-3}$

2. Divide $\dfrac{2x^3-8x^2-4x-6}{x-3}$.

Solution: (By synthetic division)

$$\begin{array}{r|rrrr} 3 & 2 & -8 & -4 & -6 \\ +) & \downarrow & 6 & -6 & -30 \\ \hline & 2 & -2 & -10, & -36 \end{array}$$

↑ Remainder

Ans: $\dfrac{2x^3-8x^2-4x-6}{x-3}=2x^2-2x-10-\dfrac{36}{x-3}$

3. Divide $\dfrac{8x^3+2x^2-51}{4x-7}$.

Solution: (By synthetic division)

$$\frac{8x^3+2x^2-51}{4x-7}=\frac{1}{4}\cdot\frac{8x^3+2x^2-51}{x-\frac{7}{4}}$$

$$\begin{array}{r|rrrr} \frac{7}{4} & 8 & +2 & +0 & -51 \\ +) & & +14 & +28 & +49 \\ \hline & 8 & +16 & +28, & -2 \end{array}$$

↑ Remainder

$\therefore \dfrac{8x^3+2x^2-51}{4x-7}$

$=\dfrac{1}{4}(8x^2+16x+28-\dfrac{2}{x-\frac{7}{4}})$

$=\dfrac{1}{4}(8x^2+16x+28-\dfrac{2}{\frac{4x-7}{4}})$

$=\dfrac{1}{4}(8x^2+16x+28-\dfrac{8}{4x-7})$

$=2x^2+4x+7-\dfrac{2}{4x-7}$. Ans.

2-19 Remainder and Factor Theorems

We already know : $\frac{7}{2} = 3 + \frac{1}{2}$ where "1" is the remainder.

$\therefore\ 7 = 2\ (3 + \frac{1}{2}) = 2 \cdot 3 + 1$

The above example shows the following fact:

Dividend = divisor × quotient + remainder.

In general, when a polynomial $p(x)$ is divided by $x - c$, $P(x)$ is the dividend, $x - c$ is the divisor, $Q(x)$ is the quotient, R is the remainder, we have:

$P(x) = Q(x)(x - c) + R$

Let $x = c$

$P(c) = Q(c) \cdot (c - c) + R$

$\therefore\ P(c) = R$ -------------- Called **Remainder Theorem.**

Remainder Theorem: The remainder on dividing $P(x)$ by $x - c$ is $P(c)$.

Factor Theorem: If and only if $P(c) = 0$ (i.e. no remainder), then $x - c$ is a factor of $p(x)$ and c is a root of the equation $P(x) = 0$.

This is an easier way to find the value of polynomial $P(x)$ at $x = c$. We apply the **remainder theorem** and use synthetic division to find the remainder on dividing $P(x)$ by $x - c$. Then , the remainder is $P(c)$.

We apply the **factor theorem** to identify and find possible rational roots of a polynomial equation $P(x) = 0$. If $P(c) = 0$, then c is a root of $P(x) = 0$.

Examples

1. Find the value of $P(x) = 2x^3 - 8x^2 - 4x - 6$ when $x = 3$.

Solution:

Method 1): Direct substitution.

$P(x) = 2x^3 - 8x^2 - 4x - 6$

$P(3) = 2(3)^3 - 8(3)^2 - 4(3) - 6$

$= 54 - 72 - 12 - 6$

$= -36.$

Ans: $P(3) = -36$.

Method 2): Remainder theorem & Synthetic Division.

$$\begin{array}{r|rrrr} 3 & 2 & -8 & -4 & -6 \\ +) & & 6 & -6 & -30 \\ \hline & 2 & -2 & -10, & -36 \end{array}$$

(Remainder)

Ans: $P(3) = -36$.

Examples

2. Divide $\frac{2x^3-8x^2-4x-6}{x-3}$, find remainder.

Solution:

Method 1: Common division.

$$\begin{array}{r} 2x^2-2x\ -10 \\ x-3\overline{)2x^3-8x^2-4x-6} \\ -)\ 2x^3-6x^2 \\ \hline -2x^2-4x-6 \\ -)\ -2x^2+6x \\ \hline -10x-6 \\ -)\ -10x+30 \\ \hline -36 \end{array}$$

(Remainder)

Method 2: Synthetic division:

$$\begin{array}{r} 3|\ 2-8-4-6 \\ +)\quad 6-6-30 \\ \hline 2-2-10,-36 \end{array}$$

(Remainder)

Method 3: Remainder theorem.

$P(x)=2x^3-8x^2-4x-6$

$P(3)=2(3)^3-8(3)^2-4(3)-6$

$=54-72-12-6$

$=-36$ (Remainder)

3. Is $x-3$ a factor of

$P(x)=2x^3-8x^2-4x-6$?

Solution: $P(x)=2x^3-8x^2-4x-6$

$P(3)=2(3)^3-8(3)^2-4(3)-6$

$=54-72-12-6$

$=-36\neq 0.$

Ans: $x-3$ is not a factor of $P(x)$.

(3 is not a root of $P(x)=0$.)

4. Is $x+1$ a factor of

$P(x)=x^3+6x^2+11x+6$?

Solution:

$P(x)=x^3+6x^2+11x+6$

$P(-1)=(-1)^3+6(-1)^2+11(-1)+6$

$=-1+6-11+6$

$=0$

Ans: $x+1$ is a factor of $P(x)$.

(-1 is a root of $P(x)=0$.)

5. Is $x-\sqrt{3}$ a factor of

$P(x)=x^3-3x^2-3x+9$?

Solution:

$P(x)=x^3-3x^2-3x+9$

$P(\sqrt{3})=(\sqrt{3})^3-3(\sqrt{3})^2-3(\sqrt{3})+9$

$=3\sqrt{3}-9-3\sqrt{3}+9$

$=0$

Ans: $x-\sqrt{3}$ is a factor of $P(x)$.

($\sqrt{3}$ is a root of $P(x)=0$.)

6. Solve $x^3-4x^2+8x-5=0$ if one root is 1.

Solution: By synthetic division:

$$\frac{x^3-4x^2+8x-5}{x-1}=x^2-3x+5$$

$x^3-4x^2+8x-5=0$

$(x-1)(x^2-3x+5)=0$

$x-1=0$, $x^2-3x+5=0$

$x=1,\ x=\frac{3\pm\sqrt{9-20}}{2}=\frac{3\pm\sqrt{11}i}{2}$

Ans: $x=1$ and $\frac{3\pm\sqrt{11}i}{2}$

2-20 Solving Polynomial Equations with Degree n

Polynomial Function: It is an algebraic function in the following form: $y = p(x)$

$$p(x) = a_n x^n + a_{n-1}x^{n-1} + \cdots + a_1 x + a_0 .$$

(Where a_n, a_{n-1}, ·····, a_1, a_0 are real numbers and n is a nonnegative integer.)

Finding the zeros of a polynomial function is to find any value of x that makes $p(x) = 0$.
Therefore, finding the zeros of a polynomial function is to solve the following polynomial equation. $a_n x^n + a_{n-1}x^{n-1} + \cdots + a_1 x + a_0 = 0$

The **real zeros** of a polynomial function $y = p(x)$ (the real roots of a polynomial equation $p(x) = 0$) are also the x – intercepts of the graph of $y = p(x)$ on the coordinate plane.

Rules for finding the zeros of a polynomial function with degree n ($n \geq 3$):

1) **Fundamental Theorem of Polynomial**
   Every polynomial function of positive degree n has exactly n zeros.
2) **Conjugate Pairs Theorem:** a and b are real numbers, $b \neq 0$.
   If $a + bi$ is a zero of polynomial function, then $a - bi$ is also a zero.
3) **Newton's Rule (Rational Zeros Theorem):** It is used to find the rational zeros of a polynomial function $p(x) = a_n x^n + a_{n-1}x^{n-1} + \cdots + a_1 x + a_0$.
   a) If $a_n = 1$ and r is an integer zero, then r must be a factor of a_0.
   b) If $a_n \neq 1$ and $\frac{b}{a}$ is a zero (a, b are integers, $x = \frac{b}{a}$ and $ax - b = 0$), then b must be a factor of a_0, a must be a factor of a_n.

To solve a polynomial equation in degree n, we find certain roots by Newton's Rule and factor it as a product of its factors in degree 1 and quadratic forms.
If a polynomial $p(x)$ has the factor $(x-c)^n$, we say that c is a zero of **multiplicity** n of $p(x)$, or a root of multiplicity n of $p(x) = 0$. It has an x – intercept at $x = c$.

Examples

1. Solve $x(x+3)^2(x-1) = 0$.

Solution:

$x = 0$ | $(x+3)^2 = 0$, $x = -3$ | $x - 1 = 0$, $x = 1$

Ans: $x = 0, 1,$
and -3 (multiplicity 2, or a double root)

2. If two roots of a quadratic equations are -3 and 1, find such an equation.

Solution:

The equation is:
$(x+3)(x-1) = 0$
$\therefore x^2 + 2x - 3 = 0$ is the equation. Ans.

3. Solve $x^3 + 6x^2 + 11x + 6 = 0$.

Solution: By Newton's Rule: (factors of 6)
Possible roots: $\pm 1, \pm 2, \pm 3, \pm 6$
By factor theorem:
$P(-1) = (-1)^3 + 6(-1)^2 + 11(-1) + 6 = 0$
$\therefore x + 1$ is a factor.
By synthetic division:

$$\frac{x^3 + 6x^2 + 11x + 6}{x+1} = x^2 + 5x + 6$$

$x^3 + 6x^2 + 11x + 6 = 0$
$(x+1)(x^2 + 5x + 6) = 0$
$(x+1)(x+2)(x+3) = 0$
Ans: $x = -1, -2$, and -3

Finding Complex Zeros of Polynomial Functions with Degree n

We have learned how to find the real zeros of a polynomial function. Here, we will learn how to find the **complex zeros** in the form $a+b\,i$ of a polynomial function with degree n.
In the complex number $a+b\,i$, it is a real number if $b=0$.
Therefore, Finding the complex zeros of a polynomial function requires finding all zeros of the form $a+b\,i$ including the real zeros ($b=0$).
We have also learned to solve a polynomial equation in degree n by factoring it as a product of its factors in degree 1 and quadratic forms.
In the real number system, a quadratic equation is said to be **no solution** (no zeros) if the equation has no real number solution.
In the complex number system, every quadratic equation has a solution (zeros), either real or complex.
In the complex number system, every polynomial function $p(x)$ of degree $n \geq 1$ can be factored into n linear factors of the form:

$p(x)=a_n(x-r_1)(x-r_2)\cdots\cdots(x-r_n)$: $r_1, r_2, \cdots\cdots, r_n$ are the zeros, either real or complex.

Fundamental Theorem of Algebra: In the complex number system, a polynomial function $p(x)$ of degree $n \geq 1$ has at least one zero.

In the complex number system, every polynomial function of degree $n \geq 1$ has exactly n zeros, either real or complex.

Conjugate Pairs Theorem: In the complex number system, if $a+b\,i\ (b \neq 0)$ is a complex zero of a polynomial function $p(x)$ whose coefficients are real numbers, then the conjugate $a-b\,i$ is also a complex zero of $p(x)$.

Examples

4. Finding all zeros of the function $f(x)=x^4-5x^3+3x^2+19x-30$.

Solution:

Using the Newton's Rule (Rational Zeros Theorem), we can find two rational zeros of $f(x)$, 3 and –2.

Using synthetic division, we have:

$$f(x)=x^4-5x^3+3x^2+19x-30$$
$$=(x-3)(x+2)(x^2-4x+5)$$

To find zeros(roots), let $f(x)=0$.

$$(x-3)(x+2)(x^2-4x+5)=0$$

We have the following four zeros(roots):

$$x=3,\ -2$$
$$x=\frac{-(-4)\pm\sqrt{(-4)^2-4\cdot 1\cdot 5}}{2(1)}=\frac{4\pm\sqrt{-4}}{2}=\frac{4\pm 2i}{2}=2\pm i$$

Ans: $x=3$, –2, and $2\pm i$.

The Relationships between Roots and Coefficients

The standard form of the quadratic equation $ax^2 + bx + c = 0$ ($a \neq 0$) can be written as:

$$x^2 + \frac{b}{a}x + \frac{c}{a} = 0 \ (a \neq 0)$$

If r_1 and r_2 are the roots (solutions) of the quadratic equation, we can write the equation in the form $(x - r_1)(x - r_2) = 0$. We have the equation $x^2 - (r_1 + r_2)x + r_1 r_2 = 0$.
Therefore, the relationships between the roots and coefficients of a quadratic equation are:

$$\textbf{1)}\ r_1 + r_2 = -\frac{b}{a} \quad ; \quad \textbf{2)}\ r_1 \cdot r_2 = \frac{c}{a}$$

We can use the above two relationships to find a quadratic equation having the given roots.
If $r_1, r_2, r_3, ..., r_n$ are roots of a polynomial equation $p(x) = a_x x^n + a_{n-1}x^{n-1} + \cdots + a_1 x + a_0 = 0$, then **the relations between roots and coefficients of a polynomial equation are:**

1) $r_1 + r_2 + r_3 + \cdots + r_n = -\dfrac{a_{n-1}}{a_n}$ **2)** $r_1 \cdot r_2 \cdot r_3 \cdots r_n = (-1)^k \dfrac{a_0}{a_n}$, k is the number of roots.

3) $r_1 r_2 r_3 \cdots r_k + r_1 r_2 r_4 \cdots r_k + \cdots\cdots = (-1)^k \dfrac{a_{n-k}}{a_n}$, k is the number of roots.

Examples

1. Find the quadratic equation whose roots are 1 and -3.

Solution: $r_1 + r_2 = 1 + (-3) = -2 = -\dfrac{b}{a}$; $\dfrac{b}{a} = 2$

$r_1 \cdot r_2 = 1 \cdot (-3) = -3 = \dfrac{c}{a}$; $\dfrac{c}{a} = -3$ $\therefore x^2 + 2x - 3 = 0$ is the equation. Ans.

2. Find the quadratic equation whose roots are $\dfrac{-1+\sqrt{13}}{3}$ and $\dfrac{-1-\sqrt{13}}{3}$.

Solution: $r_1 + r_2 = \dfrac{-1+\sqrt{13}}{3} + \dfrac{-1-\sqrt{13}}{3} = -\dfrac{2}{3} = -\dfrac{b}{a}$; $\dfrac{b}{a} = \dfrac{2}{3}$

$r_1 \cdot r_2 = \dfrac{-1+\sqrt{13}}{3} \cdot \dfrac{-1-\sqrt{13}}{3} = \dfrac{(-1)^2 - (\sqrt{13})^2}{9} = \dfrac{-12}{9} = -\dfrac{4}{3} = \dfrac{c}{a}$; $\dfrac{c}{a} = -\dfrac{4}{3}$

$x^2 + \dfrac{2}{3}x - \dfrac{4}{3} = 0$ $\therefore 3x^2 + 2x - 4 = 0$ is the equation. Ans.

3. Given the equation $2x^4 - 5x^3 + 5x - 2 = 0$, find: a) the sum and the product of the roots.
b) the sum of the product of the roots, take 3 at a time. (Hint: $x = 1, -1, 2, \frac{1}{2}$)

Solution:

a) $r_1 + r_2 + r_3 + r_4 = -\dfrac{a_{n-1}}{a_n} = -\dfrac{-5}{2} = \dfrac{5}{2}$; $r_1 r_2 r_3 r_4 = (-1)^4 \dfrac{a_0}{a_n} = (-1)^4 \dfrac{-2}{2} = -1$

b) $r_1 r_2 r_3 + r_1 r_2 r_4 + r_1 r_3 r_4 + r_2 r_3 r_4 = (-1)^3 \dfrac{a_{n-3}}{a_n} = -\dfrac{5}{2}$. (Hint: $a_{n-2} = 0$)

2-21 Polynomial Functions and Graphs

We have already learned how to graph the most commonly used polynomial functions in algebra. Understanding the basic shapes of the following graphs will help us to analyze and graph more complicated polynomial functions.

1. $f(x) = 0$. It is a zero function. The graph is the x-axis.
2. $f(x) = c$. It is a constant function. The graph is a horizontal line with $y-$intercept c.
3. $f(x) = x$. It is an identity function. The graph is a straight line passing through the origin.
4. $f(x) = mx + b$. It is a linear function. The graph is a straight line with slope m and $y-$intercept b.
5. $f(x) = x^2$. It is a quadratic function. It is a parabola. The graph is symmetric with respect to the $y-$axis.
6. $f(x) = x^3$. It is a cubic function. The graph is symmetric with respect to the origin.

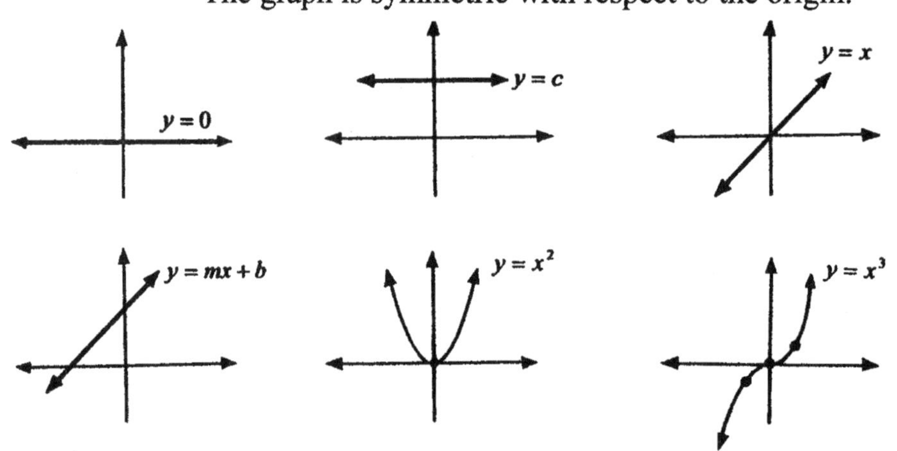

Graphing Power Functions of the form $f(x) = x^n$

As n increases, the graph tends to be flat near the origin and closer to the $x-$axis in the interval $[-1, 1]$. The graph increases rapidly when $x < -1$ or $x > 1$.

A. $f(x) = x^n$ if n is even. **B.** $f(x) = x^n$ if n is odd.

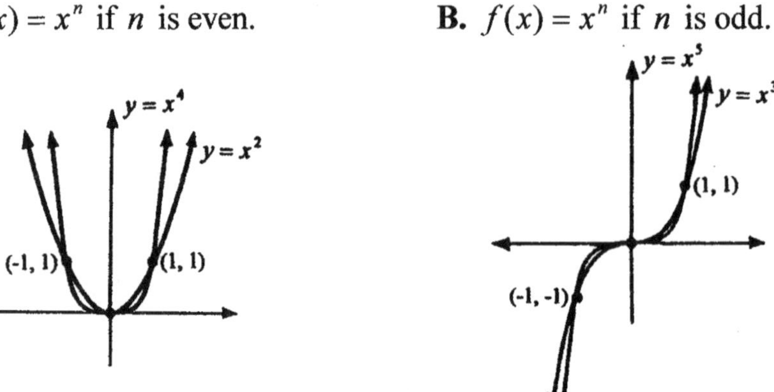

Transformations of Graphs (Shifts, Reflects, Stretches, and Compressions)

Knowing the graphs of common functions $y = f(x)$ can help us to graph a wide variety of polynomial functions by transformations (shifts, reflects, stretches, compressions). Shift and reflection translate only the position of the graph. Stretch and compression distort the basic shape of the original graph.

The transformations of the original graph of $y = f(x)$ have the following types:

1. $y = f(x)+c$. It moves (vertical shift) the original graph c units upward.
2. $y = f(x)-c$. It moves (vertical shift) the original graph c units downward.
3. $y = f(x+c)$. It moves (horizontal shift) the original graph c units to the left.
4. $y = f(x-c)$. It moves (horizontal shift) the original graph c units to the right.
5. $y = -f(x)$. It reflects the original graph over the $x-axis$.
6. $y = f(-x)$. It reflects the original graph over the $y-axis$.
7. $y = c\ f(x)$. It stretches vertically if $c > 1$ (narrower). It compresses vertically if $0 < c < 1$ (wider).
8. $y = f(cx)$. It compresses horizontally if $c > 1$ (narrower) It stretches horizontally if $0 < c < 1$ (wider).

Examples (Show how the graphs are transformed with the above types.)

1. Graph: $y = x+3$ and $y = x-3$.

Solution:

$y = x+3$ It moves the graph of $y = x$, 3 units upward.

$y = x-3$ It moves the graph of $y = x$, 3 units downward.

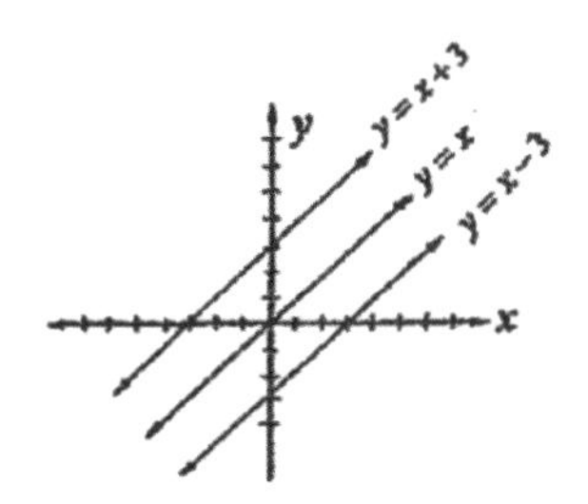

2. Graph: $y = x^2+1$ and $y = (x+1)^2$.

Solution:

$y = x^2+1$ It moves the graph of $y = x^2$, 1 unit upward.

$y = (x+1)^2$ It moves the graph of $y = x^2$, 1 unit to the left.

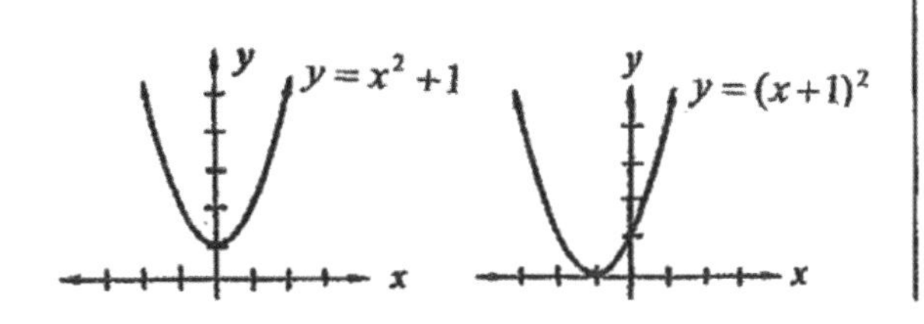

3. Graph: $y = (x-2)^2-3$.

Solution:

It moves the graph of $y = x^2$, 2 units to the right and 3 units downward.

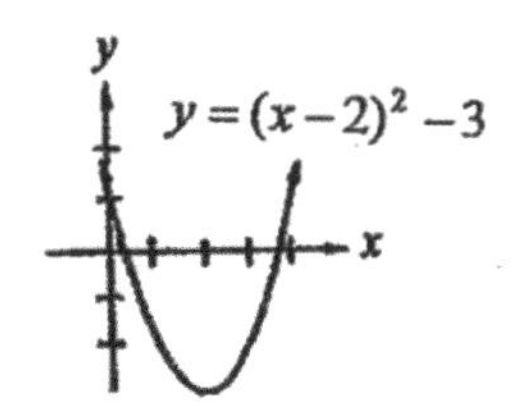

4. Graph: $y = -x^3$ and $y = (-x)^3$

Solution:

$y = -x^3$ It reflects the graph of $y = x^3$ over the $x-$axis.

$y = (-x)^3$ It reflects the graph of $y = x^3$ over the $y-$axis.

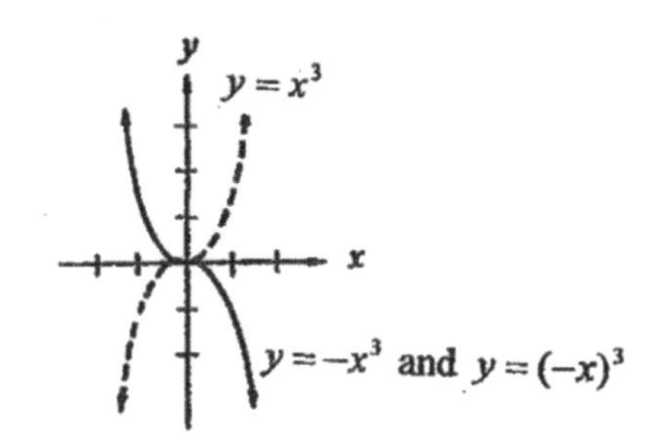

Examples

5. Graph: $y = 2x^2$ and $y = \frac{1}{2}x^2$.

   Solution:

   $y = 2x^2$. It stretches the graph of $y = x^2$ vertically from (2, 4) to(2, 8) and give a narrower parabola.
   $y = \frac{1}{2}x^2$. It compresses the graph of $y = x^2$ vertically from (2, 4) to (2, 2) and gives a wider parabola.

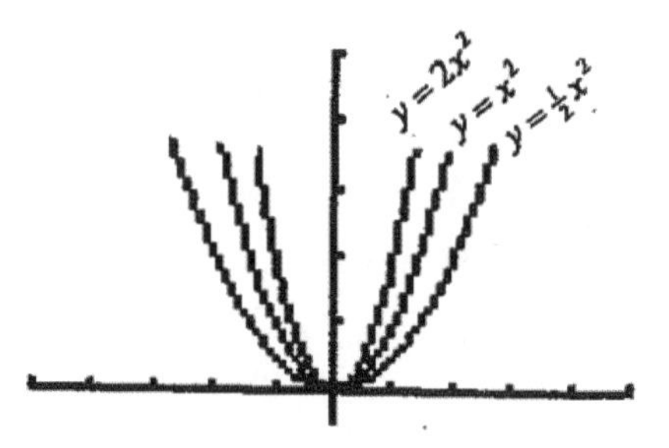

6. Graph: $y = (2x)^2$ and $y = (\frac{1}{2}x)^2$.

   Solution:

   $y = (2x)^2$. It compresses the graph of $y = x^2$ horizontally from (2, 4) to (1, 4) and give a narrower parabola.
   $y = (\frac{1}{2}x)^2$. It stretches the graph of $y = x^2$ horizontally from (2, 4) to (4, 4) and give a wider parabola.

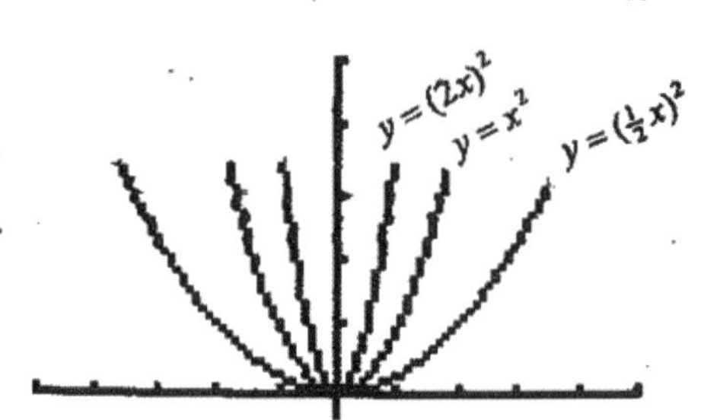

Graphing polynomial functions of higher degree

1. The graph of every polynomial function $f(x) = a_n x^n + a_{n-1}x^{n-1} + \cdots + a_1 x + a_0$ is a smooth and continuous curve. Its turning points are rounded. It has no holes, gaps, or sharp turns.
2. The points at which a graph changes direction are called **turning points**. The turning points are the **local maxima or minima (relative maxima or minima)** of the function. The maximum number of turning points of the graph is $n-1$.
3. The maximum number of real zeros (x – intercepts) is n.
4. If r is a repeated zero of **even** multiplicity (the sign of the function does not change from one side to the other side of r), the graph **touches** the x – intercept at r.
5. If r is a repeated zero of **odd** multiplicity (the sign of the function changes from one side to the other side of r), the graph **crosses** the x – intercept at r.
6. **End Behavior**: If n is even and the **leading coefficient** is positive ($a_n > 0$), the graph rises to the left and right. If n is even and the leading coefficient is negative ($a_n < 0$), the graph falls to the left and right.
7. **End Behavior**: If n is odd and the **leading coefficient** is positive ($a_n > 0$), the graph falls to the left and rises to the right. If n is odd and the leading coefficient is negative ($a_n < 0$), the graph rises to the left and falls to the right.

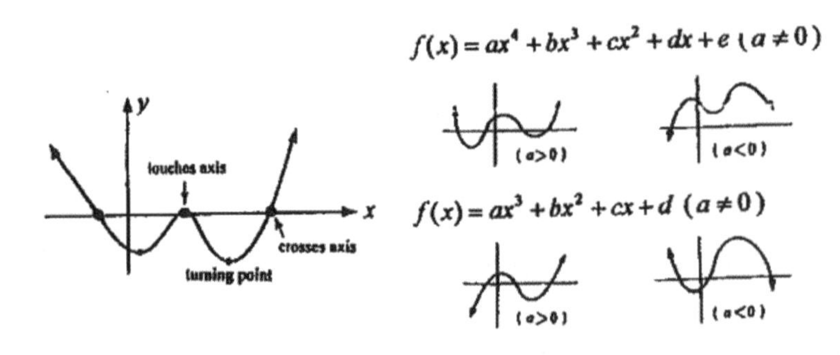

Steps for graphing a polynomial function $y = f(x)$

1. Solve the equation $f(x) = 0$ for real zeros (x – intercepts).
2. Decide whether the graph crosses or touches the x – axis at each x – intercept.
3. Decide the maximum number of turning points.
4. Apply the Leading Coefficient Test to decide whether the graph rises or falls. (Optional)
5. Test each interval between two x – intercepts to find each interval on which the graph is either above the x – axis or below the x – axis.
6. Plot a few additional points and connect the points with a smooth and continuous curve.

Examples

7. Graph: $f(x) = x^3 - 8x^2 + 20x - 16$.

Solution:

Factor the function and solve for real zeros:

$$f(x) = (x-2)^2(x-4) = 0$$

The x – intercepts are 2 (multiplicity 2) and 4.
The graph touches the x – axis at 2 and crosses the x – axis at 4.
It has at most two turning points.
Test each interval: The two x – intercepts divide the graph into 3 intervals.

$x < 2$, $f(x)$: $+ \cdot - = -$ below x – axis
$2 < x < 4$, $f(x)$: $+ \cdot - = -$ below x – axis
$x > 4$, $f(x)$: $+ \cdot + = +$ above x – axis

Compute a few additional points:

x	1	1.5	2.5	3	3.5	4.5
$f(x)$	–3	–0.63	–0.38	–1	–1.13	3.13

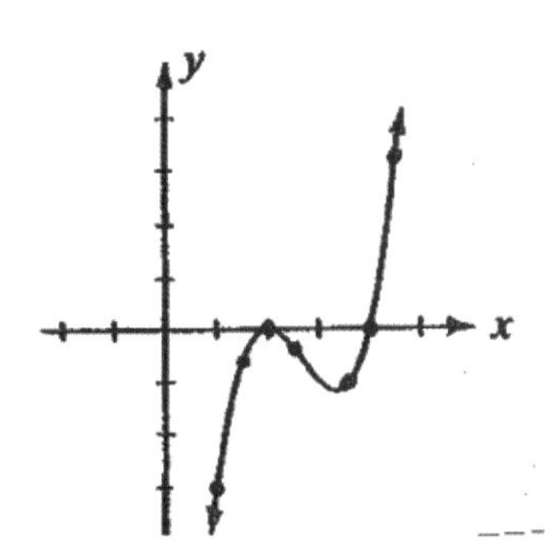

Hint: In calculus, we can find the location of the turning points, a local maximum at $x = 2$ and a local minimum at $x = 3.3$.

8. Graph: $f(x) = x^3(x-4)^2$.

Solution:

$$f(x) = x^3(x-4)^2 = 0$$

The x – intercepts are 0 (multiplicity 3) and 4 (multiplicity 2).
The graph crosses the origin and touches the x – axis at 4
It has at most four turning points.
Test each interval: (3 intervals)

$x < 0$, $f(x)$: $- \cdot + = -$ below x – axis
$0 < x < 4$, $f(x)$: $+ \cdot + = +$ above x – axis
$x > 4$, $f(x)$: $+ \cdot + = +$ above x – axis

Compute a few additional points:

x	–1	1	2	3	5
$f(x)$	–25	9	32	27	125

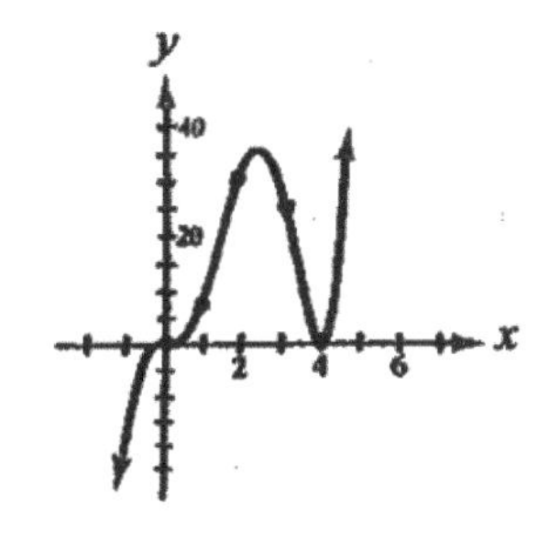

Graphing Techniques: Transformations

The graph of a function $y = f(x)$ is given. Graph each of the following functions using the techniques of transformations.

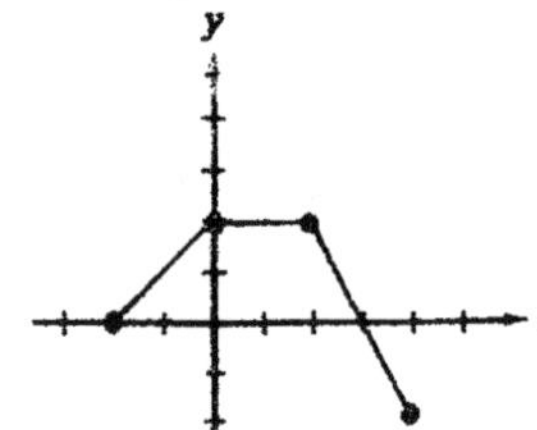

1. $y = -f(x)$ **2.** $y = f(-x)$ **3.** $y = f(x)+2$ **4.** $y = f(x+2)$
5. $y = f(x+2)-3$ **6.** $y = 2f(x)$ **7.** $y = f(2x)$ **8.** $y = f(\frac{1}{2}x)$
9. $y = f(|x|)$

1. $y = -f(x)$
Reflect about the *x-axis*.

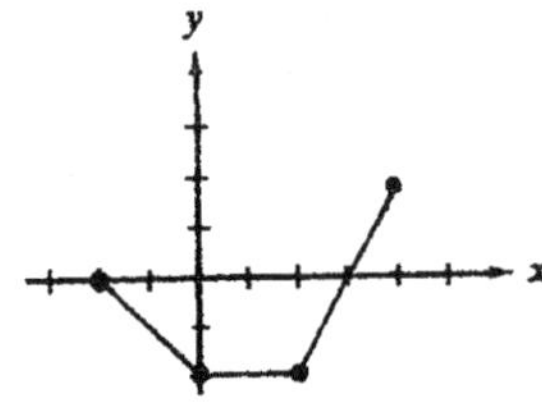

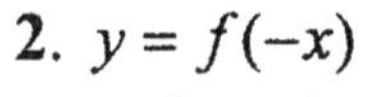
2. $y = f(-x)$
Reflect about the *y-axis*.

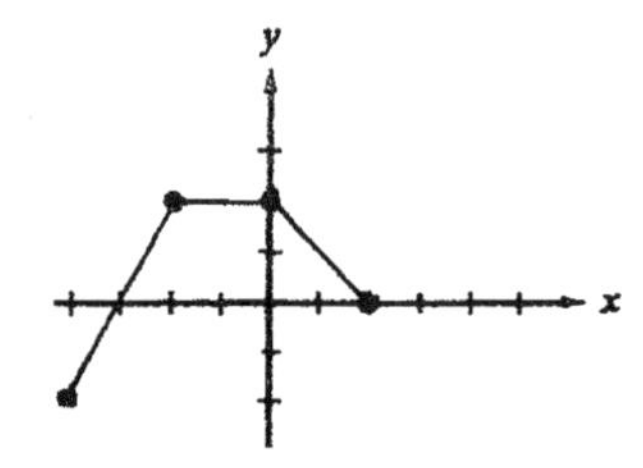

3. $y = f(x)+2$
Vertical shift up 2 units

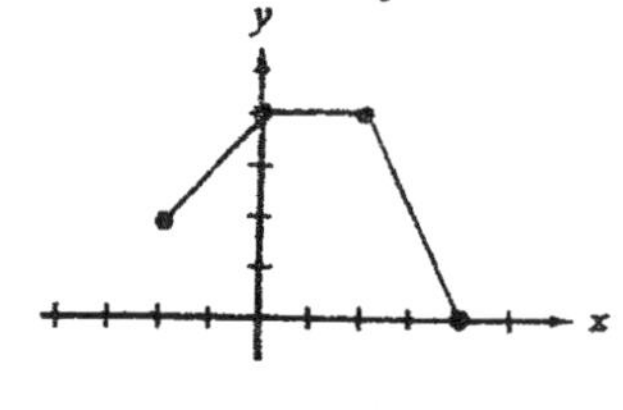

4. $y = f(x+2)$
Horizontal shift left 2 units.

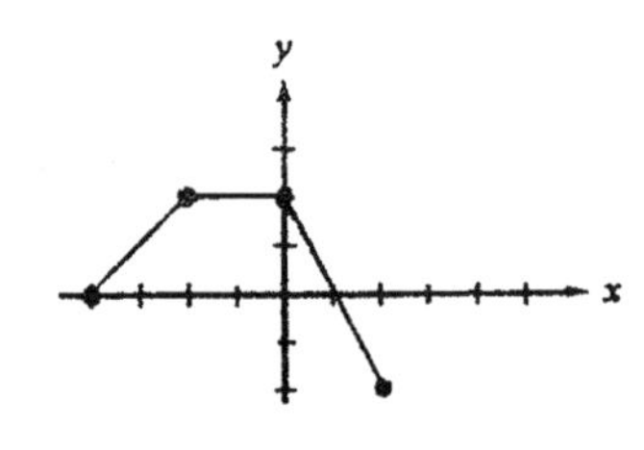

5. $y = f(x+2)-3$
Horizontal shift left 2 units.
Vertical shift down 3 units.

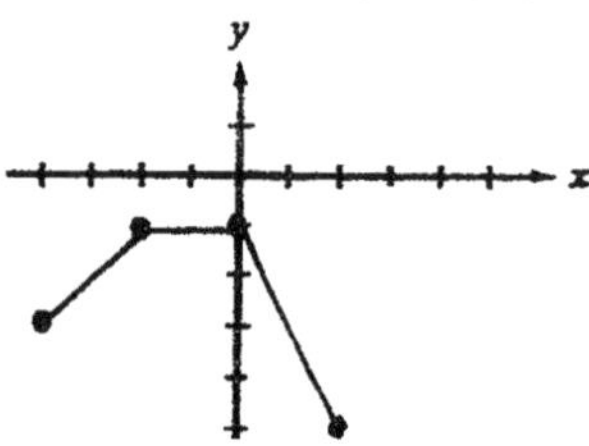

6. $y = 2f(x)$
Stretching vertically by multiplying *y*-value by 2.

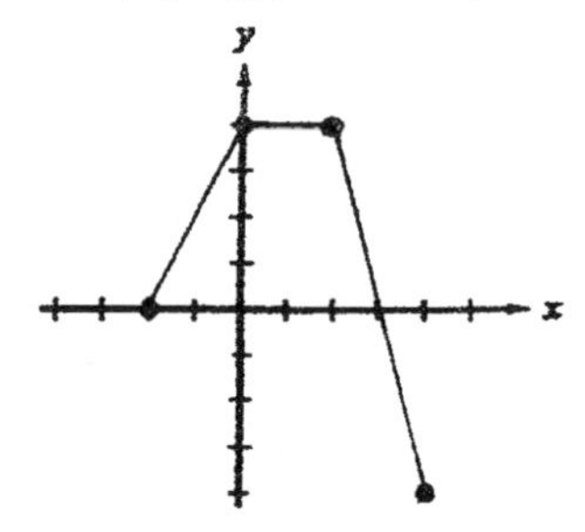

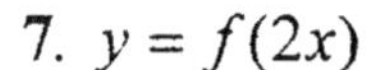
7. $y = f(2x)$
Compressing vertically.
Multiplying *x*-value by $\frac{1}{2}$.

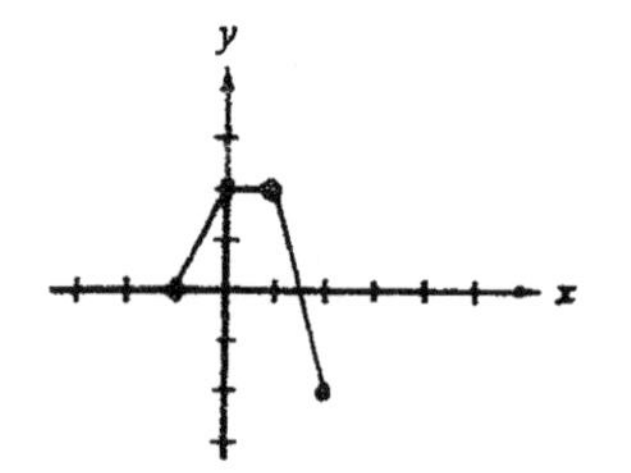

8. $y = f(\frac{1}{2}x)$
Stretching horizontally.
Multiplying *x*-value by 2.

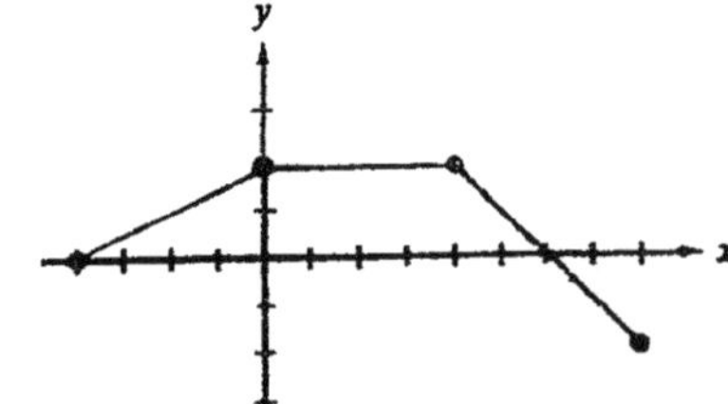
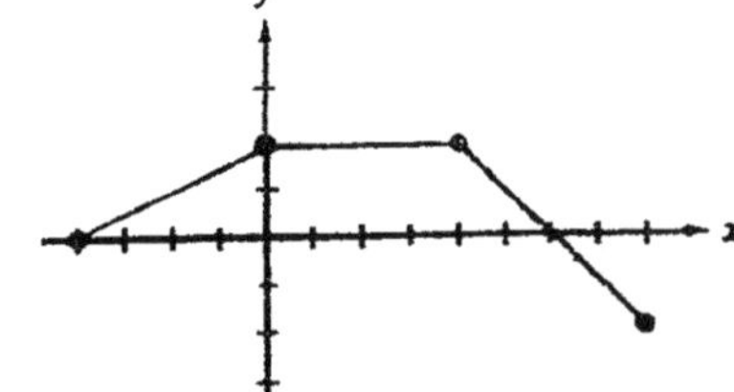

9. $y = f(|x|)$
See Hint below.

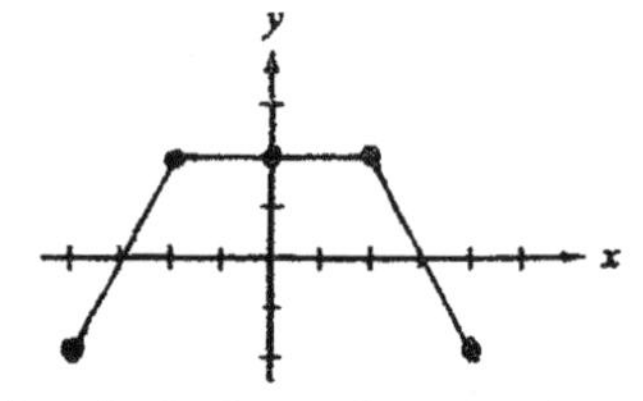

Hint: **1.** Graphing $y = |f(x)|$, we graph $y = f(x)$ and reflect the part that is below the *x*-axis through the *x-axis*. (See Example 1, Page 101)

2. Graphing $y = f(|x|)$, we graph $y = f(x)$ and replace the graph on the left of the *y-axis* by reflecting the part that is on the right of the *x-axis* about the *y-axis*.

Additional Examples

Examples

1. Graph $f(x)=|x^2-4|$.

Solution:

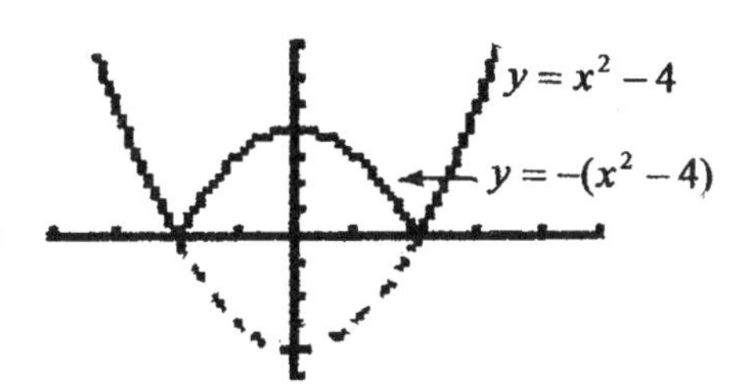

1. If $x^2-4\geq 0$, then $y=x^2-4$.
   We have $y=x^2-4$ when $x\leq -2$ or $x\geq 2$.

2. If $x^2-4\leq 0$, then $y=-(x^2-4)$
   We have $y=-(x^2-4)$ when $-2\leq x\leq 2$.

Hint: Graphing $y=|f(x)|$, we graph $y=f(x)$ and reflect the part that is below the x-axis through the x-axis.

2. An open box with a square bottom is to be made from a cardboard that has a surface area 432 square centimeters. Find the length of one side of the bottom for which the volume of the box is maximum.

Solution:

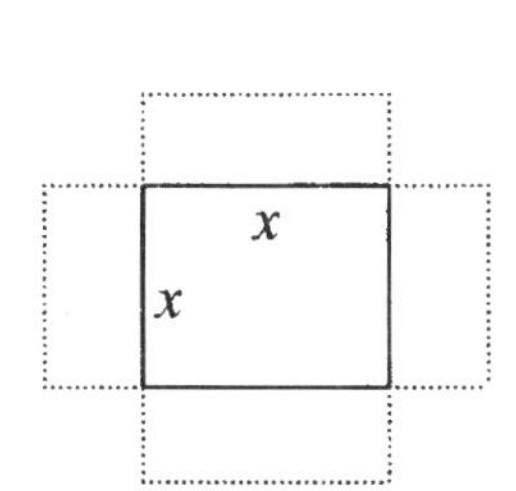

$$\text{Volume } V(x)=x^2\cdot\frac{432-x^2}{4x}$$

$$\therefore\ V(x)=-\frac{x^3}{4}+108x$$

x	11	**12**	13
$V(x)$	855.25	**864**	854.75

$\therefore x=12\ cm$. Ans.

Hint: $V=12\times 12\times 6=864\ cm^3$ is a maximum.

We can easily find the answer by the method in calculus.

$$\frac{dv}{dx}=\frac{d}{dx}(-\frac{x^3}{4}+108x)=-\frac{3}{4}x^2+108=0$$

$$3x^2=432\quad\therefore x=12\ cm.\ \text{Ans.}$$

2-22 Polynomial Inequalities and Absolute Values

We have learned how to solve inequalities in one variable of degree 1, and how to solve absolute-value equations and inequalities.
To solve a polynomial inequality which can be factored into factors of degree less than two, we find the critical points and test each interval between two sets of critical points. If the original inequality is satisfied by testing the interval, the interval is the answer. A sign graph is useful in the process.

Examples

1. Solve $x^2 + 2x - 15 > 0$.

Solution:
$$x^2 + 2x - 15 > 0$$
$$(x+5)(x-3) > 0$$
$$\text{Let } y = (x+5)(x-3) = 0$$

We have the **critical points**: $x = -5$ and 3

Test each interval between two sets of critical points if the inequality is satisfied. (Choose a convenient x – value in each interval.)

$y = (x+5)(x-3) > 0$ (It is positive.)

$x < -5$, $- \cdot - = +$ yes

$-5 < x < 3$, $+ \cdot - = -$ no

$x > 3$, $+ \cdot + = +$ yes

+ − +
−5 0 3

Ans: $x < -5$ or $x > 3$. (disjunction inequality)

2. Solve $x^3 + x^2 - 2x \le 0$.

Solution:
$$x^3 + x^2 - 2x \le 0$$
$$x(x^2 + x - 2) \le 0$$
$$x(x+2)(x-1) \le 0$$
$$\text{Let } y = x(x+2)(x-1) = 0$$

We have the **critical points**: $x = -2,\ 0,\ 1$

Test each interval between two sets of critical points if the original inequality is satisfied: (Choose a convenient x – value in the interval.)

$y = x(x+2)(x-1) \le 0$. (It is negative.)

$x < -2$, $- \cdot - \cdot - = -$ yes

$-2 < x < 0$, $- \cdot + \cdot - = +$ no

$0 < x < 1$, $+ \cdot + \cdot - = -$ yes

$x > 1$, $+ \cdot + \cdot + = +$ no

− + − +
−2 0 1

The critical points ($x = -2,\ 0,\ 1$) also satisfy the original inequality.

Ans: $x \le -2$ or $0 \le x \le 1$. (disjunction inequality)

Solving polynomial inequalities with Absolute Values

We have learned how to solve simple inequalities involving absolute values.
To solve a polynomial inequality with absolute value, we may use the following facts:

1. $|x|^2 = x^2$. **2.** If $x^2 > a$, then $|x| > \sqrt{a}$. **3.** If $x^2 < a$, then $|x| < \sqrt{a}$.

Examples

3. Solve $|x+2| < 5$.

Solution:

Method 1: $|x+2| < 5$

$-5 < x+2 < 5$

$\therefore -7 < x < 3$. Ans.

Method 2: $|x+2| < 5$

$(x+2)^2 < 5^2$

$x^2 + 4x + 4 < 25$

$x^2 + 4x - 21 < 0$

$(x+7)(x-3) < 0$

We have the critical points:

$x = -7$ and 3

Test: $y = (x+7)(x-3) < 0$

$x < -7$, $- \cdot - = +$ no

$-7 < x < 3$, $+ \cdot - = -$ yes

$x > 3$, $+ \cdot + = +$ no

Ans: $-7 < x < 3$

4. Solve $|x^2 - 2x| < x$.

Solution:

$(x^2 - 2x)^2 < x^2$

$x^4 - 4x^3 + 4x^2 - x^2 < 0$

$x^4 - 4x^3 + 3x^2 < 0$

$x^2(x^2 - 4x + 3) < 0$

$x^2 - 4x + 3 < 0$

$(x-3)(x-1) < 0$

We have the critical points:

$x = 3$ and 1

Test: $y = (x-3)(x-1) < 0$

$x < 1$, $- \cdot - = +$ no

$1 < x < 3$, $- \cdot + = -$ yes

$x > 3$, $+ \cdot + = +$ no

Ans: $1 < x < 3$

5. Solve $|x^2 + 4x + 4| > 5$.

Solution:

$|(x+2)^2| > 5$

$(x+2)^2 > 5$

$|x+2| > \sqrt{5}$

$x + 2 < -\sqrt{5}$ or $x + 2 > \sqrt{5}$

Ans: $x < -1 - \sqrt{5}$ or $x > -2 + \sqrt{5}$

6. Solve $|x^2 + 6x + 9| < 25$

Solution:

$|(x+3)^2| < 25$

$(x+3)^2 < 25$

$|x+3| < 5$

$-5 < x + 3 < 5$

Ans: $-8 < x < 2$

Solving inequalities if factoring is not possible, we apply the method of completing the square.

Examples

7. Solve $x^2+6x+9<25$.

Solution:

$$(x+3)^2<25$$
$$|x+3|<5$$
$$-5<x+3<5$$
Ans: $-8<x<2$.

(Hint: Same answer as example 6)

8. Solve $x^2+2x+1\ge\frac{9}{4}$.

Solution:

$$(x+1)^2\ge\frac{9}{4}$$
$$|x+1|\ge\frac{3}{2}$$
$$x+1\le-\frac{3}{2} \text{ or } x+1\ge\frac{3}{2}$$
$$x\le-\frac{3}{2}-1 \qquad x\ge\frac{3}{2}-1$$
Ans: $x\le-\frac{5}{2}$ or $x\ge\frac{1}{2}$

9. Solve $x^2-3x+1\le0$.

Solution:

Completing the square:

$$(x-\tfrac{3}{2})^2-\tfrac{9}{4}+1\le0$$
$$(x-\tfrac{3}{2})^2-\tfrac{5}{4}\le0$$
$$(x-\tfrac{3}{2})^2\le\tfrac{5}{4}$$
$$|x-\tfrac{3}{2}|\le\tfrac{\sqrt5}{2}$$
$$-\tfrac{\sqrt5}{2}\le x-\tfrac{3}{2}\le\tfrac{\sqrt5}{2}$$
Ans: $\frac{3-\sqrt5}{2}\le x\le\frac{3+\sqrt5}{2}$

10. Solve $2x^2-x-5\ge0$.

Solution:

Completing the square:

$$2(x^2-\tfrac{1}{2}x)-5\ge0$$
$$2[(x-\tfrac{1}{4})^2-\tfrac{1}{16}]-5\ge0$$
$$2(x-\tfrac{1}{4})^2-\tfrac{1}{8}-5\ge0$$
$$2(x-\tfrac{1}{4})^2\ge\tfrac{41}{8},\ (x-\tfrac{1}{4})^2\ge\tfrac{41}{16}$$
$$|x-\tfrac{1}{4}|\ge\tfrac{\sqrt{41}}{4}$$
$$x-\tfrac{1}{4}\le-\tfrac{\sqrt{41}}{4} \text{ or } x-\tfrac{1}{4}\ge\tfrac{\sqrt{41}}{4}$$
Ans: $x\le\frac{1-\sqrt{41}}{4}$ or $x\ge\frac{1+\sqrt{41}}{4}$.

11. Solve $3x^2-6x\le4$.

Solution:

Completing the square:

$$3(x^2-2x)\le4$$
$$3[(x-1)^2-1]\le4$$
$$3(x-1)^2-3\le4$$
$$3(x-1)^2\le7$$
$$(x-1)^2\le\tfrac{7}{3}$$
$$|x-1|\le\sqrt{\tfrac{7}{3}}$$
$$-\sqrt{\tfrac{7}{3}}\le x-1\le\sqrt{\tfrac{7}{3}}$$
Ans: $1-\sqrt{\frac{7}{3}}\le x\le1+\sqrt{\frac{7}{3}}$

12. Solve $|x^2-2x-5|>2$.

Solution:

$$x^2-2x-5<-2 \text{ or } x^2-2x-5>2$$
Completing the squares:

$(x-1)^2-1-5<-2$	$(x-1)^2-1-5>2$
$(x-1)^2<4$	$(x-1)^2>8$
$\|x-1\|<2$	$\|x-1\|>\sqrt8$

$$-2<x-1<2,\ x-1<-\sqrt8 \text{ or } x-1>\sqrt8$$
$$-1<x<3,\ x<1-\sqrt8 \text{ or } x>1+\sqrt8$$
Ans: $x<1-\sqrt8$, or $-1<x<3$,

or $x>1+\sqrt8$

2-23 Inverse Functions

One to One function: A function f is a one-to-one function if and only if for each range value there corresponds exactly one domain value. An increasing (or decreasing) function is a one-to-one function. A horizontal line intersects the graph of a one-to-one function in at most one point. Note that, both $y = x^3$ and $y = x^2$ are functions. However, $y = x^3$ is a one-to-one function, $y = x^2$ is not a one-to-one function.

Inverse of a function: The inverse of a function f is obtained by interchanging the variables x and y in f. The inverse of f has the set of ordered pairs (y, x).

Inverse Functions: If $y = f(x)$ is a one-to-one function, a function $y = g(x)$ is the inverse function of f **if and only if** $f(g(x)) = g(f(x)) = x$.

$y = x^2$ has an inverse $x = y^2$ (or $y = \pm\sqrt{x}$) which is not a function.

For the function $y = f(x)$ to have an inverse function, f must be one-to-one.

The inverse function of the function f is denoted by f^{-1}, which is read " f inverse ".

If the functions $f(x)$ and $g(x)$ are inverses (functions) of each other, then:

1. the ordered pairs (x, y) of one equation are obtained by interchanging the variables x and y in the other equation.
2. their graphs are **mirror image** of each other with respect to the line $y = x$.
3. If $f(x)$ and $g(x)$ are inverses, then they have the property : $f(g(x)) = g(f(x)) = x$

   Example: $f(x) = x + 2$ and $g(x) = x - 2$ are inverse functions of each other.

   $f(g(x)) = f(x-2) = x - 2 + 2 = x$; $g(f(x)) = g(x+2) = x + 2 - 2 = x$

A function f must have an inverse (a relation) which can be a function or not a function. In certain textbooks, they conventional state that "this function has no inverse". It means "this function has no inverse function".

Examples

1. Let $f(x) = x + 2$, find $f^{-1}(x)$.

   Solution: $y = x + 2$

   Interchange x and y: $x = y + 2$

   We have: $y = x - 2$

   $\therefore f^{-1}(x) = x - 2$. Ans.

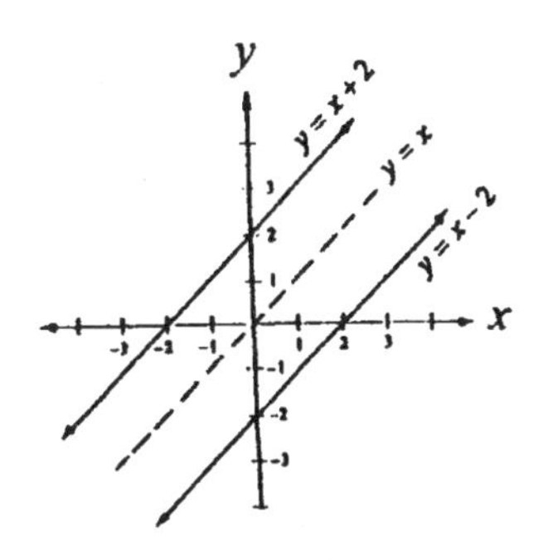

2. Let $f(x) = \frac{2}{3}x + 1$, find $f^{-1}(x)$.

   Solution: $y = \frac{2}{3}x + 1$

   Interchange x and y: $x = \frac{2}{3}y + 1$

   We have: $y = \frac{3}{2}x - \frac{3}{2}$

   $\therefore f^{-1}(x) = \frac{3}{2}x - \frac{3}{2}$. Ans.

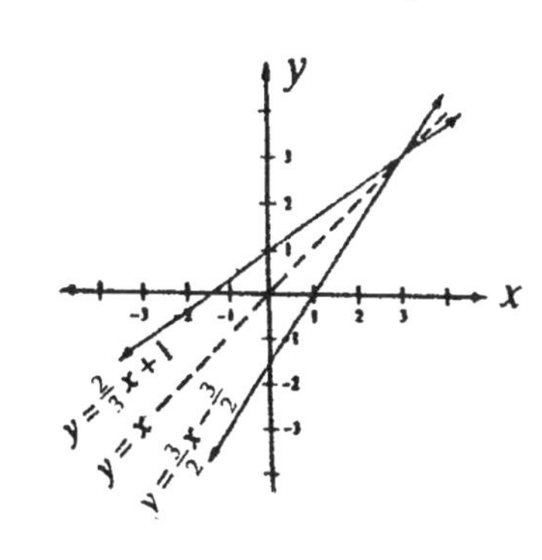

Examples

3. Let $f(x) = x^2$, find the inverse. If $f(x)$ has no inverse function, so state.

Solution:

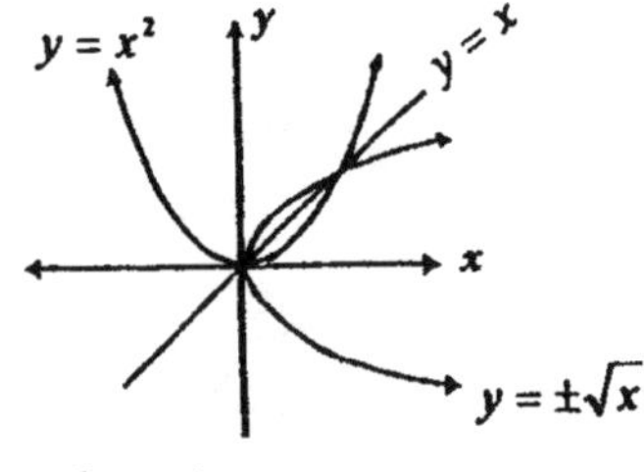

$$y = x^2$$

Interchange x and y:

We have: $x = y^2$ (It is a relation, not a function.)

$$y = \pm\sqrt{x}$$

Ans: $f(x)$ has an inverse $y = \pm\sqrt{x}$. The inverse is not a function.

4. Let $f(x) = x^2$ and $x \le 0$, find $f^{-1}(x)$.

Solution:

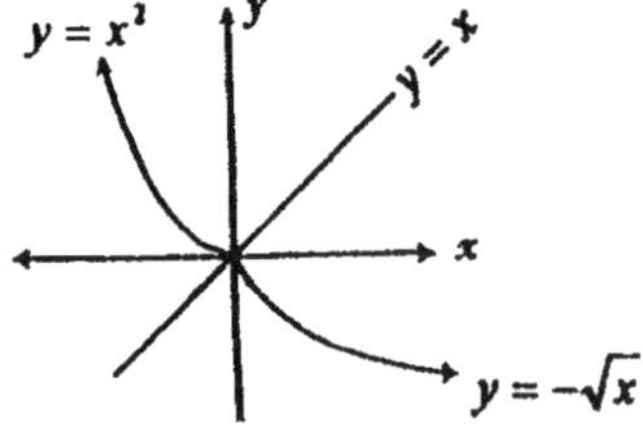

$$y = x^2 \text{ and } x \le 0$$

Interchange x and y:

We have: $x = y^2$ and $y \le 0$ (It is a function.)

$$y = -\sqrt{x} \text{ where } x \ge 0$$

Ans: $f(x)$ has an inverse $f^{-1}(x) = -\sqrt{x}$ where $x \ge 0$.

5. If $f(x) = \sqrt{x+1}$, find the inverse function by limiting the domain of $f^{-1}(x)$.

Solution:

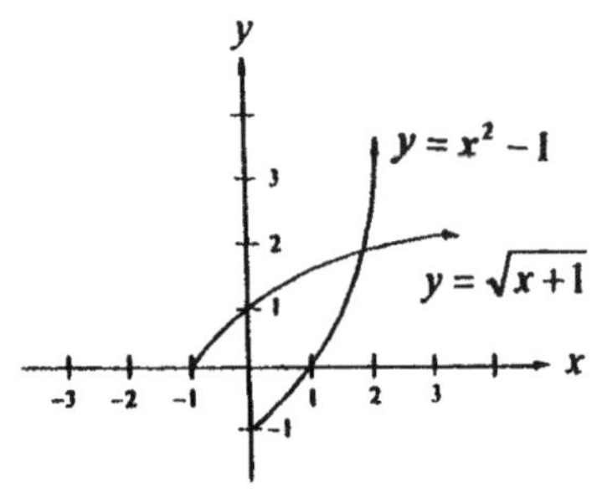

$$y = \sqrt{x+1}$$

$$y^2 = x+1 \text{ and } x \ge -1$$

Interchange x and y:

$x^2 = y+1$ and $y \ge -1$ (it is a function)

$$y = x^2 - 1 \text{ and } x \ge 0$$

Therefore $f^{-1}(x) = x^2 - 1$ where $x \ge 0$.

6. If $f = \{(3,1),(4,2),(5,2)\}$, find the inverse function by limiting the domain of f.

Solution:

The inverse {(1, 3), (2, 4), (2, 5)} is not a function.

The domain of the original function must be limited to “3 and 4” or “3 and 5”

Therefore $f^{-1} = \{(1,3),(2,4)\}$ by limiting the domain of f to 3 and 4.

Or $f^{-1} = \{(1,3),(2,5)\}$ by limiting the domain of f to 3 and 5.

7. Are the function $f(x) = \frac{2}{3}x + 1$ and $g(x) = \frac{3}{2}x - \frac{3}{2}$ inverses of each other ?

Solution:

$$f(g(x)) = \tfrac{2}{3}(\tfrac{3}{2}x - \tfrac{3}{2}) + 1 = x - 1 + 1 = x$$

$$g(f(x)) = \tfrac{3}{2}(\tfrac{2}{3}x + 1) - \tfrac{3}{2} = x + \tfrac{3}{2} - \tfrac{3}{2} = x$$

$f(g(x)) = g(f(x)) = x$ $\therefore f(x)$ and $g(x)$ are inverses of each other.

2-24 Composition of Functions

When we apply one function after another function's values, such as $f(g(x))$, is called " **the composition of** the function f with the function g ". It is denoted as $f \circ g$.

$f(g(x))$ indicates that the domain of $f \circ g$ is all x-values **such that x is in the range of** g and g **is in the domain of** f.

The composition of f with g is: $(f \circ g)(x) = f(g(x))$
The composition of g with f is: $(g \circ f)(x) = g(f(x))$
The composition of a function $f(x)$ with its inverse $f^{-1}(x)$ is equal to x.

$$(f \circ f^{-1})(x) = f(f^{-1}(x)) = x, \quad (f^{-1} \circ f)(x) = f^{-1}(f(x)) = x$$

Examples

1. If $f(x) = 3x^2 - 6$ and $g(x) = x^2 - 4$, find $(f \circ g)(x)$ and $(g \circ f)(x)$.
Solution:

$$(f \circ g)(x) = f(g(x)) = 3(g(x))^2 - 6 = 3(x^2 - 4)^2 - 6 = 3(x^4 - 8x^2 + 16) - 6$$
$$= 3x^4 - 24x^2 + 42$$
$$(g \circ f)(x) = g(f(x)) = (f(x))^2 - 4 = (3x^2 - 6)^2 - 4 = 9x^4 - 36x^2 + 36 - 4$$
$$= 9x^4 - 36x^2 + 32$$

2. Given $f(x) = \frac{2}{3}x + 1$ and $f^{-1}(x) = \frac{3}{2}x - \frac{3}{2}$, find $f \circ f^{-1}$ and $f^{-1} \circ f$.
Solution:

$$f \circ f^{-1} = f(f^{-1}(x)) = \tfrac{2}{3}(\tfrac{3}{2}x - \tfrac{3}{2}) + 1 = x - 1 + 1 = x$$
$$f^{-1} \circ f = f^{-1}(f(x)) = \tfrac{3}{2}(\tfrac{2}{3}x + 1) - \tfrac{3}{2} = x + \tfrac{3}{2} - \tfrac{3}{2} = x$$

3. If $f(g(x)) = 2x - 8$ and $f(x) = 2x$, find $g(x)$ and $g(2)$.
Solution:

$$f(g(x)) = 2(g(x)) = 2x - 8 \quad \therefore g(x) = x - 4 \text{ and } g(2) = 2 - 4 = -2$$

4. If $f(g(x)) = 2x - 8$ and $g(x) = x - 4$, find $f(x)$.
Solution:

$$f((g(x)) = 2x - 8 = 2(x - 4) = 2g(x) \quad \therefore f(x) = 2x$$

5. If $f(g(x)) = 2x - 1$ and $f(x) = x - 4$, find $g(x)$.
Solution:

$$f(g(x)) = g(x) - 4 = 2x - 1 \quad \therefore g(x) = 2x + 3$$

6. If $f(g(x)) = 2x - 1$ and $g(x) = 2x + 3$, find $f(x)$.
Solution:

$$f(g(x)) = 2x - 1 = (2x + 3) - 4 = g(x) - 4 \quad \therefore f(x) = x - 4$$

2-25 Greatest Integer Functions

The greatest-integer function is the most common type of the **step functions.**
The greatest-integer function is in the form:

$f(x) = \text{int}(x)$ or $y = [x] = i$, i is an integer and $i \le x < i + 1$

(i is the greatest integer that is less than or equal to the real number x.)

We have:

int(1)=1 or $[1] = 1$, int(2.7)=2 or $[2.7] = 2$, $[3] = 3$, $[1.2] = 1$, $[2.8] = 2$, $[0.4] = 0$, $[0.9] = 0$, $[\frac{7}{3}] = 2$, $[-1] = -1$, $[-2] = -2$, $[-3] = -3$, $[-0.5] = -1$, $[-1.5] = -2$, $[-2.1] = -3$, $[-3.2] = -4$, $[-3.5] = -4$, $[-3.9] = -4$.

To graph $f(x) = \text{int}(x)$, we plot several points. For values of $-3 \le x < -2$, the value of $f(x)$ is -3. The x-intercepts are in the interval [0, 1). A solid dot is used to indicate the value of $f(x) = -3$ at $x = -3$. The graph has an infinite number of line segments (or steps) and suddenly steps from one value to another without continuity.
The domain of $f(x) = \text{int}(x)$ is all real numbers and its range is all integers.

Examples:

1. Graph $y = [x]$.

Solution:

$-3 \le x < -2$, $y = -3$
$-2 \le x < -1$, $y = -2$
$-1 \le x < 0$, $y = -1$
$0 \le x < 1$, $y = 0$
$1 \le x < 2$, $y = 1$
$2 \le x < 3$, $y = 2$
$3 \le x < 4$, $y = 3$

2. Graph $y = [2x]$.

Solution:

$-1.5 \le x < -1$, $y = -3$
$-1 \le x < -0.5$, $y = -2$
$-0.5 \le x < 0$, $y = -1$
$0 \le x < 0.5$, $y = 0$
$0.5 \le x < 1$, $y = 1$
$1 \le x < 1.5$, $y = 2$
$1.5 \le x < 2$, $y = 3$

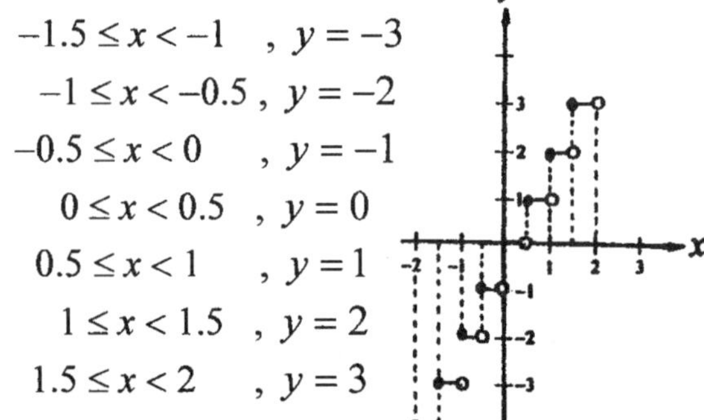

3. The cost of using a long-distance call is \$1.25 for the first minutes and \$0.35 for each additional minute or portion of a minute.

a) Use the greatest integer function to write the cost (c) for a call lasting t minutes.
b) Find the cost of a call lasting 16 minutes and 15 seconds.
c) Graph the function.

Solution:

a) $c(t) = 1.25 - 0.35\,[1 - t]$

b) $c(16\tfrac{15}{60}) = 1.25 - 0.35\,[1 - 16.25]$
$= 1.25 - 0.35\,[-15.25]$
$= 1.25 - 0.35(-16)$
$= 1.25 + 5.60 = \$6.85$

c)

Hint: $c(t) = 1.25 + 0.35[t - 1]$ is incorrect.

2-26 Square Root and Cubic Root Functions

We have learned how to graph the quadratic functions ($y = cx^2$) in Section 5-3 and the cubic functions ($y = cx^3$) in Section 5-7.
In this section we will learn to graph two types of radical functions, **the square root** and **cubic root functions**.
Understanding the basic shapes of the following two graphs will help us to analyze and graph more complicated square root and cubic root functions. (See page 150)

The graph of $y = \sqrt{x}$

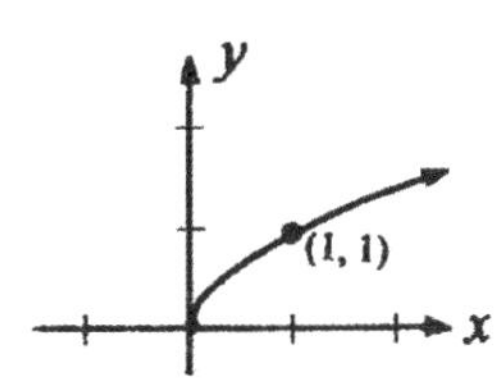

Starts at (0, 0) and passes through the point (1, 1).
Domain: $x \geq 0$. Range: $y \geq 0$
It is the upper half of the parabola $x = y^2$ having a horizontal axis.

The graph of $y = \sqrt[3]{x}$

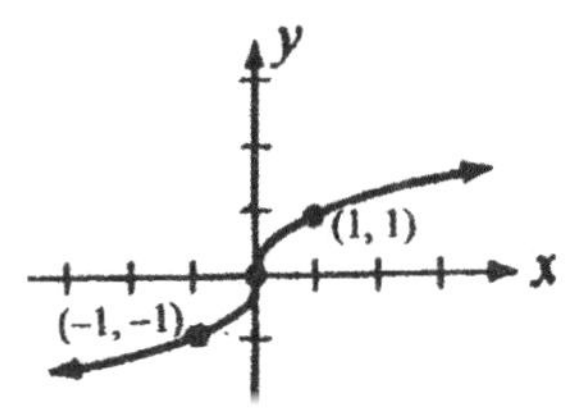

Passes through (0, 0) and the points (−1, −1) and (1, 1).
Domain: x are all real numbers.
Range: y are all real numbers.

1. The graph of $y = a\sqrt{x}$ starts at (0, 0) and passes through the point (1, a).
2. The graph of $y = a\sqrt[3]{x}$ passes through (0, 0) and the points (-1, $-a$) and (1, a).

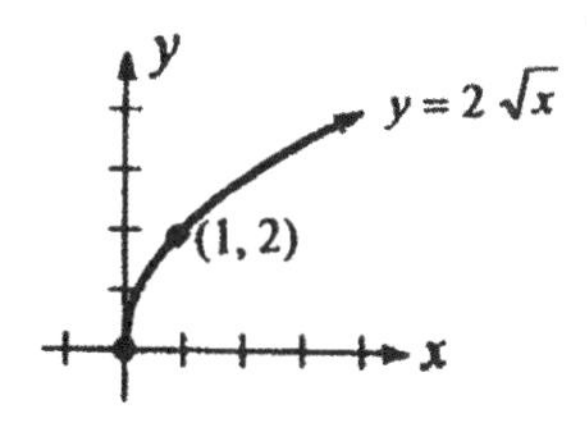

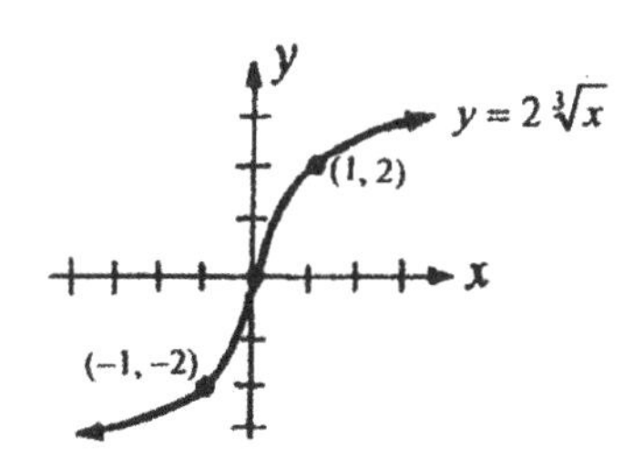

3. The graph of $y = a\sqrt{x-h} + k$ moves (shifts) the graph of $y = a\sqrt{x}$ by h units to the right and k units upward.
4. The graph of $y = a\sqrt[3]{x-h} + k$ moves (shifts) the graph of $y = a\sqrt[3]{x}$ by h units to the right and k units upward.

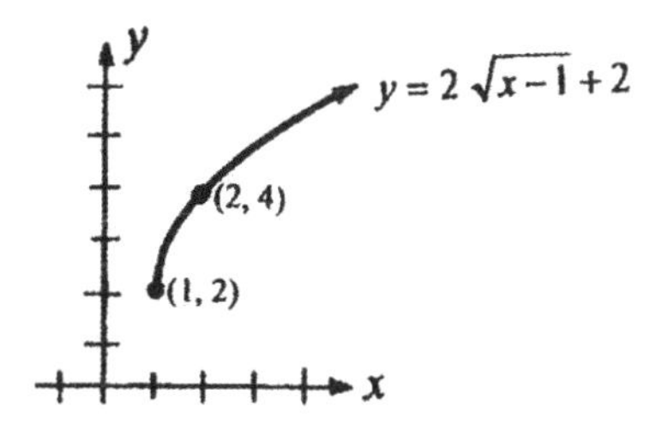

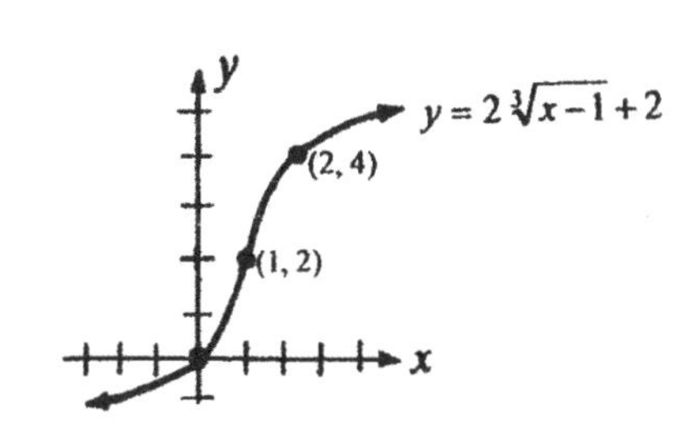

2-27 Operations with Rational Expressions

1) To add or subtract rational expressions (algebraic fractions) having the same denominators, we combine their numerators.

$$\frac{b}{a}+\frac{c}{a}=\frac{b+c}{a} \quad ; \quad \frac{b}{a}-\frac{c}{a}=\frac{b-c}{a}$$

2) To add or subtract rational expressions (algebraic fractions) having different denominators, we rewrite each fraction having the least common denominator (**LCD**), and then combine the resulting fractions. **LCD** is the least common multiple (**LCM**) of their denominators.

$$\frac{c}{a}+\frac{d}{b}=\frac{bc}{ab}+\frac{ad}{ab}=\frac{bc+ad}{ab} \quad ; \quad \frac{c}{a}-\frac{d}{b}=\frac{bc}{ab}-\frac{ad}{ab}=\frac{bc-ad}{ab}$$

3) To multiply two rational expressions (algebraic fractions), we multiply their numerators and multiply their denominators. We can multiply first and then simplify, or simplify first and then multiply.

$$\frac{x^2}{3y}\cdot\frac{6}{5x}=\frac{{}^{2}\cancel{6}\cancel{x^2}^{1}}{{}_{5}\cancel{15}\cancel{x}y}=\frac{2x}{5y} \; ; \quad \text{Or: } \frac{\cancel{x^2}^{1}}{{}_{1}\cancel{3}y}\cdot\frac{\cancel{6}^{2}}{5\cancel{x}}=\frac{2x}{5y}$$

4) To divide two rational expressions (algebraic fractions), we multiply the reciprocal of the divisor.

$$\frac{4x}{y}\div\frac{8x^2}{3y}=\frac{{}^{1}\cancel{4x}}{\cancel{y}}\cdot\frac{3\cancel{y}}{{}_{2}\cancel{8x^2}^{1}}=\frac{3}{2x}$$

Examples

1. $\frac{5a+3}{4}-\frac{a-1}{4}=\frac{5a+3-(a-1)}{4}=\frac{5a+3-a+1}{4}=\frac{4a+4}{4}=\frac{\cancel{4}(a+1)}{\cancel{4}}=a+1$

2. $\frac{1}{2x^2}-\frac{5}{6x}=\frac{3}{6x^2}-\frac{5x}{6x^2}=\frac{3-5x}{6x^2}, x\neq 0$

3. $\frac{7}{x-2}+\frac{2}{2-x}=\frac{7}{x-2}-\frac{2}{x-2}=\frac{5}{x-2}, x\neq 2$

4. $\frac{-2}{x^2-5x+6}+\frac{2}{x-3}=\frac{-2}{(x-3)(x-2)}+\frac{2}{x-3}=\frac{-2}{(x-3)(x-2)}+\frac{2\cancel{(x-2)}}{(x-3)\cancel{(x-2)}}$

$=\frac{-2+2(x-2)}{(x-3)(x-2)}=\frac{-2+2x-4}{(x-3)(x-2)}=\frac{2x-6}{(x-3)((x-2)}=\frac{2\cancel{(x-3)}}{\cancel{(x-3)}(x-2)}=\frac{2}{x-2}, x\neq 3, 2$

Examples (Assume that no variable has a value for which the denominator is zero.)

5. $\frac{x^2}{4}\div\frac{x}{2}=\frac{x^2}{4}\cdot\frac{2}{x}=\frac{x}{2}$

6. $\frac{a-3}{a}\cdot\frac{a^3}{a^2-9}=\frac{\cancel{a-3}}{a}\cdot\frac{a^3}{(a+3)\cancel{(a-3)}}=\frac{a^2}{a+3}$

7. $\frac{a^2-5a+6}{a^2}\div\frac{a^2-9}{a}=\frac{a^2-5a+6}{a^2}\cdot\frac{a}{a^2-9}=\frac{(a-2)\cancel{(a-3)}}{a^2}\cdot\frac{a}{(a+3)\cancel{(a-3)}}=\frac{a-2}{a(a+3)}$

8. $\frac{x^2-4x-5}{x}\div\frac{x^2+x}{x-5}=\frac{x^2-4x-5}{x}\cdot\frac{x-5}{x^2+x}=\frac{\cancel{(x+1)}(x-5)}{x}\cdot\frac{x-5}{x\cancel{(x+1)}}=\frac{(x-5)^2}{x^2}$

2-28 Solving Rational Equations

If a rational equation consists of one fraction equal to another (it is a proportion), then the **cross products** are equal.
To solve a rational equation consisting of two or more fractions on one side, we multiply both sides of the equation by their **least common denominator (LCD)**.

Example 1:

Solve $\frac{2}{x} = \frac{4}{5}$.

Solution:

$$2 \cdot 5 = 4 \cdot x$$

$$10 = 4x$$

$$\therefore x = \tfrac{10}{4} = 2\tfrac{1}{2}. \text{ Ans.}$$

Example 2:

Solve $\frac{x}{3} + \frac{x}{4} = 7$.

Solution: LCD = 12

$$12\left(\frac{x}{3} + \frac{x}{4}\right) = 12(7)$$

$$4x + 3x = 84$$

$$7x = 84$$

$$\therefore x = 12. \text{ Ans.}$$

Example 3:

Solve $\frac{3}{x} - \frac{1}{2x} = \frac{1}{3}$.

Solution: LCD = $6x$

$$6x\left(\frac{3}{x} - \frac{1}{2x}\right) = 6x\left(\frac{1}{3}\right)$$

$$18 - 3 = 2x$$

$$15 = 2x$$

$$\therefore x = \tfrac{15}{2} = 7\tfrac{1}{2}. \text{ Ans.}$$

To solve a rational equation, we always test each root in the original equation. A root is **not permissible** if it makes the denominator of the original equation equal to 0.

Example 4: Solve $\frac{18}{x^2 - 9} = \frac{x}{x - 3}$.

Solution:

$$18(x-3) = x(x^2 - 9)$$

$$18(\cancel{x-3}) = x(x+3)(\cancel{x-3})$$

$$18 = x(x+3)$$

$$18 = x^2 + 3x$$

$$0 = x^2 + 3x - 18$$

$$x^2 + 3x - 18 = 0$$

$$(x+6)(x-3) = 0$$

$x = -6$ | $x = 3$ **(not permissible)** $\therefore x = -6$. Ans.

If the final statement is a "**true**" statement, the equation has roots for all real numbers.
If the final statement is a "**false**" statement, the equation has no root.

Example 5: Solve $\frac{4}{2x - 2} = \frac{2}{x - 1}$.

Solution:

$$4(x-1) = 2(2x-2)$$

$$4x - 4 = 4x - 4$$

$0 = 0$ **(true)**

Ans: The equation has roots for all real numbers except 1.

Example 6: Solve $\frac{4}{2x+1} = \frac{2}{x}$.

Solution:

$$4x = 2(2x+1)$$

$$4x = 4x + 2$$

$0 = 2$ **(false)**

Ans: No solution.

2-29 Solving Word Problems with Rational Equations

Rational equations (fractional equations) are often used in solving word problems.

Examples:

1. One fifth of a number is 7 less than two thirds of the number. Find the number.

Solution:

Let $n =$ the number

The equation is:

$$\frac{n}{5}+7=\frac{2n}{3}$$

LCD = 15

$$3n+105=10n$$

$$105=7n \quad \therefore n=15$$

Ans: 15.

2. The sum of two numbers is 48 and their quotient is $\frac{3}{5}$. Find the numbers.

Solution:

Let $n = 1^{st}$ number

$48-n=2^{nd}$ number

The equation is: $\frac{n}{48-n}=\frac{3}{5}$

$$5n=3(48-n)$$

$$5n=144-3n$$

$$8n=144 \quad \therefore n=18$$

Ans: 18 and 30.

3. If two pounds of beef cost \$5.80, how much do 8 pounds cost ?

Solution:

Let $x =$ cost for 8 pounds

The equation is:

$$\frac{5.80}{2}=\frac{x}{8}$$

$$5.80(8)=2x$$

$$46.40=2x \quad \therefore x=23.20$$

Ans: \$23.20.

4. A bus uses 4.5 gallons of gasoline to travel 75 miles. How many gallons would the bus use to travel 95 miles ?

Solution:

Let $x =$ the number of gallons

The equation is: $\frac{75}{4.5}=\frac{95}{x}$

$$75x=4.5(95)$$

$$75x=427.5 \quad \therefore x=5.7$$

Ans: 5.7 gallons.

5. John can finish a job in 5 hours. Steve can finish the same job in 4 hours. How many hours do they need to finish the job if they work together ?

Solution:

Let $x =$ number of hours needed to do the job together

John can do $\frac{1}{5}$ of the job per hour.

Steve can do $\frac{1}{4}$ of the job per hour.

The equation is: $\frac{x}{5}+\frac{x}{4}=1$

$$4x+5x=20, \quad 9x=20$$

$\therefore x=2\frac{2}{9}$ Ans: $2\frac{2}{9}$ hours.

6. A 16-gallon salt-water solution contains 25% pure salt. How much water should be added to produce the solution to 20% salt ?

Solution:

Let $x =$ gallons of water added

Pure salt in the original solution:

$$16\times 25\%=16\times 0.25=4 \text{ gal.}$$

The equation is:

$$\frac{4}{16+x}=20\%, \quad \frac{4}{16+x}=\frac{1}{5}$$

$$20=16+x \quad \therefore x=4$$

Ans: 4 gallons of water.

2-30 Rational Functions and Graphs

A **rational expression (algebraic fraction)** is a quotient (ratio) of two polynomials.
A **rational (fractional) function** is a function with the quotient (ratio) of two polynomials.
$y = \frac{1}{x}$ is the most popular and simplest rational function. Its graph is a **hyperbola**.

The graph of a rational function has a **vertical asymptote** at $x = a$ if the values for y increase (or decrease) without bound as x approaches a, either from the right or from the left.
The graph of a rational function has a **horizontal asymptote** at $y = b$ if the values for y approach b as x increases (or decreases) without bound.
The graph of a rational function may have several vertical asymptotes, but at most one horizontal asymptote.

Rules for graphing a rational function:

$$y = f(x) = \frac{a_n x^n + a_{n-1} x^{n-1} + \cdots\cdots + a_0}{b_m x^m + b_{m-1} x^{m-1} + \cdots\cdots + b_0}$$

1. Find $x-$intercept (let $y = 0$) and $y-$intercept (let $x = 0$).
2. Find the vertical asymptotes by setting the denominator equal to zero.
3. If $n < m$, then $y = 0$ (it is the $x-axis$) is a horizontal asymptote.
4. If $n > m$, then there is no horizontal asymptote.
5. If $n = m$, then $y = \frac{a_n}{b_m}$ is a horizontal asymptote.
6. If n is more than m only by 1 degree, then there is a slant (oblique) asymptote which can be determined by dividing the denominator into the numerator.
7. Complete the missing portions of the graph by plotting a few additional points.
8. The graphs of all rational functions of the form $y = \frac{ax + a_0}{bx + b_0}$ are hyperbolas.

Examples

1. Graph $y = \dfrac{1}{x}$.

Solution:

1. $f(-x) = -f(x)$. It is symmetric about the origin.
2. There is no $x-$intercept and $y-$intercept.
3. $x = 0$ (It is the $y-axis$) is a vertical asymptote.
4. The degree of the numerator is less than the degree of denominator. Therefore $y = 0$ (It is the $x-axis$) is a horizontal asymptote.
5. The domain and range are all nonzero real numbers.

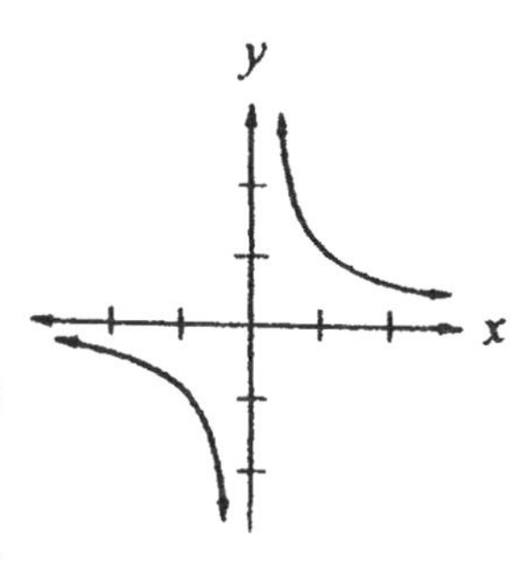

Examples

2. Graph $y = \dfrac{3}{x-2}$.

Solution:

1. Let $x = 0$, $y = -\frac{3}{2}$, $(0, -\frac{3}{2})$ is the $y-$intercept.
2. $x = 2$ is a vertical asymptote.
3. The degree of the numerator is less than the degree of the denominator. Therefore, $y = 0$ (it is the $x-axis$) is a horizontal asymptote.
4. The domain is all real numbers except 2. The range is all nonzero real numbers.

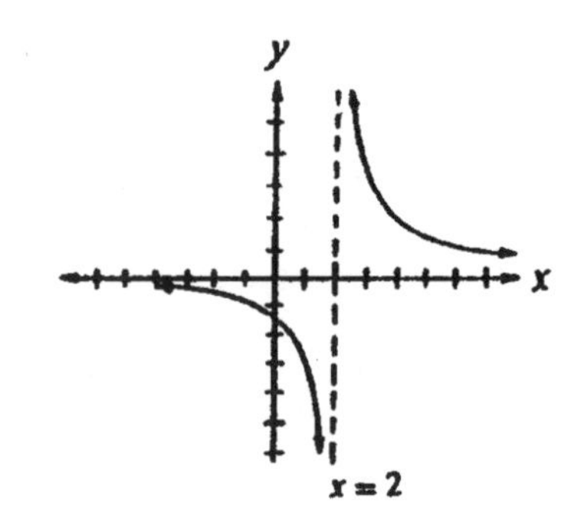

3. Graph $y = \dfrac{1}{x^2}$.

Solution:

1. $f(-x) = f(x)$. It is symmetric about the $y-axis$.
2. There is no $x-$intercept and $y-$intercept.
3. $x = 0$ (it is the $y-axis$) is a vertical asymptote.
4. $y = 0$ (it is the $x-axis$) is a horizontal asymptote.
5. The domain is all nonzero real numbers. The range is all real numbers of $y > 0$.

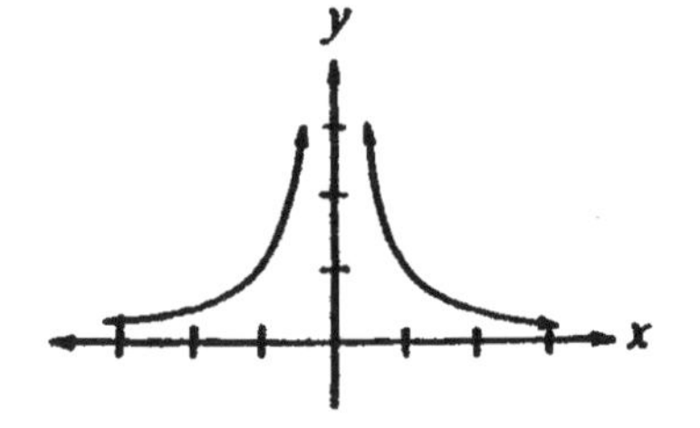

4. Graph $y = \dfrac{1}{(x-1)^2} - 4$.

Solution:

1. Let $x = 0$, then $y = -3$. (0, –3) is the $y-$intercept.
   Let $y = 0$, then $x = \frac{1}{2}$ and $1\frac{1}{2}$. $(\frac{1}{2}, 0)$ and $(1\frac{1}{2}, 0)$ are the $x-$intercepts.
2. The line $x = 1$ is a vertical asymptote.
3. The line $y = -4$ is a horizontal asymptote because y approaches –4 as x increases (or decreases) without bound.
4. The domain is all real numbers except 1. The range is all real numbers of $y > -4$.

Other method: It shifts the graph of $y = \frac{1}{x^2}$ by 1 unit to the right and 4 units downward.

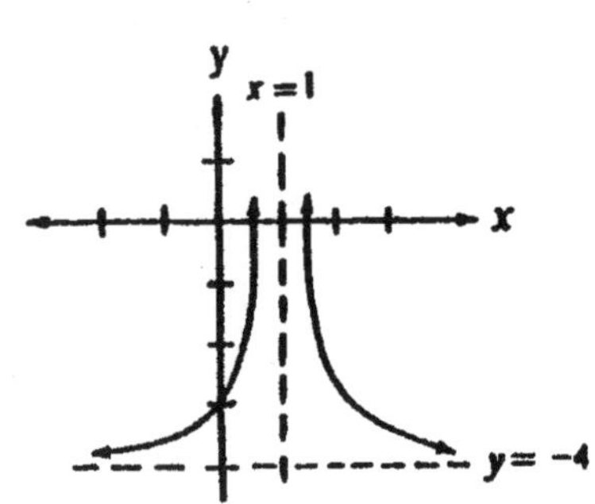

Examples

5. Graph $y = \dfrac{2}{x^2 - 2x + 2}$.

Solution:

1. Let $x = 0$, then $y = 1$. (0, 1) is the $y-$intercept.
2. $x^2 - 2x + 2 = x^2 - 2x + 1 + 1 = (x-1)^2 + 1 \neq 0$.
   There is no vertical asymptote.
3. The degree of the numerator is less than the degree of the denominator. Therefore, $y = 0$ is the horizontal asymptote.
4. $y = \dfrac{2}{(x-1)^2 + 1} > 0$. The graph is above the $x-axis$.
5. $f(0) = f(2)$. The graph is symmetric about the line $x = 1$.

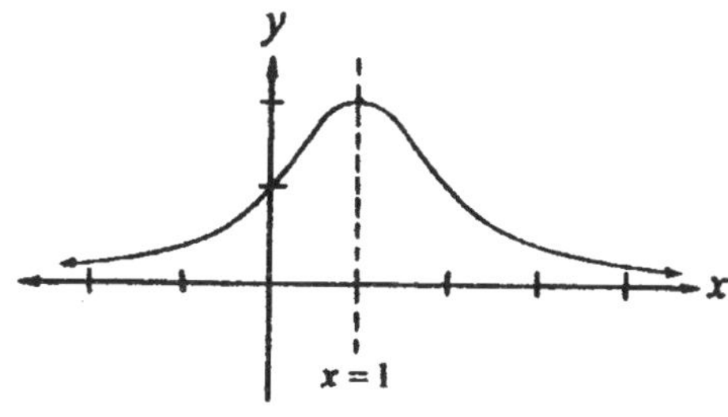

6. Graph $y = \dfrac{x^2 - 7x + 12}{x - 3}$.

Solution:

1. $y = \dfrac{x^2 - 7x + 12}{x - 3} = \dfrac{(x-3)(x-4)}{x-3} = x - 4$, $x \neq 3$
2. The value of y at $x = 3$ is undefined. The graph of the function is a straight line that has an open circle (hole) at $(3, -1)$.

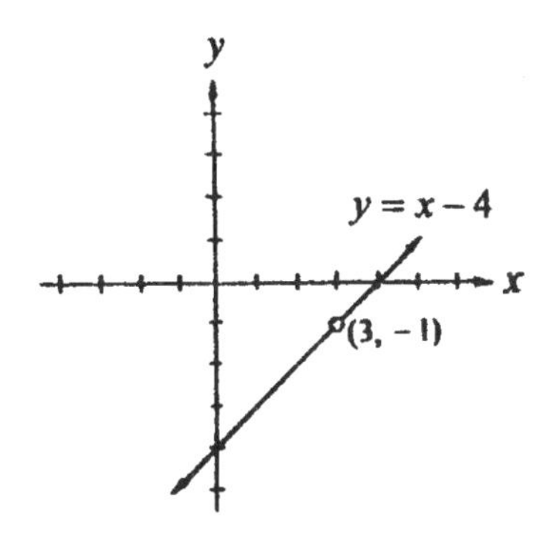

7. Graph $y = \dfrac{x - 3}{x^2 + x - 2}$.

Solution:

$$y = \frac{x-3}{x^2 + x - 2} = \frac{x-3}{(x+2)(x-1)}.$$

1. Let $x = 0$, then $y = \frac{-3}{-2} = \frac{3}{2}$. $(0, \frac{3}{2})$ is the $y-$intercept.
   Let $y = 0$, then $x = 3$. (3, 0) is the $x-$intercept.
2. Two vertical asymptotes: $x = -2$ and $x = 1$.
3. One horizontal asymptote: $y = 0$
4. Critical points: $x = -2, 1, 3$
   $-\infty < x < -2$, y is negative.
   $-2 < x < 1$, y is positive.
   $1 < x < 3$, y is negative.
   $3 < x < \infty$; y is positive.

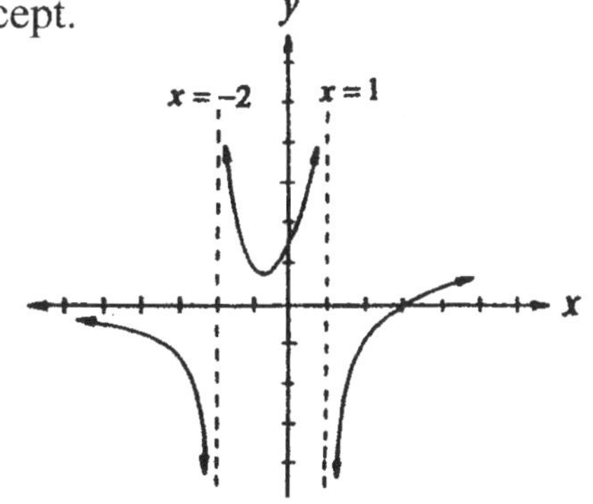

Examples

8. Graph $y = \frac{1}{|x|}$.

Solution:

If $x > 0$, $y = \frac{1}{x}$.

If $x < 0$, $y = -\frac{1}{x}$.

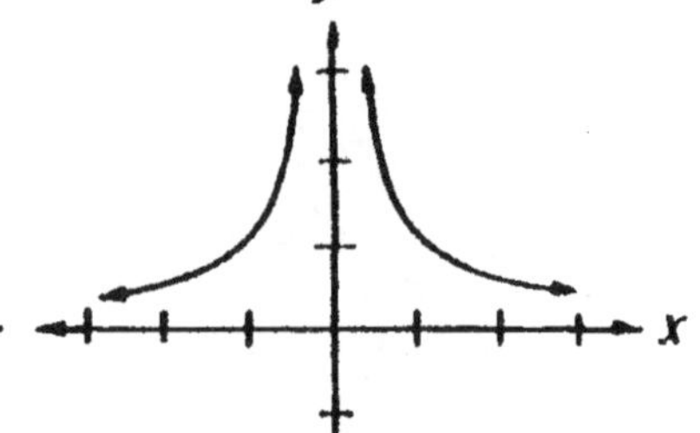

1. $f(-x) = f(x)$. It is symmetric about the $y-axis$.
2. There is no $x-$intercept and $y-$intercept.
3. $x = 0$ is a vertical asymptote.
4. $y = 0$ is a horizontal asymptote.
5. $y > 0$.

9. Graph $y = \frac{x^2}{x+1}$.

Solution:

1. Let $x = 0$, then $y = 0$. It passes the origin. There is no $y-$intercept.
   Let $y = 0$, then $x = 0$. It passes the origin. There is no $x-$intercept.
2. $x = -1$ is the vertical asymptote.
   There is no horizontal asymptote.
3. Dividing the denominator into the numerator, we have:

   $$y = \frac{x^2}{x+1} = x - 1 + \frac{1}{x+1}.$$

   $y = x - 1$ is the slant asymptote.
   (oblique asymptote)

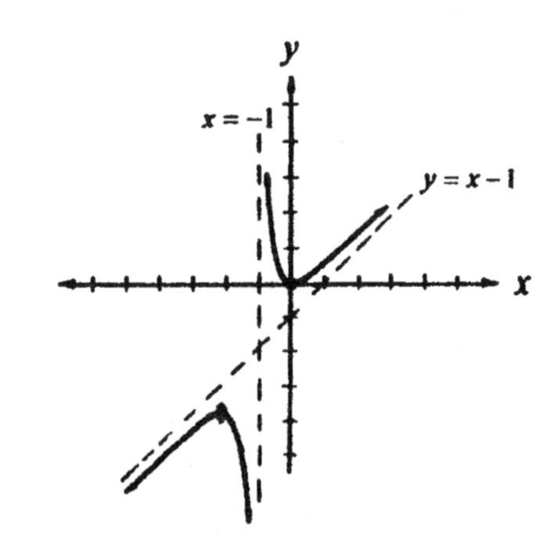

In an English class, the teacher asked the kids to write a letter to his(her) mother regarding what he (she) did today in school. The teacher noticed that John was writing very slowly.
Teacher: John, why do you write so slowly ?
John: My mom could not read fast.

2-31 Rational Inequalities and Absolute Values

Since we don't know whether its variable is positive or negative, a rational inequality in which a variable appears in a denominator can change signs of inequality if we multiply (or divide) each side by its common denominator. Therefore, we rewrite the inequality with **0** on the right side and consider two cases. **Testing the intervals** of the critical points is a good method to identify the answers.

Examples

1. Solve $\frac{3}{x} < 5$.

Solution: **Method 1:** $\frac{3}{x} - 5 < 0$, $\therefore \frac{3-5x}{x} < 0$

$x > 0$ and $3 - 5x < 0$	$x < 0$ and $3 - 5x > 0$
$-5x < -3$	$-5x > -3$
$x > \frac{3}{5}$	$x < \frac{3}{5}$
$\therefore x > \frac{3}{5}$	$\therefore x < 0$

Ans. $x < 0$ or $x > \frac{3}{5}$

Method 2: $\frac{3-5x}{x} < 0$. It is negative.

Critical points: $x = 0,\ \frac{3}{5}$

Test intervals:

$x < 0$, $\frac{(+)}{(-)} = -$ yes

$0 < x < \frac{3}{5}$, $\frac{(+)}{(+)} = +$ no

$x > \frac{3}{5}$, $\frac{(-)}{(+)} = -$ yes

(-) (+) (-)
0 3/5

Ans: $x < 0$ or $x > \frac{3}{5}$

2. Solve $\frac{x+1}{x} > 3$.

Solution:

$\frac{x+1}{x} - 3 > 0$, $\frac{x+1-3x}{x} > 0$

$\therefore \frac{1-2x}{x} > 0$ It is positive.

Critical points: $x = 0,\ \frac{1}{2}$

Test intervals:

$x < 0$, $\frac{(+)}{(-)} = -$ no

$0 < x < \frac{1}{2}$, $\frac{(+)}{(+)} = +$ yes

$x > \frac{1}{2}$, $\frac{(-)}{(+)} = -$ no

Ans: $0 < x < \frac{1}{2}$

(-) (+) (-)
0 1/2 1

We have learned how to solve simple inequalities involving absolute values. To solve a rational inequality involving absolute value, we use this fact:

$$|x|^2 = x^2$$

3. Solve $\left|\frac{x-2}{4x+1}\right| \geq 1$.

Solution:

$$|x-2| \geq |4x+1|$$
$$(x-2)^2 \geq (4x+1)^2$$
$$x^2 - 4x + 4 \geq 16x^2 + 8x + 1$$
$$-15x^2 - 12x + 3 \geq 0$$
$$5x^2 + 4x - 1 \leq 0$$
$$(5x-1)(x+1) \leq 0$$

$5x - 1 \geq 0$ and $x + 1 \leq 0$	$5x - 1 \leq 0$ and $x + 1 \geq 0$
$x \geq \frac{1}{5}$, $x \leq -1$	$x \leq \frac{1}{5}$, $x \geq -1$
Not permissible	$\therefore -1 \leq x \leq \frac{1}{5}$, $x \neq -\frac{1}{4}$.

(We can also test the intervals to find the answer.)

Ans: $-1 \leq x \leq \frac{1}{5}$, $x \neq -4$

2-32 Inverse Variations

We learned that a linear equation in the form $y = kx$ is called an equation having a **direct variation** with x and y.
An equation in the form $y = \frac{k}{x}$ (where $x \neq 0$), or $xy = k$ is called an equation having an **inverse variation** with x and y. k is a nonzero constant. k is called the **constant of variation** or **constant of proportionality**. For example, in the equation $y = \frac{3}{x}$, we say that y varies inversely as x, or y is inversely proportional to x. The graph of $y = \frac{3}{x}$, or $xy = 3$ is a **hyperbola.**
If (x_1, y_1) and (x_2, y_2) are two ordered pairs of an equation having an inverse variation defined by $y = \frac{k}{x}$, we have: $x_1y_1 = k$ and $x_2y_2 = k$. $\therefore x_1y_1 = x_2y_2$
In inverse variation, the variables that vary may involve powers (exponents).

Note that the equation $y = \frac{3}{x-1}$ does not show x and y as an inverse variation because the products of its ordered pairs x_1y_1 and x_2y_2 are not equal (not proportional).
However, the equation $y = \frac{3}{x-1}$ shows $(x-1)$ and y as an inverse variation $[y(x-1) = 3]$.

Combined Variation: The combination of direct and inverse variations.

In science, there are many word problems which involve the concept of inverse variations. The gravitational force (F) attracted each other between any two objects in the universe is given by $F = G\frac{m_1m_2}{r^2}$ (**Newton's Law of Universal Gravitation**).

Examples

1. If y varies inversely as x, and if $y = 6$ when $x = 2$, find the constant of variation.
   Solution: Let $xy = k$ $\therefore k = xy = 2 \cdot 6 = 12$. Ans.
2. If y varies inversely as x, and if $y = 6$ when $x = 2$, find y when $x = 3$.
   Solution: Let $xy = k$ $\therefore k = xy = 2 \cdot 6 = 12$, $xy = 12$ is the equation.
   When $x = 3$, $y = \frac{12}{x} = \frac{12}{3} = 4$. Ans.
3. If (x_1, y_1) and (x_2, y_2) are ordered pairs of the same inverse variation, find y_1.
   $x_1 = 3$, $y_1 = ?$, $x_2 = 12$, $y_2 = 8$
   Solution: Let $x_1y_1 = x_2y_2$, $3y_1 = 12 \cdot 8$, $3y_1 = 96$ $\therefore y_1 = 32$. Ans.
   Or: $k = x_2y_2 = 12 \cdot 8 = 96$ $\therefore x_1y_1 = 96$, $y_1 = \frac{96}{x_1} = \frac{96}{3} = 32$. Ans.

4. The number of days needed to build a house varies inversely as the number of workers working on the job. It takes 140 days for 5 workers to finish the job. If the job has to be finished in 100 days, how many workers are needed ?
   Solution: Let $d_1 = 140$, $w_1 = 5$, $d_2 = 100$, $w_2 = ?$
   We have: $d_1w_1 = d_2w_2$, $140 \cdot 5 = 100 \cdot w_2$ $\therefore w_2 = \frac{700}{100} = 7$ workers. Ans.

2-33 Exponential Functions and Graphs

Definition of an exponential function:

An exponential function $f(x)$ with a constant base a is a function of the form

$$f(x) = a^x \quad \text{or} \quad y = a^x$$

where $a > 0$, $a \neq 1$, and x is any real number. It is a one-to-one function.

Graphs of $y = a^x$ and $y = a^{-x}$:

1. The graph of $y = a^x$ is an increasing function. y increases as x increases. The graph lies above the $x-$axis and passes through the point (0, 1) and (1, a). The $x-$axis ($y = 0$) is a horizontal asymptote to the graph as $x \to -\infty$.
2. The graph of $y = a^{-x}$ is a decreasing function. y decreases as x increases. The graph lies above the $x-$axis and passes through the point (0, 1) and (1, $\frac{1}{a}$). The $x-$axis ($y = 0$) is a horizontal asymptote to the graph as $x \to \infty$.
3. The graph of $y = a^{-x}$ is the reflection of the graph of $y = a^x$ through the $y-$axis.

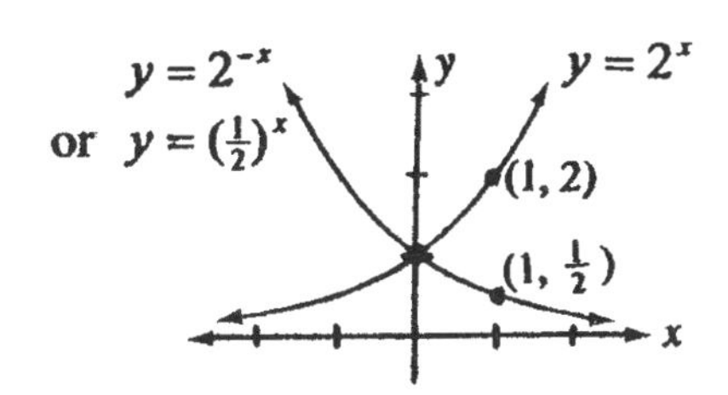

Domain: All real numbers $(x \in R)$

Range: All positive numbers ($y > 0$)

Graphing exponential functions using transformations:

1. The graph of $y = 3^{x-1}$ shifts the graph of $y = 3^x$, 1 unit to the right.
2. The graph of $y = 3^x - 1$ shifts the graph of $y = 3^x$, 1 unit downward.
3. The graph of $y = -3^x$ reflects the graph of $y = 3^x$ over the $x-$axis.
4. The graph of $y = 3^{-x}$ reflects the graph of $y = 3^x$ over the $y-$axis.

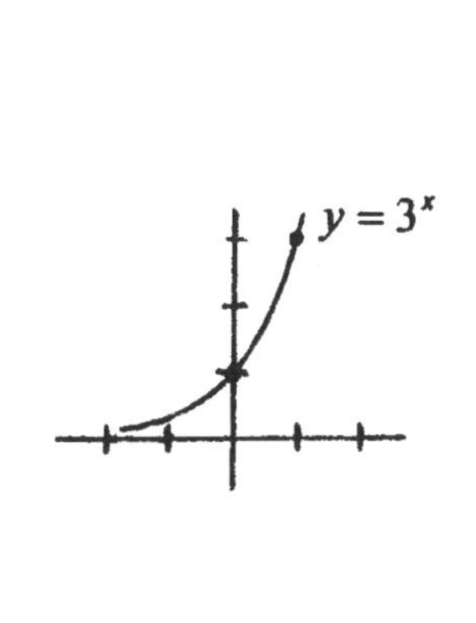

1.

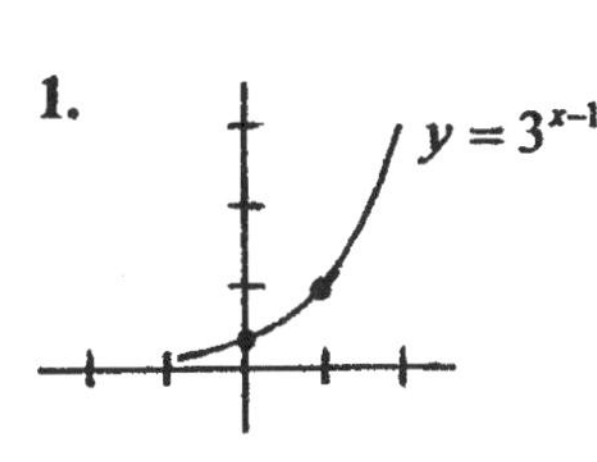

3.

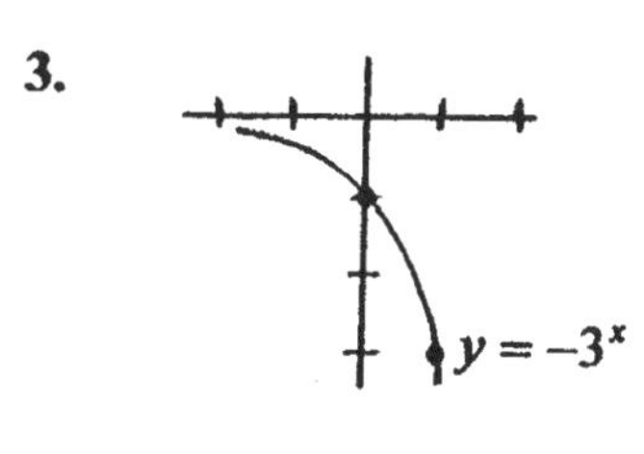

2.

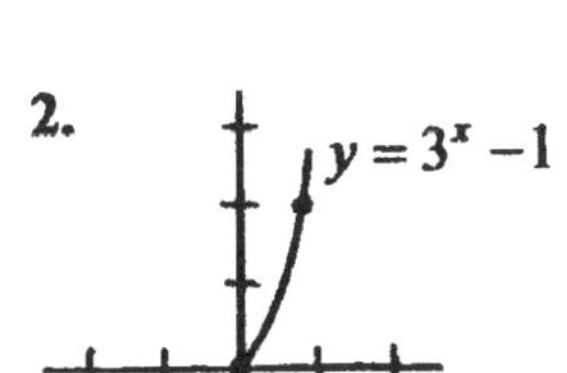

4.

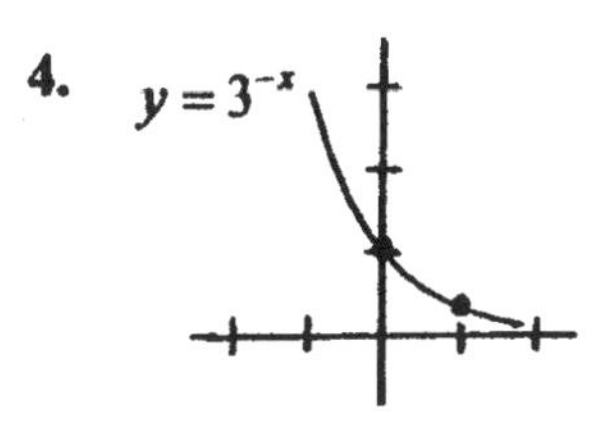

2-34 Solving Exponential Equations

Equations that contains the terms a^x, $a > 0$, $a \neq 1$ are called **exponential equations.** There are different ways to solve various types of exponential equations.

Laws of Exponents (Rules of Powers)

If a and b are real numbers, m and n are positive integers, $a \neq 0$, $b \neq 0$, then:

1) $a^0 = 1$ **2)** $a^m \cdot a^n = a^{m+n}$ **3)** $\dfrac{a^m}{a^n} = a^{m-n}$ **4)** $\dfrac{a^m}{a^n} = \dfrac{1}{a^{n-m}}$

5) $(a^m)^n = a^{mn}$ **6)** $(ab)^m = a^m b^m$ **7)** $a^{-m} = \dfrac{1}{a^m}$ **8)** $\dfrac{1}{a^{-m}} = a^m$

9) $\left(\dfrac{b}{a}\right)^m = \dfrac{b^m}{a^m}$ **10)** $\left(\dfrac{b}{a}\right)^{-m} = \left(\dfrac{a}{b}\right)^m$ **11)** $a^{\frac{1}{m}} = \sqrt[m]{a}$ **12)** $a^{\frac{n}{m}} = (\sqrt[m]{a})^n$

To solve **exponential equations**, we apply the laws of exponents.

1. Rewrite the equation using the same base on both sides.
2. Make a Substitution
3. Take logarithms on both sides

To solve an exponential equation, we must check each solution and discard solutions that are not permissible.

Examples

1. Solve $3^{x+1} = 81$.
Solution:

$$3^{x+1} = 81$$
$$3^{x+1} = 3^4$$
$$x + 1 = 4$$
$$\therefore x = 3. \text{ Ans.}$$

2. Solve $3^{x+1} = 9^{2x-1}$.
Solution:

$$3^{x+1} = 9^{2x-1}$$
$$3^{x+1} = (3^2)^{2x-1}$$
$$3^{x+1} = 3^{4x-2}$$
$$x + 1 = 4x - 2$$
$$3 = 3x$$
$$\therefore x = 1. \text{ Ans.}$$

3. Solve $4^x - 3 \cdot 2^x - 4 = 0$
Solution:

$$2^{2x} - 3 \cdot 2^x - 4 = 0$$

Let $a = 2^x$ (make a substitution):

$$a^2 - 3a - 4 = 0$$
$$(a-4)(a+1) = 0$$
$$a - 4 = 0 \text{ or } a + 1 = 0$$

$a - 4 = 0$	$a + 1 = 0$
$a = 4$	$a = -1$
$2^x = 4$	$2^x = -1$
$x = 2$	not permissible

Ans: $x = 2$

Examples

4. Solve $3^{4x} = 81$.

Solution:

$$3^{4x} = 81$$

$$3^{4x} = 3^4$$

$$4x = 4$$

$$\therefore x = 1. \text{ Ans.}$$

5. Solve $3^{4x+1} = (\frac{1}{9})^{x-1}$.

Solution:

$$3^{4x+1} = (\tfrac{1}{9})^{x-1}$$

$$3^{4x+1} = (9^{-1})^{x-1} = 9^{-x+1} = (3^2)^{-x+1}$$

$$3^{4x+1} = 3^{-2x+2}, \quad 4x+1 = -2x+2$$

$$6x = 1 \quad \therefore x = \tfrac{1}{6}. \text{ Ans.}$$

6. Solve $x^{2/3} = 16$.

Solution:

$$(x^{2/3})^{3/2} = 16^{3/2}$$

$$x = (\sqrt{16})^3$$

$$\therefore x = 4^3 = 64. \text{ Ans.}$$

7. Solve $0.25^x = 4$

Solution:

$$0.25^x = 4$$

$$0.25^x = (\tfrac{1}{4})^{-1}$$

$$0.25^x = (0.25)^{-1} \quad \therefore x = -1. \text{ Ans.}$$

On the airplane, a three-year old boy asks his mother.

Boy: Where are going ?

Mother: We are going to grandma's home.

Boy: Why are so many people going to Grandma's home ?

2-35 Systems of Equations and Inequalities

The graph of a linear equation $ax + by = c$ is a **straight line.**

The graphs of two linear equations (two straight lines) in the same coordinate plane must show one of the following:

1) Intersect in one point (they are consistent and independent).
2) Intersect in no point (parallel, they are inconsistent).
3) Intersect in unlimited number of points (coincide, they are consistent and dependent).

System of equations: Two or more equations with the same variables.
System of inequalities: Two or more inequalities with the same variables.
A solution to a system of equations in two variables is the intersection point (x, y) that satisfies both equations.
A system of equations is also called a **system of simultaneous equations.**

There are two algebraic methods for solving systems of equations.

1) The substitution method
2) The elimination method (or **The combination method)**

We may choose any method of the above two methods to find the solution of a system. In general, it is good to choose the **substitution method** if the coefficient of a variable is 1 or -1.
The main idea of the **elimination method** is to eliminate one variable by **adding or subtracting** the two equations to obtain a new equation with only one variable.

Examples

1. Solve $\begin{cases} 2x - y = 4 \cdots\cdots\cdots ① \\ 2x + 5y = 16 \cdots\cdots ② \end{cases}$

Solution:

Method 1: (Substitution Method)

From equation ①, we have $y = 2x - 4$
Substitute $y = 2x - 4$ in equation ②, we have:

$$2x + 5y = 16$$
$$2x + 5(2x - 4) = 16$$
$$2x + 10x - 20 = 16$$
$$12x = 36$$
$$\therefore x = 3$$

Substitute $x = 3$ in equation ①:

$$2x - y = 4$$
$$2(3) - y = 4 \;,\; 6 - y = 4 \quad \therefore y = 2$$

Ans: $x = 3$, $y = 2$ or $(3, 2)$

Method 2: (Combination Method)

equation ② – equation ①
We have: (x will be eliminated.)

$$\begin{array}{r} \cancel{2x} + 5y = 16 \\ -)\ \cancel{2x} - y = 4 \\ \hline 6y = 12 \end{array}$$
$$\therefore y = 2$$

Substitute $y = 2$ in equation ①

$$2x - y = 4$$
$$2x - 2 = 4$$
$$2x = 6$$
$$\therefore x = 3$$

Ans: $x = 3$, $y = 2$ or $(3, 2)$

Sometimes it is necessary to multiply one or both of the equations by a number to obtain the same coefficients(or differ in sign) for one of the variables. Then, we can eliminate this variable. If the final statement is a "**true statement**", the system of equations has unlimited number of solutions. The system is **consistent** and **dependent**.
If the final statement is a "**false statement**", the system of equations has no solution. The system is **inconsistent**.

Examples

2. Solve $\begin{cases} 3x - 2y = 10 & \cdots\cdots ① \\ 5x + 4y = 13 & \cdots\cdots ② \end{cases}$

Solution:

$$\begin{array}{lr} ① \times 2: & 6x - 4y = 20 \\ ②: & +)\ 5x + 4y = 13 \\ \hline & 11x \quad = 33 \end{array}$$

$$\therefore x = 3$$

Substitute $x = 3$ in equation ①:

$$3x - 2y = 10$$
$$3(3) - 2y = 10$$
$$9 - 2y = 10$$
$$-2y = 1$$
$$\therefore y = -\tfrac{1}{2}$$

Ans: $x = 3$, $y = -\frac{1}{2}$
or $(3, -\frac{1}{2})$.

3. Solve $\begin{cases} 3x - 5y = 12 & \cdots\cdots ① \\ 4x - 2y = \frac{20}{3} & \cdots\cdots ② \end{cases}$

Solution:

$$\begin{array}{lr} ① \times 4: & 12x - 20y = 48 \\ ② \times 3: & -)\ 12x - \ 6y = 20 \\ \hline & -14y = 28 \end{array}$$

$$\therefore y = -2$$

Substitute $y = -2$ in ①:

$$3x - 5y = 12$$
$$3x - 5(-2) = 12$$
$$3x + 10 = 12$$
$$3x = 2$$
$$\therefore x = \tfrac{2}{3}$$

Ans: $x = \frac{2}{3}$, $y = -2$
or $(\frac{2}{3}, -2)$.

4. Solve $\begin{cases} 4x - 2y = 7 & \cdots\cdots ① \\ 10x - 5y = 6 & \cdots\cdots ② \end{cases}$

Solution:

$$\begin{array}{lr} ① \times 5: & 20x - 10y = 35 \\ ② \times 2: & -)\ 20x - 10y = 12 \\ \hline & 0 = 23 \text{ (false)} \end{array}$$

Ans: No solution
(They are parallel)

5. Solve $\begin{cases} 4x + 6y = 9 & \cdots\cdots ① \\ 12x + 18y = 27 & \cdots\cdots ② \end{cases}$

Solution:

$$\begin{array}{lr} ① \times 3: & 12x + 18y = 27 \\ ②: & -)\ 12x + 18y = 27 \\ \hline & 0 = 0 \text{ (true)} \end{array}$$

Ans: Unlimited number of solutions
(they coincide).

Graphing Linear inequalities

In a coordinate plane, the graph of a linear inequality in two variables is an open or close half-plane.

For example, the graph of the linear equation $x+2y=4$ (a straight line) separates the coordinate plane into two open half-planes.

The upper open half-plane is the region of the inequality $x+2y>4$.

The lower open half-plane is the region of the inequality $x+2y<4$.

$x+2y \geq 4$ and $x+2y \leq 4$ are called the close half-planes.

$x+2y=4$ is the associated equation of each linear inequality.

To find out which half-plane region (upper or lower) is the region of each linear inequality, we use the origin (0, 0) to test it. Simply substitute (0, 0) in the inequality and see if (0, 0) is included in the region of the inequality.

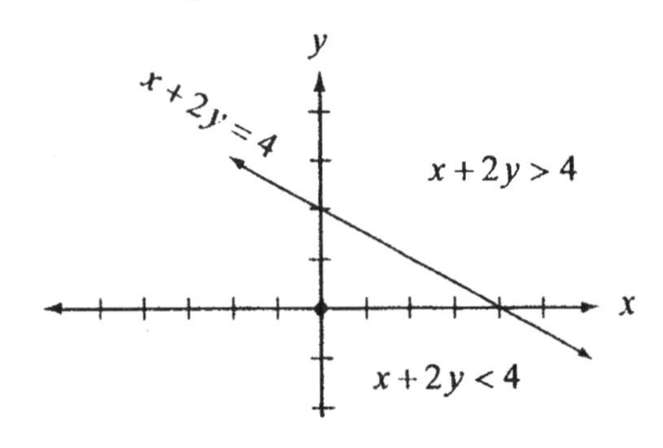

Substitute (0, 0) in $x+2y<4$

$0<4$ (True)

Therefore, (0, 0) is included in the region of $x+2y<4$.

Examples (Hint: Graph the boundary line first.)

6. Graph $4x+5y<20$.

Solution:

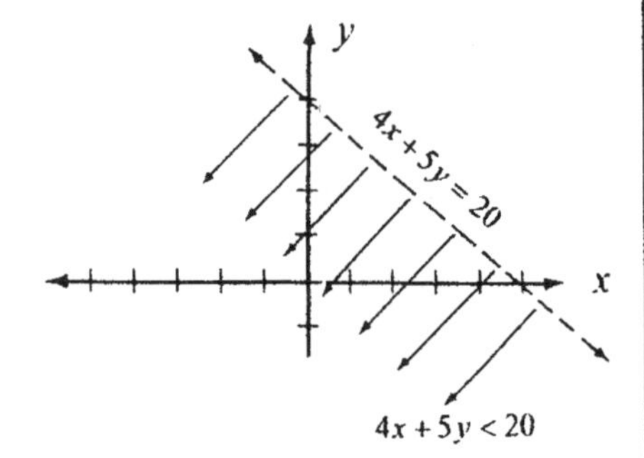

Its graph covers the lower region of the half-plane below the line. The boundary line $4x+5y=20$ is not included. The line is dashed.

7. Graph $4x+5y \geq 20$.

Solution:

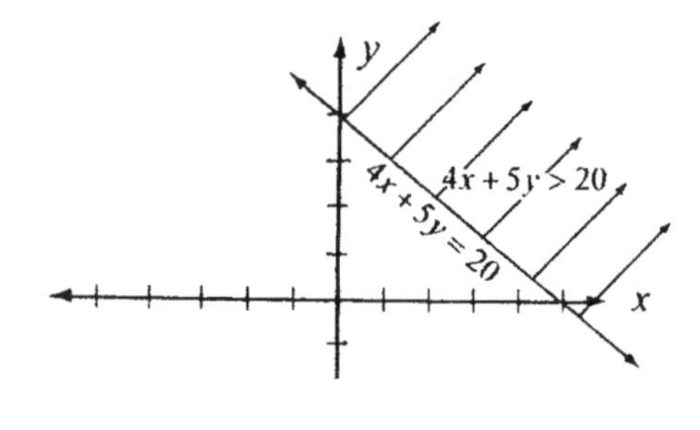

Its graph covers the upper region of the half-plane and the region on the line. The boundary line $4x+5y=20$ is included. The line is solid.

Solving a system of two linear inequalities

To solve a system of two linear inequalities, we graph the two boundary lines. Use a solid boundary line to represent ≤ or ≥ and a dashed boundary line to represent < or >.
The solution of the system is the intersection of two overlap (combined) regions of the combined graph. Every point in the intersection region satisfies both inequalities.

Example

8. Solve $\begin{cases} 4x+5y \ge 20 \\ 4x-5y<10 \end{cases}$

Solution:

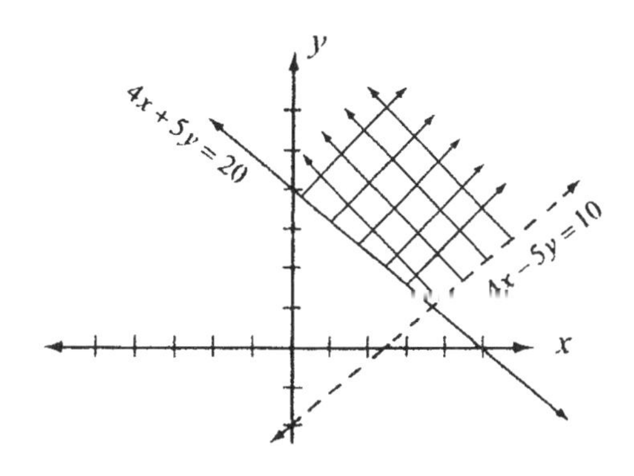

Its graph covers the overlap region above and on the line $4x+5y=20$ and above the line $4x-5y=10$.

Solving a system of three linear equations

Solving a system of three linear equations containing three variables, we eliminate one variable at a time by two of the three equations to obtain a system of two equations in two variables.

Example

9. Solve $\begin{cases} x+y+z=0 & \cdots\cdots ① \\ 2x-y-2z=6 & \cdots\cdots ② \\ x-y-3z=8 & \cdots\cdots ③ \end{cases}$

Solution:

① + ② $\quad 3x - \not{z} = 6 \quad \cdots\cdots ④$

② − ③ $\quad x + \not{z} = -2 \quad \cdots\cdots ⑤$

④ + ⑤ $\quad 4x = 4$

$\therefore x = 1$

Substitute $x=1$ in ④

$3(1)-z=6$

$\therefore z=-3$

Substitute $x=1$ and $z=-3$ in ①

$1+y-3=0$

$\therefore y=2$

Ans. $x=1,\ y=2,\ z=-3$

2-36 Piecewise-Defined Functions

The piecewise-defined function is a function described by two or more equations over a specified domain (the x-values).

$$f(x)=\begin{cases} x^2-2, & x\le 0 \\ x-1, & x>0 \end{cases}$$

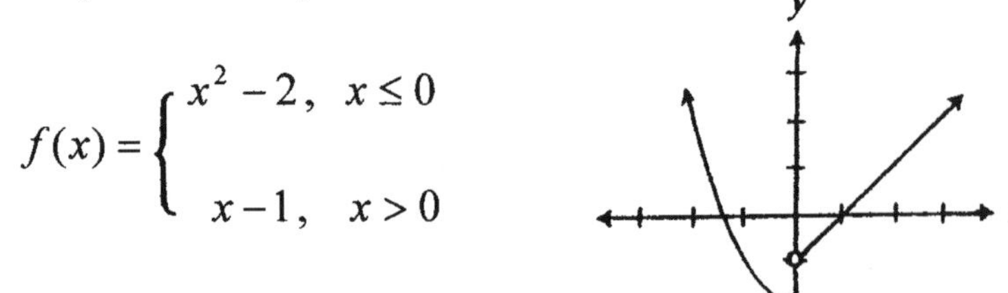

A close circle "•" shows "included". An open circle "o" shows "not included".

Examples

1. Evaluate the function when $x=-2$, 0, 1, and 2.

$$f(x)=\begin{cases} x^2-2, & x\le 0 \\ x-1, & x>0 \end{cases}$$

Solution:

$f(x)=x^2-2$ when $x=-2$: $f(-2)=(-2)^2-2=2$.

$f(x)=x^2-2$ when $x=0$: $f(0)=0^2-2=-2$.

$f(x)=x-1$ when $x=1$: $f(1)=1-1=0$.

$f(x)=x-1$ when $x=2$: $f(2)=2-1=1$.

2. A publishing company pays income tax at a rate of 30% on the first income \$50,000, and all income over \$50,000 is taxed at a rate of 40%.

a) Write a piecewise-defined function $T(x)$ that shows the total tax on an income of x dollars.

b) What is the total tax on an income \$95,000 ?

Solution:

a) If $0<x\le 50000$, $T(x)=0.30x$

If $x>50000$, $T(x)=0.30(50{,}000)+0.40(x-50{,}000)$

$=15{,}000+0.40x-20{,}000$

$=0.40x-5{,}000$

We have the piecewise-defined function:

$$f(x)=\begin{cases} 0.30x & \text{if } 0<x\le 50{,}000 \\ 0.40x-5000 & \text{if } x>50{,}000 \end{cases}$$

b) $T(95{,}000)=0.40(95{,}000)-5{,}000=\$33{,}000$..

2-37 Recursive Functions for Sequences

In Chapter 1, we have learned how to find the nth term of a sequence. The nth term (a_n) of a sequence can be considered as a function of its position n.

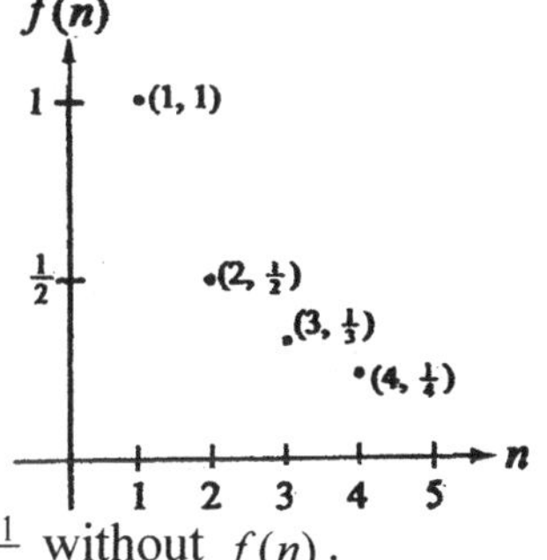

For example, the sequence $1, \frac{1}{2}, \frac{1}{3}, \frac{1}{4}, \cdots\cdots, \frac{1}{n}, \cdots\cdots$ can be represented as a function $f(n) = \frac{1}{n}$ for every positive integer n.

$a_1 = f(1) = 1$, $a_2 = f(2) = \frac{1}{2}$, $a_3 = f(3) = \frac{1}{3}$, $a_n = f(n) = \frac{1}{n}$, ····

Therefore, we can graph a sequence on the coordinate plane.
All points on the graph whose x-coordinates are positive integers.
In order to find any term of a sequence easily, we use this rule $a_n = \frac{1}{n}$ without $f(n)$.

There are two ways to represent a sequence as a function: **Explicit Rule** and **Recursive Rule.**
For example, the nth term of the arithmetic sequence 4, 7, 10, 13, ····· can be represented in two different rules:

1. Explicit Rule for the nth term is: $a_n = a_1 + (n-1)d = 4 + (n-1) \cdot 3$

The explicit rule is $a_n = 1 + 3n$.

2. Recursive Rule for the nth term is: $a_n = a_{n-1} + d = a_{n-1} + 3$

The recursive rule is $a_1 = 4$, $a_n = a_{n-1} + 3$

Recursive rule is used to find the nth term of a sequence if the first few terms of the sequence is given.
When the pattern or rule of a sequence is known, we can also write the sequence by the recursive formula for the nth term by placing braces without writing the entire sequence.

The sequence $\{a_n\} = \{2n\}$ represents the sequence 2, 4, 6, 8, 10, ····· , $2n$, ····· .

The sequence $\{b_n\} = \left\{\begin{matrix} n & \text{if } n \text{ is even} \\ n^2 & \text{if } n \text{ is odd} \end{matrix}\right\}$ represents the sequence 1, 2, 9, 4, ····· .

Examples

1. Write the nth term for the geometric sequence with $a_1 = 2$ and $r = 3$.

a. explicit rule b. recursive rule

Solution:

a. The explicit rule for the nth term is: $a_n = a_1 r^{n-1} = 2 \cdot 3^{n-1}$ Ans.

b. The recursive rule for the nth term is: $a_1 = 2$, $a_n = 3 \cdot a_{n-1}$ Ans.

2. Write the recursive rule for the sequence 2, 2, 4, 8, 32, 256, ·····

Solution:

The recursive rule is: $a_1 = 2$, $a_2 = 2$, $a_n = a_{n-2} \cdot a_{n-1}$ Ans.

3. Write the first six terms of the recursive sequence: $a_1 = 1$, $a_2 = 1$, $a_{n+2} = a_n + a_{n+1}$.

Solution:

$a_1 = 1$, $a_2 = 1$, $a_3 = 2$, $a_4 = 3$, $a_5 = 5$, $a_6 = 8$

The sequence is 1, 1, 2, 3, 5, 8, ······ . Ans. (It is called Fibonacci Sequence.)

Examples

4. Write the first five terms of the sequence: $a_1 = 1$, $a_{n+1} = (n+1)a_n$.

Solution:

$a_1 = 1$, $a_2 = 2a_1 = 2 \cdot 1 = 2$, $a_3 = 3a_2 = 3 \cdot 2 = 6$

$a_4 = 4a_3 = 4 \cdot 6 = 24$, $a_5 = 5a_4 = 5 \cdot 24 = 120$

The sequence is 1, 2, 6, 24, 120, ······ . Ans.

(It is called Factorial Sequence. The nth term of the sequence is $n!$)

5. Write the first four terms of the sequence: $\{t_n\} = \left\{\left(\frac{1}{2}\right)^n\right\}$.

Solution:

$t_1 = \left(\frac{1}{2}\right)^1$, $t_2 = \left(\frac{1}{2}\right)^2$, $t_3 = \left(\frac{1}{2}\right)^3$, $t_4 = \left(\frac{1}{2}\right)^4$ The sequence is $\frac{1}{2}, \frac{1}{4}, \frac{1}{8}, \frac{1}{16}$, ······ . Ans.

6. Write the first five terms of the sequence: $\{a_n\} = \{2^{n-1}\}$.

Solution:

$a_1 = 2^{1-1}$, $a_2 = 2^{2-1}$, $a_3 = 2^{3-1}$, $a_4 = 2^{4-1}$, $a_5 = 2^{5-1}$

The sequence is 1, 2, 4, 8, 16, ······ . Ans.

7. In a supermarket, marketing survey shows that the following information is always true within a year.

1) There are 9,300 customers in the beginning of this year.
2) 20% of the customers will go to the nearby supermarkets.
3) 1800 new customers will come to shop in this supermarket.

a. Write a recursive rule for the number a_n of the customers at the beginning of the nth year.

b. How many customers will shop in this supermarket at the beginning of 5th year ?

c. After several years, find the number of the customers which will be stable on shopping at this supermarket.

Solution: **a.** The recursive rule is: $a_1 = 9300$, $a_n = 0.8a_{n-1} + 1800$

b. $a_2 = 0.8(9300) + 1800 = 9240$, $a_3 = 0.8(9240) + 1800 = 9192$

$a_3 = 0.8(9192) + 1800 = 9153$, $a_5 = 0.8(9192) + 1800 = 9122$

We can use a graphing calculator to find a_5.

c. Continue pressing "ENTER" on the calculator, the number of customers will be stable by 9000 customers at the beginning of 33rd year.

Press: 9300→ENTER→0.8→×→2nd→ANS→ENTER

```
9300
                9300
0.8*Ans+1800
                9240
                9192
              9153.6
             9122.88
```

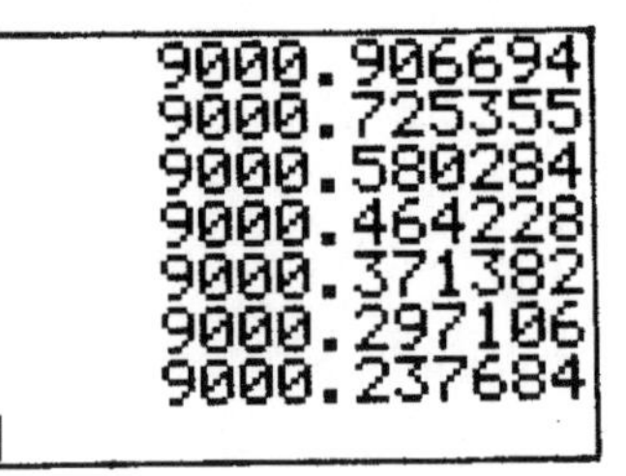

2-38 Logarithmic Functions and Graphs

We have learned in the preceding section that the exponential function $y = a^x$, for $a > 0$ and $a \neq 1$, is a one-to-one function.
Since every one-to-one function has an inverse function, $y = a^x$ has the inverse function by the equation: $x = a^y$, $a > 0$ and $a \neq 1$

In order to solve the equation $x = a^y$ for the exponent y in terms of x, we create the word **logarithm** for the exponent y:

$y = \log_a x$ Read " y is the logarithm of x to the base a. ".

We simply say that $\log_2 8$ represents " 2 to what power gives 8. ". Therefore, $\log_2 8 = 3$.
$y = \log_a x$ is the equivalent logarithmic form of $x = a^y$.
$y = \log_a x$ and $y = a^x$ are inverse functions of each other (see graphs on section 7~1).
The domains and ranges of $y = \log_a x$ and $y = a^x$ are interchanged.

Definition of a logarithmic function:

A logarithmic function $f(x)$ to a constant base a is a function of the form

$$f(x) = \log_a x \quad \text{or} \quad y = \log_a x$$

where $a > 0$, $a \neq 1$, and x is any real number. It is a one-to-one function.

Graphs of $y = \log_a x$**:**

1. The function is an increasing function if $a > 1$, y increases as x increases. The $y-$axis ($x = 0$) is a vertical asymptote to the graph as $x \to 0$.
2. The function is a decreasing function if $0 < a < 1$, y decreases as x increases. The $y-$axis ($x = 0$) is a vertical asymptote to the graph as $x \to 0$.
3. The graph lies to the right of $y-$axis and passes through the point (1, 0)

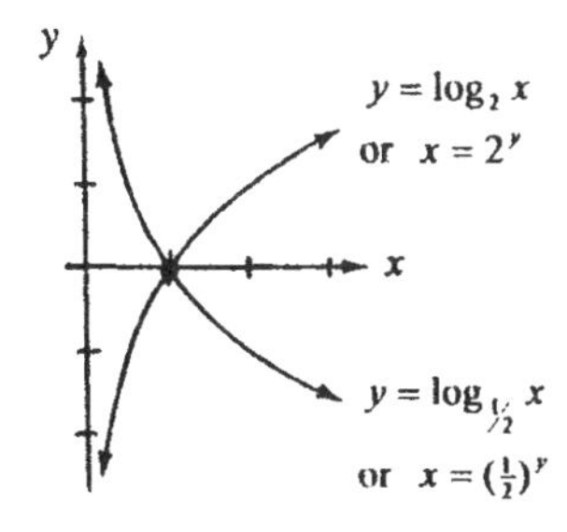

Domain: All positive real numbers ($x > 0$)
Range: All real numbers ($y \in R$)

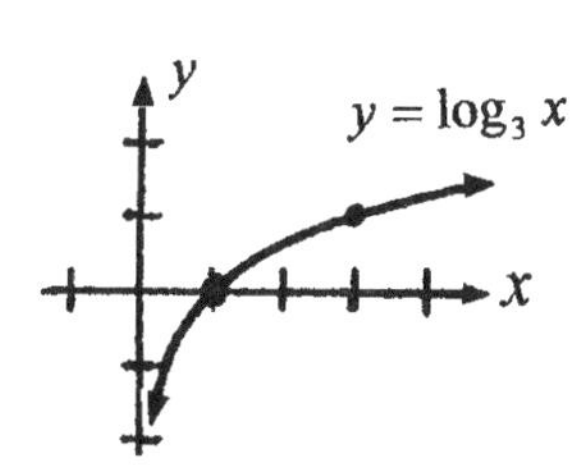

Graphing logarithmic functions using transformations

1. The graph of $y = \log_3(x-1)$ shifts the graph of $y = \log_3 x$, 1 unit to the right.
2. The graph of $y = \log_3 x - 1$ shifts the graph of $y = \log_3 x$, 1 unit downward.
3. The graph of $y = -\log_3 x$ reflects the graph of $y = \log_3 x$ over the $x-$axis.
4. The graph of $y = \log_3(-x)$ reflects the graph of $y = \log_3 x$ over the $y-$axis.

2-39 The Number e and Natural Logarithms

When we graph the exponential function $y = a^x$, where $a > 0$ and $a \neq 1$, we notice the following two facts:

1. The slope (m) of the tangent to $y = 2^x$ through the point $p(0, 1)$ is less than 1.

2. The slope (m) of the tangent to $y = 3^x$ through the point $p(0, 1)$ is more than 1.

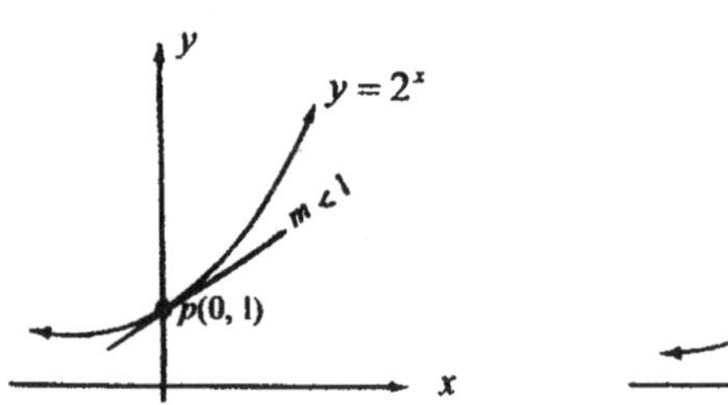

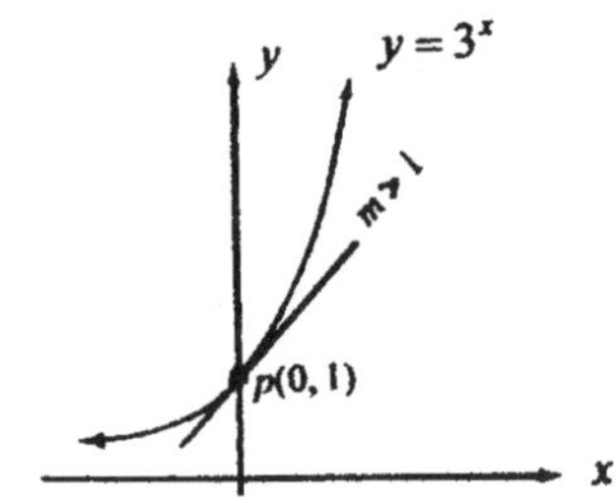

Therefore, we expect that there must be a number a, $2 < a < 3$, so that the slope of the tangent to $y = a^x$ through the point $p(0, 1)$ is exactly equal to 1. This number is designed by the letter e. The number e is called the **Euler number**.
The number e is one of the most famous numbers found in the history of mathematics. It is as famous as π and i. In advanced mathematics, we have found e from the following function in limit notation:

$$e = \lim_{n \to \infty}(1 + \tfrac{1}{n})^n . \qquad e = 2.71828\cdots\cdots. \text{ It is an irrational number.}$$

$x = e^y$ is the inverse function of $y = e^x$. The **natural logarithmic function** of $x = e^y$ is written by $y = \log_e x$. (Note that : $y = \log_e x$ and $x = e^y$ are equivalent.)
Natural logarithms are abbreviated by ln, with the base understood to be e.

$y = \ln x \rightarrow$ read as "y **is the logarithm of** x **to the base** e".

$y = \ln x$ and $y = e^x$ are inverses each other. The graph of $y = \ln x$ and $y = e^x$ are reflection of each other about the line $y = x$.
The domains and ranges of the two functions are interchanged.

$y = e^x \rightarrow x \in R,\ y > 0.$

$y = \ln x$ or $x = e^y \rightarrow y \in R,\ x > 0.$

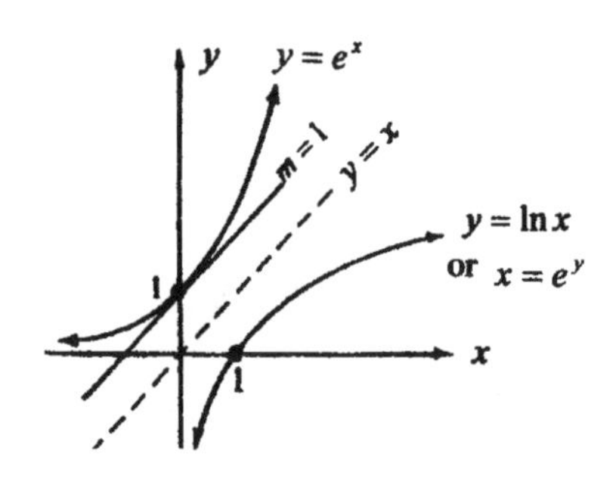

Natural logarithms are widely used in calculus and science. All laws of logarithms and the formulas for the antilogarithms that we have learned apply to the natural logarithms as well.

The function $y = e^x$ is called the **natural exponential function**. Its graph is shown below.

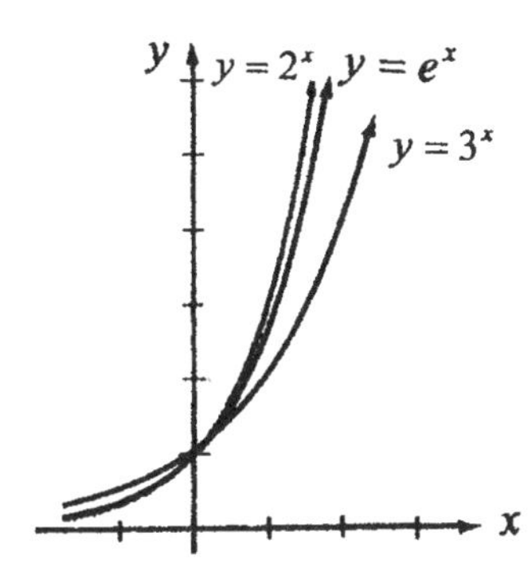

Most calculators have the key " e^x " to evaluate the exponential function for a given x-value and the key " LN " to evaluate the natural logarithms. The symbol for the antilogarithm of x to the base e is antiln x.

$$\text{If } \ln x = 0.6931, \text{ then } x = anti\ln 0.6931 = e^{0.6931} \approx 2.$$

Examples

1. $e^2 = (2.7182\cdots)^2 = 7.389$

2. $e^3 = (2.7182\cdots)^3 = 20.086$

3. $e^{1.2} = (2.7182\cdots)^{1.2} = 3.320$

4. $e^{5.8} = (2.7282\cdots)^{5.8} = 330.30$

5. $e^{-1.2} = \dfrac{1}{e^{1.2}} = \dfrac{1}{3.320} = 0.301$

6. $e^{-0.04} = \dfrac{1}{e^{0.04}} = \dfrac{1}{1.041} = 0.961$

7. $e^8 \cdot e^5 = e^{8+5} = e^{13}$

8. $3e^5 \cdot 4e^2 = 12e^{5+2} = 12e^7$

9. $\dfrac{12e^{10}}{2e^7} = 6e^{10-7} = 6e^3$

10. $\dfrac{12e^7}{2e^{10}} = 6e^{7-10} = 6e^{-3} = 6/e^3$

11. $(3e^5)^2 = 3^2 e^{5(2)} = 9e^{10}$

12. $(2e^{-4x})^3 = 2^3 e^{-4x\cdot 3} = 8e^{-12x} = 8/e^{12x}$

13. Given $\ln 2 = 0.693$, $\ln 3 = 1.099$, and $\ln 5 = 1.609$, Find:

$\ln 20 = \ln(4\times 5) = \ln 4 + \ln 5 = \ln 2^2 + \ln 5 = 2\ln 2 + \ln 5 = 2(0.693) + 1.609 = 2.995$.

$\ln 0.2 = \ln\frac{1}{5} = \ln 1 - \ln 5 = 0 - 1.609 = -1.609$.

$\ln 81 = \ln 3^4 = 4\ln 3 = 4(1.099) = 4.396$.

$$\log_{0.1} 3 = \frac{\ln 3}{\ln 0.1} = \frac{\ln 3}{\ln 1 - \ln 10} = \frac{\ln 3}{0 - (\ln 2 + \ln 5)} = \frac{1.099}{-(0.693 + 1.609)} = -0.477.$$

$$\log_{\sqrt{2}} \sqrt{5} = \frac{\ln\sqrt{5}}{\ln\sqrt{2}} = \frac{\frac{1}{2}\ln 5}{\frac{1}{2}\ln 2} = \frac{\ln 5}{\ln 2} = \frac{1.609}{0.693} = 2.322.$$

14. If $\ln N = 0.231$, find N.
Solution:
$N = anti\ln(0.231) = e^{0.231} = 1.26$.

15. If $\ln x = 0.492$, find x.
Solution:
$x = anti\ln(0.492) = e^{0.492} = 1.636$

2-40 Solving Logarithmic Equations

Equations that contains the terms $\log_a x$, where a is a positive real number, with $a \neq 1$, are called **logarithmic equations**.
There are different ways to solve various types of logarithmic equations.

Laws of Logarithms (Rules of Logarithms)

For all positive real numbers a, p, q with $a \neq 1$, and n is any real number:

1) $\log_a 1 = 0$ **2)** $\log_a a = 1$ **3)** $\log_a a^n = n$ **4)** $a^{\log_a n} = n$

5) $\log_a pq = \log_a p + \log_a q$ **6)** $\log_a \frac{p}{q} = \log_a p - \log_a q$ **7)** $\log_a p^n = n\log_a p$

8) Formula for Changing Base: $\log_a n = \dfrac{\log_b n}{\log_b a}$ where $b \neq 1$

In practice, we use $b = 10$. Therefore, $\log_a n = \dfrac{\log_{10} n}{\log_{10} a}$

To solve **logarithmic equations**, we apply the laws of logarithms.

1. If write the exponential form using the same base on both side is not possible, take the logarithms on both sides.
2. Change the equation to exponential form if it has only a single logarithm.
3. Combine it as a single logarithm if it has multiple logarithms.

To solve a logarithmic equation, we must check each solution in the original equation and discard solutions that are not permissible.

Examples

1. Solve $3^x = 20$.

Solution:

$$\log 3^x = \log 20$$
$$x\log 3 = \log 20$$
$$\therefore x = {}^{\log 20}/_{\log 3} = {}^{1.301}/_{0.477} = 2.727. \text{ Ans.}$$

2. Solve $\log_a 16 = 4$.

Solution:

$$a^4 = 16$$
$$a = 2 \text{ or } -2$$
($a = -2$ is not permissible.)
$$\therefore a = 2. \text{ Ans.}$$

3. Solve $\log_4 x + \log(x-6) = 2$.

Solution:

$$\log_4 x(x-6) = 2$$
$$x(x-6) = 4^2$$
$$x^2 - 6x - 16 = 0$$
$$(x-8)(x+2) = 0$$
$$x = 8 \text{ or } -2$$
($x = -2$ is not permissible.)
$$\therefore x = 8. \text{ Ans.}$$

4. Solve $\ln x + \ln(x-2) = 1$

Solution:

$$x(x-2) = e$$
$$x^2 - 2x - e = 0$$
$$x = \frac{2 \pm \sqrt{4+4e}}{2} = \frac{2 \pm 2\sqrt{1+e}}{2}$$
$$= 1 \pm \sqrt{1+e} \approx 2.3. \text{ Ans.}$$

2-41 Parametric Equations and Graphs

Parametric equations are the equations to express a relationship between x and y in terms of a third variable (called a **parameter**). We write:

$$x = f\ (\mathrm{t}) \quad \text{and} \quad y = f(\mathrm{t})$$

We say that the two equations form a set of parametric equations. Each value of t determines a point $(x,\ y)$ in the plane. The graph of the parametric equations is the set of points in the plane when t takes on all possible real numbers ($\mathrm{t} \in \mathrm{R}$).

Parametric equation is useful to trace the movement or orientation of the graph of the equation. Each point (x, y) on the graph is determined from a value chosen for t.

The parametric equations of a line:

$$\begin{cases} x = x_1 + at \\ y = y_1 + bt \end{cases}$$
. If $a = 0$, the line is vertical. If $b = 0$, the line is horizontal.

Example 1. Find the parametric equation of the line passing the points (–2, 5) and (4, 8).
Solution:

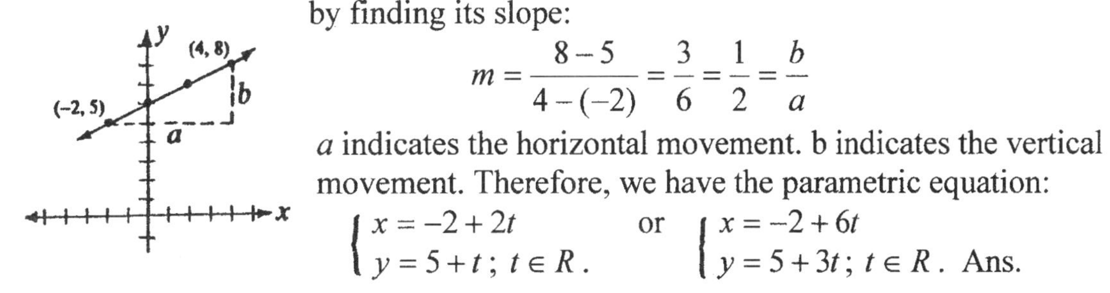

We select (–2, 5) as $(x_1,\ y_1)$ and trace the movement of the line by finding its slope:

$$m = \frac{8-5}{4-(-2)} = \frac{3}{6} = \frac{1}{2} = \frac{b}{a}$$

a indicates the horizontal movement. b indicates the vertical movement. Therefore, we have the parametric equation:

$$\begin{cases} x = -2 + 2t \\ y = 5 + t;\ t \in R. \end{cases} \quad \text{or} \quad \begin{cases} x = -2 + 6t \\ y = 5 + 3t;\ t \in R. \end{cases} \quad \text{Ans.}$$

When $t = 0$, we have (–2, 5); $t = 1$, we have (0, 6); $t = 2$, we have (2, 7); $t = 3$, we have (4, 8), and so on. Plotting and connecting these points, we have the graph of the line. There are many parametric equations for this line for $\frac{b}{a} = \frac{1}{2} = \frac{2}{4} = \frac{3}{6} = \frac{4}{8} = \cdots\cdots$.

To graph a pair of parametric equations, we eliminate the parameter to obtain the rectangular equation by solving for it in one of the equations and then substituting in the other. We may need to adjust the domain of the resulting rectangular equations to match the restriction of the original parametric equations.

Example 2. Graph the parametric equations $\begin{cases} x = t^2 \\ y = 2t^2 + 3;\ t \in R \end{cases}$

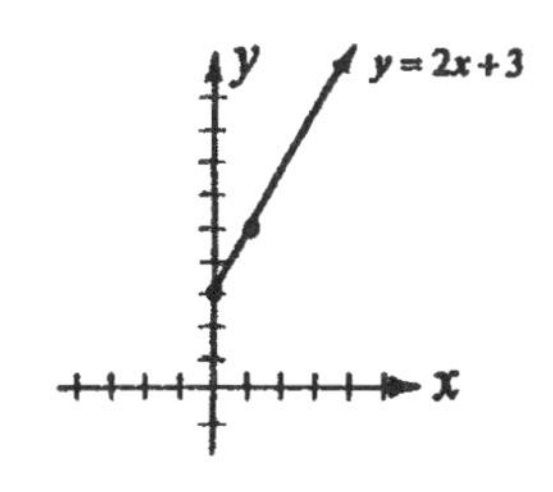

Solution: Since $t^2 \geq 0$, we have $x \geq 0$ and $y \geq 3$.

Substituting $x = t^2$ into the second equation, we have the rectangular equation:

$y = 2x + 3,\ x \geq 0$. Ans.

Orientation of the graph of a parametric equation

Parametric equation is very useful to provide the information about the orientation (the direction) of the graph. The graph is oriented by the increasing values of the parameter.
We indicate an orientation by placing arrowheads on or alongside the graph.

Example 3: Graph the parametric equations $x = t^2 - 4$, $y = 2t^2 - 2$; $t \in R$. Indicate the orientation.

Solution: $t^2 = x + 4$, $x \geq -4$

Substitute into the second equation:

$y = 2(x+4) - 2$

We have the rectangular equation:

$y = 2x + 6$, $x \geq -4$

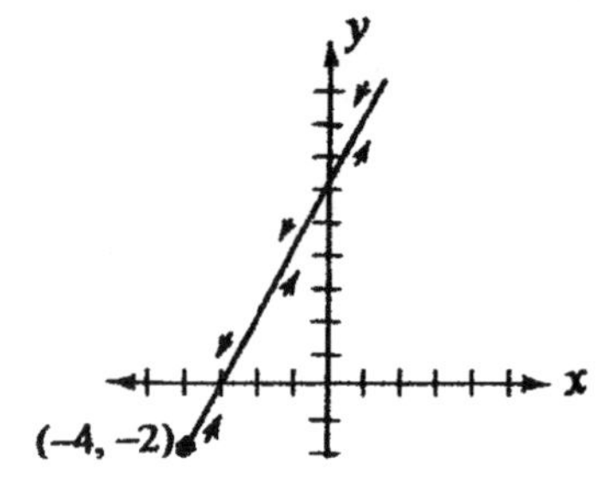

If $t = -3$, we have (5, 16). If $t = 0$, we have (–4, –2). If $t = 3$, we have (5, 16).
As t increases from $-\infty$ to 0, $p(x, y)$ slants downward.
As t increases from 0 to ∞, $p(x, y)$ slants upward.
(Hint: The infinite symbol, ∞, indicates the value continues without end.)

Example 4. Graph the parametric equations $x = 4t^2 - 5$, $y = 2t + 3$; $t \in R$. Indicate the orientation.

Solution:

$y = 2t + 3$, we have: $t = \frac{y-3}{2}$

Substitute into the first equation: $x = 4(\frac{y-3}{2})^2 - 5$

We have: $(y-3)^2 = x + 5$; $x \geq -5$

or $y = \pm\sqrt{x+5} + 3$; $x \geq -5$

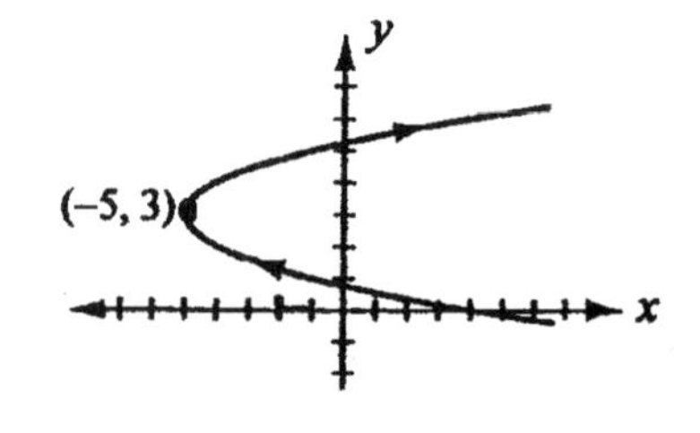

The graph is the entire parabola. The arrowheads on the graph indicate the direction of the curve as t increases from $-\infty$ to $+\infty$.

In a toy store, a five-year old girl wants her father to buy her an expensive toy.
Father: No. it is too expensive.
Girl: Dad, I like it, Honey and Sweetheart.
Father: Why are you calling me " Honey and Sweetheart" ?
Girl: Because whenever mom wants to buy a new dress, she calls you Honey, or Sweetheart.

2-42 Limits and Continuity

The concept of limits and continuity is very important to the study of calculus. Suppose we sketch the graph of the following function by direct substitution:

$$f(x) = \frac{x^2 - 1}{x - 1}, \quad x \neq 1$$

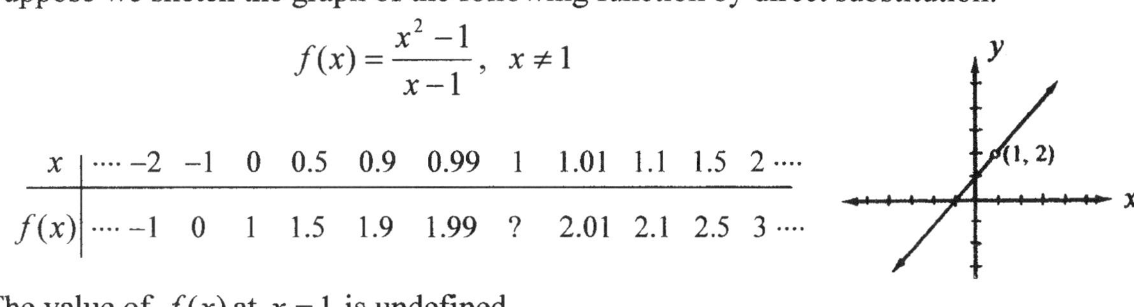

x	.... −2	−1	0	0.5	0.9	0.99	1	1.01	1.1	1.5	2 ....
$f(x)$	.... −1	0	1	1.5	1.9	1.99	?	2.01	2.1	2.5	3 ....

The value of $f(x)$ at $x = 1$ is undefined.
Since x cannot equal to 1, we are not sure what the value of $f(x)$ is at $x = 1$.
The graph of $f(x)$ is a straight line that has a hole at the point (1, 2).
The point (1, 2) is not part of the graph.

To find the value of $f(x)$ near the undefined point at $x = 1$, we could use the values of x that approaches 1 from the left, and the values of x from the right. We estimate that the value of $f(x)$ moves close to 2 when x approaches 1 from either the right or the left. We say that " the limit of $f(x)$ is 2 as x approaches 1 ", and we write the result in limit notation: $\lim_{x \to 1} f(x) = 2$

Definition of a limit: If the values of a function $f(x)$ approach or equal L as the values of x approaches c, we say that $f(x)$ has limit L as x approaches c.
We write: $\lim_{x \to c} f(x) = L$.
Read " The limit of $f(x)$ equals L as x approaches c.

Some functions do not have a limit as $x \to c$. If a limit of a function exists, it is unique. In other words, a function cannot have two different limits as $x \to c$.

By reducing the function, we have $f(x) = \frac{x^2 - 1}{x - 1} = \frac{(x+1)\cancel{(x-1)}}{\cancel{x-1}} = x + 1, \; x \neq 1$.

Therefore, for all points other than $x = 1$, $\lim_{x \to 1} \frac{x^2 - 1}{x - 1} = \lim_{x \to 1}(x + 1) = 2$.

We use the following ways to evaluate the limit of $f(x)$ as $x \to c$:

1. Evaluated by direct substitution.
2. If $f(x)$ is undefined at $x = c$, we try to reduce $f(x)$ and find a new function that agrees with $f(x)$ for all x other than c.

One-side limits: $\lim_{x \to c^+} f(x) = L$ represents the limit of $f(x)$ from the right. It is the limit of $f(x)$ as x approaches c from values greater than c.

$\lim_{x \to c^-} f(x) = L$ represents the limit of $f(x)$ from the left. It is the limit of $f(x)$ as x approaches c from values less than c.

$\lim_{x \to c} f(x) = L$ **exists only and only if** $\lim_{x \to c^+} f(x) = L$ **and** $\lim_{x \to c^-} f(x) = L$.

Infinite Limits: $\lim_{x \to c} f(x) = \infty$ means that $f(x)$ increases without bound as x approaches c.

$\lim_{x \to c} f(x) = -\infty$ means that $f(x)$ decreases without bound as x approaches c.

Since infinite " ∞ " is not a number, we may say that "the limit does not exist".

Limits of the greatest integer functions: If n is an integer, then $\lim_{x \to n} [x]$ does not exist.

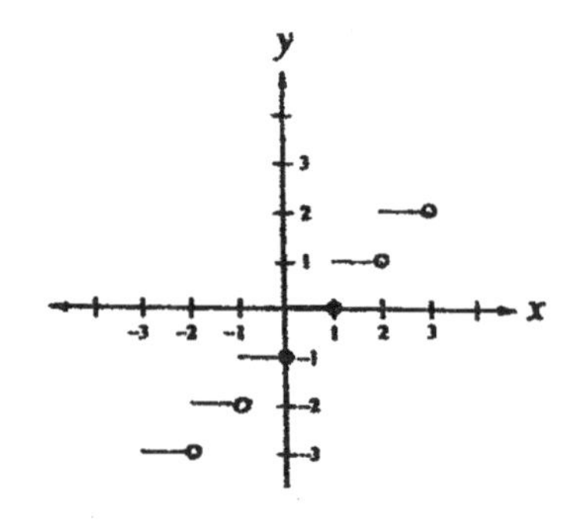

Example: $\lim_{x \to 2^+} [x] = \lim_{x \to 2^+} 2 = 2$

$\lim_{x \to 2^-} [x] = \lim_{x \to 2^-} 1 = 1$

Therefore, $\lim_{x \to 2} [x]$ does not exist.

If the degree of $p(x)$ is greater than the degree of $q(x)$, then $\lim_{x \to \infty} \frac{p(x)}{q(x)} = \pm \infty$.

If the degree of $p(x)$ is less than the degree of $q(x)$, then $\lim_{x \to \infty} \frac{p(x)}{q(x)} = 0$.

To find the limit of a quotient of polynomials, we divide the highest power of x and use the fact that $\lim_{x \to \infty} \frac{1}{x} = 0$.

$$\lim_{x \to \infty} \frac{-x^3}{2 - x^2} = \lim_{x \to \infty} \frac{-1}{\frac{2}{x^3} - \frac{1}{x}} = \frac{-1}{0 - 0} = \frac{-1}{0} = -\infty.$$

The limit does not exist in this example.

A high school boy is playing the violin.
His girlfriend sits next to him.
Boy: Do you like the song I played ?
Girl: Yes. It makes me miss my father.
Boy: Was your father a musician ?
Girl: No. He was a sawmill worker.

Examples

1. $\lim_{x\to 2} 3 = 3$ **2.** $\lim_{x\to 2} x^3 = 2^3 = 8$ **3.** $\lim_{x\to 1}(x^2+x+2) = 1^2+1+2 = 4$ **4.** $\lim_{x\to -2}|x| = 2$

5. $\lim_{x\to 1}\dfrac{x^2+x-2}{x-1} = \lim_{x\to 1}\dfrac{(x+2)(x-1)}{x-1} = \lim_{x\to 1}(x+2) = 1+2 = 3$

6. $\lim_{x\to -4}\dfrac{x^2+2x-8}{x+4} = \lim_{x\to -4}\dfrac{(x+4)(x-2)}{x+4} = \lim_{x\to -4}(x-2) = -4-2 = -6$

7. $\lim_{x\to 2}\dfrac{x-2}{x^3-2x^2+3x-6} = \lim_{x\to 2}\dfrac{x-2}{x^2(x-2)+3(x-2)} = \lim_{x\to 2}\dfrac{\cancel{x-2}}{\cancel{(x-2)}(x^2+3)} = \lim_{x\to 2}\dfrac{1}{x^2+3} = \dfrac{1}{7}$

8. $\lim_{x\to 0}\dfrac{x}{x^3} = \lim_{x\to 0}\dfrac{1}{x^2} = \dfrac{1}{0} = \infty$ (undefined). The limit does not exist.

$f(x)$ increases without bound as x approaches 0 from either the right or the left. It has no limit as $x \to 0$.

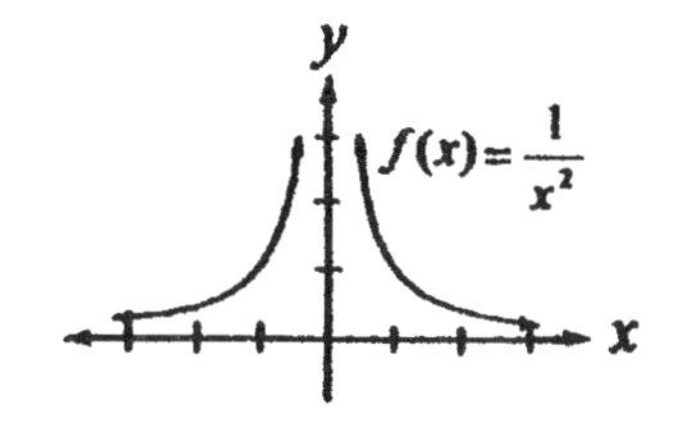

9. $\lim_{x\to 0}\dfrac{\sqrt{x+4}-2}{x} = \lim_{x\to 0}\dfrac{(\sqrt{x+4}-2)(\sqrt{x+4}+2)}{x(\sqrt{x+4}+2)} = \lim_{x\to 0}\dfrac{(x+4)-4}{x(\sqrt{x+4}+2)} = \lim_{x\to 0}\dfrac{\cancel{x}}{\cancel{x}(\sqrt{x+4}+2)}$

$= \lim_{x\to 0}\dfrac{1}{\sqrt{x+4}+2} = \dfrac{1}{\sqrt{0+4}+2} = \dfrac{1}{4}$

10. $\lim_{x\to 2^+}\sqrt{x-2} = \sqrt{2-2} = 0$. It is a parabola $y^2 = x-2$ with vertex (2, 0).

11. Evaluate $\lim_{x\to 0}\dfrac{|x|}{x}$.

Solution: $\lim_{x\to 0^+}\dfrac{|x|}{x} = \lim_{x\to 0^+}\dfrac{x}{x} = \lim_{x\to 0^+} 1 = 1$

$\lim_{x\to 0^-}\dfrac{|x|}{x} = \lim_{x\to 0^-}\dfrac{-x}{x} = \lim_{x\to 0^-}(-1) = -1$

Therefore, the limit does not exist.

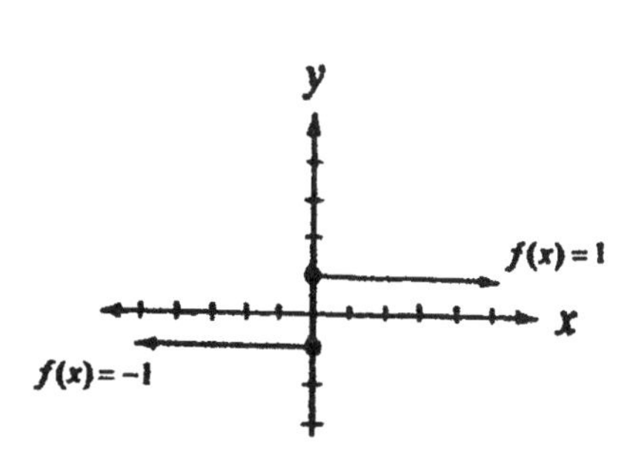

12. Evaluate $\lim_{x\to 1} f(x)$, Where $f(x) = \begin{cases} 2x^2, & -1 \le x \le 1 \\ -x+3, & 1 \le x \le 3 \end{cases}$

Solution:

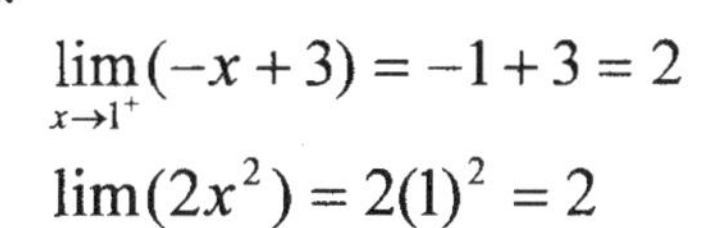

$\lim_{x\to 1^+}(-x+3) = -1+3 = 2$

$\lim_{x\to 1^-}(2x^2) = 2(1)^2 = 2$

Therefore, $\lim_{x\to 1} f(x) = 2$

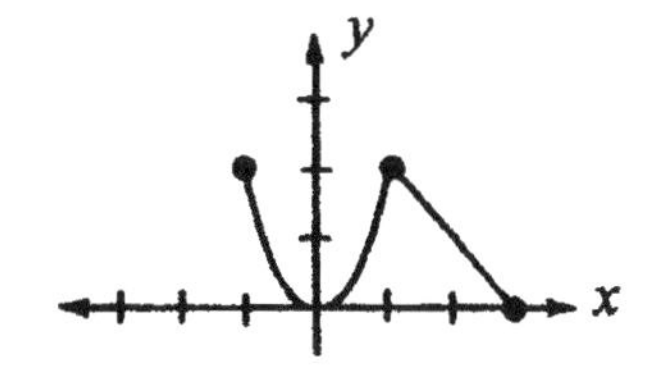

Continuity

A function $f(x)$ is said to be continuous at $x = c$ if its graph is unbroken (no holes) at $x = c$. If a function is not continuous at $x = c$, we say that it is discontinuous at $x = c$.

A function $f(x)$ is said to be continuous on the closed interval $[a, b]$ if it is continuous on the interval $a \le x \le b$.

A function $f(x)$ that is continuous on the entire real numbers $(-\infty, +\infty)$ is called a continuous function.

Definition of Continuity: A function $f(x)$ is said to be continuous at $x = c$ if the following three conditions are satisfied:

1. $f(x)$ is defined. **2.** $\lim\limits_{x \to c} f(x)$ exists. **3.** $\lim\limits_{x \to c} f(x) = f(c)$.

Continuity on a closed interval: If a function $f(x)$ is continuous on a closed interval $[a, b]$, we say that it is continuous at every interior point, continuous on the right at a, and continuous on the left at b.

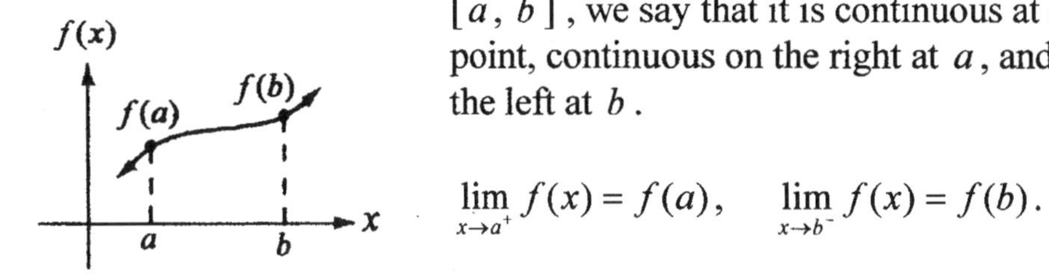

$\lim\limits_{x \to a^+} f(x) = f(a)$, $\lim\limits_{x \to b^-} f(x) = f(b)$.

A discontinuity of a function $f(x)$ at $x = c$ is called removable if $f(x)$ can be made continuous by redefining $f(x)$ at $x = c$.

Intermediate Value Theorem: If $f(x)$ is continuous on the closed interval $[a, b]$ and k is any number between $f(a)$ and $f(b)$, then there is at least one number c between a and b such that $f(c) = k$.

The Intermediate Value Theorem states that if a function is continuous on a closed interval, there is no hole in its graph.

If there are two numbers a and b in the function $f(x)$, such that $f(a)$ is negative and $f(b)$ is positive, then the function $f(x)$ must have at least one number c between a and b such that $f(c) = 0$. Its graph must cross the $x-axis$ at $x = c$.
The equation $f(x) = 0$ has a real root at $x = c$.

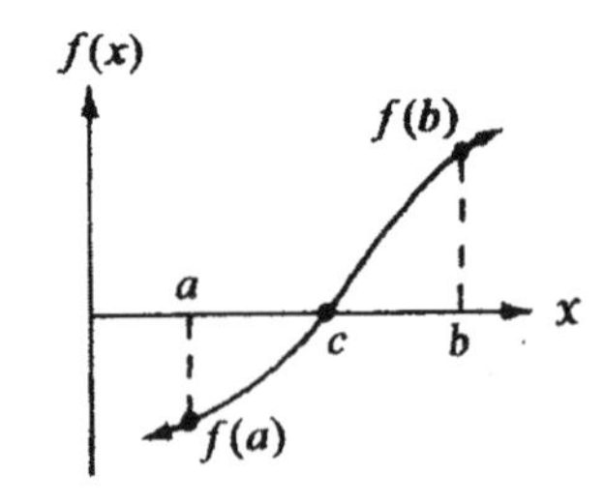

2-43 Slope and Derivative of a Curve

In basic Algebra, we learned how to find the slope of a straight line. The slope of a straight line is used to describe the rate of change when a line rises or falls, $\Delta y/\Delta x$.
The slope is the same at every point on a straight line (a linear function).
The slope is different from point to point on a curve (a non-linear function).

There are many applications of rates of change in science and our real life.
For example, the scientists and astronauts want to calculate the velocity and acceleration of a space shuttle at every point and time on its elliptical orbit. First, we need to find the slope of the elliptical orbit at every point.

In Calculus, we start to find the slope (rates of change) of the curve at any point.
The process to find the slope of the curve at a point is called **differentiation.**

The slope of the curve at a point is the slope of the **tangent line** of the curve at the point.
In the process of differentiation, we derive the slope formula of the function $y = f'(x)$.
The derived function is called the **derivative** of $y = f(x)$ at x.

To find the slope of the tangent line of the curve at a point, we use the following **Newton's Four-Step Process of Differentiation:**

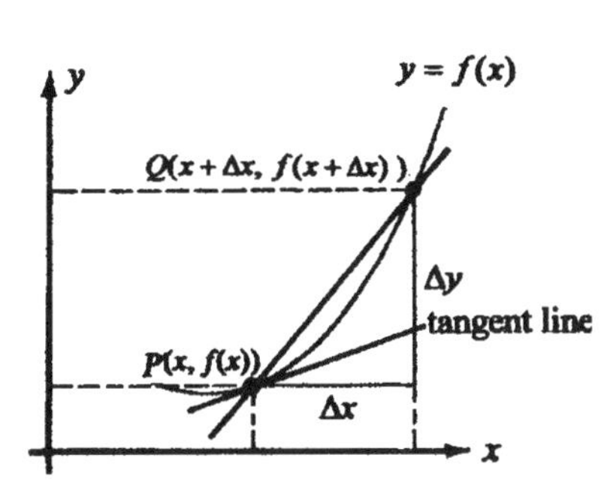

1. Evaluate: $f(x+\Delta x)$
2. Subtract: $\Delta y = f(x+\Delta x) - f(x)$
3. Divide by Δx : $\dfrac{\Delta y}{\Delta x} = \dfrac{f(x+\Delta x) - f(x)}{\Delta x}$
4. Take the limit: m (slope) $= \lim\limits_{\Delta x \to 0} \dfrac{f(x+\Delta x) - f(x)}{\Delta x}$

Δx and Δy are **changes** in x and y, or **increments** of x and y. Symbol Δ is read "Delta".

$\Delta x/\Delta y$ is the **average rate of change** of $f(x)$ from point P to point Q. It is the slope of the secant line PQ passing through P (the point of tangent) and any nearby point Q. Therefore, the slope (m) of the secant line PQ is given by:

m(secant line) $= \dfrac{\Delta y}{\Delta x} = \dfrac{f(x+\Delta x) - f(x)}{\Delta x}$. It is called **difference quotient** for $f(x)$ at x.

When the point Q moves closer to the point P, we obtain better approximation of the slope of the tangent line at P. When Δx approaches 0 ($\Delta x \to 0$), we can find the exact slope of the tangent line at point P.

The expression in Step 4 is called the derivative of the function $y = f(x)$ at x.
The notations to describe the derivative are y', $f'(x)$, or $\dfrac{dy}{dx}$.

Definition of Derivative of a Function:

The derivative of the function $y = f(x)$ at x is given by:

$$y' = \frac{dy}{dx} = \lim_{\Delta x \to 0} \frac{\Delta y}{\Delta x} = \lim_{\Delta x \to 0} \frac{f(x+\Delta x) - f(x)}{\Delta x} \text{ if the limit exists.}$$

The other form to express the derivative of a function:

The derivative of the function $y = f(x)$ at c is given by:

$$y' = \frac{dy}{dx} = \lim_{\Delta x \to 0} \frac{\Delta y}{\Delta x} = \lim_{x \to c} \frac{f(x) - f(c)}{x - c} \text{ if the limit exists.}$$

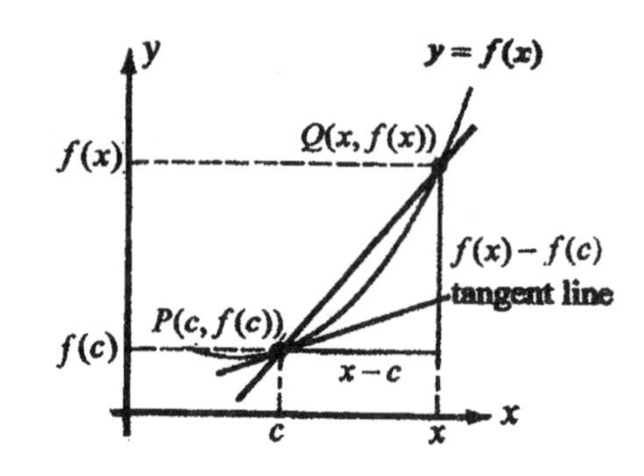

Examples

1. Find the slope of the line $y = 3x - 5$ at the point (2, 1).

Solution:

$$m = \lim_{\Delta x \to 0} \frac{f(x+\Delta x) - f(x)}{\Delta x} = \lim_{\Delta x \to 0} \frac{[3(x+\Delta x) - 5) - (3x - 5)]}{\Delta x} = \lim_{\Delta x \to 0} \frac{3\Delta x}{\Delta x} = \lim_{\Delta x \to 0} 3 = 3. \text{ Ans.}$$

2. Find the slope of the function $y = x^2$ at the point (1, 1) and the point (2,4).

Solution:

$$m = \lim_{\Delta x \to 0} \frac{f(x+\Delta x) - f(x)}{\Delta x} = \lim_{\Delta x \to 0} \frac{(x+\Delta x)^2 - x^2}{\Delta x} = \lim_{\Delta x \to} \frac{x^2 + 2 \cdot x \cdot \Delta x + (\Delta x)^2 - x^2}{\Delta x}$$

$$= \lim_{\Delta x \to 0} \frac{2 \cdot x \cdot \Delta x + (\Delta x)^{2\,1}}{\Delta x} = \lim_{\Delta x \to 0} (2x + \Delta x) = 2x$$

Therefore, the slope at the point (1, 1): $m = 2(1) = 2$. At (2, 4): $m = 2(2) = 4$. Ans.

Applying the Four-Step Process of differentiation, we can prove and obtain the following **General Formulas of Differentiation**:

1. If c is a constant, then $\frac{d}{dx}[c] = 0$	**4.** $\frac{d}{dx}[f(x) \pm g(x)] = f'(x) \pm g'(x)$
2. $\frac{d}{dx}[x^n] = nx^{n-1}$	**5.** $\frac{d}{dx}[f(x)g(x)] = f(x)g'(x) + g(x)f'(x)$
3. $\frac{d}{dx}[cf(x)] = cf'(x)$	**6.** $\frac{d}{dx}\left[\frac{f(x)}{g(x)}\right] = \frac{g(x)f'(x) - f(x)g'(x)}{[g(x)]^2}$

Fortunately, we use the above formulas to calculate the derivatives directly and simply without the tedious process by using the limit definition of the derivative.

Examples: **1.** $\frac{d}{dx}(2) = 0$ **2.** $\frac{d}{dx}(12) = 0$ **3.** $\frac{d}{dx}(x) = 1x^{1-1} = x^0 = 1$ **4.** $\frac{d}{dx}x^2 = 2x^{2-1} = 2x$

5. $\frac{d}{dx}x^5 = 5x^{5-1} = 5x^4$ **6.** $\frac{d}{dx}(4x^{10}) = 4 \cdot \frac{d}{dx}(x^{10}) = 4 \cdot 10x^{10-1} = 40x^9$

Examples

7. $\frac{d}{dx}(x^{-3}) = -3 \cdot x^{-3-1} = -3x^{-4} = -\frac{3}{x^4}$

8. $\frac{d}{dx}\left(\frac{2}{x}\right) = \frac{d}{dx}(2x^{-1}) = 2\frac{d}{dx}(x^{-1}) = 2(-1)x^{-2} = -\frac{2}{x^2}$

9. $\frac{d}{dx}(-\frac{3}{5x^4}) = -\frac{3}{5}\frac{d}{dx}(x^{-4}) = -\frac{3}{5} \cdot -4x^{-5} = \frac{12}{5x^5}$

10. $\frac{d}{dx}(3\pi\, x^4) = 3\pi \frac{d}{dx}(x^4) = 3\pi \cdot 4x^3 = 12\pi\, x^3$

11. $\frac{d}{dx}(\sqrt{x}) = \frac{d}{dx}(x^{1/2}) = \frac{1}{2}x^{-1/2} = \frac{1}{2\sqrt{x}}$

12. $\frac{d}{dx}(x^{-2/3}) = -\frac{2}{3} \cdot x^{-5/3} = -\frac{2}{3x^{5/3}}$

13. $\frac{d}{dx}(5x^2 + 3x - 6) = \frac{d}{dx}(5x^2) + \frac{d}{dx}(3x) + \frac{d}{dx}(-6) = 10x + 3 + 0 = 10x + 3$

14.
$$\frac{d}{dx}[(2x^2 - 3x)(4x+5)] = (2x^2 - 3x)\frac{d}{dx}(4x+5) + (4x+5)\frac{d}{dx}(2x^2 - 3x)$$
$$= (2x^2 - 3x)(4) + (4x+5)(4x-3) = (8x^2 - 12x) + (16x^2 + 8x - 15)$$
$$= 24x^2 - 4x - 15$$

15.
$$\frac{d}{dx}[(2 + x^{-1})(x-2)] = (2 + x^{-1})\frac{d}{dx}(x-2) + (x-2)\frac{d}{dx}(2 + x^{-1})$$
$$= (2 + x^{-1})(1) + (x-2)(-x^{-2})$$
$$= 2 + \frac{1}{x} - \frac{x-2}{x^2} = \frac{2x^2 + x - x + 2}{x^2} = \frac{2x^2 + 2}{x^2}$$

16.
$$\frac{d}{dx}\left(\frac{x-1}{x+1}\right) = \frac{(x+1)\frac{d}{dx}(x-1) - (x-1)\frac{d}{dx}(x+1)}{(x+1)^2} = \frac{(x+1)(1) - (x-1)(1)}{(x+1)^2} = \frac{2}{(x+1)^2}$$

17.
$$\frac{d}{dx}\left(\frac{x^2 + 3x - 2}{2x - 1}\right) = \frac{(2x-1)\frac{d}{dx}(x^2 + 3x - 2) - (x^2 + 3x - 2)\frac{d}{dx}(2x-1)}{(2x-1)^2}$$
$$= \frac{(2x-1)(2x+3) - (x^2 + 3x - 2)(2)}{(2x-1)^2} = \frac{4x^2 + 4x - 3 - 2x^2 - 6x + 4}{(2x-1)^2}$$
$$= \frac{2x^2 - 2x + 1}{(2x-1)^2}$$

18. Find the slope of the line $y = 3x - 5$ at the point (2, 1).
Solution:
$$m = \frac{d}{dx}(3x - 5) = 3.$$ Ans.

19. Find the slope of the function $f(x) = x^3 - 2x$ at the point (1, −1).
Solution:
$$m = \frac{d}{dx}(x^3 - 2x) = 3x^2 - 2 = 3(1^2) - 2 = 1.$$ Ans.

Notes

SAMPLE QUESTIONS

CHAPTER 2: ALGEBRA and FUNCTIONS

1. A rectangle has width x and length y. Its perimeter is 64. Write an equation to represent its perimeter.
Solution:

$2(x+y)=64$. Ans.

2. A rectangle has width 20 and length x. Its area is 210. Write an equation to represent its area.
Solution:

$20x = 210$. Ans.

3. John rented a car for one day. The car rental costs $60 plus 0.35 a mile. He paid $260 when he returned the car. The mileage that he traveled is x. Write an equation to represent the total rental charges.
Solution:

$60 + 0.35x = 260$. Ans.

4. For a field trip to the museum, each bus costs b dollars and holds 45 passengers. Two teachers must be on each bus. Each student and each teacher must pay x dollars for admission to the museum. Write an expression for the total cost, in dollars, for 235 students to the field trip.
Solution:

Each bus can take 43 students and 2 teachers.
We needs at least 6 buses and 12 teachers.
Total passengers are 247.
The total cost for the trip is: $247x + 6b$. Ans.

5. If $f(x) = 2x+1$, $g(x) = \frac{1}{2}x - \frac{1}{2}$,
Find $f(g(x))$ and $g(f(x))$.
Solution:

$f(g(x)) = 2(\frac{1}{2}x - \frac{1}{2}) + 1 = x - 1 + 1 = x$.

$g(f(x)) = \frac{1}{2}(2x+1) - \frac{1}{2} = x + \frac{1}{2} - \frac{1}{2} = x$.

Ans.

6. If $f(g(x)) = x$ and $f(x) = 2x+1$, Find $g(x)$.
Solution:

$f(g(x)) = 2g(x) + 1$

$f(g(x)) = x$

$\therefore\ 2g(x) + 1 = x,\ \ g(x) = \frac{1}{2}x - \frac{1}{2}$. Ans.

7. The profit, in dollars, of a certain product from now to next few years is estimated by using the function p, defined by $p(x) = -0.2x^3 + 0.54x^2 + 2.6x + 5.9$, where x is the number of years from the beginning of this year. What is the maximum profit estimated for the product during next few years ?
Solution:

We can use a graphing calculator to graph the function and find the maximum.
(Read the book "A-Plus Notes for Algebra Page 186 and Page 192)

Graph the function and change the view window:
Press: **WINDOW**

$x_{min} = -2,\ \ x_{max} = 10,\ \ x_{scl} = 1$

$y_{min} = -2,\ \ y_{max} = 20,\ \ y_{scl} = 1$

Calculate the maximum"
Press: **2nd → CALC → 4: maximum →ENTER**

Move the cursor on the maximum point: $p_{max.} \approx \$13.19$ at $x \approx 3.11$

SAMPLE QUESTIONS

CHAPTER 2: ALGEBRA and FUNCTIONS

1. Given $f(x) = 3x^2 - 4$. Find the value of each function.

a. $f(1)$ b. $f(-3)$ c. $f(2m)$ d. $2f(m)$

Solution:

a. $f(1) = 3(1)^2 - 4 = -1$. Ans.

b. $f(3) = 3(-3)^2 - 4 = 23$. Ans.

c. $f(2m) = 3(2m)^2 - 4 = 12m^2 - 4$. Ans.

d. $2f(m) = 2(3m^2 - 4) = 6m^2 - 8$. Ans.

2. If $f(x)$ is a linear function, $f(0) = a$, $f(1) = 15$, and $f(2) = b$, what is the value of $a + b$?

Solution:

Let $f(x) = cx + d$

We have $f(0) = d = a$

$f(1) = c + d = 15$

$f(2) = 2c + d = b$

$a + b = d + 2c + d = 2c + 2d$

$= 2(c + d) = 2(15) = 30$. Ans.

3. Solve $\frac{r(r^2-1)+r^2-5}{r^2-1} = \frac{r^2+r+2}{r+1}$.

Solution:

$\frac{r^3-r+r^2-5}{(r+1)(r-1)} = \frac{r^2+r+2}{r+1}$, $r^3+r^2-r-5 = (r-1)(r^2+r+2)$

$r^3 - r + r^2 - 5 = r^3 + r^2 + 2r - r^2 - r - 2$

$r^2 - 5 = 2r - 2$

$r^2 - 2r - 3 = 0$

$(r-3)(r+1) = 0$

$r = 3$ | $r = -1$ Not permissible

Ans: $r = 3$

4. If a line passes through the origin and is perpendicular to the line $y = -\frac{1}{3}x + b$, where b is a constant. The point of intersection of these two lines is $(a, a-3)$. What is the value of a ?

Solution:

The slope of the line is 3, and passes through the origin (0, 0).
The equation of the line is $y = 3x$.
The line passes through the point $(a, a-3)$

$a - 3 = 3a$

$\therefore a = -\frac{3}{2}$. Ans.

SAMPLE QUESTIONS

CHAPTER 2: ALGEBRA and FUNCTIONS

(Newton's Law of Free-Falling Objects)
Isaac Newton (1642-1727, Creator of Calculus)

1. The distance (d) in feet that an object falls in t seconds is given by the formula $d = 16t^2$. Find the distance of an object falls in 5 seconds.
Solution: $d = 16t^2 = 16 \cdot 5^2 = 16 \cdot 25 = 400$ feet. Ans.

2. The height (h) in feet that an object falls from its initial height h_0 in t seconds is given by the formula $h = h_0 - 16t^2$.

a) If the initial height of a ball is 1,000 feet, find the height of the ball drops in 5 seconds.
b) How many seconds will it take the ball to reach the ground ?

Solution:

a) $h = h_0 - 16t^2 = 1000 - 16 \cdot 5^2 = 1000 - 16 \cdot 25 = 1000 - 400 = 600$ feet. Ans.

b)
$$h = 1000 - 16t^2$$
$$0 = 1000 - 16t^2$$
$$16t^2 = 1000$$
$$t^2 = \tfrac{1000}{16} \qquad \therefore t = \sqrt{\tfrac{1000}{16}} = \sqrt{62.5} \approx 7.91 \text{ seconds. Ans.}$$

3. A rocket is fired upward with an initial speed of 160 feet per second. The height (h) of the rocket is given by the formula $h = vt - 16t^2$, where v is the initial speed and t is the time in seconds.

a) Find the height of the rocket reaches in 3 seconds after being fired upward.
b) When will the rocket be at a height of 384 feet ?
c) How many seconds will it take to reach the ground again ?
d) What will be the maximum height the rocket reaches ?

Solution:

a) $h = vt - 16t^2 = 160(3) - 16(3)^2 = 480 - 144 = 336$ feet. Ans.

b)
$$h = vt - 16t^2$$
$$384 = 160t - 16t^2$$
$$16t^2 - 160t + 384 = 0$$
Divide each side by 16:
$$t^2 - 10t + 24 = 0$$
$$(t-4)(t-6) = 0$$
$\therefore t =$ 4 and 6 seconds. Ans.

c)
$$h = vt - 16t^2$$
$$h = 160t - 16t^2$$
Let $h = 0$
$$0 = 160t - 16t^2$$
$$16t^2 - 160t = 0$$
Factor out the GCF:
$16t(t-10) = 0 \quad \therefore t = 10$ seconds. Ans.

d) It is the time halfway between the starting time and the time to reach the ground again. Therefore, it reaches the maximum height when $t = 5$.
$\therefore h = vt - 16t^2 = 160(5) - 16(5)^2 = 800 - 400 = 400$ feet. Ans.

····· **Continued on next page** ·····

4. A ball is thrown upward from the top of a tower 96 feet high with an initial upward speed of 80 feet per second. The height of the ball is given by the formula $h = 96 + vt - 16t^2$, where v is the initial speed and t is the time in seconds. How many seconds will it take the ball to reach the ground again ?
Solution:

$h = 96 + vt - 16t^2$

$h = 96 + 80t - 16t^2$

Let $h = 0$

$0 = 96 + 80t - 16t^2$

$16t^2 - 80t - 96 = 0$

$t^2 - 5t - 6 = 0$

$(t+1)(t-6) = 0$

$\therefore t = 6$ seconds. Ans.

5. An arrow is shot upward with an initial speed of 58.8 meter per second. The height of the arrow is given by the formula $h = vt - 4.9t^2$, where v is the initial speed and t is the time in seconds. How many seconds will it take the arrow to reach the ground again ? What will be the maximum height the arrow reaches ?
Solution:

$h = vt - 4.9t^2 = 58.8t - 4.9t^2$

Let $h = 0$, $58.8t - 4.9t^2 = 0$

$4.9t(12 - t) = 0$

$\therefore t = 12$ seconds. Ans.

$\therefore$ Max. height $h = 58.8(6) - 4.9(6)^2$

$= 352.8 - 176.4 = 176.4$ meters. Ans.

6. A rocket is fired upward from the top of a tower 20 feet. The height of the rocket after t seconds is given by the function $h(t) = c - (d - 2t)^2$, where c and d are positive numbers. The rocket reaches the maximum height of 120 feet at $t = 5$. What will be the height, in feet, of the rocket at $t = 1$?

Solution:

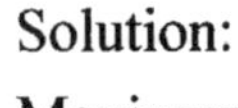

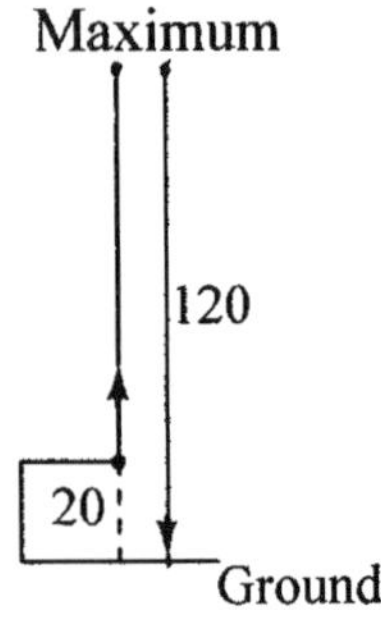

$h(t) = c - (d - 2t)^2$

First, we need to find the values of c and d from the given information.

At $t = 0$, $h = 20$, $h(0) = c - d^2 = 20$

At $t = 5$, $h = 120$, $h(5) = c - (d - 10)^2 = 120$

$c - (d^2 - 20d + 100) = 120$

We have: $c - d^2 + 20d - 100 = 120$ and $c - d^2 = 20$

$20 + 20d - 100 = 120$

$20d = 200 \quad \therefore d = 10$

$c - d^2 = 20, \quad c - 100 = 20 \quad \therefore c = 120$

We have the function:

$h(t) = 120 - (10 - 2t)^2$

At $t = 1$, $h(1) = 120 - (10 - 2)^2 = 56$ feet. Ans.

SAMPLES QUESTIONS

CHAPTER 2: ALGEBRA and FUNCTIONS

1. A rectangle has width x and length y. Its perimeter is 64. Write an equation to represent its perimeter.
   Solution:
   $$2(x+y)=64.\ \text{Ans.}$$

2. A rectangle has width 20 and length x. Its area is 210. Write an equation to represent its area.
   Solution:
   $$20x=210.\ \text{Ans.}$$

3. John rented a car for one day. The car rental costs \$60 plus 0.35 a mile. He paid \$260 when he returned the car. The mileage that he traveled is x. Write an equation to represent the total rental charges.
   Solution:
   $$60+0.35x=260.\ \text{Ans.}$$

4. For a field trip to the museum, each bus costs b dollars and holds 45 passengers. Two teachers must be on each bus. Each student and each teacher must pay x dollars for admission to the museum. Write an expression for the total cost, in dollars, for 235 students to the field trip.
   Solution:
   Each bus can take 43 students and 2 teachers.
   We needs at least 6 buses and 12 teachers.
   Total passengers are 247.
   The total cost for the trip is: $247x+6b$. Ans.

5. If $f(x)=2x+1$, $g(x)=\frac{1}{2}x-\frac{1}{2}$, Find $f(g(x))$ and $g(f(x))$.
   Solution:
   $$f(g(x))=2(\tfrac{1}{2}x-\tfrac{1}{2})+1=x-1+1=x.$$
   $$g(f(x))=\tfrac{1}{2}(2x+1)-\tfrac{1}{2}=x+\tfrac{1}{2}-\tfrac{1}{2}=x.$$
   Ans.

6. If $f(g(x))=x$ and $f(x)=2x+1$, Find $g(x)$.
   Solution:
   $$f(g(x))=2g(x)+1$$
   $$f(g(x))=x$$
   $$\therefore\ 2g(x)+1=x,\ g(x)=\tfrac{1}{2}x-\tfrac{1}{2}.\ \text{Ans.}$$

7. The profit, in dollars, of a certain product from now to next few years is estimated by using the function p, defined by $p(x)=-0.2x^3+0.54x^2+2.6x+5.9$, where x is the number of years from the beginning of this year. What is the maximum profit estimated for the product during next few years ?
   Solution:
   We can use a graphing calculator to graph the function and find the maximum.
   (Read the book "A-Plus Notes for Algebra Page 186 and Page 192)

   Graph the function and change the view window:
   Press: **WINDOW**
   $$x_{\min}=-2,\quad x_{\max}=10,\quad x_{scl}=1$$
   $$y_{\min}=-2,\quad y_{\max}=20,\quad y_{scl}=1$$
   Calculate the maximum"
   Press: **2nd → CALC → 4: maximum →ENTER**

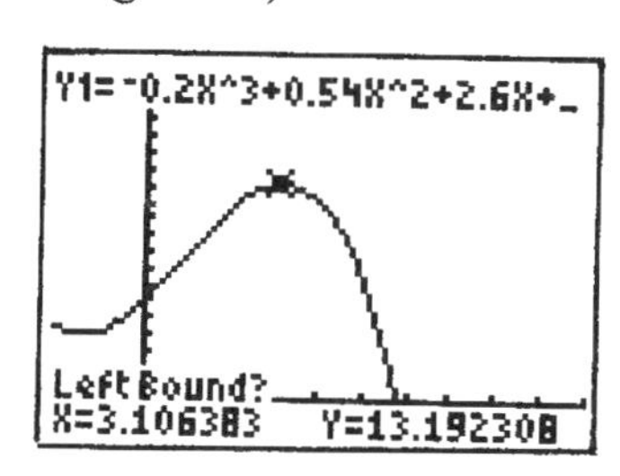

   Move the cursor on the maximum point: $p_{\max}\approx \$13.19$ at $x\approx 3.11$

SAMPLE QUESTIONS

CHAPTER 2: ALGEBRA and FUNCTIONS

DOMAIN and RANGE

The domain of a function $y = f(x)$ is the collection of all input values (x-values). The range of a function is the collection of all output values (y-values). For each x-value, there is exactly one y-value.

When a function is defined by an equation, its domain is restricted to **real numbers** for which the function can be evaluated. Otherwise, it must be excluded from the domain, such as:

1. In the Square-Root Function $y = \sqrt{x}$, the radicand x of a square root is always nonnegative. Both the domain and range are all nonnegative real numbers (greater than or equal to 0).
2. In the Absolute-Value Function $y = |x|$, the domain is all real numbers. The range is all nonnegative real numbers (greater than or equal to 0) because the absolute value of a number is never negative.
3. In the Rational Function $y = 1/x$, it is undefined if $x = 0$. A value that makes a division by zero must be excluded from the domain. Both the domain and range are all nonzero real numbers. $y = 0$ (the x-axis) is the horizontal asymptote.

The graph of $y = \sqrt{x}$

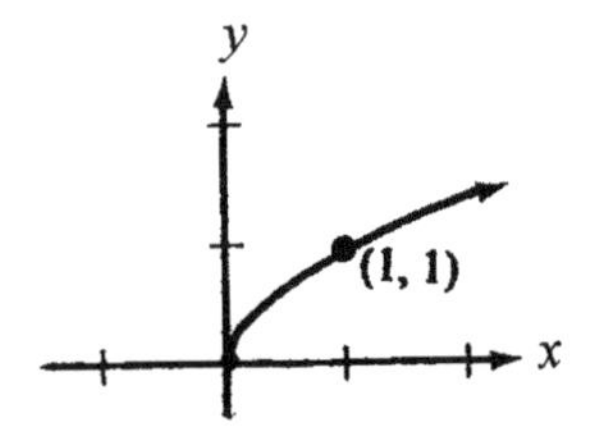

The graph of $y = |x|$

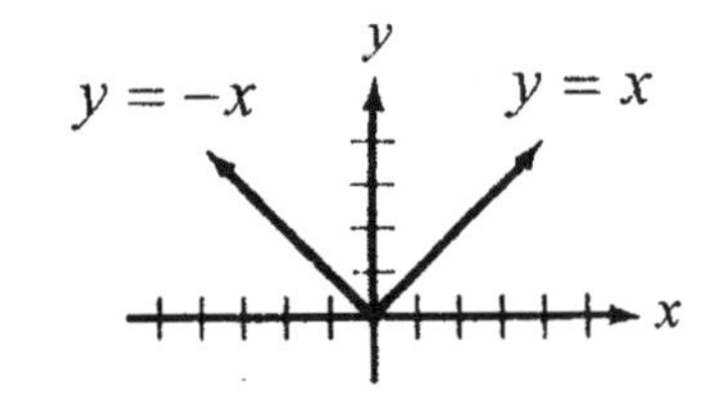

The graph of $y = \dfrac{1}{x}$

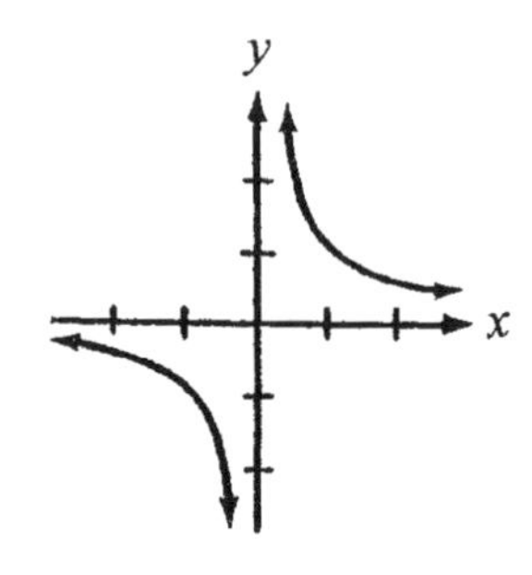

Find the domain and range of each function.

1. $y = \sqrt{x} + 2$
   D = All real numbers greater than or equal to 0
   R = All real numbers greater than or equal to 2
2. $y = \sqrt{x+3}$
   D = All real numbers greater than or equal to −3
   R = All real numbers greater than or equal to 0
3. $y = 5 + \sqrt{x-2}$
   D = All real numbers greater than or equal to 2
   R = All real numbers greater than or equal to 5
4. $y = |x-2|$
   D = All real numbers
   R = All real numbers greater than or equal to 0
5. $y = |x| - 2$
   D = All real numbers
   R = All real numbers greater than or equal to −2
6. $y = -|x-2|$
   D = All real numbers
   R = All real numbers less than or equal to 0

Find the domain and range of each function

7. $y = \dfrac{1}{x-2}$

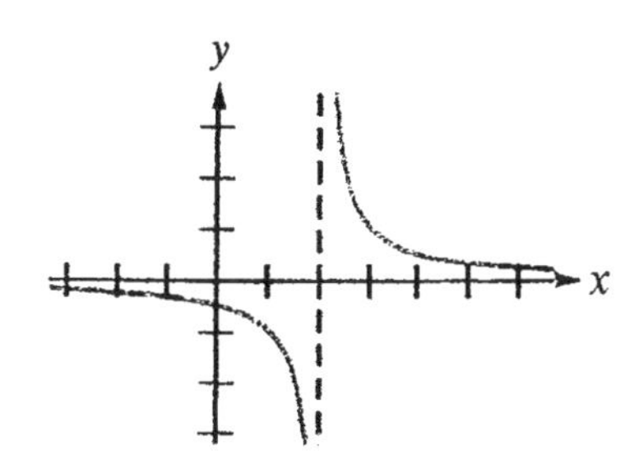

$y = 0$ is the horizontal asymptote.

D = All real numbers except 2
R = All nonzero real numbers

8. $y = \dfrac{x-3}{x-2}$

Dividing the denominator into the numerator, we have:

$$y = \frac{x-3}{x-2} = 1 - \frac{1}{x-2}$$

$y = 1$ is the horizontal asymptote.

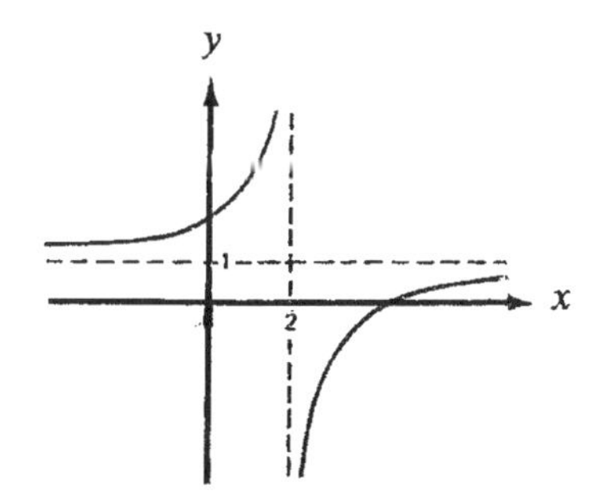

D = All real numbers except 2
R = All real numbers except 1

9. $y = \dfrac{x+1}{x+2}$

Dividing the denominator into the numerator, we have:

$$y = \frac{x+1}{x+2} = 1 - \frac{1}{x+2}$$

$y = 1$ is the horizontal asymptote.

D = All real numbers except -2
R = All real numbers except 1

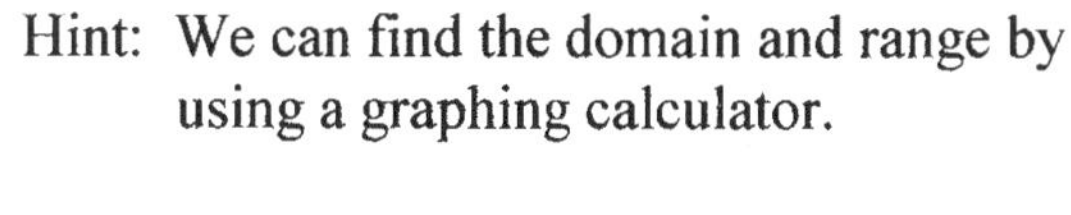
Hint: We can find the domain and range by using a graphing calculator.

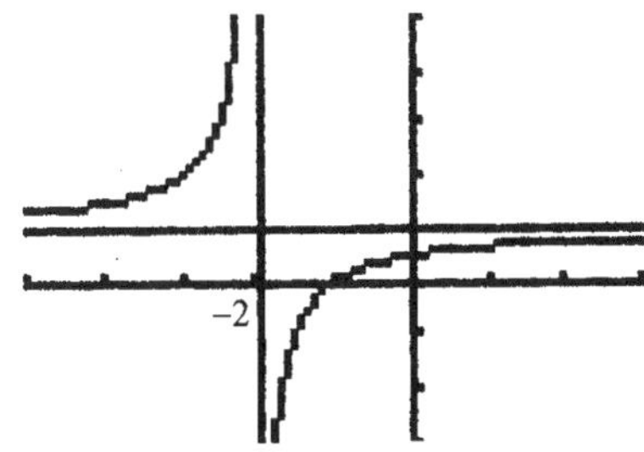

Notes

CHAPTER THREE

Geometry and Measurements

3-1 Basic Terms in Geometry

Geometry is the study of the properties and mutual relations of points, lines, planes, and solids in space.

There are many applications of geometry in science and our real life, such as archaeology, navigation, astronomy, housing and gardening.

The famous ancient mathematicians who contributed in finding geometric facts and theorems are Pythagoras, Plato, Aristotle, and Euclid.
The Greek mathematician Euclid (300 B.C.) is the first person to find and organize basic geometric principles.

Geometry includes **Plane (Euclidean) Geometry**, **Coordinate (Analytic) Geometry**, and **Solid (Analytic) Geometry.**

Plane (Euclidean) Geometry deals with figures in **plane** (or two-dimensions).
Coordinate (Analytic) Geometry deals with the solutions by **algebraic analysis**.
Solid (Analytic) Geometry deals with figures in **space** (or three-dimensions).

In the next few sections, we will start with some basic terms in geometry.

1. Undefined Terms: The terms that we all know what they mean fundamentally and intuitively by their descriptions. These terms are not necessary to be defined, but they can be described. However, their descriptions are not considered as definitions. We can make definitions of the other terms by using these undefined terms.
Point, line, and plane are undefined terms.

2. Defined Terms: The terms that we make common definitions by using the words that are previously defined or are easy understood by the geometric relationships. A good definition of a term gives a clear and correct meaning about it. Angle and triangle are defined terms.

3-2 Postulates and Theorems

In the study of geometry, we analyze the relationships among the geometric facts and principles, and then make statements and conclusions.
The term "**principle**" includes the geometrical statements such as definitions, postulates, and theorems.

Postulates (or Axioms): They are statements that we assume to be true without proof. Since not all statements in geometry can be proved, we need some basic assumptions (hypotheses) as a beginning to develop other statements. Postulates are basic assumptions or hypotheses. We need not prove a postulate because we are no doubt and accept them as true. By using the postulates, we can later develop other statements that we will prove to be true.

Theorems: They are statements that can be proved to be true.
To prove a theorem, we use deductive reasoning. **Deductive reasoning** is a logical reasoning from given facts (definitions, postulates, and previously proven theorems) to prove a desired conclusion.. We can not use a new theorem to prove a previous one. Otherwise, the logical sequence is violated.
If-then statements are common used in theorem. The " if-part " is the hypothesis taken as a condition, and the " then-part " is the conclusion.
A **corollary** is a theorem that follows directly from a postulate or another theorem, and can be used as a fact in proofs.

There are many postulates and theorems in geometry. Some basic statements are presented below.

Postulates:

1. Every line contains at least two distinct points.
2. Every plane contains at least three distinct, noncollinear points.
3. Two distinct points determine one and only one straight line.
4. Three distinct noncollinear points determine one and only one plane.
5. The line containing two distinct points in a plane lies in the plane.
6. Two intersecting planes have at least two points in common.
7. The length of a segment is the shortest distance between two points.
8. A segment has one and only one midpoint.
9. An angle has one and only one bisector.

Theorems:

1. If two distinct lines intersect, then their intersection is a point.
2. If two distinct planes intersect, then their intersection is a line.
3. The sum of the measures of the angles of a triangle is 180°.
4. Corollary: The measure of each angle of an equilateral triangle is 60°.
5. Pythagorean Theorem: In any right triangle, the square of the length of the hypotenuse equals the sum of the squares of the lengths of the two legs.

Methods of proof in Geometry

1. **Inductive Reasoning:** It is a logical reasoning from a pattern of a few examples to reach a desired conclusion by assuming the pattern will never end.

   Example: Consider the sum of the product of four consecutive integers and 1.

$$1 \cdot 2 \cdot 3 \cdot 4 + 1 = 25 = 5^2$$
$$2 \cdot 3 \cdot 4 \cdot 5 + 1 = 121 = 11^2$$
$$3 \cdot 4 \cdot 5 \cdot 6 + 1 = 361 = 19^2$$
$$4 \cdot 5 \cdot 6 \cdot 7 + 1 = 841 = 29^2$$

   It shows that the sum of four consecutive integers and 1 will always be a perfect square. On the evidence of this pattern, we can conclude this result (a perfect square) will never end. This is an example of inductive reasoning.
   Inductive reasoning is not a valid method of proof and the statement often can be proved by other method. See proof of this example in the book: " A-Plus Notes for Beginning Algebra ", page 354.

2. **Deductive Reasoning:** It is a logical reasoning from given facts (definitions, postulates, and previously proven theorems) to prove a desired conclusion.

   Example 1: Prove the intersection of two distinct planes is a line.
   Proof: Assume the following postulates are true:
   1. Two intersecting planes have at least two points in common.
   2. The line containing two distinct points in a plane lies in the plane.
   3. Two distinct points determine one and only one straight line.

   Since the line lies in both planes, the intersection is a line.

   Example 2: The Base Angles Theorem states that if two sides of a triangle are congruent, then the angles opposite those sides are congruent.
   Prove the bisector of the vertex angle of an isosceles triangle will intersect the base to form two right angles.
   Proof:

   $\overline{AB} \cong \overline{AC}$ (Given)
   $< B \cong < C$ (**Basic Angle Theorem**)
   $< 1 \cong < 2$ ($\overline{AD}$ is a bisector)
   $\overline{AD} \cong \overline{AD}$ (Common sides)
   We have $\Delta ABD \cong \Delta ACD$ (SAS Postulate)
   Since $< 3 \cong < 4$ and $m < 3 + m < 4 = 180$
   Therefore $m < 3 = m < 4 = 90$. Ans. (Hint: It means $\overline{AD} \perp \overline{BC}$.)

   Hint: An angle bisector divides an angle into two congruent angles.
   The symbol $\perp$ is read as "is perpendicular to".
   SAS Postulate: If two sides and the included angle of one triangle are congruent to the corresponding parts of the second triangle, then the two triangles are congruent.

Note: Observation, Measurement, and **Experimentation** are not valid methods of proof.

3-3 Points, Lines, Planes, and Angles

A **point** indicates position only. It is represented by a dot " • " and is named by a single capital letter near the dot. A point has no size (length, width, or height).

> **Collinear points** are points that lie on the same line.
> **Noncollinear points** are points that do not lie on the same line.
> **Coplanar points** are points that lie in the same plane.
> **Noncoplanar points** are points that do not lie in the same plane.

A **line** is represented by a set of continuous points that extend without end in two opposite directions and is named by two capital letters of any points on the line and drawing a " — " over the letters. A line has infinite length but no width or height.

A **ray** is the part of a line that extends without end from its one endpoint.

A **line segment** of a line consists of two endpoints with all the points between them. It is customary to use " $m\overline{AB}$ " or " AB " to express the length of line segment $\overline{AB}$.

The **midpoint** on a line segment is the point that bisects (two equal halves) the line segment.

A **plane** is a flat surface that has length and width but no height. The length and width of a plane can be extended without end in all directions. A plane is represented by a four-sided figure and is named by placing a capital letter at one of its corners.

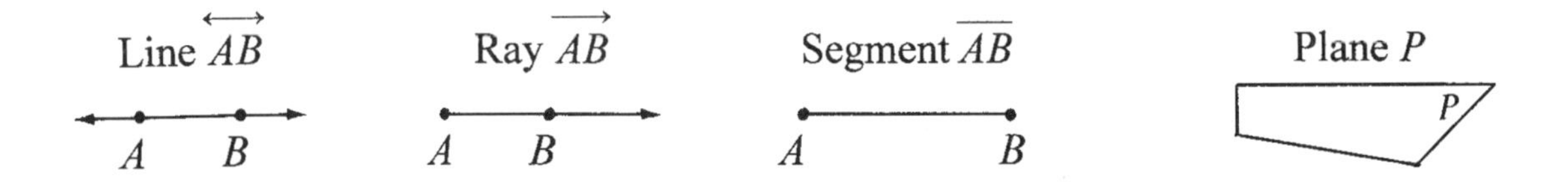

An **angle** is the figure formed by two noncollinear rays having the same endpoint and is named by the symbol " < " or " ∢ ". The rays are the sides of the angle and the endpoint is the vertex. The angle with vertex B below may be named in different ways as long as it does not cause any confusion, such as $< ABC$, $< CBA$, $< B$, or < 1. We use a protractor to measure an angle. In using a protractor, we place the center of the protractor at the vertex of the angle and line up on ray of the angle with the "0" mark on the protractor.
No matter how large or small the two sides are, the size of the angle would not be changed.

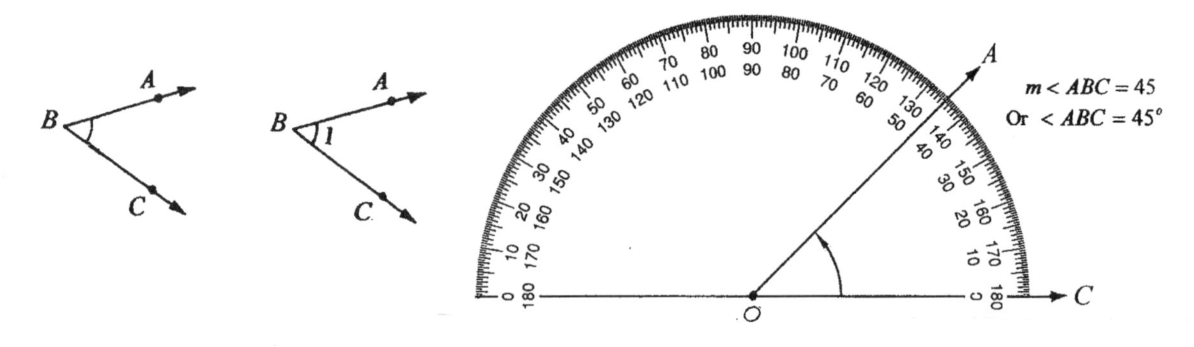

The measure of an angle is the number of degrees or radians it contains. If an angle $<B$ measures 45 degrees, we write $m<B=45$ or $<B=45^o$. An angle is often denoted by α **(alpha)**, β **(belta)**, θ **(theta)**, γ **(gamma)**, or ϕ **(phi)**.

An **acute angle** is an angle whose measure is less than 90°.
A **right angle** is an angle whose measure is 90° and is denoted by a square.
An **obtuse angle** is an angle whose measure is more than 90° and less than 180°.
A **straight angle** is an angle whose measure is 180°.
A **reflex angle** is an angle whose measure is more than 180° and less than 360°.

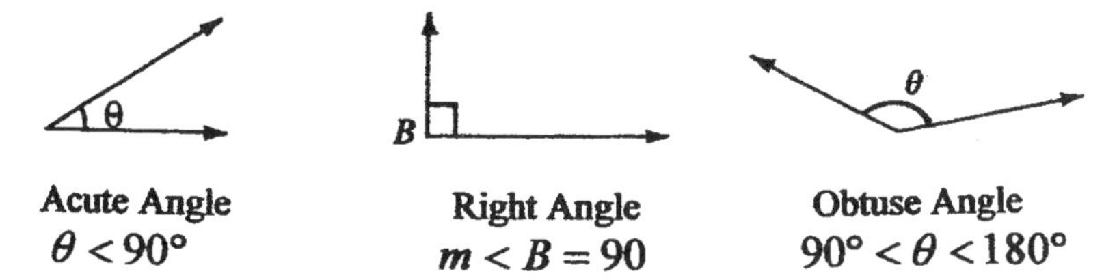

Acute Angle $\theta<90°$ **Right Angle** $m<B=90$ **Obtuse Angle** $90°<\theta<180°$

Complementary Angles are two angles whose measures have a sum of 90°.
Supplementary Angles are two angles whose measures have a sum of 180°.
Two angles form a linear pair are supplementary.

Vertical Angles are two nonadjacent angles when two lines or line segments intersect.
Vertical angles are equal.

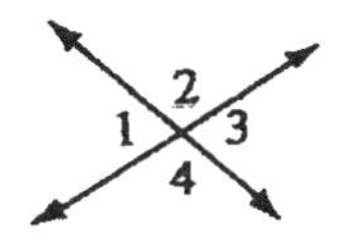

Linear Pairs: <1 and <2, <2 and <3, <3 and <4, <1 and <4

Vertical Angles: <1 and <3, <2 and <4

$m<1=m<3$

$m<2=m<4$

> Using a phone, a four-year old boy calls his mother who is working in a foreign country.
> Son: Mom, are you coming home today ?
> Mother: No, I am very far away at the other side of the earth.
> Son: Where is the other side of the earth ?
> Mother: Earth is like a basketball. You are on top of the ball. I am at the bottom.
> Son: It must be very hard for you with your head upside-down when you drink water.

Examples

1: Find the measure of each angle of the figure.

a. $<CBE$ **b.** $<CBF$ **c.** $<CBG$
d. $<ABH$ **e.** $<DBH$ **f.** $<PBD$
g. $<PBF$ **h.** $<ABC$

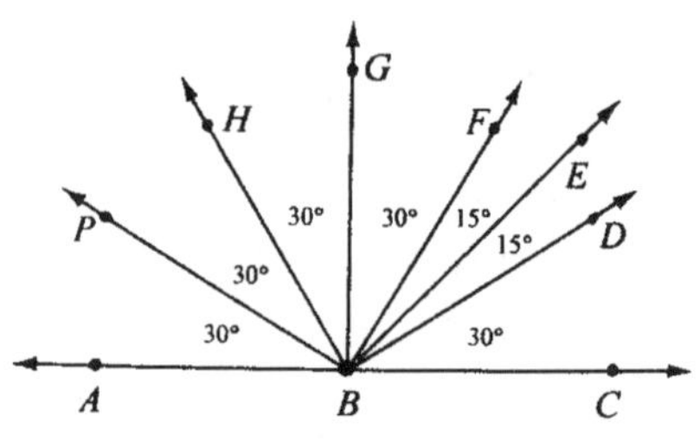

Solution:

a. $<CBE=45°$ b. $<CBF=60°$ c. $<CBG=90°$ d. $<ABH=60°$
e. $<DBH=90°$ f. $<PBD=120°$ g. $<PBF=90°$ h. $<ABC=180°$. Ans.

2: Identify each angle as acute, right, or obtuse.

a.
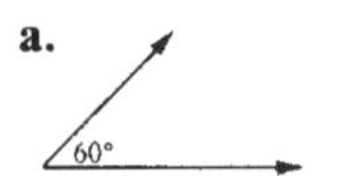

b.
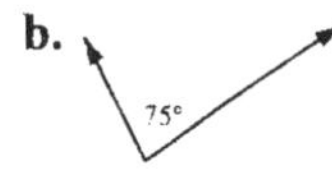

c.
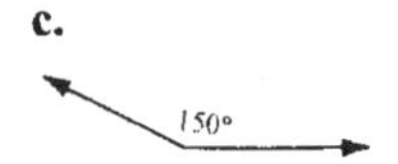

d.
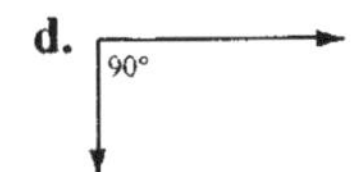

e.
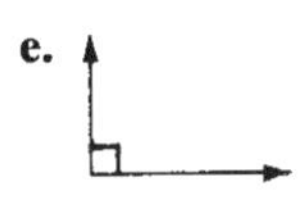

Solution: a. Acute b. Acute c. Obtuse d. Right e. Right. Ans.

3: The measure of an angle is $30°$.

a. Find the measure of its supplement. b. Find the measure of its complement.

Solution:

a. Measure of the supplement: $180°-30°=150°$. Ans.
b. Measure of the complement: $90°-30°=60°$. Ans.

4: The measure of an angle is $30°$ larger than its supplement. Find the measure of the angle and the measure of its supplement.

Solution:

Let $x=$ the measure of the angle
$x-30=$ the measure of its supplement
We have the equation:

$$x+x-30=180$$
$$2x=210$$
$x=105° \rightarrow$ The measure of the angle. Ans.
$x-30=75° \rightarrow$ The measure of its supplement. Ans.

A patient said to his dentist.
Patient: Why are you charging me \$300 ?. Last time, you charged me only \$100.
Dentist: Two patients in the waiting room ran away when they heard you screaming.

3-4 Parallel Lines and Angles

Perpendicular Lines: Two lines (rays, segments) are perpendicular if they intersect to form a right ($90°$) angle. The symbol is " $\perp$ ".

A line (or ray, segment) is called a **perpendicular bisector** when it is perpendicular to the segment at the midpoint.

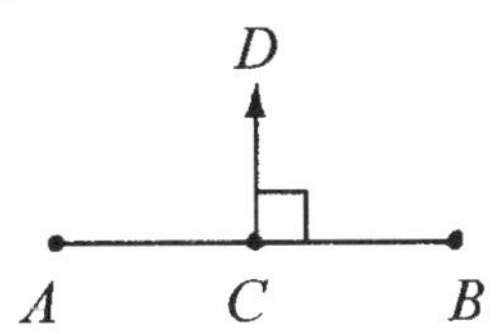

$\overrightarrow{CD} \perp \overline{AB}$ and $\overline{AC} = \overline{CB}$

$\overrightarrow{CD}$ is the perpendicular bisector of $\overline{AB}$.

Parallel Lines: Two lines are parallel if they will never intersect. The symbol is " // ".

If two parallel lines are cut by a transversal, the transversal will form eight angles (4 **interior angles** and 4 **exterior angles**).
Corresponding angles are two angles in corresponding positions relative to the two lines and the transversal. **Alternate angles** are nonadjacent angles on opposite sides of a transversal.

Among the eight angles, we have the following special relationships:

1. The **corresponding angles** are congruent.
2. The **alternate interior angles** are congruent.
3. The **alternate exterior angles** are congruent.

In geometry, **congruent figures** have the same shape and size. The symbol is " $\cong$ ".

Lines *m* and *n* cut by transversal k, $m // n$, we have:

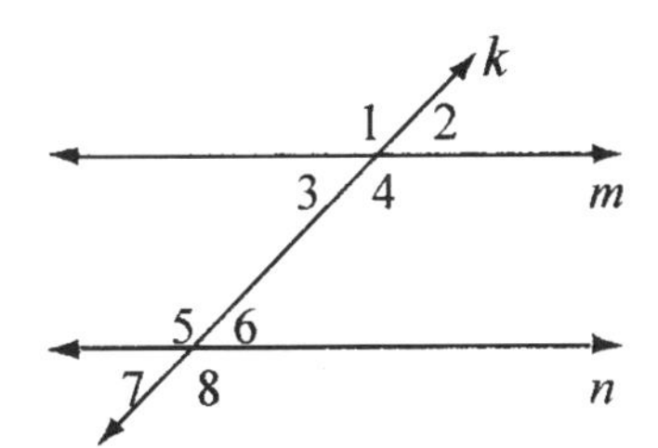

Corresponding angles: $<1 \cong <5$, $<2 \cong <6$, $<3 \cong <7$, $<4 \cong <8$

Alternate interiors angles: $<3 \cong <6$, $<4 \cong <5$

Alternate exterior angles: $<2 \cong <7$, $<1 \cong <8$

Example 1: If $<1 = 135°$ on the above figure, find each measure.

Solution:

$<2 = 45°$, $<3 = 45°$, $<4 = 135°$, $<5 = 135°$
$<6 = 45°$, $<7 = 45°$, $<8 = 135°$. Ans.

Example 2: If line *f* and line *g* are parallel, find the sum of measures of $<a$ and $<b$.

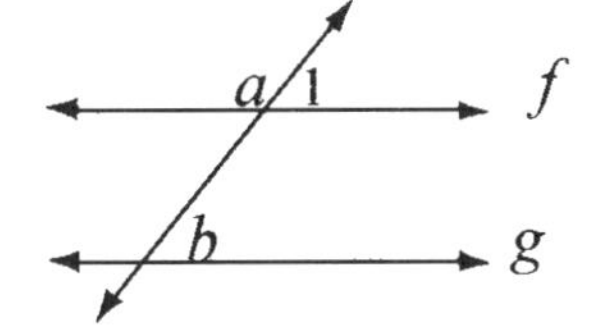

Solution:

$m < a + m < 1 = 180$

$< b = < 1$ (corresponding angles)

$\therefore\ m < a + m < b = 180$. Ans. (Hint: $a + b = 180°$)

Fred's Theorem: When two parallel lines are cut by a transversal, two kinds of angles are formed, little angles and big angles. All of the little angles are equal. All of the big angles are equal. Any little angle plus any big angle equals $180°$.

Examples

3: If line $s // $ line t and $<2=62^\circ$,
find $m<5$ and $m<8$.
Solution:

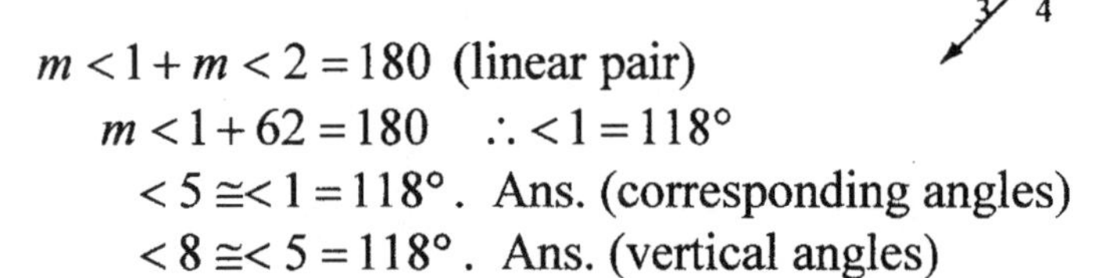

$m<1+m<2=180$ (linear pair)
$m<1+62=180 \quad \therefore <1=118^\circ$
$<5\cong<1=118^\circ$. Ans. (corresponding angles)
$<8\cong<5=118^\circ$. Ans. (vertical angles)

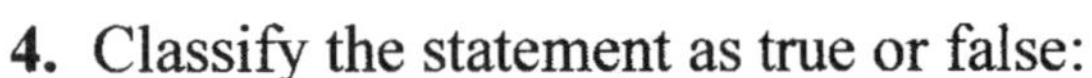

4. Classify the statement as true or false:
If $<4=135^\circ$ and $<6=45^\circ$,
then line m // line n.
Solution:

$m<3+m<4=180$ (linear pair)
$m<3+135=180 \quad \therefore <3=45^\circ$
$<2\cong<3=45^\circ$ (vertical angles)
$<6=45^\circ$ (given)
$<2\cong<6$ (They are parallel.)
The statement is **true**. Ans. (The corresponding angles are congruent.)

5. Classify the statement as true or false:
If $<4=135^\circ$ and $<7=47^\circ$,
then line t // line s.
Solution:

$<1\cong<4=135^\circ$ (vertical angles)
$m<1+m<3=180$ (linear pair)
$135+m<3=180 \quad \therefore <3=45^\circ$
$<7=47^\circ$ (given)
$m<3\neq m<7$ (They are not parallel.)
The statement is **false**. Ans. (The corresponding angles are not congruent.)

6. If line k and line p are parallel and $<1=140^\circ$, find $m<5-m<6$.

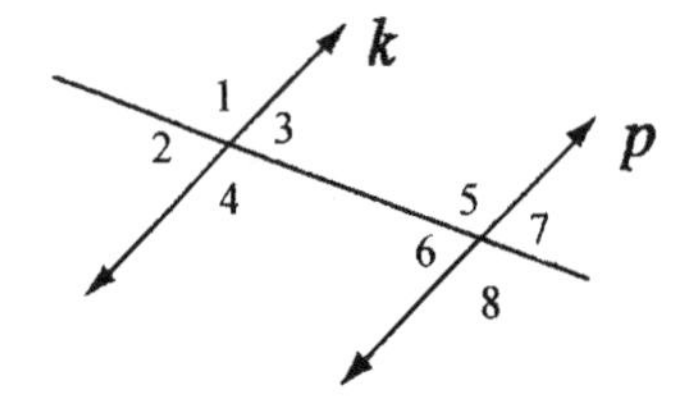

Solution:

$<5\cong<1=140^\circ$ (corresponding angles)
$m<6+m<5=180$ (linear pair)
$m<6+140=180 \quad \therefore <6=40^\circ$
$m<5-m<6=140-40=100$. Ans.

3-5 Basic Rules in Triangles

A triangle is the figure formed by connecting three noncollinear points with three line segments and is named by the symbol " Δ " . A triangle below may be named with its three letters in any order, such as ΔABC, ΔBCA, or ΔCAB. Its three sides are $\overline{AB}$, $\overline{BC}$, and $\overline{AC}$. It has three vertices A, B, and C and three angles $\angle A$, $\angle B$, and $\angle C$.
An **equilateral triangle (equiangular triangle)** is a triangle having three equal sides (angles).
An **isosceles triangle** is a triangle having at least two equal sides (called legs). An equilateral triangle is also an isosceles triangle.
A **scalene triangle** is a triangle having no equal sides or no equal measure of angels.
A **right triangle** is a triangle having a right ($90°$) angle.
An **acute triangle** is a triangle having three acute angles.
An **obtuse triangle** is a triangle having an obtuse angle.

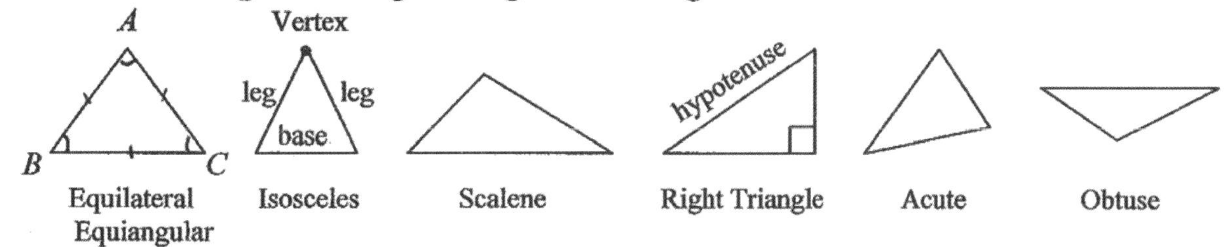

Equilateral Equiangular | Isosceles | Scalene | Right Triangle | Acute | Obtuse

Basic Rules in Triangles

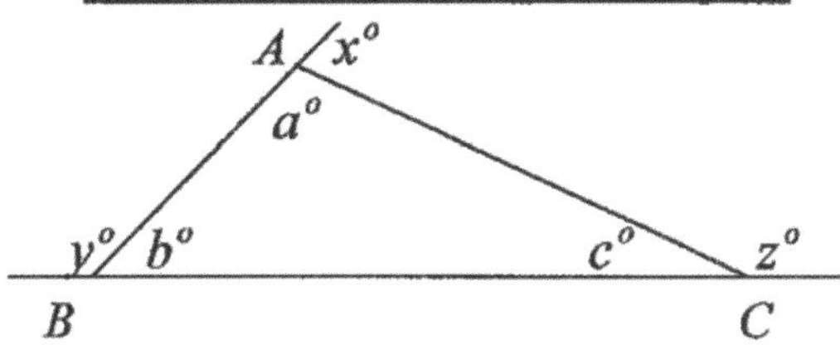

1. Triangle-Sum Theorem: The sum of the measures of interior angles of any triangle is $180°$. The sum of the measures of exterior angles of any triangle is $360°$.

$$a+b+c=180, \quad x+y+z=360$$

2. Measure of an Exterior Angle: The measure of an exterior angle of a triangle is equal to the sum of its two remote interior angles.

$$x=b+c, \quad y=a+c, \quad z=a+b$$

3. Angle-Side (Proportionality) Relationships: In every triangle, the larger angle is opposite the larger side and the smaller angle is opposite the smaller side.

If $a>b>c$, then $\overline{BC}>\overline{AC}>\overline{AB}$

4. The Triangle-Inequality Theorem (Third-Side Rule):

The sum of the lengths of any two sides is greater than the length of the third side.
The difference of the lengths of any two sides is less than the length of the third side.

$$\overline{AB}+\overline{AC}>\overline{BC}, \quad \overline{AB}+\overline{BC}>\overline{AC}, \quad \overline{AC}+\overline{BC}>\overline{AB}$$

$$\left|\overline{AB}-\overline{AC}\right|<\overline{BC}, \quad \left|\overline{AB}-\overline{BC}\right|<\overline{AC}, \quad \left|\overline{AC}-\overline{BC}\right|<\overline{AB}$$

Basic Rules in Triangles (Continued)

5. **Isosceles-Triangle Theorem:**
   If two sides of a triangle are congruent, then the two base angles are congruent.

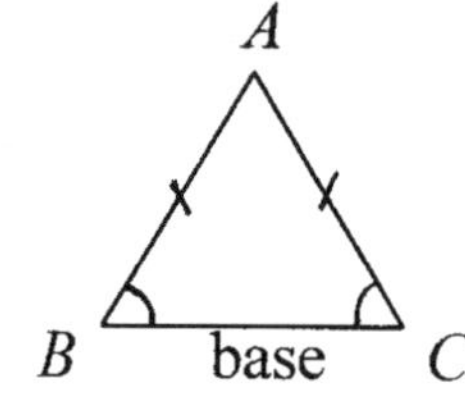

If $\overline{AB} = \overline{AC}$, then $< B \cong < C$.

6. The **median** of a triangle is a segment from one vertex to the midpoint of the opposite side. There are three medians in a triangle. The point of concurrency of three medians is the **centroid**, or **center of gravity** of the triangle. The distance between the centroid to the vertex is $\frac{2}{3}$ of the length of the median.

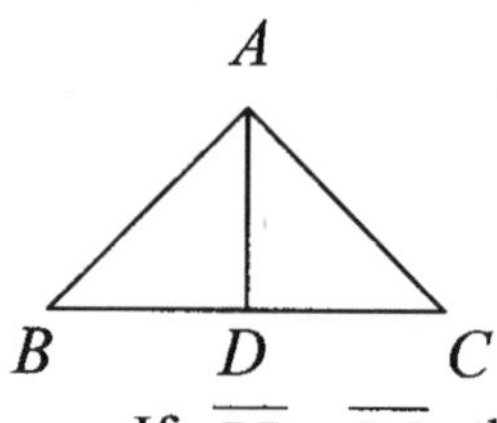

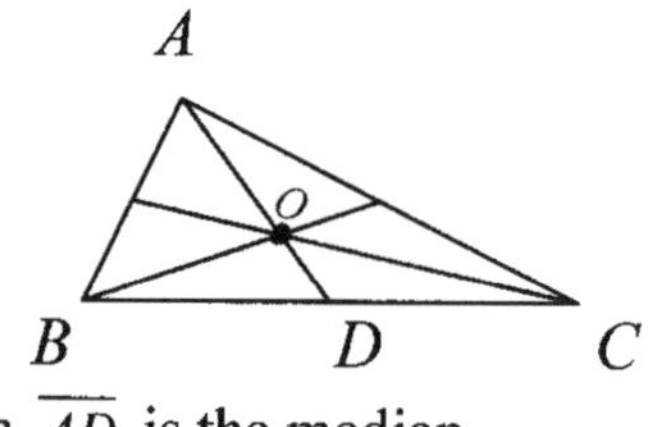

If $\overline{BD} = \overline{DC}$, then $\overline{AD}$ is the median.
Point o is the centroid.

7. The **height (altitude)** of a triangle is a segment from one vertex perpendicular to the opposite side (**the base**), or perpendicular to the line containing the opposite side.
   The height of an isosceles triangle is also the median of the triangle.
   The point of concurrency of three heights is the **orthocenter** of the triangle.

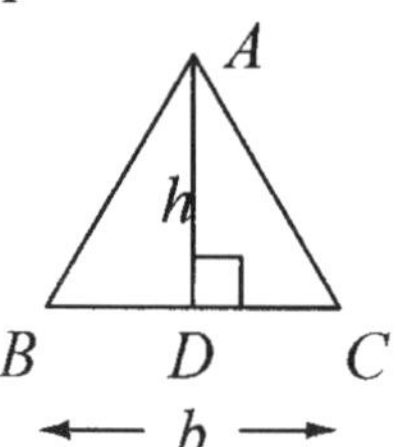

$\overline{AD}$ is the height.
$\overline{BC}$ is the base.

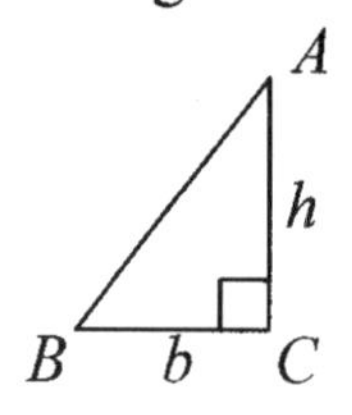

$\overline{AC}$ is the height.
$\overline{BC}$ is the base.

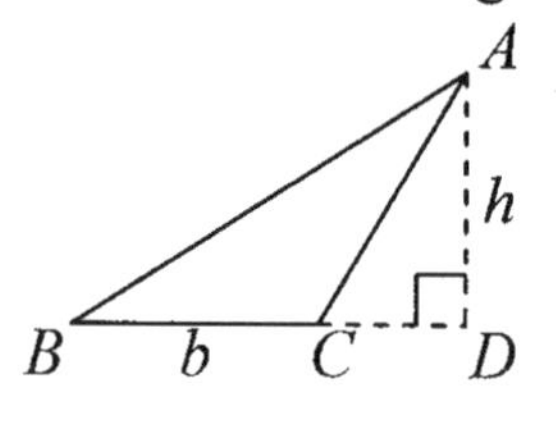

$\overline{AD}$ is the height.
$\overline{BC}$ is the base.

8. The **angle bisector** of a triangle is a segment from one side of the triangle and divides an angle into two congruent angles. There are three angle bisectors in a triangle.
   The angle bisector of an isosceles triangle is also a median and the height of the triangle.
   The point of concurrency of three angle bisectors is the **incenter** of the triangle.
   The incenter of the triangle is equidistance from the sides of the triangle. The incenter is the center of the inscribed circle.

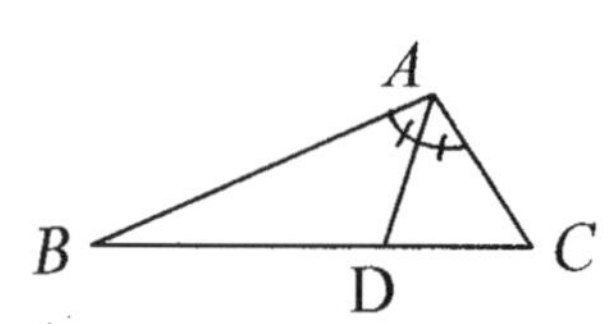

$m < BAD = m < CAD$

$\overline{AD}$ is a angle bisector of ΔABC.

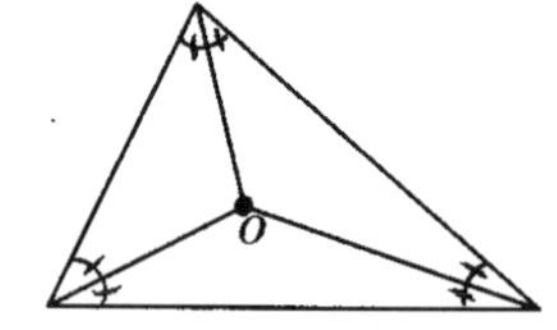

Point o is the incenter.

9. The **circumcenter** of a triangle is the point of concurrency of the perpendicular bisectors of the sides of a triangle.
   The circumcenter of a triangle is equidistance from the vertices of the triangle. The circumcenter is the center of the circumscribed circle.

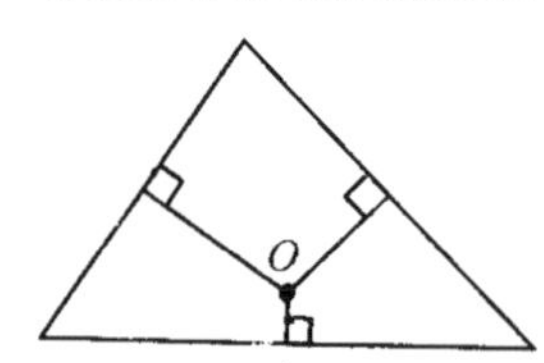

Point o is the circumcenter.

Basic Rules in Triangles (Continued)

10. Area Formula of a Triangle: The area of a triangle is one-half of (base × height). $A = \frac{1}{2}bh$.

11. Area of an Equilateral Triangle with side of length s: $A = \dfrac{s^2\sqrt{3}}{4}$

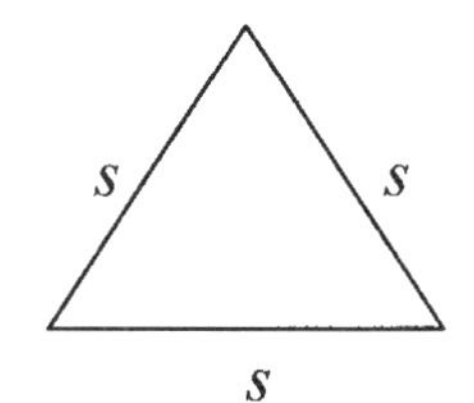

Examples

1. Find the measure of the angle x.

   Solution:

   $x + 75 + 35 = 180$

   $\therefore x = 70$. Ans.

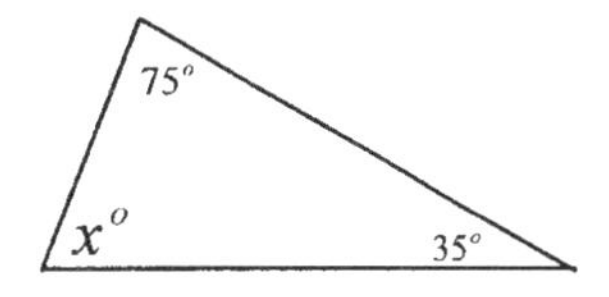

2. Find the measure of the angle a.

   Solution:

   $a = 100 + 30 = 130$. Ans.

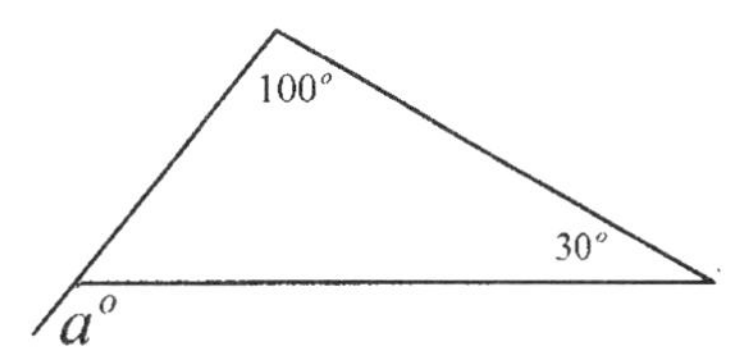

3. Identify the statement as true or false: The lengths of sides of a triangle are 2, 4, 6.

   Solution:

   False. Ans.

   (The sum of the lengths of any two sides of a triangle is greater than the length of the third side.)

4. A triangle has the lengths of sides 7, 12, and x. What are the possible values of x ?

   Solution:

   $12 - 7 < x < 12 + 7$

   $5 < x < 19$. Ans.

5. A triangle has the lengths of two sides 7 and 12. The perimeter is p. What are the possible values of p ?

   Solution:

   The length of the third side is ($p - 12 - 7$)

   $12 - 7 < p - 12 - 7 < 12 + 7$

   $5 < p - 19 < 19$

   $24 < p < 38$. Ans.

6. An isosceles triangle has the length of sides 7, 15, and x. What are the possible values of x ?

   Solution:

   A 7- 7-15 (sides) triangle is not possible.

   A 15-15-7 (sides) triangle is the only isosceles triangle.

   $\therefore x = 15$ only. Ans.

Examples

7. The sides of an equilateral triangle are each 6 inches long. Find the area.
Solution:

$$A = \frac{s^2\sqrt{3}}{4} = \frac{6^2\sqrt{3}}{4} = 9\sqrt{3} \approx 15.60 \ in.^2 \ \text{Ans.}$$

8. The area of an equilateral triangle are $9\sqrt{3}$. What is the length of its perimeter ?
Solution:

$$A = \frac{s^2\sqrt{3}}{4} = 9\sqrt{3}, \quad s^2 = 36, \quad s = 6, \quad \therefore p = 3 \times 6 = 18. \ \text{Ans.}$$

9. In the figure, what is the measure of x ?
Solution:

The sum of the measure of interior angles is 180°.

$m < 1 = 45, \quad 45 + 84 + x = 180$

$\therefore \ x = 51.$ Ans.

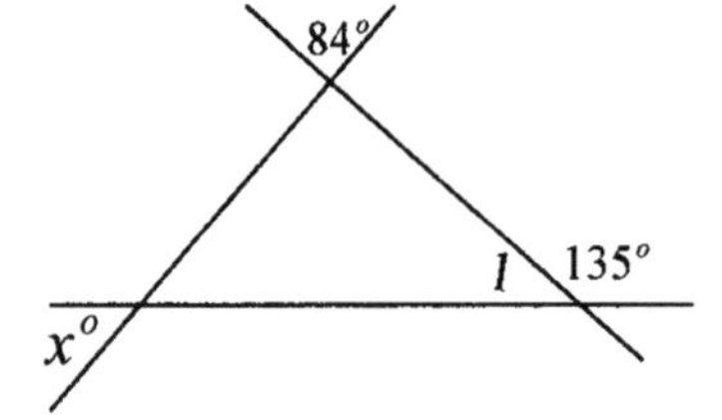

10. ΔABC is an isosceles triangle. $\overline{DB}$ and $\overline{DC}$ are two angle bisectors. $m < BDC = 140$. Find $m < A$.
Solution:

$m < DBC + m < DCB + 140 = 180$

$m < DBC = m < DCB = 20$

We have $m < B = m < C = 40$

$m < A + m < B + m < C = 180$

$m < A + 40 + 40 = 180$

$\therefore m < A = 100$. **Ans.**

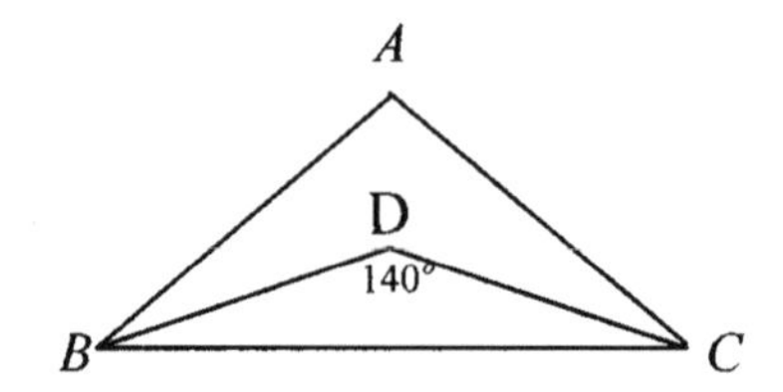

11. ΔABC is a triangle and $< A : < B : < C$ = 2:3:4. What is the measure of $< C$?
Solution:

$$m < C = 180 \times \frac{4}{2+3+4} = 180 \times \frac{4}{9} = 80. \ \text{Ans.}$$

12. A triangle has an area of 0.5. If all of the lengths of sides in this triangle are decreased as one-half of the original lengths, what is the new area ?
Solution:

We choose the original triangle with a base of 1 and a height of 1.
We have the new area

$$A = \tfrac{1}{2}bh = \tfrac{1}{2} \cdot \tfrac{1}{2} \cdot \tfrac{1}{2} = \tfrac{1}{8} = 0.125. \ \text{Ans.}$$

Other Method: If two triangles are similar, the ratio of their areas is the square of the ratio of the lengths of the corresponding sides.

$$\frac{A}{0.5} = \left(\frac{0.5}{1}\right)^2 \quad \therefore A = 0.5 \times 0.25 = 0.125. \ \text{Ans.}$$

3-6 The Pythagorean Theorem

Right angle: It is an angle with measure 90° and indicated by a small square.
Right triangle: It is a triangle having one right (90°) angle.
The hypotenuse: It is the side opposite the right angle in a triangle.
The other two sides are the **legs**.

The Pythagorean Theorem:
In any right triangle, the square of the length of the hypotenuse equals the sum of the squares of the lengths of the two legs.

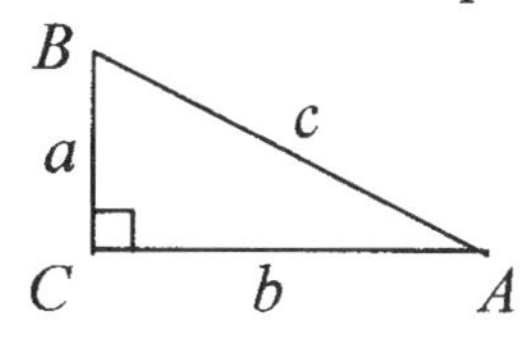

Formula: $a^2 + b^2 = c^2$

We can use the **Pythagorean Theorem** to find the length of one side of a right triangle when the lengths of the other two sides are known. To apply the Pythagorean Theorem, we need the knowledge of squares and square roots.

To determine whether or not the triangle with the given lengths of its three sides is a right triangle, we apply the Pythagorean Theorem. If the sum of the squares of the lengths of the two shorter sides is equal to the square of the length of the longest side, then it is a right triangle.

Examples

1. Find the length of the unknown side.

a)

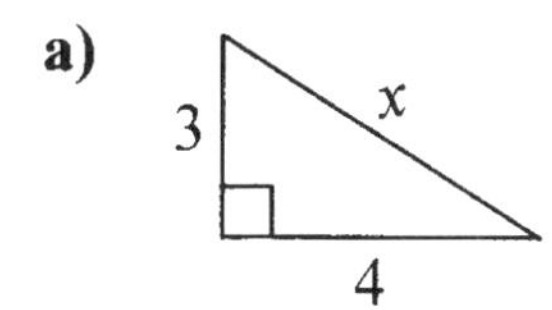

b) 4, 2, y

c) m, 12, 14

Solution:

a) $x^2 = 3^2 + 4^2$
$x^2 = 9 + 16$
$x^2 = 25$
$\therefore x = \sqrt{25} = 5$.

b) $y^2 = 2^2 + 4^2$
$y^2 = 4 + 16$
$y^2 = 20$
$\therefore y = \sqrt{20} = 4.47$.

c) $m^2 + 12^2 = 14^2$
$m^2 + 144 = 196$
$m^2 = 52$
$\therefore m = \sqrt{52} = 7.21$.

2. Determine whether or not the triangle with the given lengths of its three sides is a right triangle.

a) 6, 8, 10 **b)** 4, 5, $5\sqrt{2}$ **c)** 6, $6\sqrt{2}$, 6

Solution:

a) $6^2 + 8^2 = 100$
$10^2 = 100$
$6^2 + 8^2 = 10^2$ **Yes**

b) $4^2 + 5^2 = 41$
$(5\sqrt{2})^2 = 50$
$4^2 + 5^2 \neq (5\sqrt{2})^2$ **No**

c) $6^2 + 6^2 = 72$
$(6\sqrt{2})^2 = 72$
$6^2 + 6^2 = (6\sqrt{2})^2$ **Yes**

Examples

3. A baseball field is shaped like a square diamond. Each side is 100 feet long. Find the distance from home plate to second base.

Solution:

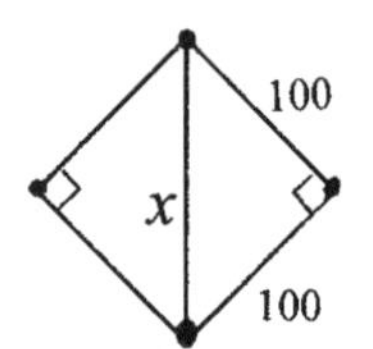

$$x^2 = 100^2 + 100^2$$
$$x^2 = 20000$$
$$\therefore x = \sqrt{20000} \approx 141.42 \text{ feet. Ans.}$$

4. The sides of an equilateral triangle are each s units long. Find the altitude(h) in terms of s.

Solution:

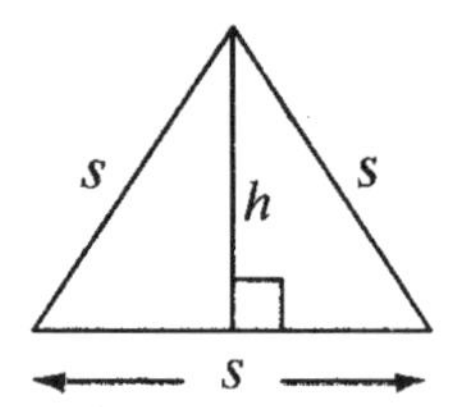

$$s^2 = h^2 + (\tfrac{1}{2}s)^2$$
$$s^2 = h^2 + \tfrac{1}{4}s^2$$
$$h^2 = \tfrac{3}{4}s^2 \quad \therefore h = \tfrac{\sqrt{3}}{2}s. \text{ Ans.}$$

5. Find the area of the triangle.

Solution:

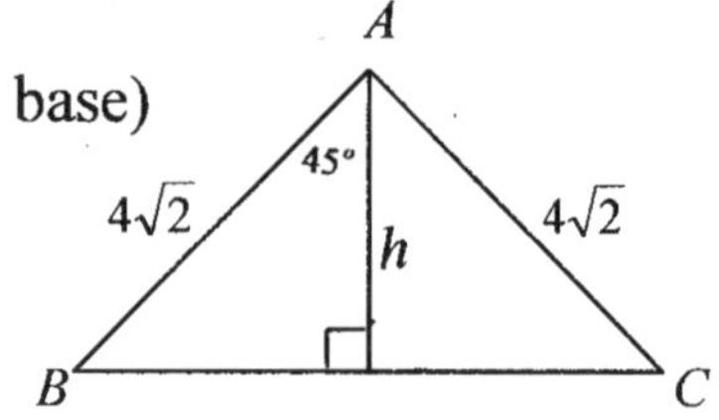

$$(BC)^2 = (4\sqrt{2})^2 + (4\sqrt{2})^2 = 32 + 32 = 64 \quad \therefore BC = 8 \text{ (the base)}$$
$$h^2 + 4^2 = (4\sqrt{2})^2$$
$$h^2 + 16 = 32 \quad \therefore h = 4$$

We have the area

$$A = \tfrac{1}{2}bh = \tfrac{1}{2}\cdot 8 \cdot 4 = 16. \text{ Ans.}$$

In an English class, the teacher asks the kids to write a letter to his (her) mother regarding what he (she) did today in school. The teacher notices that John is writing very slowly.

Teacher: John, why are you writing so slowly ?

John: My mom could not read fast.

3-7 Special Right Triangles

There are two special right triangles, **30-60-90 triangle** and **45-45-90 triangle**. The lengths of sides of each special right triangle follow a special pattern.

1. A **30-60-90 Triangle** is a triangle with angles of $30°$, $60°$, and $90°$.
   The sides of every 30-60-90 triangle will follow "$1: 2 : \sqrt{3}$ ratio".
   In a 30-60-90 triangle, the length of the shorter side (opposite side of $30°$ angle) is one-half of the length of the hypotenuse. The length of the longer side is $\sqrt{3}$ times the length of the shorter side.

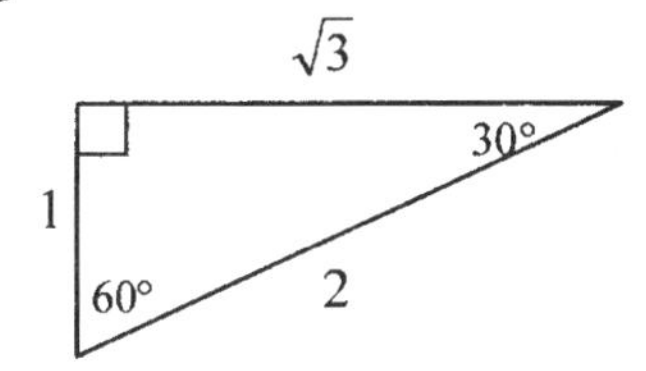

Basic 30-60-90 triangle

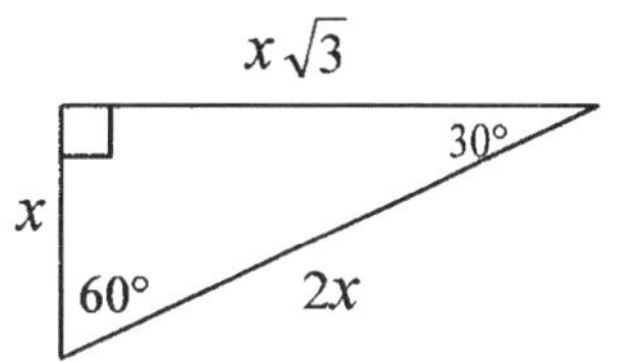

Any 30-60-90 triangle with shorter side x
(Area $= \frac{\sqrt{3}}{2}x^2$)

2. A **45-45-90 Triangle** is a triangle with angles of $45°$, $45°$, and $90°$. It is also an isosceles right triangle.
   The sides of every 45-45-90 triangle will follow "$1:1:\sqrt{2}$ ratio".
   In a 45-45-90 triangle, the two legs are always equal. The length of the hypotenuse is $\sqrt{2}$ times the length of the leg.

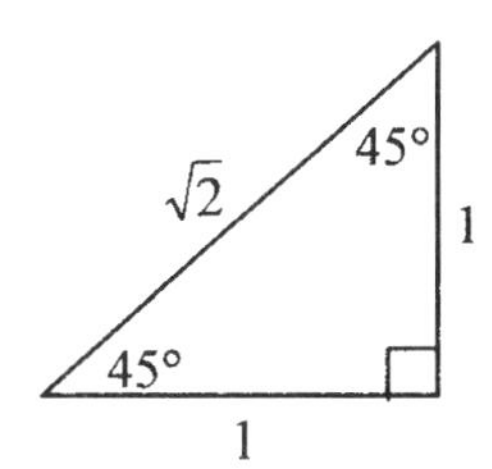

Basic 45-45-90 triangle

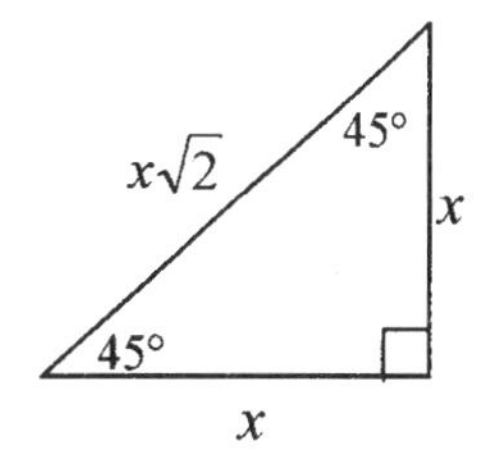

Any 45-45-90 triangle with leg x
(Area $= \frac{1}{2}x^2$)

Examples

1. The hypotenuse of a 30-60-90 triangle is 10. Find the lengths of its two sides.
   Solution:
   The shorter side is one-half of the hypotenuse = 5
   The longer side is $\sqrt{3}$ times the shorter side $= 5\sqrt{3}$. Ans.

2. The hypotenuse of a 45-45-90 triangle is 10. Find the length x of its leg.
   Solution:
   $x\sqrt{2} = 10 \quad \therefore x = \frac{10}{\sqrt{2}} = 5\sqrt{2}$. Ans.

Examples

3. Find the length of the diagonal of a square with sides of 20 inches.
Solution:

$$d^2 = 20^2 + 20^2$$
$$d^2 = 800$$
$$d = \sqrt{800} = \sqrt{400 \times 2} = 20\sqrt{2}. \text{ Ans.}$$

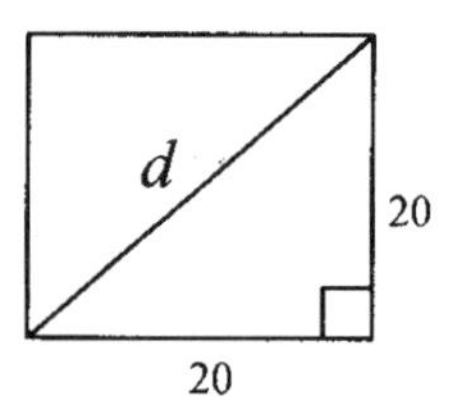

Or: It is a 45-45-90 triangle. The length of the hypotenuse is $\sqrt{2}$ times the length of the leg.
$d = 20\sqrt{2}$. Ans.

4. On the figure, Find the measures of $\overline{AD}$, $\overline{AB}$, $\overline{BD}$, and $\overline{DC}$.
Solution:

ΔADC is a 30-60-90 triangle.

$AD = \frac{1}{2}(AC) = \frac{1}{2}(20) = 10$ Ans.

$DC = 10\sqrt{3}$. Ans.

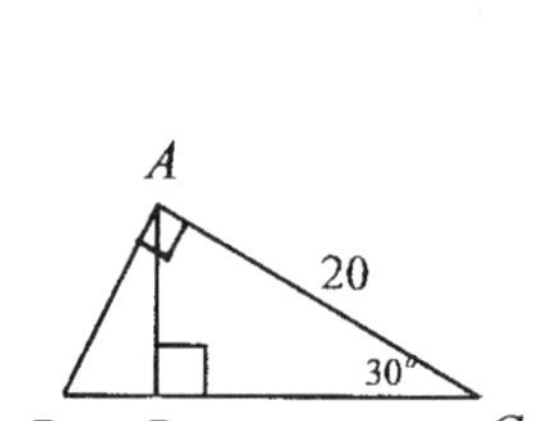

ΔABD is a 30-60-90 triangle.

$AD = \sqrt{3}(BD)$ $\therefore BD = \frac{1}{\sqrt{3}}(AD) = \frac{1}{\sqrt{3}}(10) = \frac{10\sqrt{3}}{3}$. Ans.

$AB = 2(BD) = 2 \cdot \frac{10\sqrt{3}}{3} = \frac{20\sqrt{3}}{3}$. Ans.

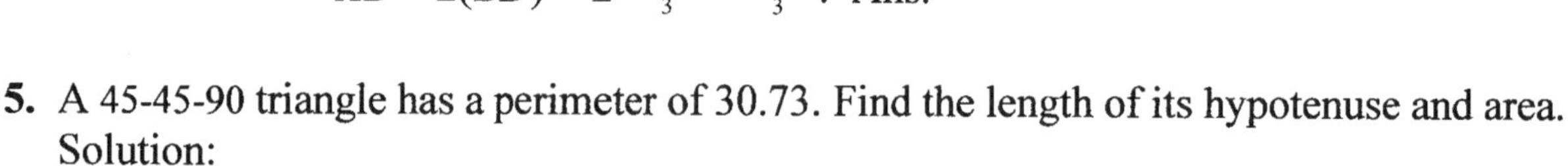

5. A 45-45-90 triangle has a perimeter of 30.73. Find the length of its hypotenuse and area.
Solution:

$$x + x + x\sqrt{2} = 30.73$$
$$(2 + \sqrt{2})x = 30.73$$
$$x = \frac{30.73}{2 + \sqrt{2}} = \frac{30.73}{3.414} = 9$$

$\therefore$ Hypotenuse $= x\sqrt{2} = 9\sqrt{2} \approx 12.73$. Ans.

Area $= \frac{1}{2}x^2 = \frac{1}{2} \cdot 9^2 = 40.5$. Ans.

6. The equilateral triangle with the side length of 12 is inscribed in the circle. Find the area of the circle.
Solution:

$m < AOC = 120$, $m < AOD = 60$

ΔAOD is a 30-60-90 triangle.

The side opposite to the 60° angle is $\sqrt{3}$ × (the side opposite to the 30° angle).

$AD = \sqrt{3}(OD)$, $OD = \frac{1}{\sqrt{3}}(AD) = \frac{1}{\sqrt{3}}(6) = 2\sqrt{3}$

$OA = 2(OD) = 2 \cdot 2\sqrt{3} = 4\sqrt{3}$

The hypotenuse $\overline{OA}$ is the radius of the circle: $r = 4\sqrt{3}$

The area of the circle: $A = \pi r^2 = \pi(4\sqrt{3})^2 = 48\pi \approx 150.72$. Ans.

3-8 Similar and Congruent Triangles

Two triangles are **similar** if the ratio of the lengths of their corresponding sides is a constant. The symbol is " $\sim$ ". The corresponding angles of two similar triangles are congruent.

AAA Similarity or AA similarity: If three angles of one triangle are congruent to the three angles of another triangle, then the triangles are similar.

If a line parallel to one side of a triangle determines the second triangle, then the second triangle is similar to the original triangle.

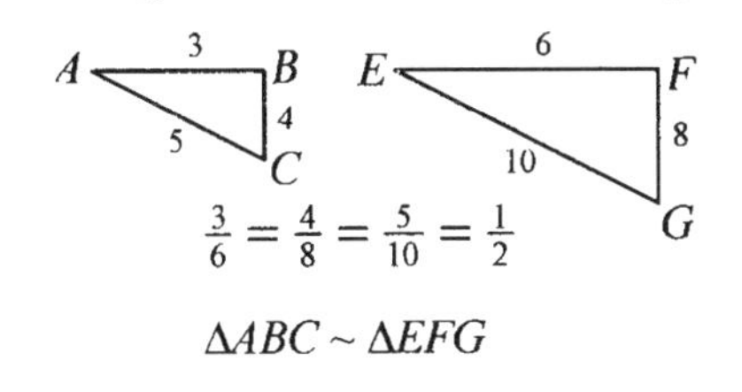

$\frac{3}{6} = \frac{4}{8} = \frac{5}{10} = \frac{1}{2}$

$\Delta ABC \sim \Delta EFG$

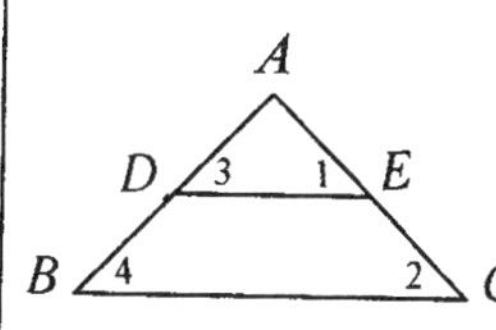

If $DE \parallel BC$, then $\angle 1 = \angle 2$ and $\angle 3 = \angle 4$. We have $\Delta ABC \sim \Delta ADE$.

Formulas: $\frac{AD}{AB} = \frac{AE}{AC}$ or $\frac{AD}{DB} = \frac{AE}{EC}$

Congruent Triangles: Triangles that have the same size and same shape are congruent. To prove two triangles are congruent, it is not necessary to show all six parts are congruent. The following rules (postulates) can simply determine and prove that two triangles are congruent.

1. **SSS Postulate (Side-Side-Side)**
   If three sides of one triangle are congruent to three sides of another triangle, then the triangles are congruent.
2. **SAS Postulate (Side-Angle-Side)**
   If two sides and the included angle of one triangle are congruent to the corresponding two sides and the included angle of another triangle, then the triangles are congruent. An included angle is the angle formed by the two sides.
3. **ASA Postulate (Angle-Side-Angle)**
   If two angles and the included side of one triangle are congruent to the corresponding two angles and the included side of another triangle, then the triangles are congruent. An included side is the side formed by the two angles.
   Since the sum of the measure of interior angles of a triangle is $180°$, we can conclude that **AAS** is also valid for congruent.

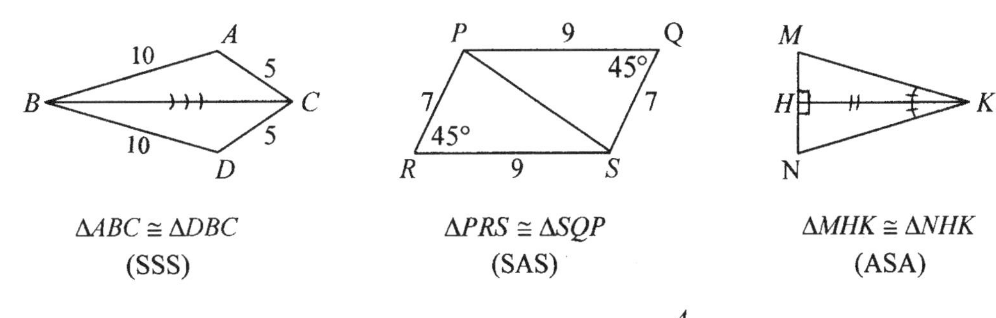

$\Delta ABC \cong \Delta DBC$ (SSS) $\Delta PRS \cong \Delta SQP$ (SAS) $\Delta MHK \cong \Delta NHK$ (ASA)

Note that **SSA is not valid for congruent.** $\Delta ABD \neq \Delta ABC$ (SSA) (Ambiguous Case)

CPCTC: Corresponding parts of congruent triangles are congruent. We can use congruent triangles to find the unknown measures of the sides and angles of the triangles.

Examples

1. In the figure, $\Delta ABC \sim \Delta CDE$, find x and y.

Solution:

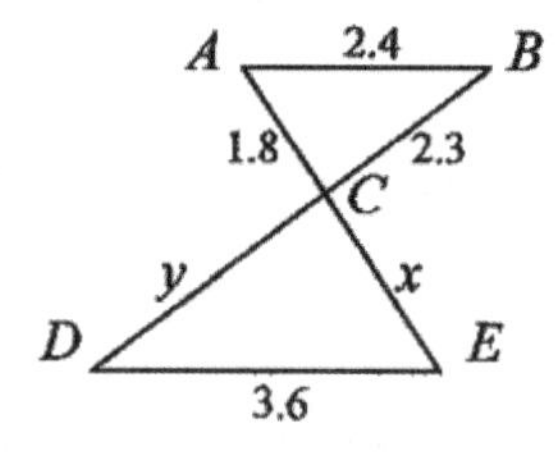

$$\frac{x}{1.8} = \frac{y}{2.3} = \frac{3.6}{2.4} = 1.5$$

Every side of ΔCDE is 1.5 times the corresponding side of ΔABC.

$\therefore x = 1.8 \times 1.5 = 2.7$. Ans

$y = 2.3 \times 1.5 = 3.45$. Ans.

2. Two triangles are similar. The lengths of sides of the large triangle is twice as long as the lengths of sides of the small triangle. If the area of the large triangle is 1, what is the area of the small triangle ?

Solution:

We choose the larger triangle with a base of 2 and a height of 1.

We have the area of the small triangle:

$A = \frac{1}{2}bh = \frac{1}{2} \cdot 1 \cdot \frac{1}{2} = \frac{1}{4} = 0.25$. Ans.

3. In the figure, $\overline{DE} \parallel \overline{BC}$. If $\overline{DB} = \frac{1}{3}\overline{AD}$, find the ratio of the area of ΔADE to the area of ΔABC.

Solution:

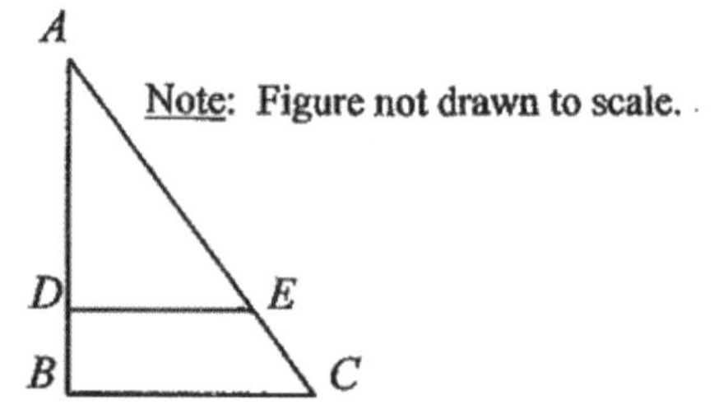

Since $\overline{DE} \parallel \overline{BC}$, $\Delta ADE \sim \Delta ABC$

We choose $AB = 4$, $AD = 3$.

If two triangles are similar, the ratio of their areas is the square of the ratio of the lengths of the corresponding sides.

$$\frac{\Delta ADE}{\Delta ABC} = \left(\frac{3}{4}\right)^2 = \frac{9}{16}$$

Ans.

4. State whether each pair of triangles is congruent by the SAS, ASA, SSS, or None.

a.

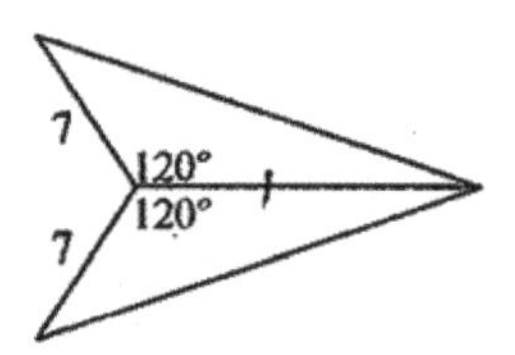

b.

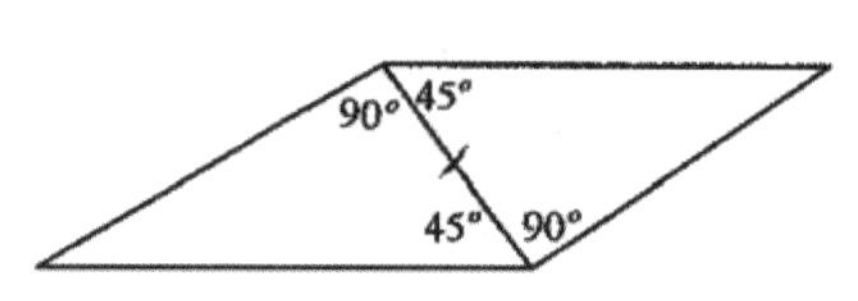

c.

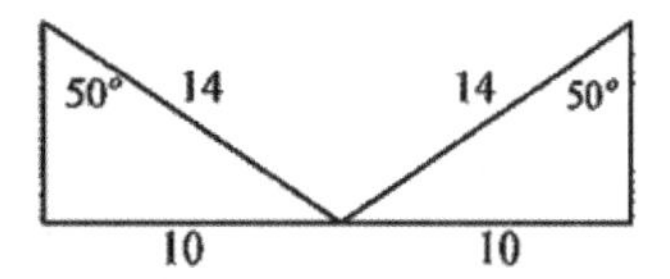

d.

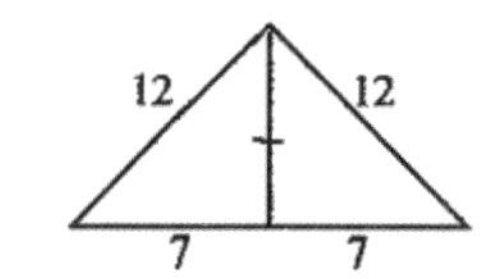

Solution:

a. SAS   b. ASA   c. None   d. SSS

3-9 Basic Rules in Polygons

Polygon: It is a flat shape formed by straight line segments. A polygon has at least three sides. Polygon are named according to the number of sides, such as:
Triangle (3), Quadrilateral (4), Pentagon (5), Hexagon (6), Heptagon (7), Octagon (8), Nonagon (9), Decagon (10), Dodecagon (12)

Regular Polygon: It is a polygon that has all equal sides and angles.
Diagonal of a polygon: It is a segment, other than a side, joining two vertices. A triangle has no diagonal.
Exterior angles of a polygon: They are the angles formed by extending each side in succession. The interior and exterior angle at each vertex form a linear pair.
Perimeter of a polygon is the sum of the lengths of its sides.
A polygon always means a convex polygon such that no line containing a side of the polygon also contain a point in the interior of the polygon.

A quadrilateral is a polygon having four sides. Each of the following quadrilaterals has a special name.

Parallelogram: A quadrilateral with both pairs of opposite sides parallel.
Rectangle: A parallelogram with four right angles.
Square: A parallelogram with four right angles and all four sides congruent.
Rhombus: A parallelogram with all four sides congruent and the diagonals are perpendicular to each other.
Trapezoid: A quadrilateral with exactly one pair of parallel sides. It includes right trapezoid and isosceles trapezoid.

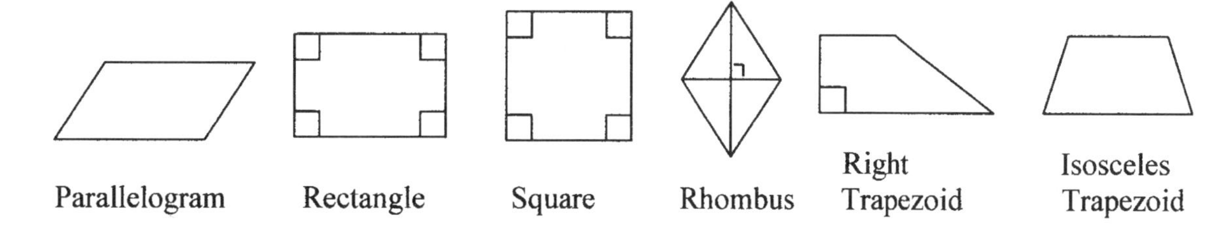

Note that rectangle, square, and rhombus are all parallelograms. Trapezoid is the only quadrilateral that is not a parallelogram.

A circle inscribed in a regular polygon is tangent to each side of the polygon.
A circle circumscribed about a regular polygon contains each vertex of the polygon.
The radius of a regular polygon is a radius of its circumscribed circle.
An **apothem** of a regular polygon is an altitude of its divided triangles. It is the radius of its inscribed circle.

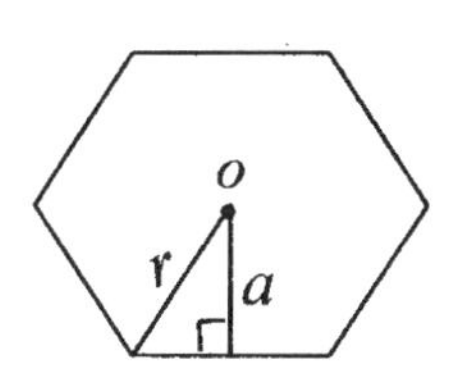

r: the radius
a: apothem

Basic Rules in Polygons

<1, <2, <3, and <4 are the interior angles.
<5, <6, <7, and <8 are the exterior angles.

1. **Interior Angle-Sum Theorem:** The sum of the measures of the interior angles of a polygon with n sides is $180(n-2)$.

   Note that a polygon with n sides can be divided into $(n-2)$ triangles.
   Therefore, the sum of the measures of the interior angles is $180(n-2)$.

   The measure of each interior angle of a regular polygon with n sides is $\frac{180(n-2)}{n}$.

2. **Exterior Angle-Sum Theorem:** The sum of the measures of the exterior angles of a polygon with n sides is 360.

   Note that a polygon with n sides has n linear pairs. The sum of the measures of the interior and exterior angles is $180 \times n$. The sum of the measures of the interior angles is $180(n-2)$. Therefore, the sum of the measures of the exterior angles is $180n - 180(n-2) = 360$.

   The measure of each exterior angle of a regular polygon with n sides is $\frac{360}{n}$.

3. **The Properties of a Parallelogram:** A parallelogram has the following properties:
   a. Opposite sides are equal.
   b. Opposite angles are congruent.
   c. Adjacent angles are supplementary (they add up to 180°).

4. **The Diagonals of a Polygon:** The total number of diagonals of a polygon with n sides can be expressed as $\frac{n(n-3)}{2}$.

   We can also use sequence to find the number of diagonals of a polygon.

n sides	4	5	6	7	8	9	10	11	12
diagonals	2	5	9	14	20	27	35	44	54

5. **Area of a Regular Polygon:**
   a. A regular polygon with n sides and side length s can be divided into n equilateral triangles having a side of length s. Apply the area of the equilateral triangle, the area of a regular polygon: $A = n \cdot \frac{s^2\sqrt{3}}{4}$
   b. The area of a regular polygon equals one-half of its apothem (a) times its perimeter (p): $A = \frac{1}{2}ap$

6. **Similarity of Polygons:** Regular polygons of the same number of sides are similar.
   The perimeters (or areas) of two similar polygons have the same ratio as their corresponding sides, as their radii, or as their apothems.

Examples

1. Find the sum of the measures of the angels of a polygon with 32 sides.
Solution:

The sum of angels is $S = 180(n-2)$

For $n = 32$, $S = 180(32-2) = 5400$. Ans.

2. A regular polygon has 10 sides. Find the measure of each angle.
Solution:

The measure of each angle of a regular polygon is $A = \frac{180(n-2)}{n}$.

For $n = 10$, $A = \frac{180(10-2)}{10} = 144$. Ans.

3. The measure of each angle of a polygon is 144. How many sides does the polygon has ?.
Solution:

$144 = \frac{180(n-2)}{n}$, $144n = 180(n-2)$, $36n = 360$

$\therefore n = \frac{360}{36} = 10$ sides. Ans.

4. How many diagonals can we draw in a decagon (10 sides) ?
Solution:

$\frac{n(n-3)}{2} = \frac{10(10-3)}{2} = 35$ diagonals. Ans.

(Hint: It is easier to find the answer by using sequence.)

5. An equilateral triangle is inscribed in a circle whose radius is 18. Find the length of the apothem (a).

Solution: $m < OBD = 30$

ΔOBD is a 30-60-90 triangle.

$OD = \frac{1}{2} OB$

$\therefore a = \frac{1}{2}(18) = 9$. Ans.

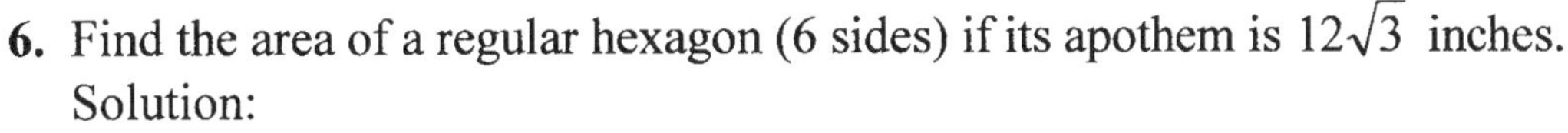

6. Find the area of a regular hexagon (6 sides) if its apothem is $12\sqrt{3}$ inches.
Solution:

$m < AOB = \frac{360}{6} = 60$

ΔAOC is a 30-60-90 triangle. $m < AOC = 30$

$AC = \frac{1}{2}r$, $a = \frac{1}{2}r \cdot \sqrt{3}$

$r = \frac{2a}{\sqrt{3}} = \frac{2 \cdot 12\sqrt{3}}{\sqrt{3}} = 24$

$AC = \frac{1}{2}r = \frac{1}{2}(24) = 12$, $AB = 24$

Perimeter $p = 24 \times 6 = 144$

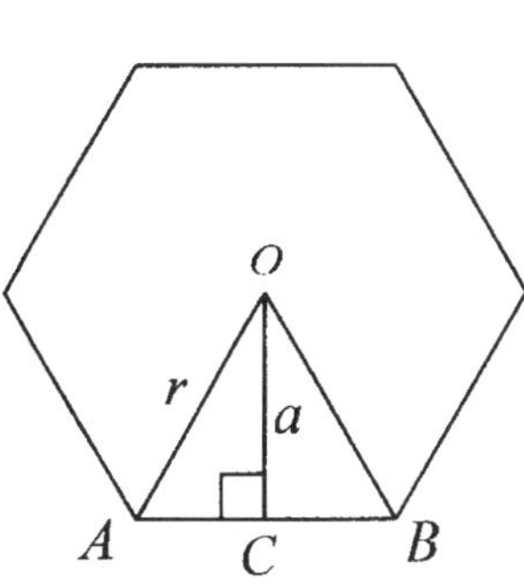

Area: $A = \frac{1}{2}ap = \frac{1}{2} \cdot 12\sqrt{3} \cdot 144 = 864\sqrt{3} \approx 1496.45\ in.^2$. Ans.

Notes

3-10 Basic Rules in Circles

Circle: It is the set of all points in a plane at a same distance (called the radius) from a fixed point (called the center).

Radius (r)**:** It is a segment joining the center of the circle with a point on the circle.

Chord: It is a segment with endpoints on the circle.

Diameter (d)**:** It is a chord passing through the center of the circle.

Circumference: It is the distance around a circle. $C = 2\pi r$ or $C = \pi d$

Concentric Circles: They are circles having the same center and radii of different lengths.

Central Angle: It is the angle having its vertex at the center of the circle.

Minor Arc: The interior of a central angle forms a minor arc on the circle.

Major Arc: The exterior of a central angle forms a major arc on the circle. We use three letters to name a major arc.

Basic Rules in Circles

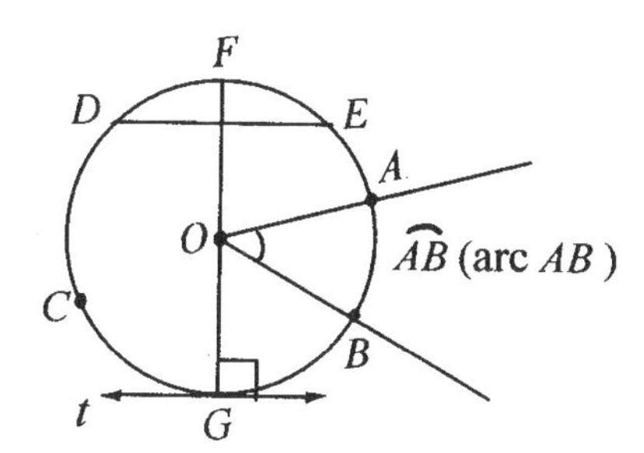

$\overline{OA}$, $\overline{OB}$, $\overline{OG}$, $\overline{OF}$ are the radii.

$\overline{FG}$ is the diameter. $\overline{DE}$ is a chord.

$\angle AOB$ is a central angle.

$\overset{\frown}{AB}$ is a minor arc. $\overset{\frown}{ACB}$ is a major arc.

Line t is tangent to the circle at G.

1. The measure of a minor arc equals the measure of its central angle. $m\angle AOB = m\overset{\frown}{AB}$

2. Two circles are congruent if their radii are congruent.

3. In the same circle, if two arcs are congruent, then their chords are congruent. In the same circle, if two chords are congruent, then their arcs are congruent. In the same circle, two chords are congruent if and only if they are equidistant from the center.

4. If a diameter of a circle is perpendicular to a chord, then the diameter bisects the chord and its arc.

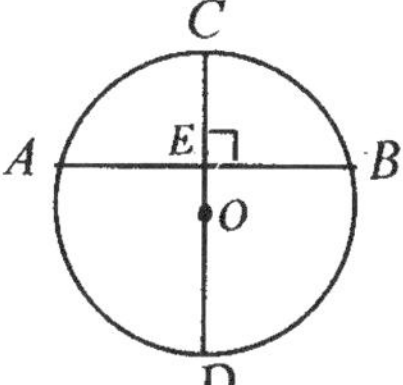

If $\overline{CD} \perp \overline{AB}$, then $\overline{AE} = \overline{EB}$, $\overset{\frown}{AC} = \overset{\frown}{CB}$.

5. If a line is tangent to a circle, then the line is perpendicular to the radius drawn to the point of tangency.

$\overline{AB} \perp \overline{OB}$, $\overline{AC} \perp \overline{OC}$

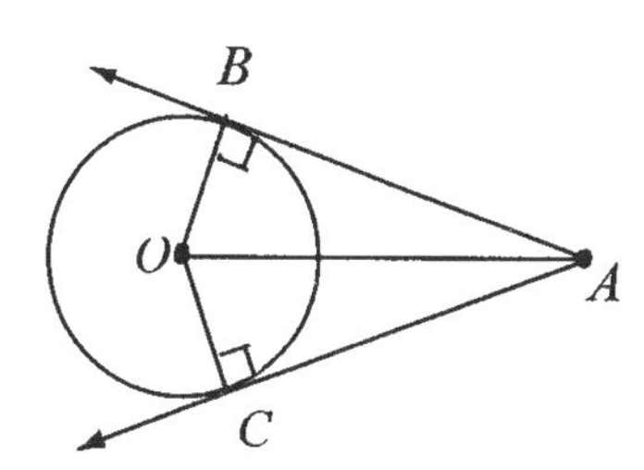

6. The two tangent segments from the same exterior point of a circle are congruent.

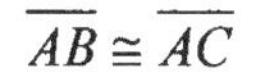

$\overline{AB} \cong \overline{AC}$

Basic Rules in Circles (Continued)

7. The measure of an inscribed angle of a circle is one-half the measure of its intercepted arc.

8. If the inscribed angles intercepts the congruent arcs, the angles are congruent.

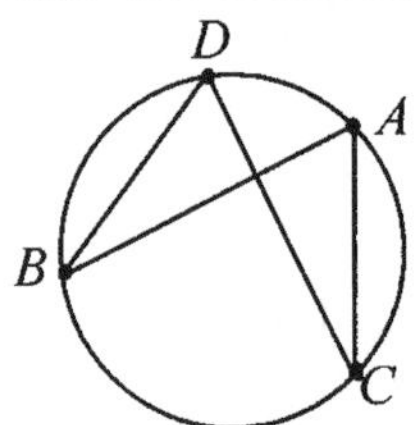

$m < BAC = \frac{1}{2} m\overset{\frown}{BC}$, $m < BDC = \frac{1}{2} m\overset{\frown}{BC}$

$m < BAC = m < BDC$

9. If an inscribed angle of a circle is inscribed in a semicircle, the inscribed angle is a right angle.

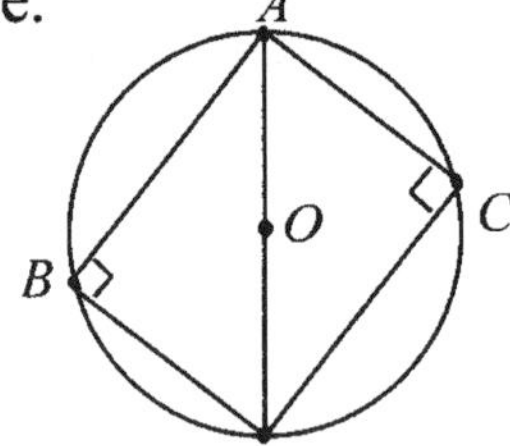

$m < ACD = 90$

$m < ABD = 90$

10. In a circle, the opposite angles of an inscribed quadrilateral are supplementary.

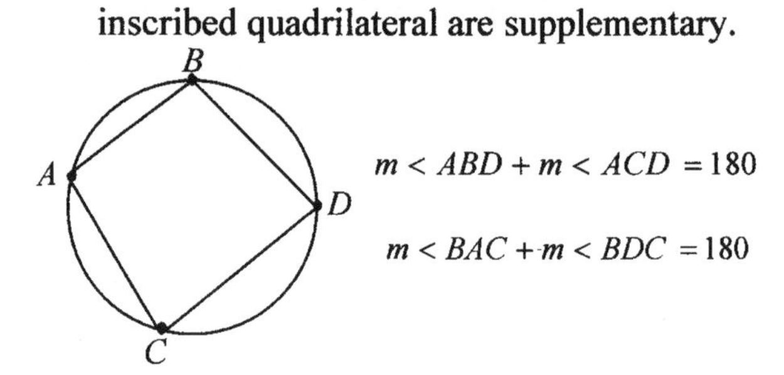

$m < ABD + m < ACD = 180$

$m < BAC + m < BDC = 180$

11. If a tangent and a secant intersect at a point on the circle, the measure of the angle formed is one-half the measure of the intercepted arc.

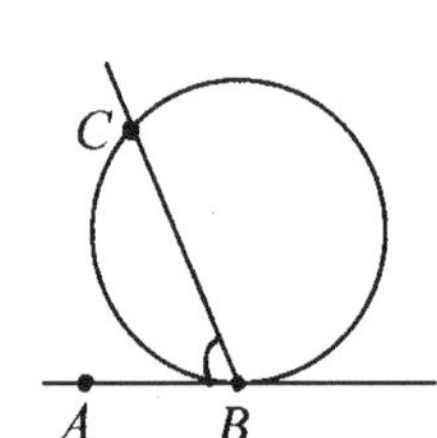

$m < ABC = \frac{1}{2} m\overset{\frown}{BC}$

12. If two secants intersect in the interior of a circle, the measure of the angle formed is one-half the sum of the measures of the intercepted arcs.

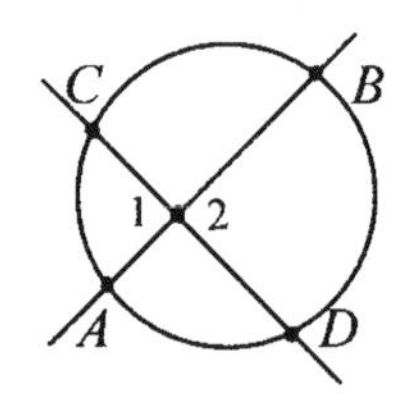

$m < 1 = m < 2$

$= \frac{1}{2}(m\overset{\frown}{AC} + m\overset{\frown}{BD})$

13. If two secants, a tangent and a secant, or two tangents intersect in the exterior of a circle, the measure of the angle formed is one-half the difference of the measures of the intercepted arcs.

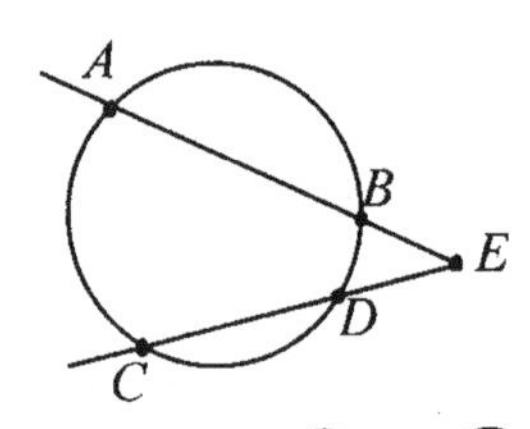

$m < E = \frac{1}{2}(m\overset{\frown}{AC} - m\overset{\frown}{BD})$

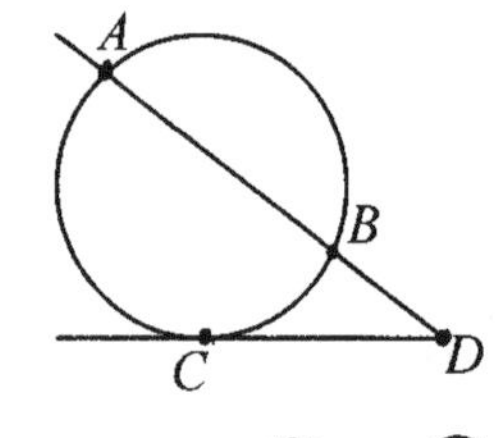

$m < D = \frac{1}{2}(m\overset{\frown}{AC} - m\overset{\frown}{BC})$

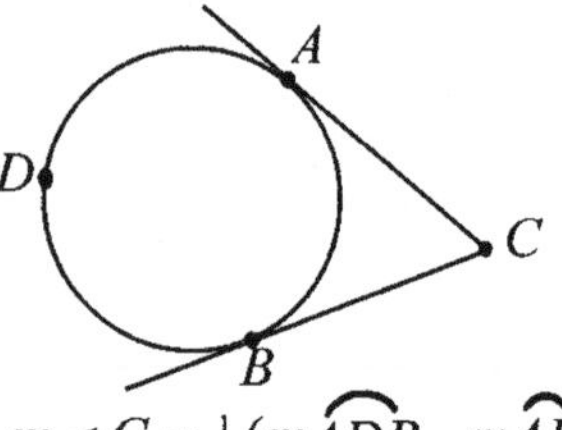

$m < C = \frac{1}{2}(m\overset{\frown}{ADB} - m\overset{\frown}{AB})$

14. If two chords intersect in the interior of a circle, the two triangles formed are similar.

15. If a tangent and a secant intersect in the exterior of a circle, the two triangles formed by the tangent and secant segments with each of the two chords are similar.

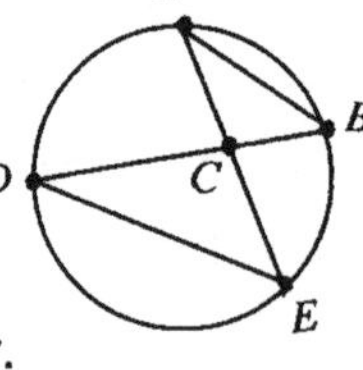

$\Delta ABC \sim \Delta CDE$

$AC \cdot CE = BC \cdot CD$

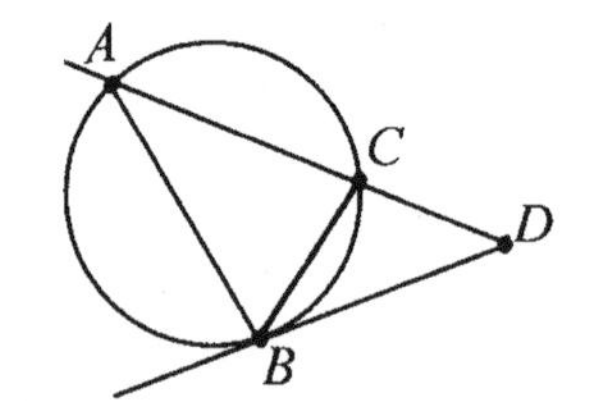

$\Delta ABD \sim \Delta CBD$

$AD : BD = BD : CD$

$(BD)^2 = AD \cdot CD$

Examples

1. In the circle, $\overline{AB}$ is the diameter, $m<C=50$, $m<D=70$. Find the measure of each numbered angle.
Solution:

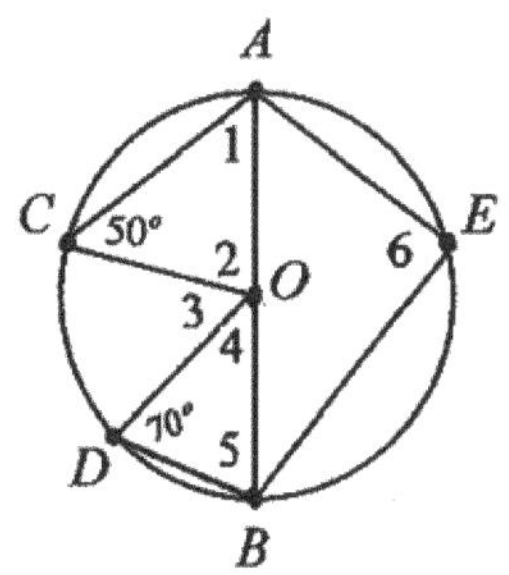

$\overline{OA}\cong\overline{OC}$, $m<1=m<C=50$
$m<2=180-50-50=80$
$\overline{OD}\cong\overline{OB}$, $m<5=m<D=70$
$m<4=180-70-70=40$
$m<3=180-80-40=60$
$m<6=90$ (Inscribed in a semicircle)

2. In the circle, $\overline{BC}$ is the diameter, the measures of the central angles are given, $m<1=45$, $m<2=60$. Find $m\widehat{AB}$, $m\widehat{AC}$, $m\widehat{AD}$, and $m\widehat{ABC}$.
Solution:

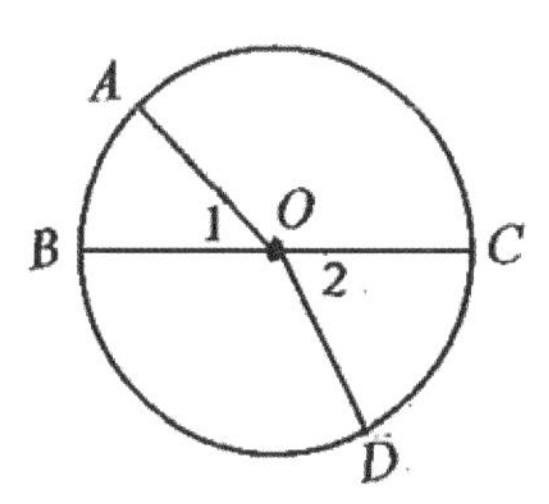

$m\widehat{AB}=m<1=45$

$m\widehat{AC}=180-45=135$

$m\widehat{AD}=m<1+m<BOD=45+120=165$

$m\widehat{ABC}=360-m<AOC=360-135=225$

3. In the circle, $\overline{AB}$ is the diameter, $\overline{CD}\perp\overline{AB}$, $CD=30$, and $OE=8$. Find the length of $\overline{CO}$.
Solution:

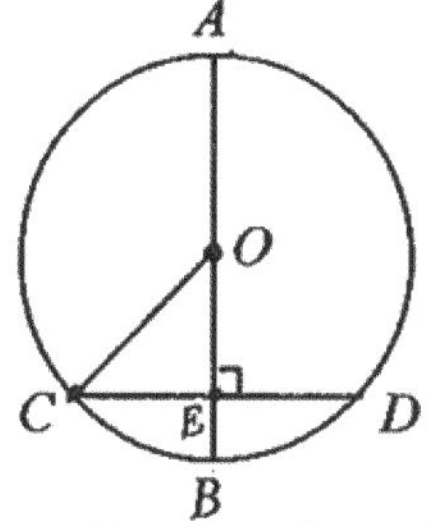

Note: Figure not drawn to scale.

$CE=\frac{1}{2}(CD)=\frac{1}{2}(30)=15$
$(CO)^2=(CE)^2+(OE)^2=15^2+8^2=289$
$CO=17$. Ans.

4. $\overline{AB}$ and $\overline{AC}$ are two tangent segments to the circle, $BO=8, AO=17$. Find AB and AC.
Solution:

ΔABO and ΔACO are right triangles and $\overline{AB}\cong\overline{AC}$.

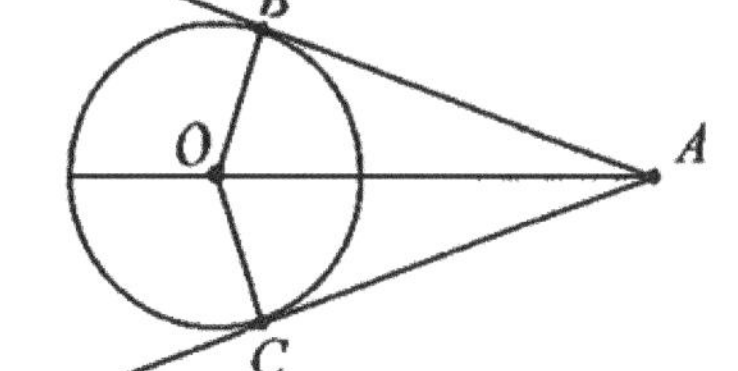

$(AB)^2+(BO)^2=(AO)^2$
$(AB)^2+8^2=17^2$
$(AB)^2+64=289$
$(AB)^2=225$
$\therefore AB=15$. Ans.
$AC=15$. Ans.

Examples

5. In the circle, find $m < M$, $m < MPN$, $m < 1$, and $m < Q$.

Solution:

$m < M = \frac{1}{2} m\widehat{OP} = \frac{1}{2}(40) = 20$

$m < MPN = \frac{1}{2} \widehat{MN} = \frac{1}{2}(100) = 50$

$m < 1 = \frac{1}{2}(m\widehat{MN} + m\widehat{OP}) = \frac{1}{2}(100 + 40) = 70$

$m < Q = \frac{1}{2}(m\widehat{MN} - m\widehat{OP}) = \frac{1}{2}(100 - 40) = 30$

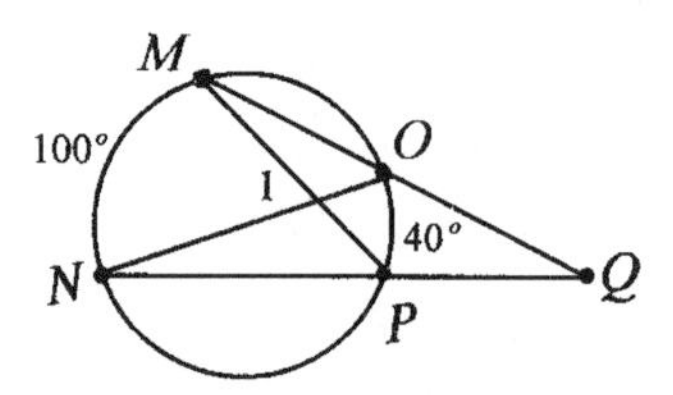

6. In the figure, quadrilateral PQSR is inscribed in the circle. Find $m < QPS$, $m < RPS$, $m < Q$, $m < R$, and $m < TPR$.

Solution:

$m < QPS = \frac{1}{2} m\widehat{QS} = \frac{1}{2}(90) = 45$

$m < RPS = \frac{1}{2} m\widehat{RS} = \frac{1}{2}(100) = 50$

$< Q$ and $< R$ are right angles.

$m < Q = m < R = 90$

$m < TPR = \frac{1}{2} m\widehat{PR} = \frac{1}{2}(180 - 100) = 40$

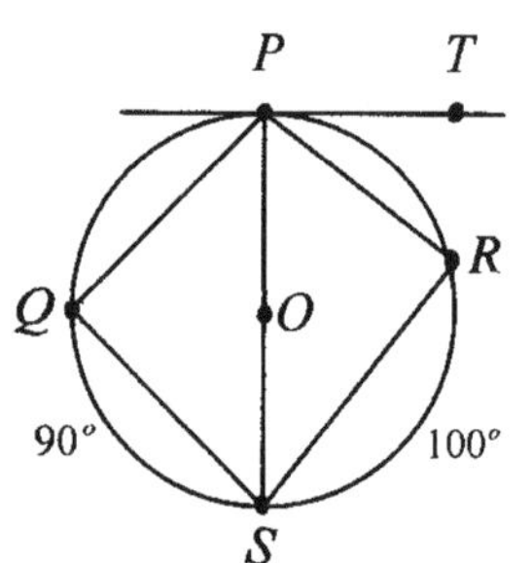

7. In the circle, chord AB and CD intersect at E. If $CE = 12$, $DE = 15$, and $AE = 18$, find BE.

Solution:

$18 \cdot BE = 15 \cdot 12$

$BE = \frac{180}{18} = 10$. Ans.

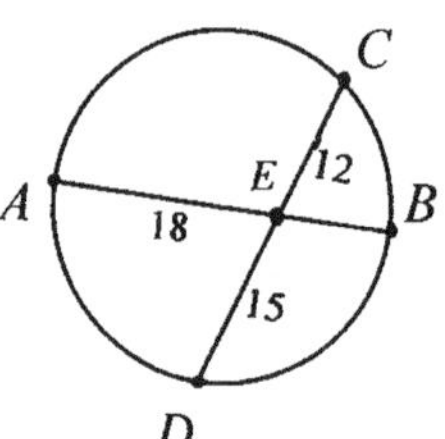

8. In the figure, $\overline{MP}$ and $\overline{QP}$ intersect at P, $MN = 16$, $NP = 9$, find QP.

Solution:

$(QP)^2 = MP \cdot NP$

$(QP)^2 = 25 \cdot 9 = 225$

$QP = 15$. Ans.

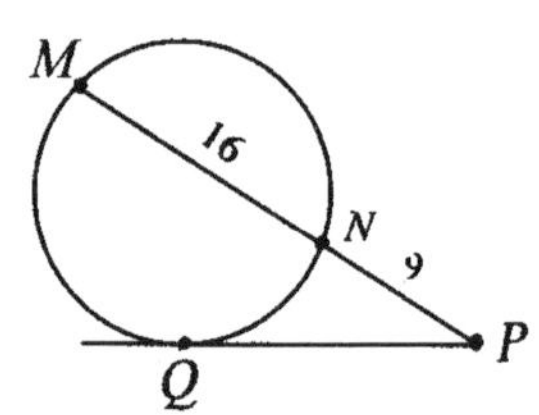

9. A right triangle is inscribed in the circle. The diameter of the circle is 10 *cm*. What would the maximum area of the triangle be ?

Solution:

The hypotenuse of a right triangle inscribed a circle is the diameter of the circle.
Therefore, the base of the triangle is 10 *cm*.
The maximum height can be the radius 5 *cm*.

Max. Area $= \frac{1}{2} bh = \frac{1}{2} \cdot 10 \cdot 5 = 25\ cm^2$. Ans.

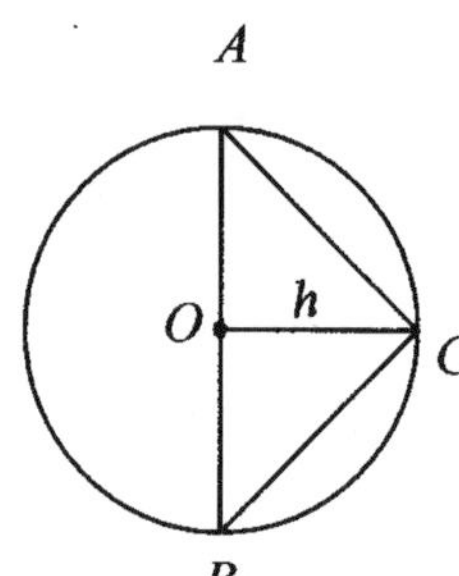

3-11 Area Formulas and Perimeters

Area is the measure of the region enclosed by a figure. Area is measured in square units, such as cm^2 (square centimeters), m^2 (square meters), $in.^2$ (square inches), and ft^2 (square feet), ······ . **Perimeter** of a polygon is the sum of the lengths of its sides.

1. Area of a Triangle: $A = \frac{1}{2}bh$

b : the base h: the height

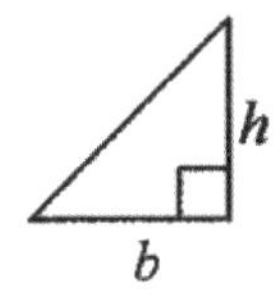

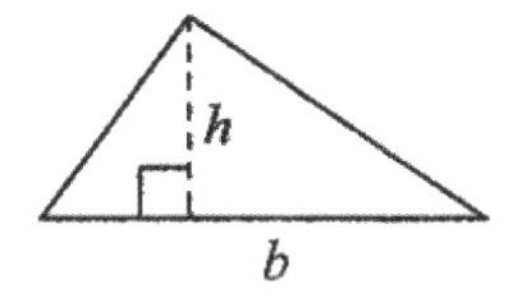

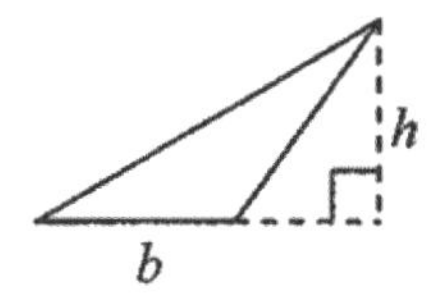

2. Area of an Equilateral Triangle: $A = \frac{s^2\sqrt{3}}{4}$

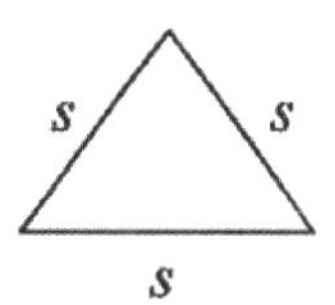

3. Area of a Square: $A = s^2$

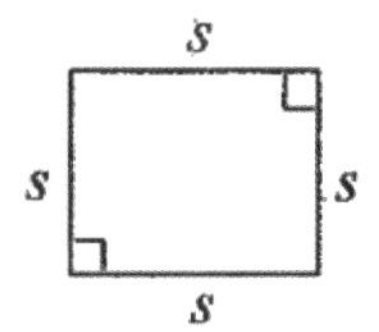

4. Area of a Rectangle: $A = \ell \cdot w$

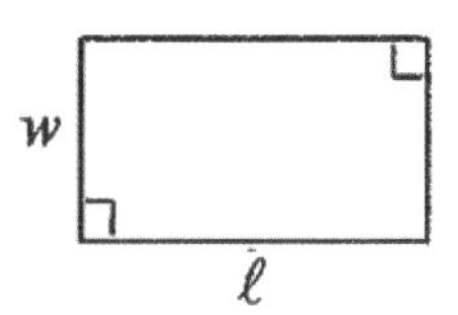

5. Area of a Parallelogram: $A = b \cdot h$

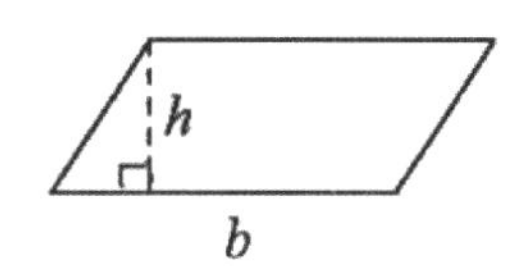

6. Area of a Trapezoid: $A = \frac{1}{2}(b_1 + b_2)h$

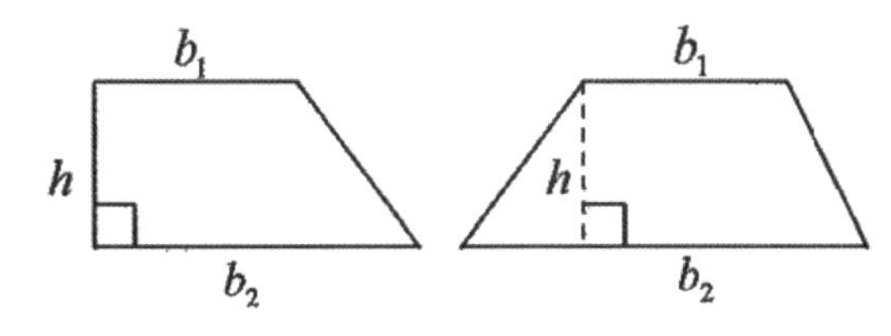

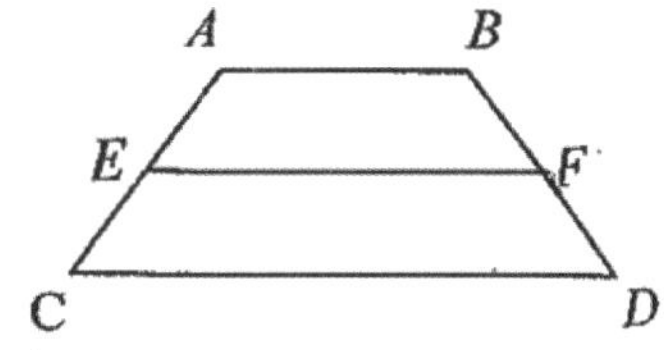

Other Formula:

$EF = \frac{1}{2}(AB + CD)$

(EF is the median.)

7. Area of a Rhombus: $A = \frac{1}{2}(d_1 d_2)$

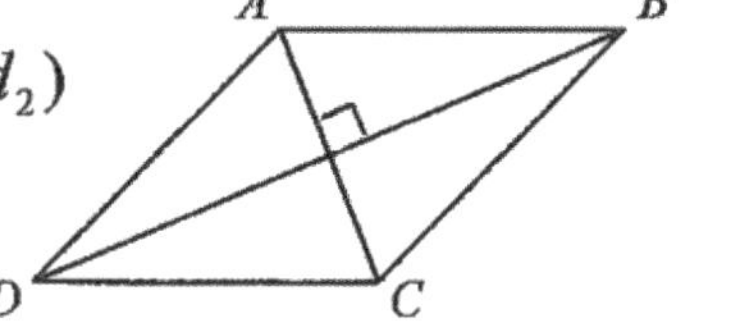

$d_1 = AC$

$d_2 = BD$

Area Formulas (Continued)

8. Area of a Circle: $A = \pi r^2$

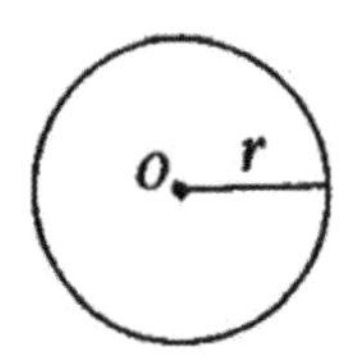

9. Area of a Sector of a Circle: $A = \pi r^2 (\frac{m}{360})$

m: measure of the arc of the sector OAB

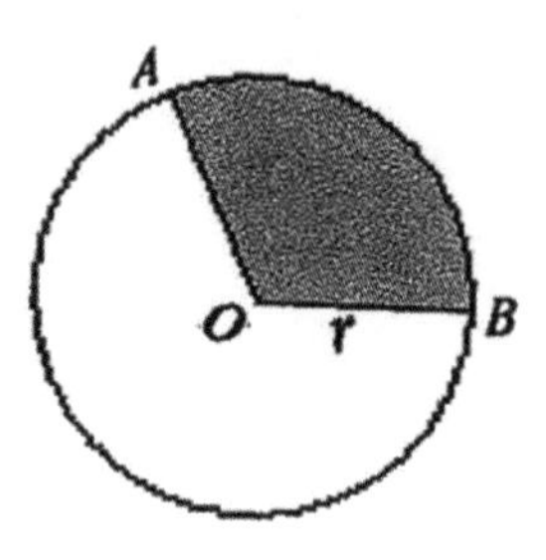

10. Area of a Regular Polygon: $A = \frac{1}{2} a p$

a: apothem
p: perimeter

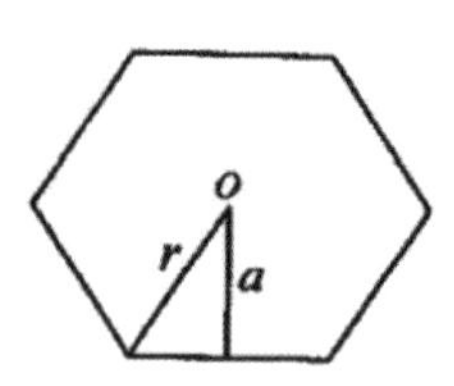

11. Area of a sector and a segment of a Circle:

A segment of a circle is the region bounded by an arc and its chord.
The area of the shaded region is:
Area of sector OAB – Area of ΔOAB

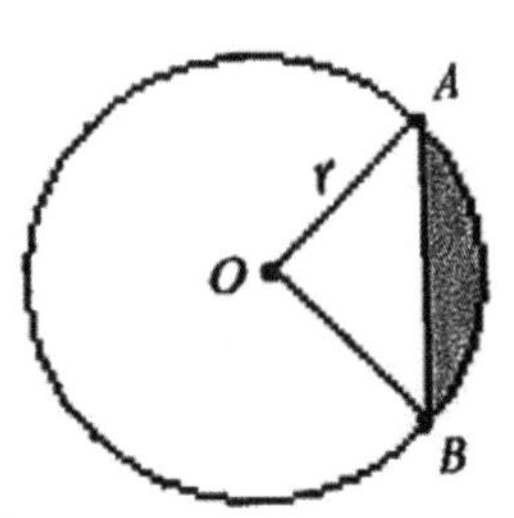

In a toy store, a five-year old girl wants her father to buy her an expensive toy.
Father: No, it is too expensive.
Girl: Dad, I like it. Honey and Sweetheart.
Father: Why are you calling me "Honey and Sweetheart" ?
Girl: Because whenever mom wants to buy a new dress, she calls you Honey, or Sweetheart.

Examples

1. In the triangle, $AC = 20$, find the area of ΔABC.
   Solution:

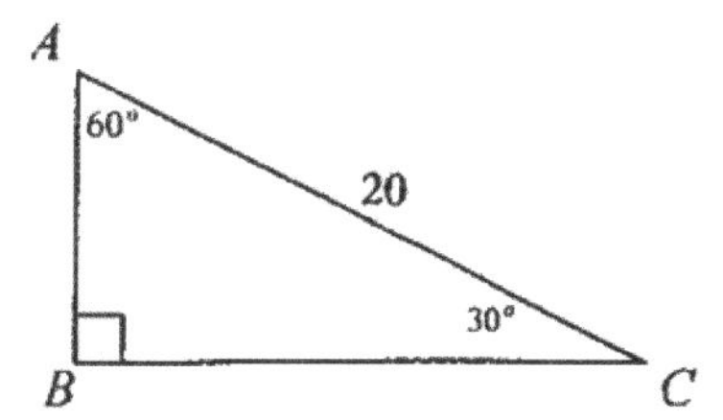

$$h = AB = \frac{1}{2}(AC) = \frac{1}{2}(20) = 10$$
$$b = BC = \sqrt{3}(AB) = \sqrt{3}(10) = 10\sqrt{3}$$
Area: $A = \frac{1}{2}bh = \frac{1}{2}(10\sqrt{3})(10) = 50\sqrt{3}$. Ans.

2. In the triangle, $AB = 20$, $BC = 50$, and $m < D = 45^o$ find the area of ΔABC.
   Solution:

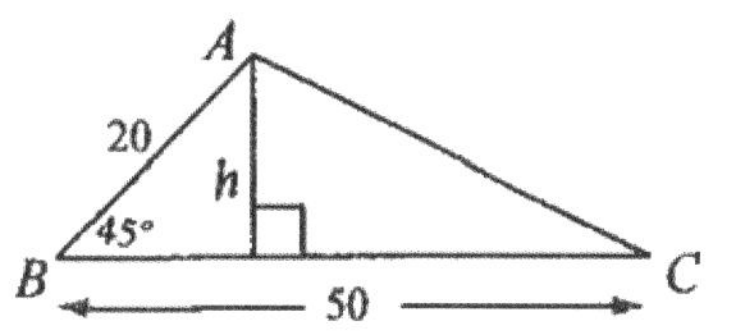

$$b = BC = 50$$
$$AB = h\sqrt{2}, \quad h = \frac{AB}{\sqrt{2}} = \frac{20}{\sqrt{2}} = 10\sqrt{2}$$
Area: $A = \frac{1}{2}bh = \frac{1}{2}(50)(10\sqrt{2}) = 250\sqrt{2}$. Ans.

3. In the figure, a square is cut out from a rectangle of length 25 and width 15. Find the area of the shaded region.
   Solution:

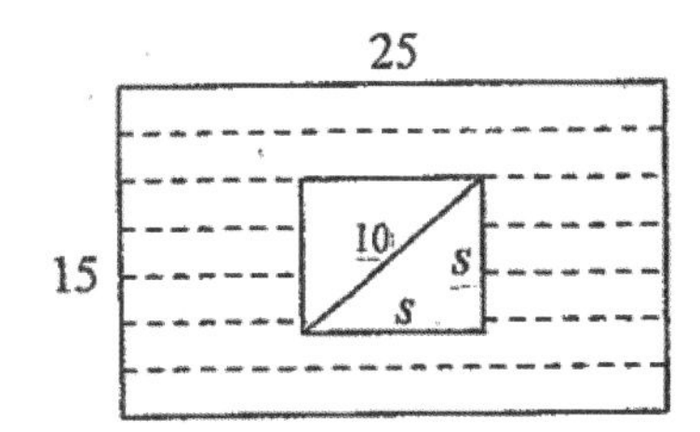

$s^2 + s^2 = 10^2$, $2s^2 = 100$ $\therefore s^2 = 50$
Area of the square $= s^2 = 50$
Area of the shaded region
$=$ Area of the rectangle $-$ Area of the square
$= 25 \cdot 15 - 50 = 325$. Ans.

4. In the figure, a rhombus is inscribed in a rectangle. Find the area of the shaded region.
   Solution:

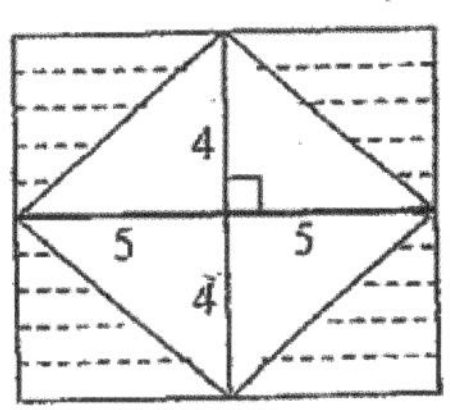

Area of the rhombus $= \frac{1}{2}(d_1 d_2) = \frac{1}{2}(10)(8) = 40$
Area of the rectangle $= l \cdot w = 10 \cdot 8 = 80$
Area of the shaded region $= 80 - 40 = 40$. Ans.

5. In the parallelogram *ABCD*, $AB = 18\,cm$, $AD = 12\,cm$, and $m < D = 60$, find the area.
   Solution:

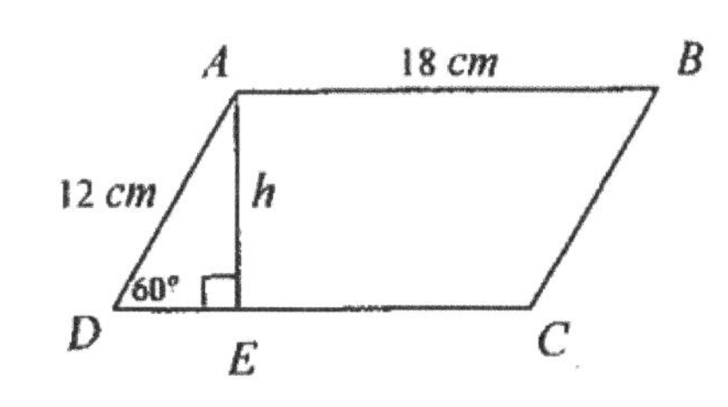

$$b = AB = 18$$
$$DE = \frac{1}{2}(AD) = \frac{1}{2}(12) = 6$$
$$h = \sqrt{3}(DE) = 6\sqrt{3}$$
Area of the parallelogram:
$A = bh = 18(6\sqrt{3}) = 108\sqrt{3}\ cm^2$. Ans.

6. In the trapezoid $MNPQ$, $MN = 18$ *in.*, $QP = 30$ *in.*, and $m\angle Q = 45$, find the area.

Solution:

$12 = \sqrt{2}h$, $h = \frac{12}{\sqrt{2}} = 6\sqrt{2}$

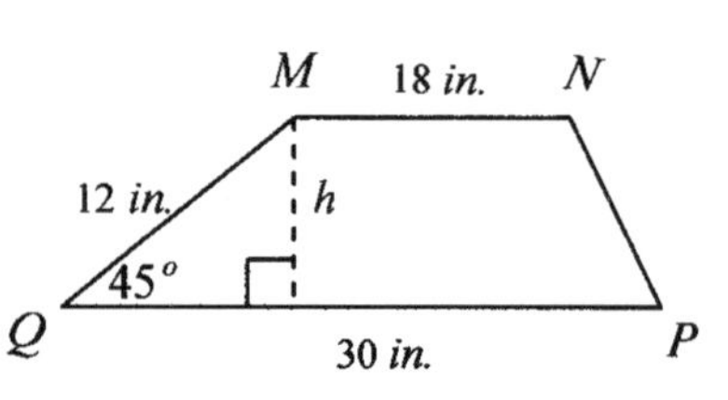

$b_1 = MN = 18$, $b_2 = QP = 30$

Area of the trapezoid:

$A = \frac{1}{2}(b_1 + b_2)h = \frac{1}{2}(18+30)(6\sqrt{2}) = 144\sqrt{2}$ $in.^2$ Ans.

7. Find the area of a regular pentagon (5 sides) with a side of 12 meters and an apothem of 3 meters.

Solution:

Perimeter of the pentagon $p = 12 \times 5 = 60$

Area of the pentagon $A = \frac{1}{2}ap = \frac{1}{2}(3)(60) = 90\ m^2$. Ans.

8. Find the area of a regular hexagon (6 sides) if its apothem is $12\sqrt{3}$ inches.

Solution:

$m\angle AOB = \frac{360}{6} = 60$, $m\angle AOC = 30$

Apothem $a = \sqrt{3}(AC) = 12\sqrt{3}$ $\therefore AC = 12$

$r = 2(AC) = 2(12) = 24$

$AB = 24$, perimeter $p = 24(6) = 144$

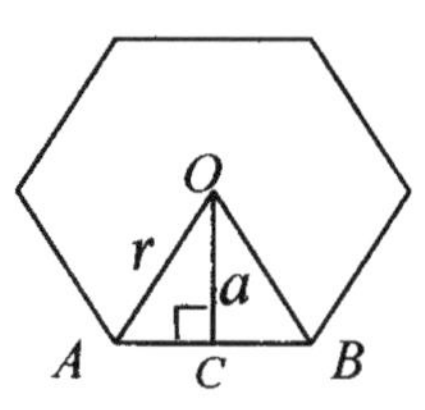

Area of the hexagon $A = \frac{1}{2}ap = \frac{1}{2}(12\sqrt{3})(144) = 864\sqrt{3}$ $in.^2$. Ans.

9. A regular hexagon (6 sides) is inscribed in the circle. The radius of the circle is 24 inches. Find the area of the shaded region

Solution:

Apothem of the hexagon $a = \sqrt{3}(\frac{1}{2}r) = \sqrt{3}(12)$

Perimeter of the hexagon $p = 24(6) = 144$

Area of the hexagon $A = \frac{1}{2}ap = \frac{1}{2}(12\sqrt{3})(144) \approx 1496.45$

Area of the circle $A = \pi r^2 = \pi(24)^2 \approx 1808.64$

Area of the shaded region $= 1808.64 - 1496.45 \approx 312.19\,in.^2$. Ans.

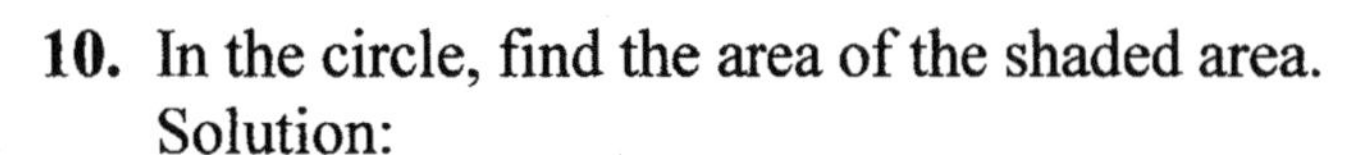

10. In the circle, find the area of the shaded area.

Solution:

Area of sector OAB $= \pi\,(8^2)(\frac{90}{360}) = 16\pi$

Area of $\Delta OAB = \frac{1}{2}bh = \frac{1}{2}(8)(8) = 32$

Area of the shaded region $= 16\pi - 32 \approx 18.27$

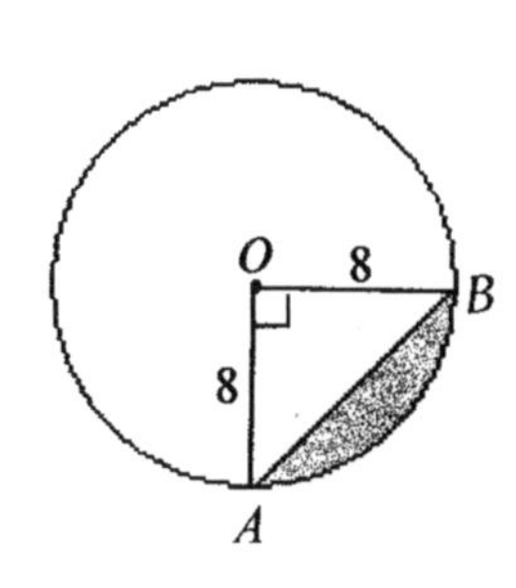

3-12 Distance and Midpoint Formulas

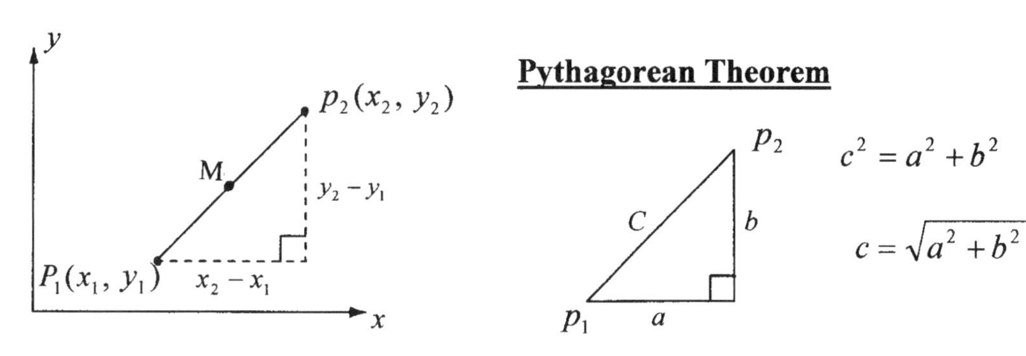

Pythagorean Theorem

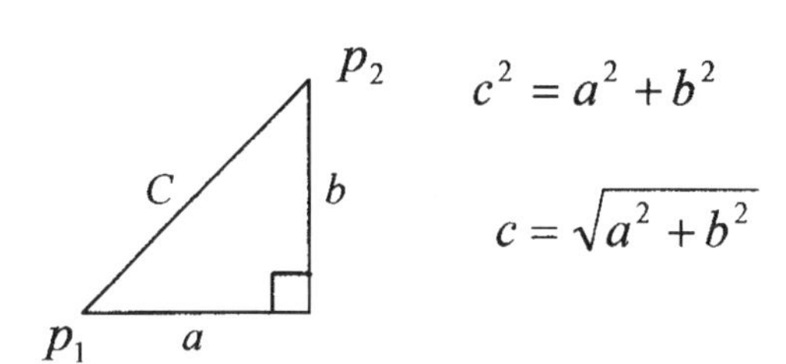

$$c^2 = a^2 + b^2$$

$$c = \sqrt{a^2 + b^2}$$

Distance Formula: The distance between two points p_1 and p_2 is

$$d = \overline{p_1 p_2} = \sqrt{(x_2 - x_1)^2 + (y_2 - y_1)^2}$$

Midpoint Formula: The midpoint between two points p_1 and p_2 is

$$M = \left(\frac{x_1 + x_2}{2}, \frac{y_1 + y_2}{2} \right)$$

Examples

1. Find the distance and midpoint between $A(-2,\ 1)$ and $B(5,\ 4)$.
Solution:

$$d = \overline{AB} = \sqrt{[5-(-2)]^2 + (4-1)^2} = \sqrt{49+9} = \sqrt{58} \approx 7.62. \text{ Ans.}$$

$$M = \left(\frac{-2+5}{2}, \frac{1+4}{2} \right) = (\ 1.5, 2.5\ ). \text{ Ans.}$$

2. Find the distance and midpoint between $(4,\ 3\sqrt{3})$ and $(2,\ -\sqrt{3})$.
Solution:

$$d = \sqrt{(2-4)^2 + (-\sqrt{3} - 3\sqrt{3})^2} = \sqrt{(-2)^2 + (-4\sqrt{3})^2} = \sqrt{4+48} = \sqrt{52} \approx 7.21. \text{ Ans.}$$

$$M = \left(\frac{4+2}{2}, \frac{3\sqrt{3} + (-\sqrt{3})}{2} \right) = (3,\ \sqrt{3}) \approx (\ 3,\ 1.73\ ). \text{ Ans.}$$

3. If $M\ (1,\ 0)$ is the midpoint of the segment $\overline{AB}$ and $B\ (-1, -2)$ is the coordinates of point B. Find the coordinates of point A.
Solution:

$$A(x, y),\quad M(1, 0),\quad B(-1, -2)$$

$$\frac{x + (-1)}{2} = 1,\quad x - 1 = 2,\quad \therefore x = 3$$

$$\frac{y + (-2)}{2} = 0,\quad y - 2 = 0,\quad \therefore y = 2$$

The coordinates of point A is $A(3,\ 2)$. Ans.

3-13 Chart of Measurements

Chart of Measurements

Length	Weight (Mass)	Volume (Capacity)
1 yard (yd) = 3 feet (ft) 1 foot (ft) = 12 inches (in.) 1 in. = 2.54 centimeters (cm) 1 mile (mi) = 5280 feet (ft) 1 mile (mi) = 1.6 km	1 ton = 2,000 pounds (lb) 1 pound (lb) = 16 ounces (oz) 1 kilogram (kg) = 2.2 lb 1 ounce (oz) = 28 grams (g)	1 gallon (gal) = 4 quarts (qt) 1 quart (qt) = 2 pints (pt) 1 pint (pt) = 2 cups (cu) 1 cup(cu) = 8 fluid ounces(fl.oz) 1 gallon (gal) = 3.8 liters (ℓ) = 128 fl.oz
1 kilometer(km)=1000 meters (m) 1 meter(m)=100 centimeters (cm) 1 meter(m)=1000 millimeters(mm)	1 kilogram (kg)=1000 grams (g) 1 gram (g)=100 centigrams (cg) 1 gram(g)=1000 milligrams (mg)	1 kiloliter ($k\ell$)=1000 liters (ℓ) 1 liter (ℓ)=100 centiliters ($c\ell$) 1 liter(ℓ)=1000 milliliters($m\ell$)

Other Measurements

1 cc.(or c.c) = 1 cubic centimeter 1 tablespoon = 3 teaspoons = 15 cc. 1 teaspoon = $\frac{1}{3}$ tablespoon = 5 cc. 1 tablespoon = $\frac{1}{2}$ fluid ounce	1 bushel = 36 liters 1 peck = 9 liters 1 bushel = 4 pecks	" hecto " means " hundred ". " deca " means " ten ". 1 hectometer and 2 decameters = 120 meters

Area of a Rectangle: $A = \ell \times w$ where ℓ is the length and w is the width

Volume of a Rectangular Solid: $V = \ell \times w \times h$ where h is the height

Area of a Parallelogram: $A = b\,h$ where b is the base and h is the height

Area of a Triangle: $A = \frac{1}{2} b\,h$ where b is the base and h is the height

Area of a Circle: $A = \pi r^2$ where r is the radius

Circumference of a Circle: $C = \pi d$ where d is the diameter

Surface Area of a Sphere: $S = 4\pi r^2$ where r is the radius

Volume of a Sphere: $V = \frac{4}{3}\pi r^3$ Where r is the radius

Area of a Right Cylinder: $A = B\,h$ Where B is the base area and h is the height

Volume of a Right Circular Cone: $V = \frac{1}{3}\pi r^2 h$ where r is the radius and h is the height

Volume of a Pyramid: $V = \frac{1}{3} B\,h$ where B is the base area and h is the height

Lateral Area of a Right Circular Cone: $S = \frac{1}{2} c\,\ell$ where c is the circumference and ℓ is the slant height

3-14 Parabolas

A conic section is a curve that can be formed by slicing a right circular cone with a plane. The figure of the intersection of the cone and the plane is a conic section.
By changing the position of the plane (horizontally or slant) , we obtain four types of conic sections, **a circle, an ellipse, a parabola**, or part of **a hyperbola**.
In this chapter, we will learn to obtain an equation for each of the conic sections by applying its geometric definition.

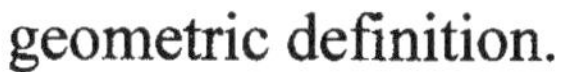

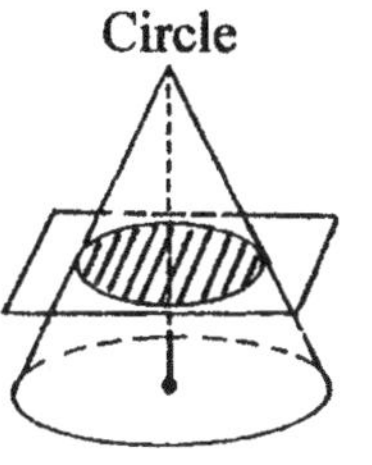

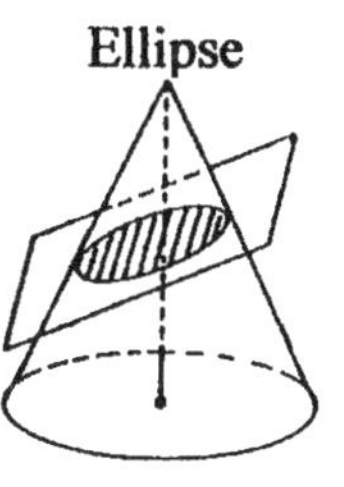

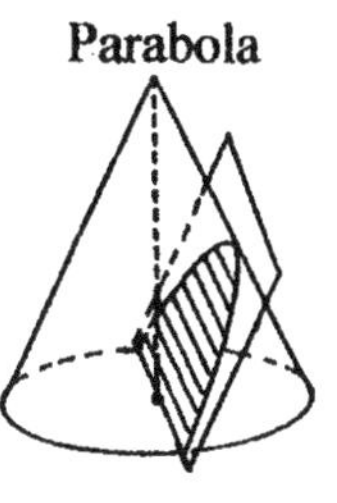

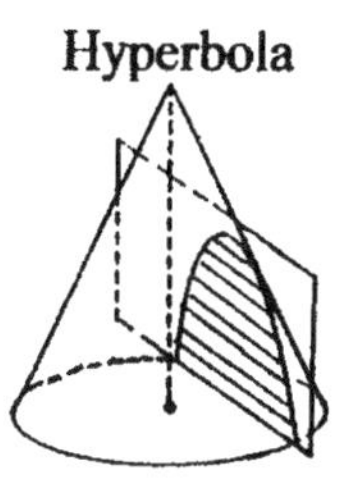

In Section 2~16, we have learned the basic equation of a parabola $y = ax^2 + bx + c \; (a \neq 0)$ and **the General Form of a Parabola**:

$$y - k = a(x - h)^2 . \text{ The vertex is } (h, k).$$

If $a > 0$, the parabola open upward. If $a < 0$, the parabola open downward.

In this section, we will learn to derive equations for parabolas by applying a general definition of a parabola and learn how to graph a parabola.

Definition of a parabola: A parabola is the set of all points equidistant from a fixed line (called the directrix) and a fixed point (called the focus).

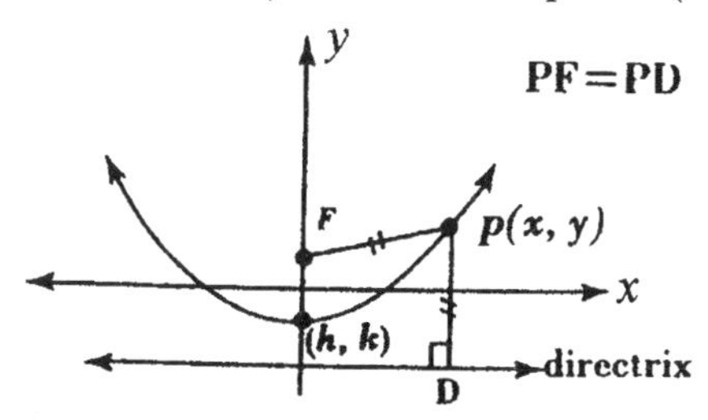

Example

1. Find the equation of a parabola whose focus is the point (0, 2) and whose directrix is the line $y = -4$.

   Solution:

$$PF = PD$$
$$\sqrt{(x-0)^2 + (y-2)^2} = \sqrt{(x-x)^2 + (y+4)^2}$$
$$x^2 + (y-2)^2 = (y+4)^2$$
$$x^2 + \cancel{y^2} - 4y + 4 = \cancel{y^2} + 8y + 16$$
$$x^2 = 12y + 12$$
$$x^2 = 12(y+1)$$
$$\therefore y + 1 = \tfrac{1}{12}x^2 \text{ (the equation)}$$

Vertex $(0, -1)$, $a = \frac{1}{12} > 0$ open upward

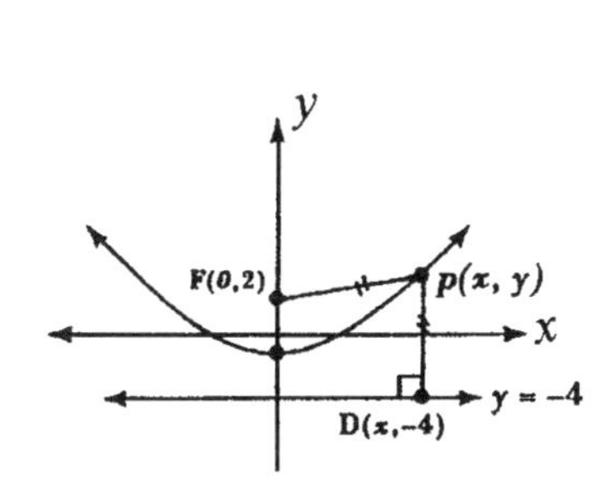

Example

2. Find the equation of a parabola whose focus is the point (0, 3) and whose directrix is the line $x = 4$.

Solution:

$$PF = PD$$
$$\sqrt{(x-0)^2+(y-3)^2} = \sqrt{(x-4)^2+(y-y)^2}$$
$$x^2+(y-3)^2 = (x-4)^2$$
$$x^2+(y-3)^2 = x^2-8x+16$$
$$(y-3)^2 = -8(x-2)$$
$$x-2 = -\tfrac{1}{8}(y-3)^2 \text{ (the equation)}$$
Vertex (2, 3), $a = -\frac{1}{8} < 0$ open to the left

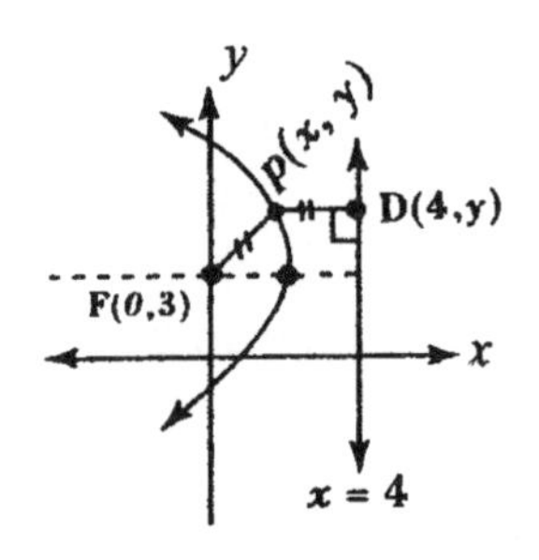

In Section 2~16, we have learned the parabolas that have a vertical axis and open upward or downward. Now, we will also learn the parabola that have a horizontal axis and open to the right or to the left. The standard form of each parabola shown below are useful to locate the vertex, the focus, the directrix, and latus rectum.

Standard Forms of Parabolas

Latus Rectum: The line segment passing through the focus and perpendicular to the axis.

1. $y^2 = 4ax$

vertex: (0, 0), open to the right
focus : $(a, 0)$
directrix: $x = -a$
latus rectum: $4a$

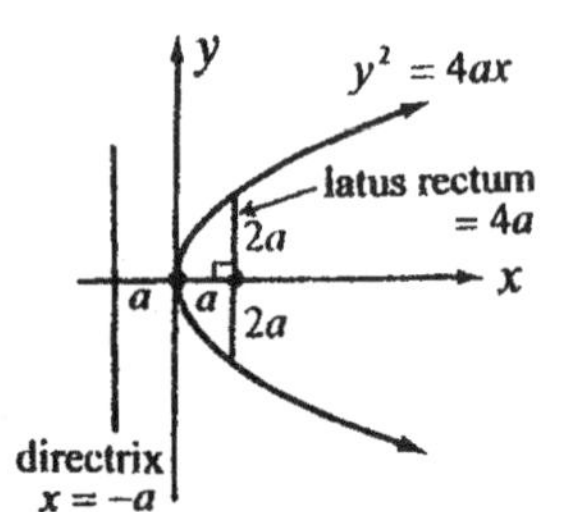

Or: $x = py^2$ where $p = \frac{1}{4a}$

2. $(y-k)^2 = 4a(x-h)$

vertex: (h, k), open to the right
focus: $(h+a, k)$
directrix: $x = h - a$
latus rectum: $4a$

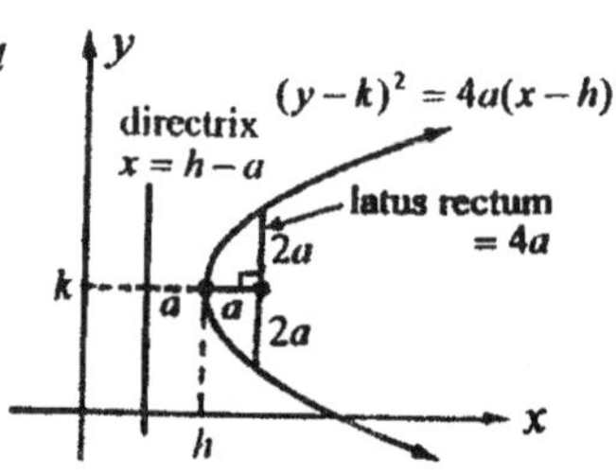

Or: $x - h = p(y-k)^2$ where $p = \frac{1}{4a}$

3. $x^2 = 4ay$

vertex: (0, 0), open upward
focus: $(0, a)$
diretrix: $y = -a$
latus rectum: $4a$

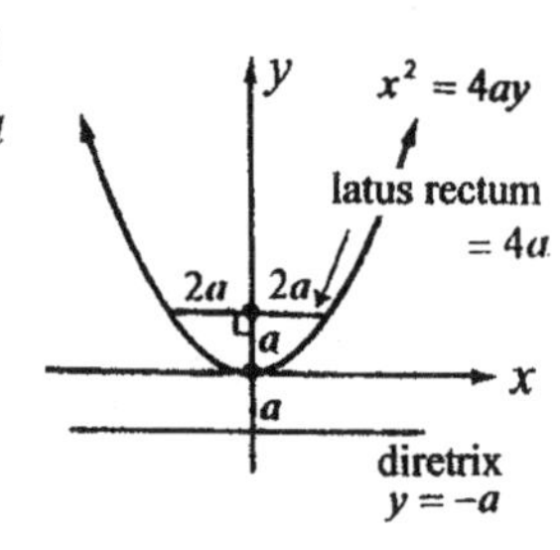

Or: $y = px^2$ where $p = \frac{1}{4a}$

4. $(x-h)^2 = 4a(y-k)$

vertex: (h, k), open upward
focus: $(h, k+a)$
diretrix: $y = k - a$
latus rectum: $4a$

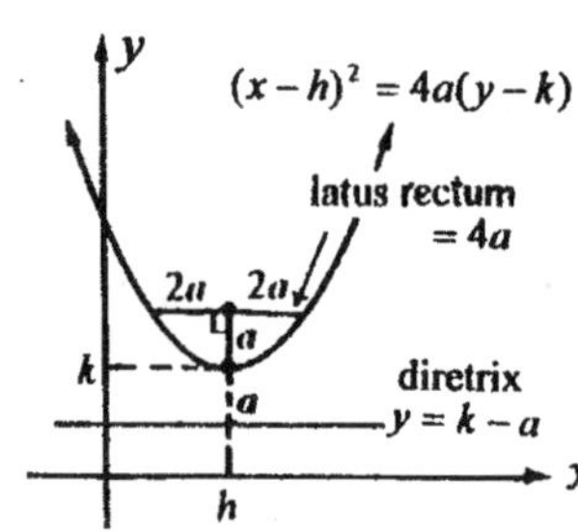

Or: $y - k = p(x-h)^2$ where $p = \frac{1}{4a}$

Standard Forms of Parabolas (Continued)

5. $y^2 = -4ax$
vertex: (0, 0), open to the left
focus: $(-a, 0)$
diretrix: $x = a$
latus rectum: $4a$

6. $(y-k)^2 = -4a(x-h)$
vertex: (h, k), open to the left
focus: $(h - a, k)$
diretrix: $x = h + a$
latus rectum: $4a$

7. $x^2 = -4ay$
vertex: (0, 0), open downward
focus: $(0, -a)$
diretrix: $y = a$
latus rectum: $4a$

8. $(x-h)^2 = -4a(y-k)$
vertex: (h, k), open downward
focus: $(h, k - a)$
diretrix: $y = k + a$
latus rectum: $4a$

Examples

1. Find an equation of the parabola with vertex (0, 0) and focus (2, 0).
Solution: Sketch and Find:
It is open to the right.
$y^2 = 4ax$ and $a = 2$
$y^2 = 4(2)x = 8x$
$\therefore y^2 = 8x$ is the equation.

2. Find an equation of the parabola with vertex (0, 0) and focus (–2, 0).
Solution: Sketch and Find:
It is open to the left.
$y^2 = -4ax$ and $a = 2$
$y^2 = -4(2)x = -8x$
$\therefore y^2 = -8x$ is the equation.

3. Find the equation of a parabola whose focus is the point (0, 4) and whose directrix is the line $y = -4$.
Solution: Sketch and find:
It is open upward. vertex (0, 0)
$x^2 = 4ay$ and $a = 4$
$x^2 = 4(4)y = 16y$
$\therefore x^2 = 16y$ is the equation.

4. Find the equation of a parabola whose focus is the point (0, 2) and whose directrix is the line $y = -4$.
Solution: Sketch and find:
It is open upward. vertex (0, –1)
$(x-h)^2 = 4a(y-k)$ and $a = 3$
$(x-0)^2 = 4(3)(y+1)$
$\therefore x^2 = 12((y+1)$ is the equation.

5. Find the vertex, focus, diretrix, and latus rectum of the parabola $x^2 + 2x + 2y - 4 = 0$.
Solution: Completing the square
$(x^2 + 2x + 4) - 4 + 2y - 4 = 0$, $(x+2)^2 + 2y - 8 = 0$
$(x+2)^2 = -2(y-4)$, open downward
$-4a = -2 \quad \therefore a = \frac{1}{2}$
vertex: $(h, k) = (-2, 4)$
focus: $(h, k - a) = (-2, \frac{7}{2})$
directrix: $y = k + a$, $y = \frac{9}{2}$
latus rectum: $4a = 2$

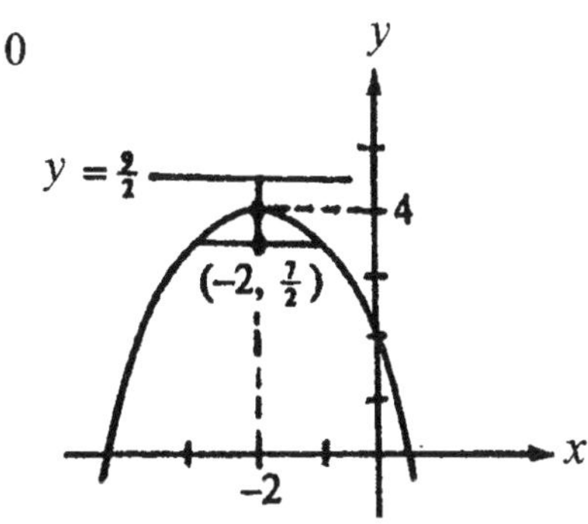

Notes

3-15 Circles

Definition of a circle: A circle is the set of all points in a plane that are a fixed distance r (called the radius) from a fixed point (called the center).

Standard Forms of Circles

1. $x^2 + y^2 = r^2$

Center (0, 0)

Radius $= r$

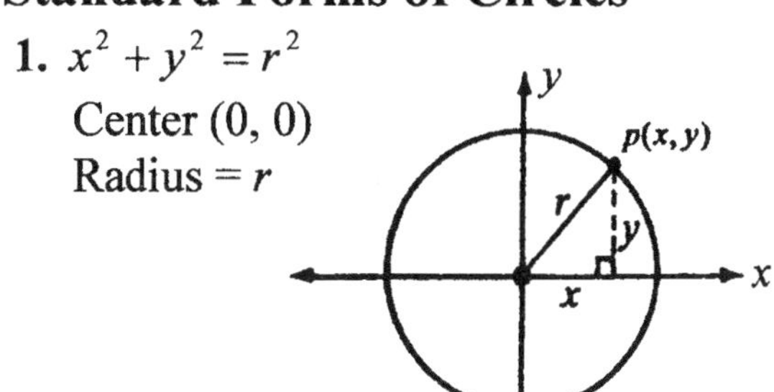

2. $(x-h)^2 + (y-k)^2 = r^2$

Center (h, k)

Radius $= r$

When the equation of a circle is given in the form $x^2 + y^2 + ax + by + c = 0$, we rewrite it to match the standard form by the method of completing the square. Then find its center and radius.

Examples

1. Find the center and radius of $x^2 + y^2 = 25$.

Solution:

$$x^2 + y^2 = 25$$

$$x^2 + y^2 = 5^2$$

$\therefore$ Center (0, 0)

Radius $r = 5$. Ans.

2. Find the equation of the circle with center (0, 0) and radius $3\sqrt{2}$.

Solution:

$$x^2 + y^2 = r^2$$

$$x^2 + y^2 = (3\sqrt{2})^2$$

$$\therefore x^2 + y^2 = 18. \text{ Ans.}$$

3. Find the equation of the circle with center (–2, 1) and radius 3.

Solution:

$$(x-h)^2 + (y-k)^2 = r^2$$

$(x+2)^2 + (y-1)^2 = 3^2$. Ans. (Standard Form)

$$x^2 + 4x + 4 + y^2 - 2y + 1 = 9$$

$x^2 + y^2 + 4x - 2y - 4 = 0$. Ans. (General Form)

4. Write the equation of a circle if the endpoints of the diameter are at (5, 4) and (–1, –2).

Solution:

$$\text{Center} = \left(\frac{5+(-1)}{2}, \frac{4+(-2)}{2}\right) = (2, 1)$$

$$\text{Radius } r = \sqrt{(5-2)^2 + (4-1)^2} = \sqrt{18}$$

$$\therefore (x-2)^2 + (y-1)^2 = 18. \text{ Ans.}$$

5. Graph $x^2 + y^2 + 4x - 8y + 11 = 0$.
Solution:

Completing the square

$$x^2 + y^2 + 4x - 8y + 11 = 0$$
$$(x^2 + 4x) + (y^2 - 8y) + 11 = 0$$
$$(x^2 + 4x + 4) - 4 + (y^2 - 8y + 16) - 16 + 11 = 0$$
$$(x+2)^2 + (y-4)^2 = 9$$
$$(x+2)^2 + (y-4)^2 = 3^2$$

$\therefore$ Center $(-2, 4)$, Radius $= 3$

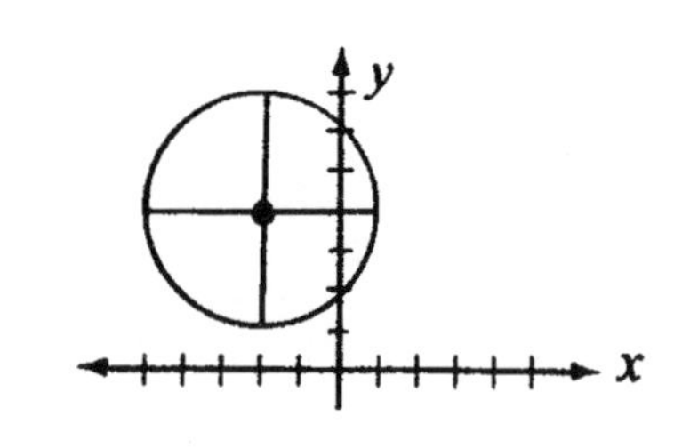

6. Find the equation of the line that is tangent to the circle $x^2 + y^2 = 34$ at point (3, 5).
Solution:

The tangent line on a circle is perpendicular to the radius of the circle.
Slope of the radius $m = \frac{5-0}{3-0} = \frac{5}{3}$
Slope of the tangent line $m_1 = -\frac{3}{5}$ passing the point (3, 5), $p(x, y)$
The equation of the tangent line:

$$\frac{y-5}{x-3} = -\frac{3}{5}$$
$$5y - 25 = -3x + 9$$
$$3x + 5y - 34 = 0. \text{ Ans.}$$

7. The line $3x + 5y - 34 = 0$ is tangent to a circle whose center is at the origin (0, 0). Find the equation of the circle.
Solution:

The tangent line on a circle is perpendicular to the radius of the circle.
Find the slope of the tangent line:

$$3x + 5y - 34 = 0$$
$$y = -\tfrac{3}{5}x + \tfrac{34}{5}, \text{ the slope } m = -\tfrac{3}{5}$$

Slope of the radius $m_1 = \frac{5}{3}$

Equation of the radius $\frac{y-0}{x-0} = \frac{5}{3}$, $5x - 3y = 0$, $x = \frac{3}{5}y$

Find the point of intersection of the tangent line and the radius:

$$y = -\tfrac{3}{5}x + \tfrac{34}{5}$$
$$y = -\tfrac{3}{5}\left(\tfrac{3}{5}y\right) + \tfrac{34}{5}$$

$\therefore y = 5$ and $x = \frac{3}{5}y = \frac{3}{5}(5) = 3$, point of intersection (3, 5)

Find the length of the radius between (0, 0) and (3, 5):

$$r = \sqrt{(3-0)^2 + (5-0)^2} = \sqrt{34}$$

Equation of the circle $x^2 + y^2 = 34$. Ans.

Notes

Notes

3-16 Symmetry and Transformations

In Chapter 2-11, we have learned how to graph a polynomial function by transformations. In this section, we will discuss the transformations of a geometric figure in the plane. Transformations can help us to design or sketch the image of a figure in arts or constructions.

In each transformation, we move the original figure by reflects, slides, or turns. Each point of the new figure corresponds (or pairs) to exactly one point of the original figure. The transformations include **Reflections**, **Translations**, **Rotations**, and **Dilations**.

Reflections, translations, and rotations are transformations that preserve size and shape. The image and the original figure are congruent figures (congruent mapping). Each of these three transformations is also called an **isometry**.

Dilations are transformations that involve enlargements or reductions. Dilation is not an isometry.

1. Symmetry & Reflections

In the **reflection**, each point of a figure reflects over from one side of a line to the opposite side. Reflection preserves its size and shape.
The reflection image and the original figure has the same distance from **the line of reflection**.
To draw the reflection image of a figure, the line of reflection is the perpendicular bisector of the line between the point and the point of its own image.

Reflect ΔABC over line k

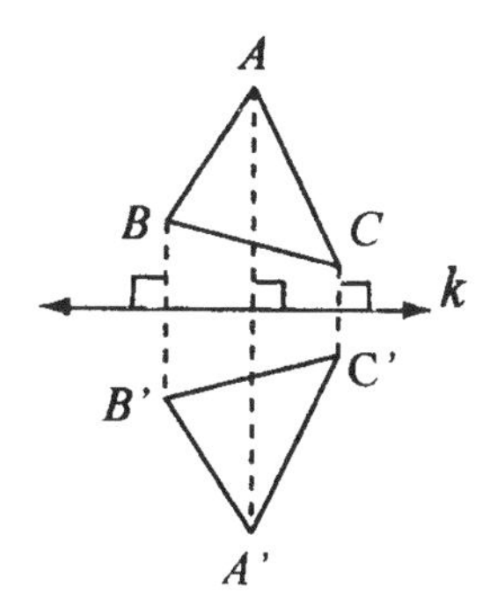

Reflect $< ABC$ over line p

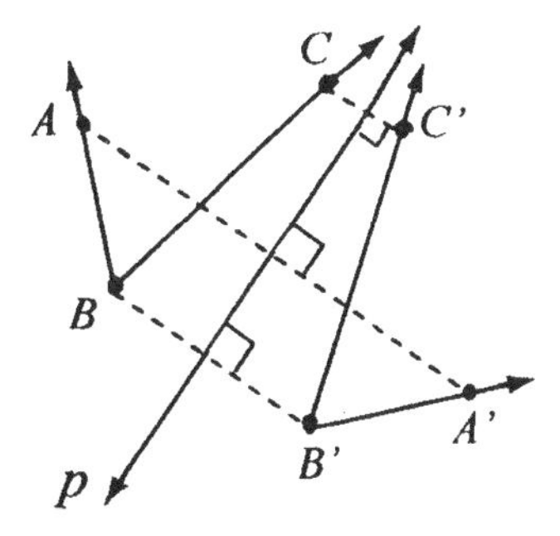

If a line can be drawn through a figure and divides the figure into two parts that are mirror images of each other, then the figure is said to have **a line of symmetry**, or **axis of symmetry**.

Polygon *ABCDEFGH* has a line k as its line of symmetry.

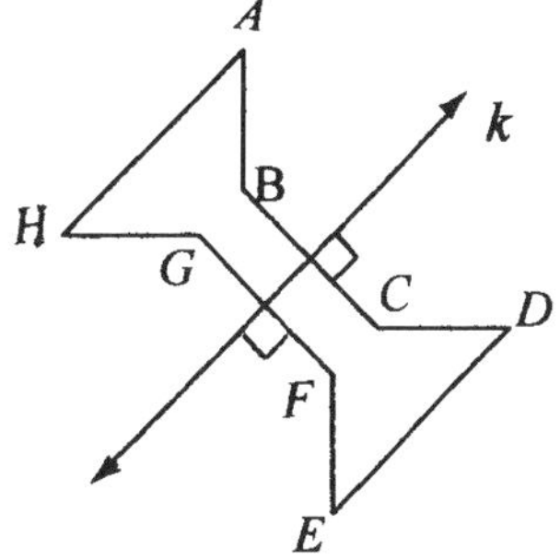

Examples

1. Reflect ΔABC over the x-axis.
   Solution:

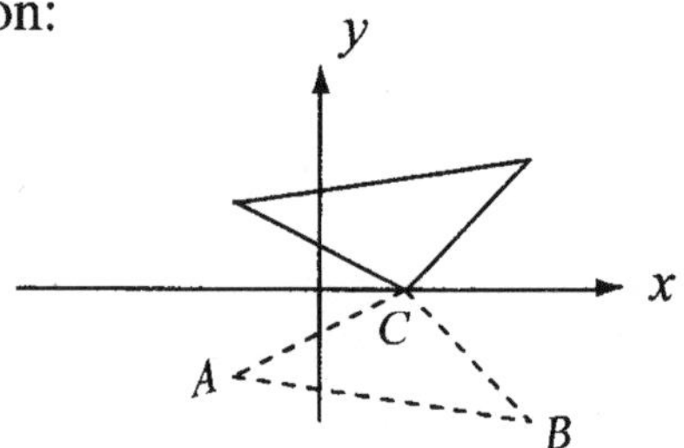

2 Reflect the quadrilateral $ABCD$ over the y-axis.
   Solution:

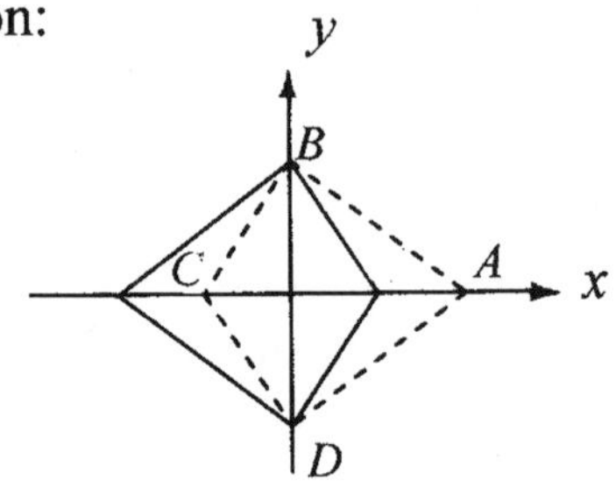

3. Draw the graph
   1. to make it symmetric with respect to the x-axis.
   2. to make it symmetric with respect to the y-axis.
   3. to make it symmetric with respect to the origin.
   4. to make it symmetric with respect to the x-axis, y-axis, and origin.

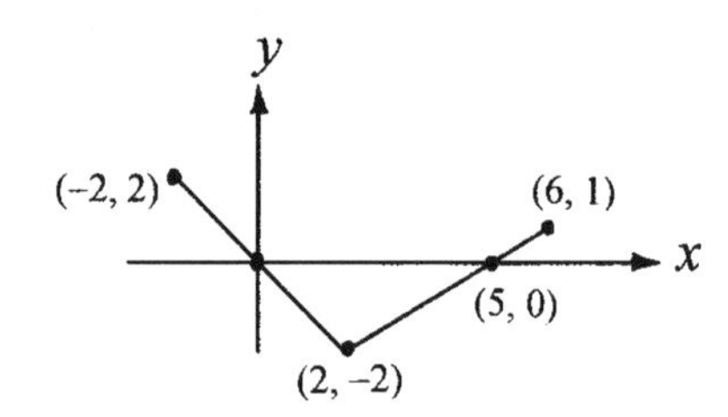

Solution:

1.

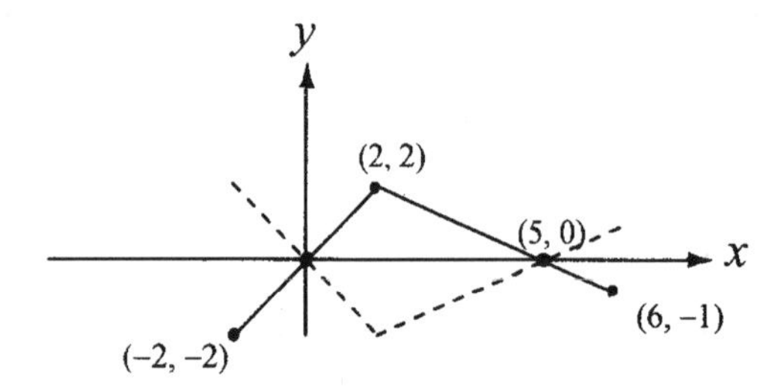

2.

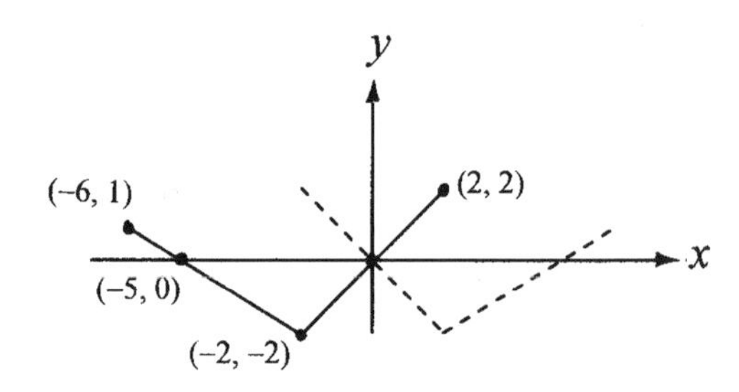

3.

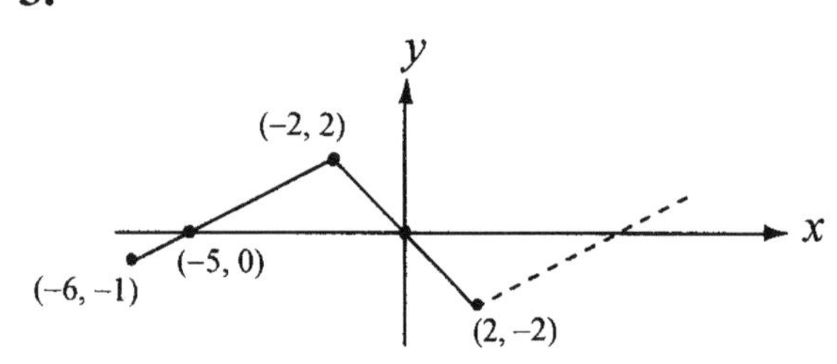

4.

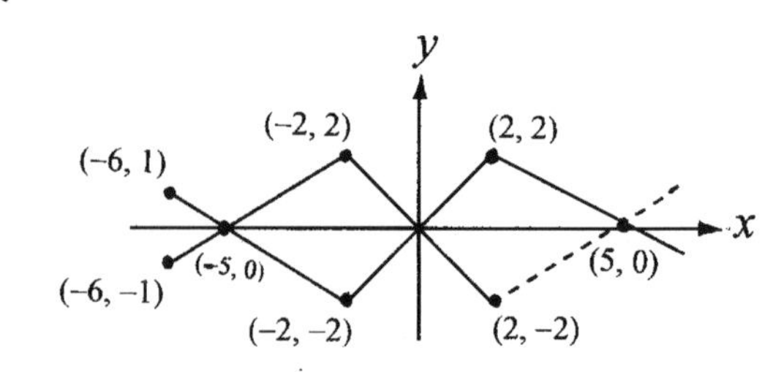

A kid said to his mother.
Kid: Did you wear contact lenses this morning ?
Mother: No. Why do you ask ?
Kid: I heard you yell to father this morning: "After 20 years of marriage, I have finally seen what kind of man you are ! ".

2. Translations

In the **translation**, each point of a figure slides the same distance in the same direction. Translation preserves its size and shape.

A composition of two reflections over two parallel lines is a translation. The distance between the original figure and the image after a translation is twice the distance between the parallel lines.

Examples

1. Translate $\overline{AB}$ 4 units to the left and and 2 units down, or by the translation vector $\vec{v} = (-4,-2)$.

   Solution:

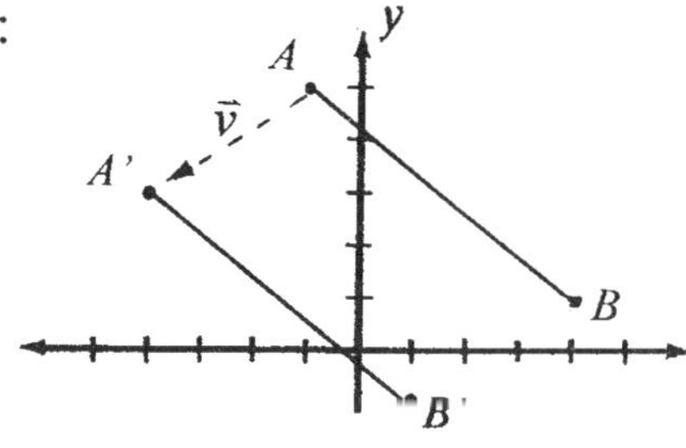

2. Lines *m* and *n* are parallel. Reflect ΔABC over line *m*, followed by a reflection over line *n*.

   Solution:

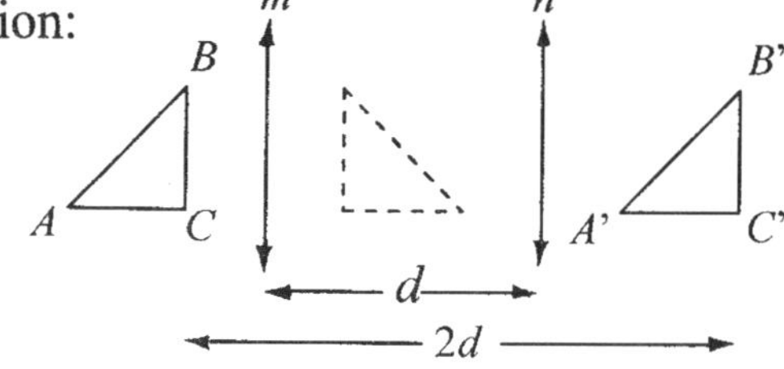

3. Rotations

In the **rotation**, each point of a figure moves along a circular path about a fixed point. Rotation preserves its size and shape. If the direction of a rotation is counterclockwise, the measure of the angle of rotation is positive. If the direction is clockwise, the measure of the angle of rotation is negative.

$\overline{AB}$ is rotated by $-90°$ (clockwise) about origin. $m < AOA' = 90$, $m < BOB' = 90$

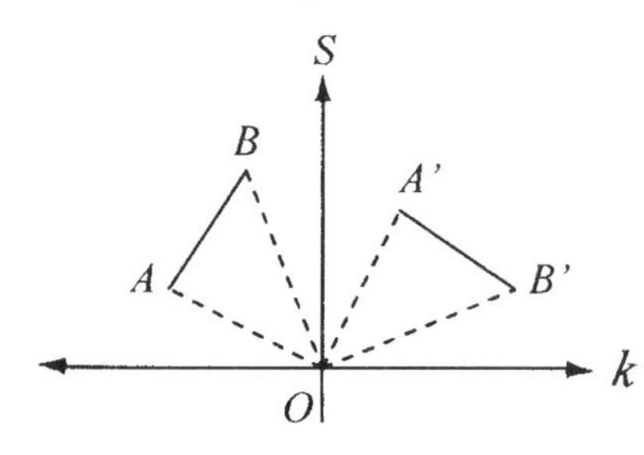

ΔABC is rotated by $-90°$ (clockwise) about the the origin.
$m < AOA' = 90$, $m < BOB' = 90$, $m < COC' = 90$

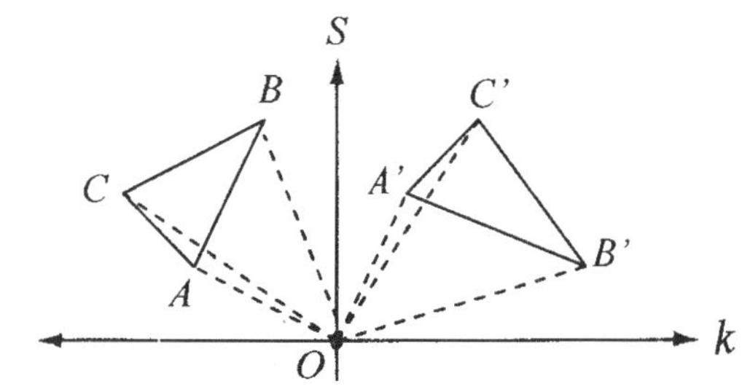

Formula for Rotation of a Point

Let $p(x, y)$ be a point in the plane, and $p'(x', y')$ be the resulting point after a rotation of angle θ with the origin as the center of rotation.

$$x = r\cos\alpha \qquad x' = r\cos(\theta + \alpha)$$

$$y = r\sin\alpha \qquad y' = r\sin(\theta + \alpha)$$

$$x' = r\cos(\theta + \alpha) = r(\cos\theta\cos\alpha - \sin\theta\sin\alpha) = x\cos\theta - y\sin\theta$$

$$y' = r\sin(\theta + \alpha) = r(\sin\theta\cos\alpha + \cos\theta\sin\alpha) = x\sin\theta + y\cos\theta$$

Formula: $\begin{cases} x' = x\cos\theta - y\sin\theta \\ y' = x\sin\theta + y\cos\theta \end{cases}$

In matrix form: $\begin{bmatrix} x' \\ y' \end{bmatrix} = \begin{bmatrix} \cos\theta & -\sin\theta \\ \sin\theta & \cos\theta \end{bmatrix} \cdot \begin{bmatrix} x \\ y \end{bmatrix}$

4. Dilations

In the dilation (or dilatation), each point of a figure is multiplied by a **scale factor** n to create a similar image. Dilation preserves the angle measure of the original figure while enlarging or reducing the size and shape. The center of dilation is the point from which the dilation is created. The perimeter of the image enlarges or reduces the original perimeter figure in the same **ratio** as the scale factor. Dilation is not an isometry.

The image A' of point A under a dilation is usually denoted by $D_n(A)$.

In the graph, $\overline{A'B'}$ is the image of $\overline{AB}$ under a dilation of $n = \frac{2}{3}$.

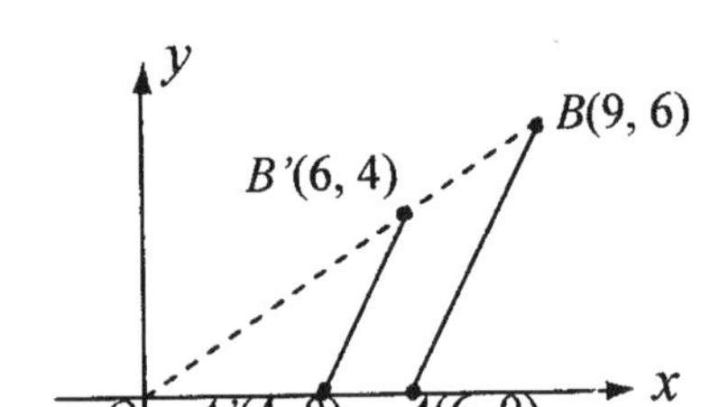

$A' = (6 \cdot \frac{2}{3},\ 0 \cdot \frac{2}{3}) = (4,\ 0)$

$B' = (9 \cdot \frac{2}{3},\ 6 \cdot \frac{2}{3}) = (6,\ 4)$

Center of dilation is (0, 0).

$\overline{A'B'} = \frac{2}{3}\left(\overline{AB}\right)$

In the graph, $\Delta OA'B'$ is the image of ΔOAB under a dilation of $n = 2$.

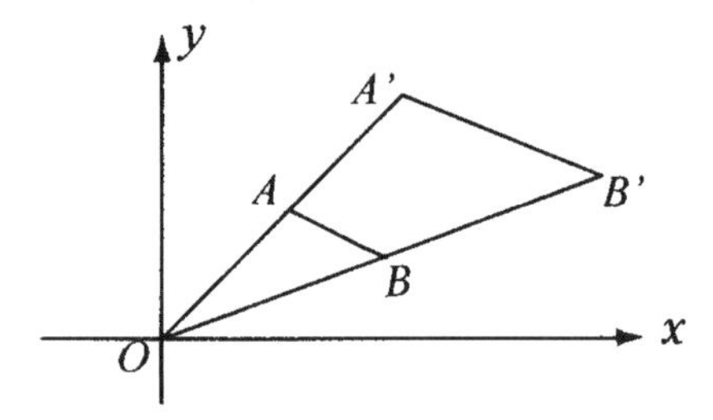

$\overline{OA'} = 2\left(\overline{OA}\right)$

$\overline{OB'} = 2\left(\overline{OB}\right)$

Center of dilation is (0, 0)

$\overline{OA'} + \overline{OB'} + \overline{A'B'} = 2\left(\overline{OA} + \overline{OB} + \overline{AB}\right)$

In the graph, enlarge ΔABC so that the perimeter of $\Delta A'B'C'$ triples the original perimeter..

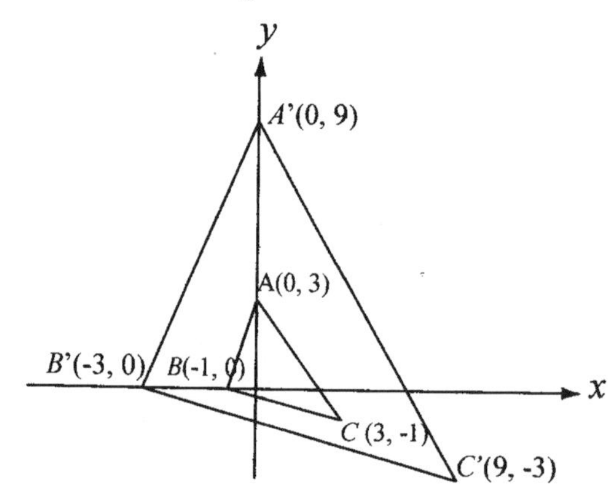

Dilation Theorem: If a closed figure F is dilated to a closed figure F' with ratio k, then the following statements are true:

1. The perimeter of F' is k times the perimeter of F.
2. The area of F' is k^2 times the perimeter of F.
3. The volume of F' is k^3 times the volume of F.

3-17 Cylinders

A **Cylinder** is a three-dimensional figure that consists of two parallel and congruent circular bases and a lateral surface. The axis of a cylinder is the segment that connects the centers of the two bases. Cylinders include **right cylinders** and **oblique cylinders**. In a right cylinder, the axis is its height (altitude).

Right Cylinder

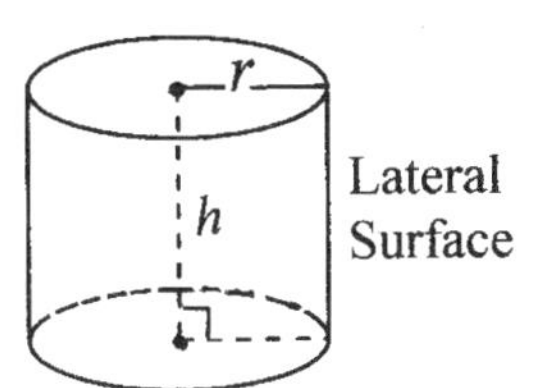

Oblique Cylinders

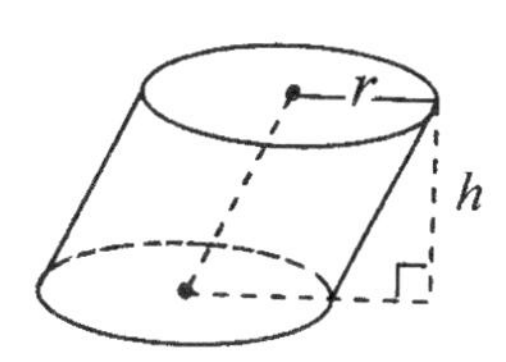

The lateral surface of a right cylinder is a rectangle. The length of the rectangle is the circumference ($C = 2\pi r$) of the circle (the base). The width of the rectangle is the height of the cylinder.

The **total surface area** (**SA**) of a right cylinder with base radius r and height h is the sum of the lateral surface area (LA) and the area of the two congruent circles.

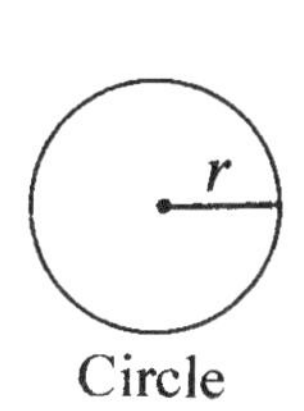

Circle

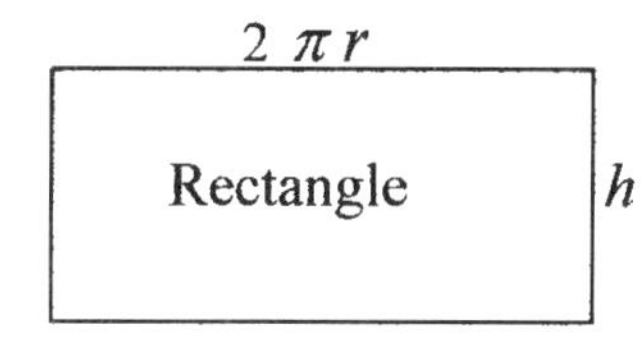

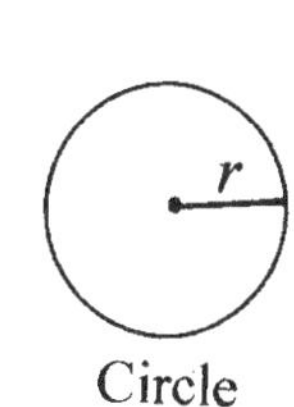

Circle

Lateral Surface Area of a Cylinder: $LA = 2\pi r h$

Total Surface Area of a Cylinder: $SA = 2(\pi r^2) + 2\pi r h$

The **volume** (V) of a right cylinder with base radius r and height h is the product of the base area (B) and the height h.

Volume of a Cylinder: $V = Bh = \pi r^2 h$

Examples

1. Find the total surface area and volume of the right cylinder of radius 5 inches and height 8 inches.

   Solution:

   Area of the bases = $2(\pi r^2) = 2\pi(5^2) = 50\pi$

   Area of the lateral surface $LA = 2\pi r h = 2\pi(5)(8) = 80\pi$

   Total surface area $SA = 50\pi + 80\pi = 130\pi$ $in.^2$ Ans.

   Volume $V = \pi r^2 h = \pi(5^2)(8) = 200\pi$ $in.^3$ Ans.

5 in.

8 in.

Examples

2. Find the volume of the oblique cylinder of radius 20 centimeters and height 30 centimeters.
Solution:

$$V = \pi r^2 h = \pi(20^2)(30) = 12{,}000\pi \ cm^3. \text{ Ans.}$$

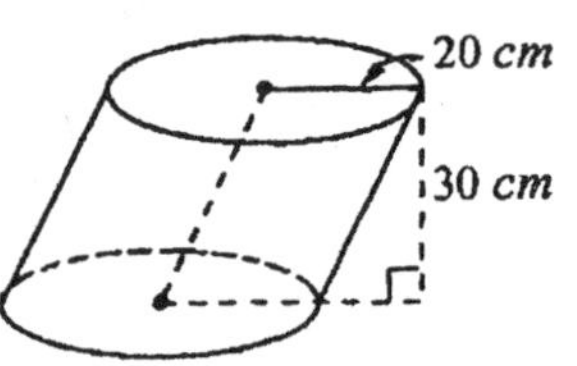

3. Find the longest line that can be drawn inside the cylinder of length 8 inches and radius 5 inches.
Solution:

$$d^2 = 10^2 + 8^2 = 164 \text{ (Pythagorean Theorem)}$$

$$d = \sqrt{164} \approx 12.81 \text{ inches. Ans}$$

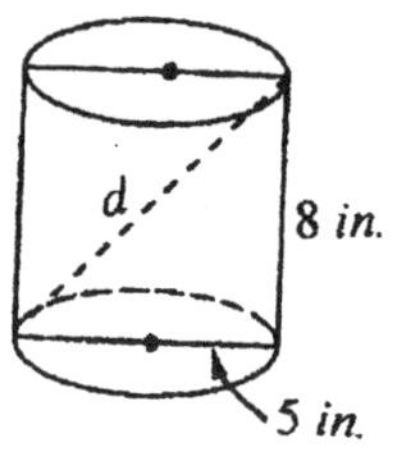

4. A water tank is shaped like a right cylinder of diameter 1.6 meters and length 4 meters. Find its volume.
Solution:

$$V = \pi r^2 h = \pi(0.8)^2(4) \approx 8.04 \ m^3. \text{ Ans.}$$

5. In the diagram, a concrete tunnel pipe is shaped as a hollowed-out right cylinder. How much concrete in cubic meters is needed to make this pipe ?
Solution:

$$r_1 = 2\,m, \quad r_2 = 1.7\,m, \quad h = 5\,m$$

$$V = \pi r_1^2 h - \pi r_2^2 h = \pi h(r_1^2 - r_2^2) = \pi(5)(2^2 - 1.7^2)$$
$$= \pi(5)(4 - 2.89) = \pi(5)(1.11) \approx 17.43 \ m^3. \text{ Ans.}$$

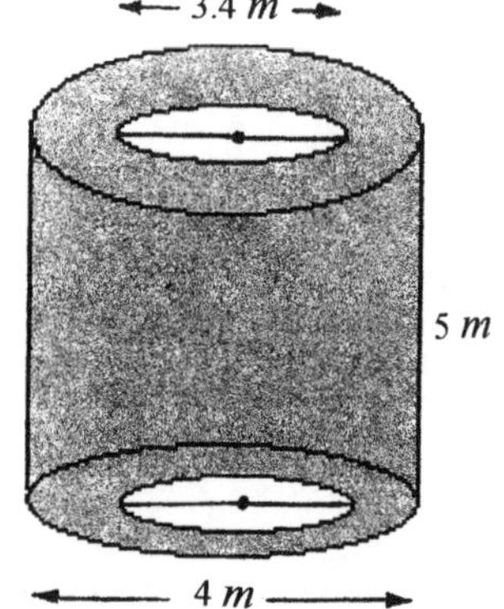

6. The volume and lateral surface area of a cylinder are equal. Find the total surface area of this cylinder with height 12 centimeters.
Solution:

$$\pi r^2 h = 2\pi r h$$

We have: $r = 2$

Total surface area: $SA = 2(\pi r^2) + 2\pi rh = 2\pi(2^2) + 2\pi(2)(12) = 56\pi \ cm^2$. Ans.

3-18 Prisms

A **prism** is a three-dimensional figure that consists of two parallel and congruent polygons as its two **faces** (called **bases**) connecting by the **lateral faces** (parallelograms). The intersections of all faces (the bases and lateral faces) are called the **edges**.

A **right prism** is a prism whose lateral edge is perpendicular to the bases.
A **regular prism** is a right prism whose bases are regular congruent polygons.
An **oblique prism** is a prism whose lateral edge is not perpendicular to the bases.

Prisms are classified by the shapes of their bases, such as **Rectangular Solids, Cubes, Triangular Prisms, Pentagonal Prisms, Hexagonal Prisms,** ⋯⋯ .

1. Rectangular Solids (Rectangular Prisms): A rectangular solid is a prism whose two bases and four lateral faces are rectangles.

Volume of a Rectangular Solid:

$V = Bh = \ell \times w \times h$; $B =$ the area of a base
$h =$ the height

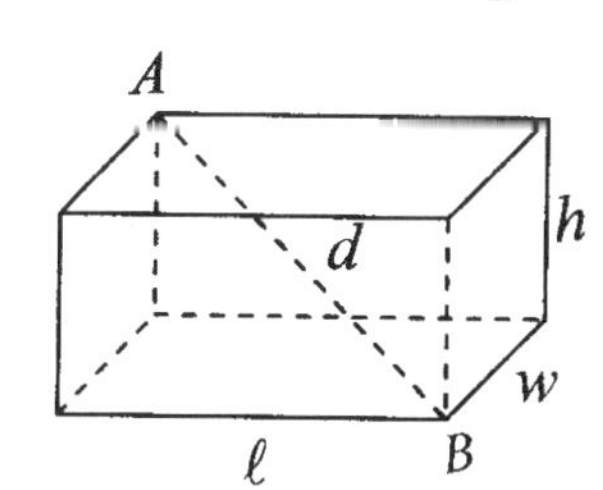

Surface Area of a Rectangular Solid:

$SA = 2(\ell h) + 2(\ell w) + 2(wh)$

$d = \overline{AB}$

The diagonal of a rectangular solid is the segment ($d = \overline{AB}$) that connects the two opposite corners through the center of the solid.

The Length of the Diagonal of a Rectangular Solid:

$d^2 = \ell^2 + w^2 + h^2$ (This is the Pythagorean Theorem in three-dimensions.)

$d = \sqrt{\ell^2 + w^2 + h^2}$

2. Cubes: A cube is a rectangular solid whose two bases and four lateral faces are squares. In a cube, the length, width, and height are all equal.

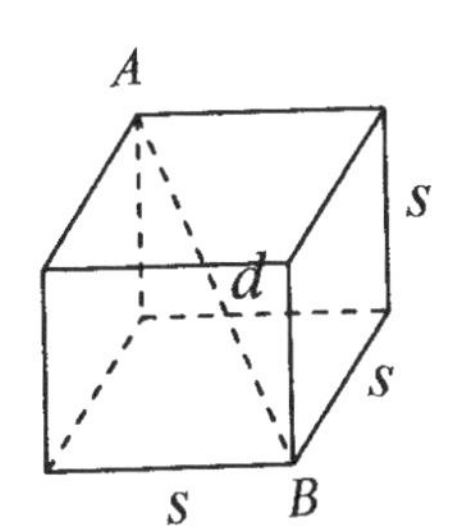

Volume of a Cube: $V = Bh = s \times s \times s = s^3$

Surface Area of a Cube: $SA = 6(s \times s) = 6s^2$

$d = \overline{AB}$

The length of the diagonal of a Cube: $d = \sqrt{s^2 + s^2 + s^2} = \sqrt{3}\, s$

3. In general, the basic formulas for any right prism are given by:

Volume of a Prism: $V = B \times h$; B = the area of a base
h = the height

Lateral Surface Area of a Right Prism: $LA = p\,h$; p = the perimeter of a base
h = the height

Surface Area of a Right Prism: SA = Area of two bases + Area of lateral faces

Examples

1. Find the volume and surface area of the rectangular solid having a length 10 inches, width 4 inches, and height 5 inches.
   Solution:
   $V = B\,h = \ell wh = (10)(4)(5) = 200\ in.^3$ Ans.
   $SA = 2(10\times5) + 2(10\times4) + 2(4\times5) = 100 + 80 + 40 = 220\ in.^2$ Ans.

2. Find the volume and surface area of the cube whose edge length is 8 *cm*.
   Solution:
   $V = B\,h = s^3 = 8^3 = 512\ cm^3$. Ans.
   $SA = 6\,s^2 = 6(8^2) = 384\ cm^2$. Ans.

3. Find the volume and surface area of the right triangular prism shown below.
   (Hint: A right triangular prism is a prism whose two bases are triangles and lateral faces are three rectangles.)
   Solution:

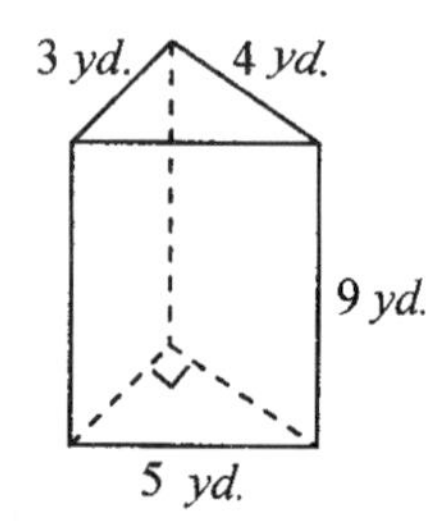

$V = B\,h = \frac{1}{2}(4)(3)\times9 = 54\ yd.^3$ Ans.

Area of two bases $= 2\left[\frac{1}{2}(4)(3)\right] = 12$
Area of three faces $= 9(5) + 9(4) + 9(3) = 108$

$SA = 12 + 108 = 120\ \ yd.^2$ Ans.

4. Find the volume of the right triangular prism shown below.
   Solution:

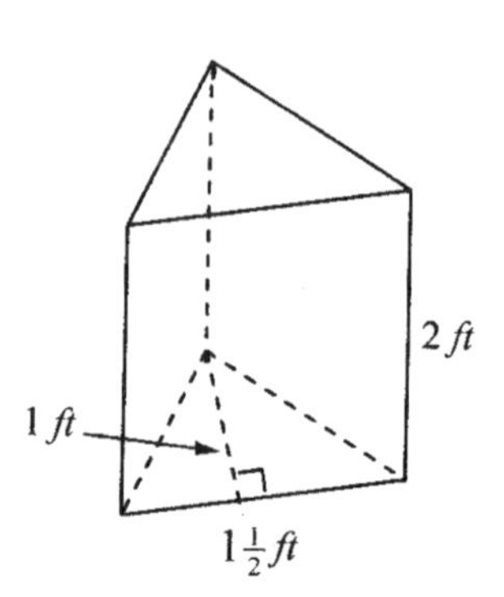

$V = B\,h = \frac{1}{2}(1\frac{1}{2})(1)\times2 = 1\frac{1}{2}\ ft^3$. Ans.

Examples

5. Find the surface area of the regular hexagonal prism shown below.
Solution:

Hint: The area of a regular polygon equals one-half of its apothem times its perimeter. (See page 178, Chapter 3-11)

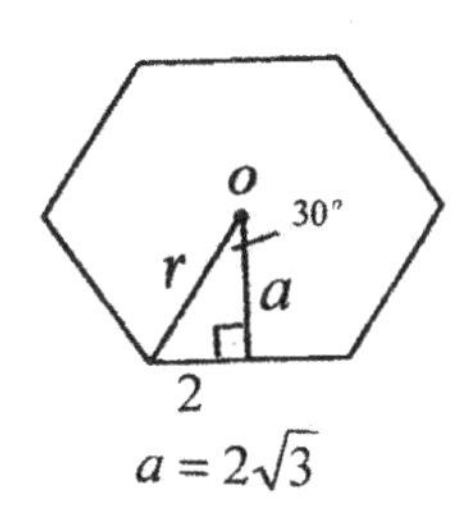

Surface area of 6 rectangles = $6(10\times 4)=240$

Area of a base = $\frac{1}{2}ap=\frac{1}{2}(2\sqrt{3})(24)=24\sqrt{3}$

$SA=240+2(24\sqrt{3})=(240+48\sqrt{3})\ cm^2$. Ans.

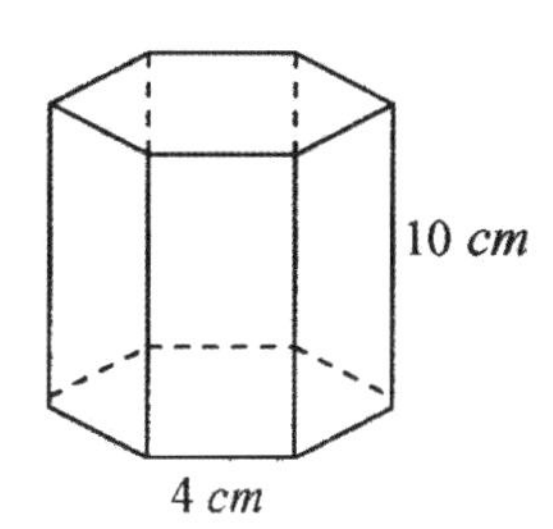

6. Find the volume of the regular hexagonal prism shown on Problem 5.
Solution:

Area of a base = $\frac{1}{2}ap=\frac{1}{2}(2\sqrt{3}\times 24)=24\sqrt{3}$

$V=B\ h=24\sqrt{3}\times 10=240\sqrt{3}\ \ cm^3$. Ans.

7. Find the length of diagonal in the rectangular solid shown below.
Solution:

$d=\overline{MN}=\sqrt{10^2+4^2+5^2}=\sqrt{100+16+25}=\sqrt{141}\approx 11.87$. Ans.

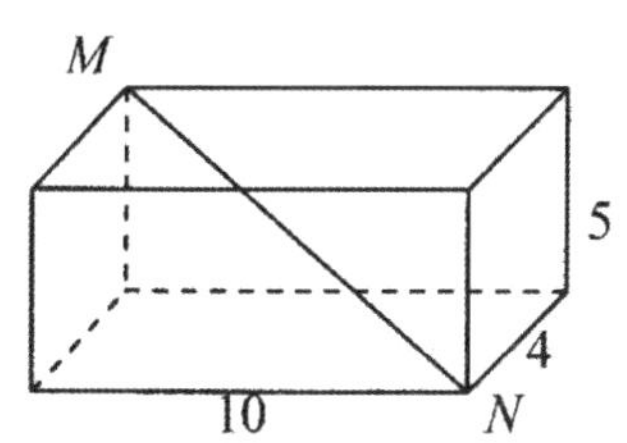

8. Find the volume of the solid made by rotating the rectangle $MNPQ$ 360° around side $\overline{MN}$.
Solution:

$V=B\ h=\pi\ r^2\ h=\ \pi(4^2)(10)=160\pi\ cm^3$. Ans.

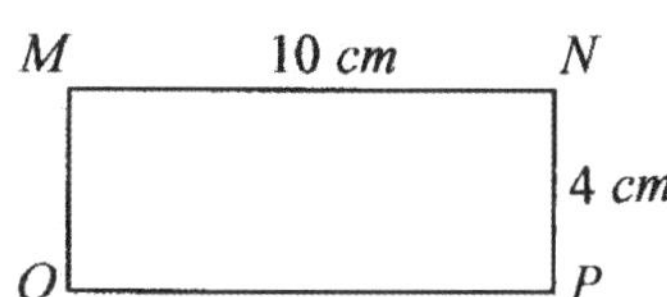

9. The volume and lateral surface area of a cube are equal. Find the edge length.
Solution:

$s^3=4s^2\quad \therefore s=4$. Ans.

10. The volume and total surface area of a cube are equal. Find the edge length.
Solution:

$s^3=6s^2\quad \therefore s=6$. Ans.

11. The surface area of a cube is 384 $in.^2$. Find the edge length.
Solution:

$6s^2=384,\quad s^2=64\quad \therefore s=8$ inches. Ans.

Examples

12. The surface area of a cube is 216. Find the longest line that can be drawn in the cube.
Solution:

$$6s^2 = 216, \quad s^2 = 36 \quad \therefore s = 6$$

The longest line that can be drawn in the cube is the diagonal of the cube.

$$d = \sqrt{6^2 + 6^2 + 6^2} = \sqrt{108} \approx 10.4. \text{ Ans.}$$

13. The volume of a cube is 64. Find the longest line that can be drawn in the cube.
Solution:

$$s^3 = 64 \quad \therefore s = 4$$

The longest line that can be drawn in the cube is the diagonal of the cube.

$$d = \sqrt{4^2 + 4^2 + 4^2} = \sqrt{64} = 8. \text{ Ans.}$$

14. If we double the side length of a cube, how much do we increase the volume of the cube ?
Solution:

$$V = s^3$$

$$V_1 = (2s)^3 = 8s^3 = 8V. \text{ Ans.}$$

The volume of the cube is multiplied by a factor of 8.
We may simply work on the side length to find the scale factor.

$$(2s)^3 = 8s^3$$

15. If the side length of a cube is halved, how much does the surface area decrease ?
Solution:

$$SA = 6s^2$$

$$(SA)_1 = 6\left(\frac{s}{2}\right)^2 = 6\left(\frac{s^2}{4}\right) = \tfrac{1}{4}(6s^2) = \tfrac{1}{4}(SA). \text{ Ans.}$$

The surface area of the cube is multiplied by a factor of $\frac{1}{4}$ (divided by 4).
We may simply work on the side length to find the scale factor.

$$(\tfrac{1}{2}s)^2 = \tfrac{1}{4}s$$

In a biology class, a student said to his teacher.
Student: I just killed five mosquitoes, three were males and two were females.
Teacher: How do you know which were males and which were males ?
Student: Three were killed on a beer bottle. Two were killed on a mirror.

3-19 Cones

A **cone** is a three-dimensional figure that consists of a circular base and a curved lateral surface tapering to a vertex so that any point on the surface is in a straight line between the circumference of the base and the vertex.

A **right (circular) cone** is a cone whose distance from the vertex to any point on the circumference of the base is a constant. An **oblique (circular) cone** is a cone whose distance from the vertex to any point on the circumference of the base is not a constant.

The **axis** of a cone is the segment that connects the vertex and the center of the base. The **height** of a cone is the length of the segment from the vertex, perpendicular to the base plane. In a right cone, the axis and the height are the same segment.

The **slant height** (s) of a cone is the distance from the vertex to a point on the edge of the base. The **lateral surface area** (**LA**) of a cone is the area of the curved surface.

In a right cone, the height, the radius, and the slant height form a right triangle.

When a plane intersects a cone, the cross section can be a circle, an ellipse, a hyperbola, or a triangle (the plane cuts through the vertex.).

Right Cone

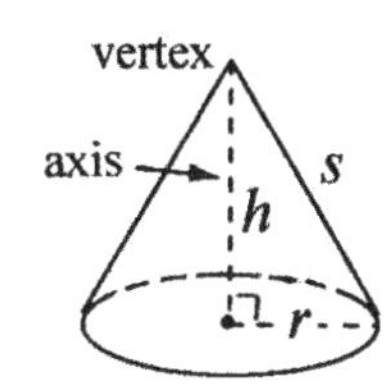

Oblique Cone

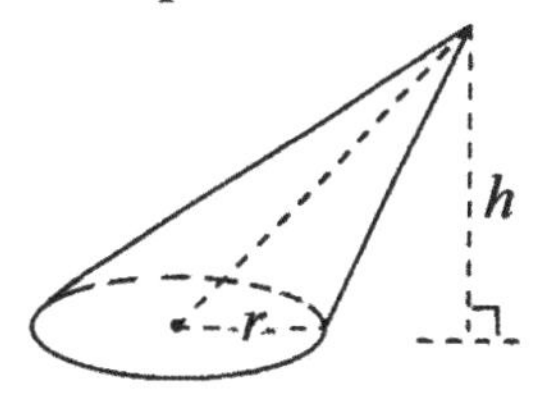

The **volume** (V) of a cone is one-third of the product of the area (B) of the base and the height h:

Volume of a Cone: $V = \frac{1}{3}Bh = \frac{1}{3}\pi r^2 h$

(Hint: It is one-third of the volume of a cylinder with same base and h.)

The **lateral surface area** (LA) of a right cone of radius r equals one-half the product of the circumference of the base and the slant height.
(Hint: As the number of sides of the base increases, a regular pyramid approaches a cone.)

Lateral Surface Area of a Right Cone: $LA = \frac{1}{2}cs = \pi r s$ s: slant height
where $c = 2\pi r$ (circumference)

The **surface area** (SA) of a right cone with base radius r and slant height s equals the sum of the lateral area and the base area.

Surface Area of a Right Cone: $SA = \pi r s + \pi r^2 = \pi r(s + r)$; s: slant height

There is no general formula for the surface area of an oblique cone. To find the surface area of an oblique cone, we use the technique in Calculus.

Examples

1. Find the volume of the cone.

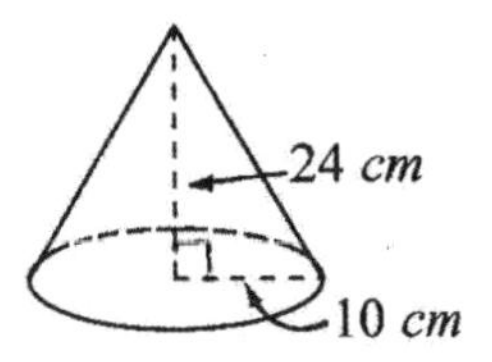

Solution:

$r = 10, \quad h = 24$

$V = \frac{1}{3}\pi r^2 h = \frac{1}{3}\pi(10^2)(24)$

$= 800\pi \ cm^3$. Ans.

2. Find the volume of the cone.

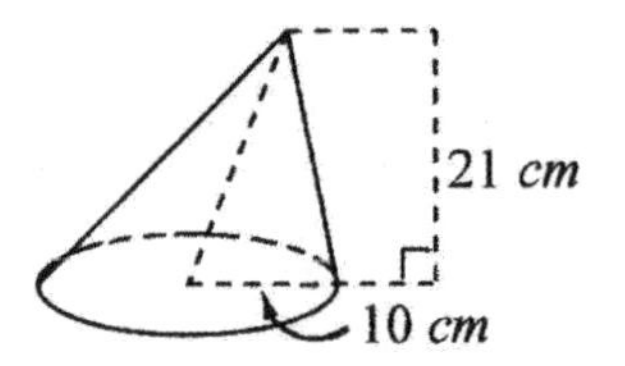

Solution:

$r = 10, \quad h = 21$

$V = \frac{1}{3}\pi r^2 h = \frac{1}{3}\pi(10^2)(21)$

$= 700\pi \ cm^3$. Ans.

3. Find the lateral area and total surface area of the cone shown on Problem 1.

Solution:

Find the slant height by the Pythagorean Theorem:

$s^2 = 24^2 + 10^2$

$s^2 = 676 \quad \therefore s = 26$

Lateral area: $LA = \pi r s = \pi(10)(26) = 260\pi \ cm^2$. Ans.

Surface area: $SA = \pi rs + \pi r^2 = \pi(10)(26) + \pi(10^2) = 360\pi \ cm^2$. Ans.

4. A right cone and a right cylinder have the same radius of base and volume. If the cylinder has a height of 9 centimeters, find the height of the cone.

Solution:

Volume of the cone = Volume of the cylinder

Let h = height of the cone , $h_1 = 9$ (height of the cylinder)

We have: $\frac{1}{3}\pi r^2 h = \pi r^2 h_1$

$\frac{1}{3}h = h_1$

$\frac{1}{3}h = 9 \quad \therefore h = 27 \ cm$. Ans.

5. Right triangle ABC is revolved about $\overline{AB}$ as an axis generating a right cone, $\overline{AB} = 7$ and $\overline{BC} = 9$. Find the volume of the cone.

Solution:

$r = 10, \quad h = 7$

Volume: $V = \frac{1}{3}\pi r^2 h = \frac{1}{3}\pi(9^2)(7) = 189\pi$. Ans.

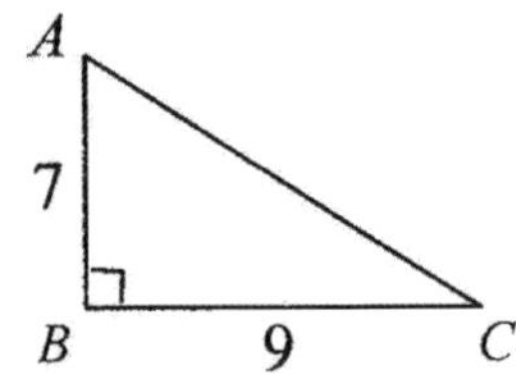

3-20 Pyramids

A **pyramid** is a three-dimensional figure that consists of a polygonal base and triangular surfaces meeting at a vertex.

A **regular pyramid** is a pyramid whose base is a regular polygon. Its height passes through the center of the base and is perpendicular to the base plane.
In a regular pyramid, the height is the **axis**.

The **slant height** (s) of a regular pyramid is the length of the height of one of its triangular surfaces. The **lateral surface area** (LA) of a pyramid is the sum of the areas of the triangular surfaces. The intersections of the lateral surfaces are the **edges**.

Pyramids are classified by the shapes of their bases, such as **Rectangular Pyramids, Triangular Pyramids, Pentagonal Pyramids, Hexagonal Pyramids,** ····· .

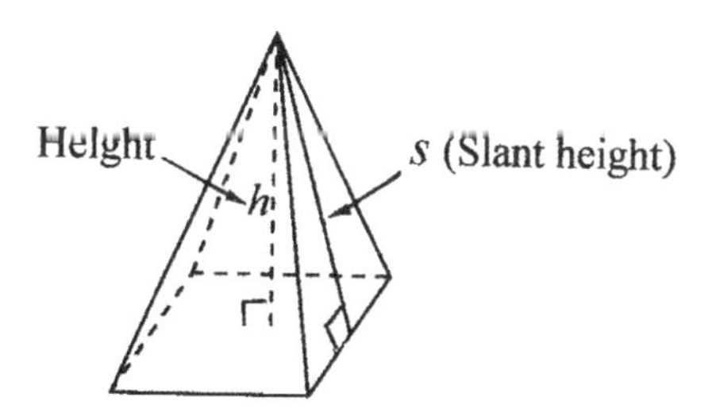

Regular Pyramid

The **volume** (V) of a pyramid is one-third of the product of the area (B) of the base and the height h.

Volume of a Pyramid: $V = \frac{1}{3}Bh$
(Hint: It is one-third of the volume of a prism with same base and h.)

The **lateral surface area** (LA) of a regular pyramid equals one-half of the product of the slant height and the perimeter of the base.

Lateral Surface Area of a Regular Pyramid: $LA = \frac{1}{2}sp$; s: slant height
p: perimeter of the base

The **surface area** (SA) of a regular pyramid with base area B and slant height s equals the sum of the lateral area and the base area.

Surface Area of a Regular Pyramid: $SA = \frac{1}{2}sp + B$; s: slant height
B: base area

There is no general formula for the surface area of a pyramid which is not regular. To find the surface area of a general pyramid, we add up the areas of all polygonal surfaces.

Examples

1. Find the volume of the pyramid whose base is a rectangle.

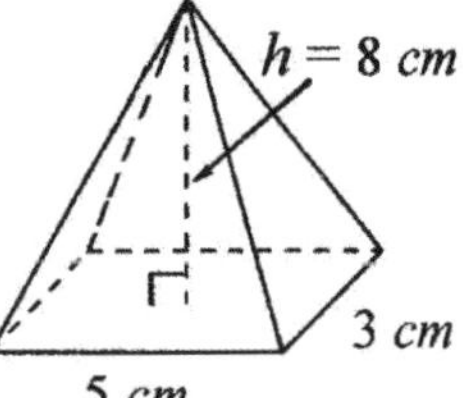

Solution:

Base area: $B = 5 \times 3 = 15$

$V = \frac{1}{3}Bh = \frac{1}{3}(15)(8) = 40\ cm^3$. Ans.

2. Find the volume of the pyramid whose base is an equilateral triangle with side length 6 inches, and height 8 inches.

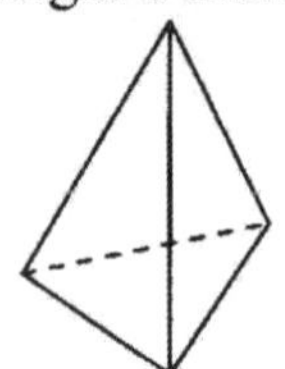

Solution:

Base area: $B = \frac{s^2\sqrt{3}}{4} = \frac{6^2\sqrt{3}}{4} = 9\sqrt{3}$

(Area of equilateral triangle)

$V = \frac{1}{3}Bh = \frac{1}{3}(9\sqrt{3})(8) = 24\sqrt{3}\ in.^3$ Ans.

3. Find the lateral surface area of the regular pyramid.

Solution:

Method 1: There are four congruent triangles. $h = s$

$LA = 4(\frac{1}{2}bh) = 4(\frac{1}{2} \cdot 4 \cdot 8) = 64\ \ cm^2$. Ans.

Method 2: $s = 8$, $p = 4(4) = 16$

$LA = \frac{1}{2}sp = \frac{1}{2}(8)(16) = 64\ \ cm^2$. Ans.

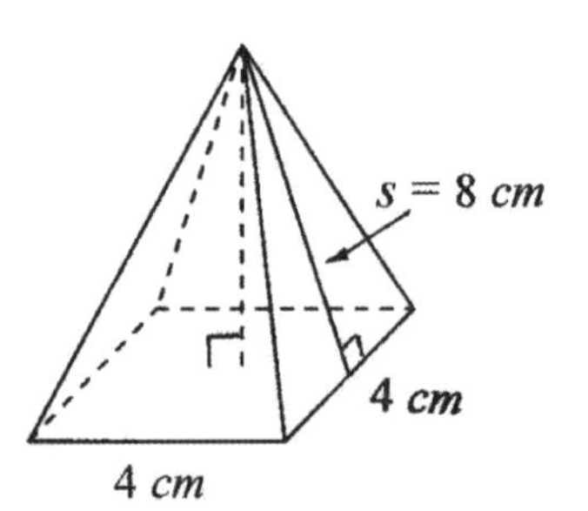

4. Find the total surface area of the regular hexagonal pyramid.

Solution:

Lateral area: $LA = \frac{1}{2}sp = \frac{1}{2}(10)(4 \times 6) = 120$

Base area: $a = 2\sqrt{3}$, $p = 4(6) = 24$

$B = \frac{1}{2}ap = \frac{1}{2}(2\sqrt{3})(24) = 24\sqrt{3}$

Surface area: $SA = (120 + 24\sqrt{3})\ \ cm^2$. Ans.

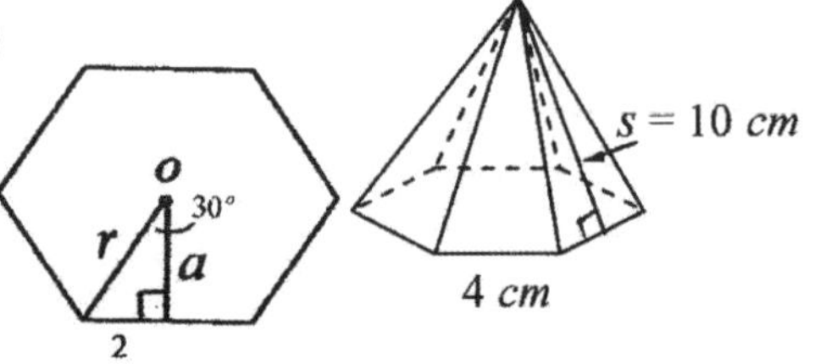

5. The base of a pyramid is a square with edge length 3, and slant height 3. Find the volume of the pyramid.

Solution:

$h^2 + (\frac{3}{2})^2 = 3^2$, $h^2 + \frac{9}{4} = 9$, $h^2 = \frac{27}{4}$ $\therefore h = \frac{3\sqrt{3}}{2}$

$V = \frac{1}{3}Bh = \frac{1}{3}(3^2)(\frac{3\sqrt{3}}{2}) = \frac{9\sqrt{3}}{2}$. Ans.

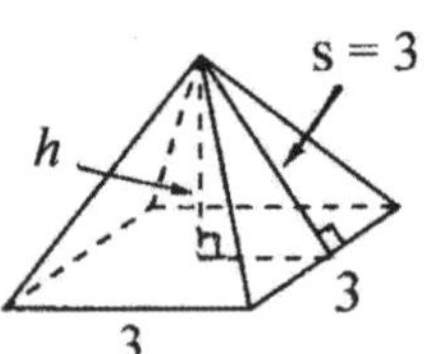

6. The base of a pyramid is a square with edge length x, and $h = 5x$. Find the volume of the pyramid in terms of h.

Solution:

Base area: $B = x^2$; $x = \frac{h}{5}$

$V = \frac{1}{3}Bh = \frac{1}{3}x^2h = \frac{1}{3}(\frac{h}{5})^2 h = \frac{1}{75}h^3$. Ans.

3-21 Spheres

A **sphere** is a three-dimensional figure having all points on the surface equally distance from the center. A sphere is a three-dimensional circle.
The **radius** of s sphere is the distance from the center to any point on the sphere.
If a plane intersects a sphere, the intersection is a circle. If a plane is tangent to a sphere, the intersection is point.

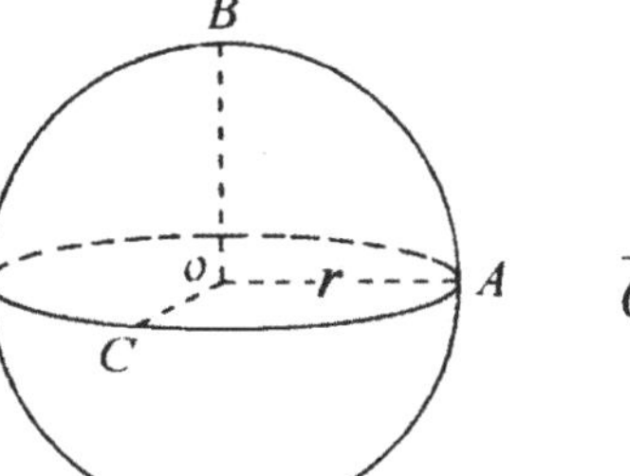

$\overline{OA} = \overline{OB} = \overline{OC} = r$ (radius)

The **volume** (V) of a sphere is given as following.

Volume of a sphere: $V = \frac{4}{3}\pi r^3$; r: radius

The **Surface Area** (SA) of a sphere is given as following.

Surface Area of a sphere: $SA = 4\pi r^2$; r: radius

Rules of Inscribed Solids

1. When a cylinder is inscribed in a sphere, the diameter of the sphere is equal to the diagonal of the rectangle formed by the heights and diameters of the cylinder.
2. When a sphere is inscribed in a cylinder, the sphere and the cylinder have the same radius.
3. When a sphere is inscribed in a cube, the diameter of the sphere is equal to the edge length of the cube.
4. When a rectangular solid or cube is inscribed in a sphere, the diagonal of the solid is equal to the diameter of the sphere.

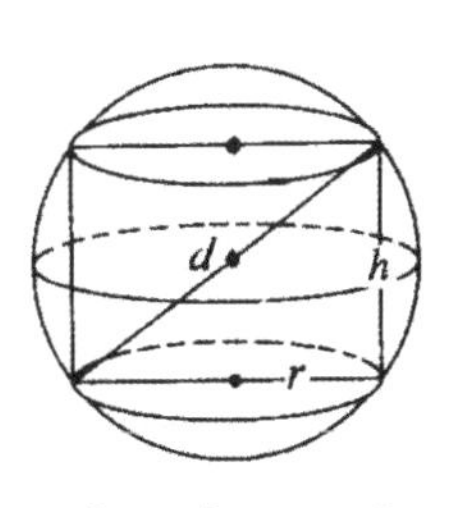

$d^2 = h^2 + (2r)^2$
h: height of the cylinder

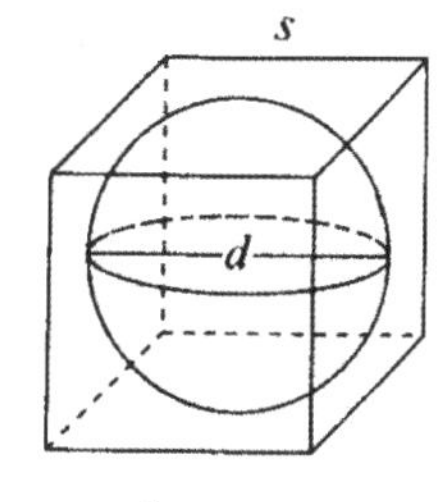

$r_1 = r_2$

$d = s$

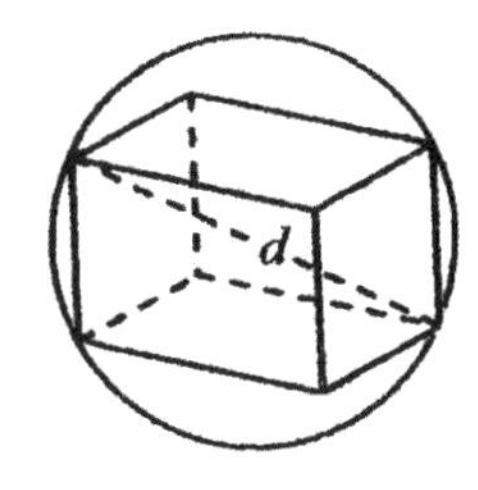

diagonal
= diameter

Examples

1. Find the volume and surface area of a sphere with a radius 6 inches.
   Solution:

$$V = \tfrac{4}{3}\pi\, r^3 = \tfrac{4}{3}\pi(6^3) = 288\,\pi\ in.^3 \text{ Ans.}$$

$$SA = 4\pi\, r^2 = 4\pi(6^2) = 144\pi\ in.^2 \text{ Ans.}$$

2. The surface area of a sphere and the volume of a cylinder are equal. What is the ratio of the radius of the sphere to the radius of the cylinder ?
   Solution:

Let r_1 = radius of the sphere
r_2 = radius of the cylinder

We have:

$$4\pi(r_1)^2 = \pi(r_2)^2(h)$$

$$\frac{(r_1)^2}{(r_2)^2} = \frac{h}{4}$$

$$\therefore \frac{r_1}{r_2} = \sqrt{\frac{h}{4}} = \frac{\sqrt{h}}{2}. \text{ Ans.}$$

3. If we double the radius of a sphere, how much do we increase the volume of the sphere ?
   Solution:

$$V = \tfrac{4}{3}\pi\, r^3$$

$$V_1 = \tfrac{4}{3}\pi(2r)^3 = \tfrac{4}{3}\pi(8r^3) = 8(\tfrac{4}{3}\pi\, r^3) = 8V. \text{ Ans.}$$

The volume of the sphere is multiplied by a factor of 8.
We may simply work on the radius to find the scale factor.

$$(2r)^3 = 8r^3$$

4. If the radius of a sphere is halved, how much does the volume decrease ?
   Solution:

$$V = \tfrac{4}{3}\pi\, r^3$$

$$V_1 = \frac{4}{3}\pi\left(\frac{r}{2}\right)^3 = \frac{4}{3}\pi\left(\frac{r^3}{8}\right) = \frac{1}{8}\left(\frac{4}{3}\pi r^3\right) = \frac{1}{8}V. \text{ Ans.}$$

The volume of the sphere is multiplied by a factor of $\frac{1}{8}$ (divided by 8).
We may simply work on the radius to find the scale factor.

$$(\tfrac{1}{2}r)^3 = \tfrac{1}{8}r^3$$

5. If the radius of a sphere is tripled, how much does the surface area increase ?
   Solution:

$$(3r)^2 = 9r^2$$

The surface area of the sphere is multiplied by a factor of 9. Ans.

Examples

6. In the figure, a sphere is inscribed in a cube. Find the ratio of the area of the sphere to the area of the cube.

Solution:

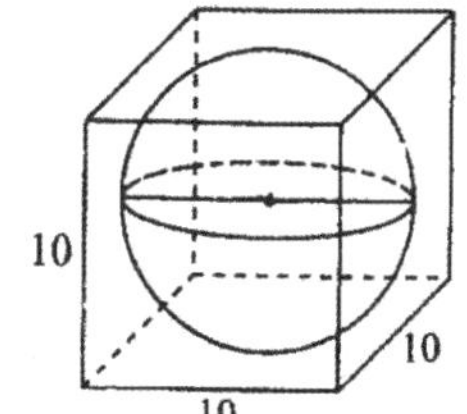

Area of the sphere: $A_1 = 4\pi\, r^2 = 4\pi(5^2) = 100\pi$

Area of the cube: $A_2 = 6s^2 = 6(10^2) = 600$

$$\frac{A_1}{A_2} = \frac{100\pi}{600} = \frac{\pi}{6} \approx 0.52. \text{ Ans.}$$

7. In the figure, a cylinder is inscribed in the sphere. The height of the cylinder is equal to its the diameter of its circular base. Find the ratio of the volume of the cylinder to the volume of the sphere.

Solution:

Volume of the cylinder: $V_1 = \pi\, r^2 h = \pi(3^2)(6) = 54\pi$

Volume of the sphere: $d^2 = 6^2 + 6^2 = 72$

$d = \sqrt{72} = 6\sqrt{2}\,,\quad r = 3\sqrt{2}$

$V_2 = \frac{4}{3}\pi\, r^3 = \frac{4}{3}\pi(3\sqrt{2})^3 = \frac{4}{3}\pi(54\sqrt{2}) = 72\sqrt{2}\,\pi$

$$\frac{V_1}{V_2} = \frac{54\pi}{72\sqrt{2}\pi} = \frac{3}{4\sqrt{2}} = \frac{3\sqrt{2}}{8} \approx 0.53. \text{ Ans.}$$

8. A cube is inscribed in a cylinder. The height of the cylinder is 12. The length of the diagonal of the cube is $8\sqrt{3}$. Find the volume of the cylinder.

Solution:

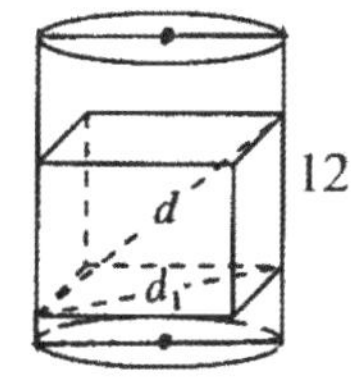

First, find the edge length s of the cube.

Length of diagonal is given: $d = 8\sqrt{3}$

$d^2 = s^2 + s^2 + s^2 = 3s^2$

$d = \sqrt{3}\, s, \quad 8\sqrt{3} = \sqrt{3}\, s \quad \therefore s = 8$

The diagonal of a face of the cube is the diameter of the circular base.

$(d_1)^2 = 8^2 + 8^2 = 128$

$d_1 = \sqrt{128} = 8\sqrt{2}$

The radius of the circular base: $r = 4\sqrt{2}$

The volume of the cylinder: $V = \pi\, r^2 h = \pi(4\sqrt{2})^2(12) = 384\pi$. Ans.

Roger was absent yesterday.
Teacher: Roger, why were you absent ?
Roger: My tooth was aching.
I went to see my dentist.
Teacher: Is your tooth still aching ?
Roger: I don't know. My dentist has it.

Notes

3-22 Ellipses

Definition of an ellipse: An ellipse is the set of all points in a plane such that for each point , the sum of the distances (called the focal radii) from two fixed points (called the foci) is a constant.

Ellipse is commonly applied to the movement of the earth. The earth's orbit around the sun is shaped as an ellipse in which the sun is located and fixed at one of its two foci.

Standard Forms of Ellipses

1. $\dfrac{x^2}{a^2}+\dfrac{y^2}{b^2}=1$

Center (0, 0)

$c^2=a^2-b^2$

Focal radii:

$PF_1+PF_2=2a$

Vertex: $(a, 0)$ and $(-a, 0)$

Co-vertex: $(0, b)$ and $(0, -b)$

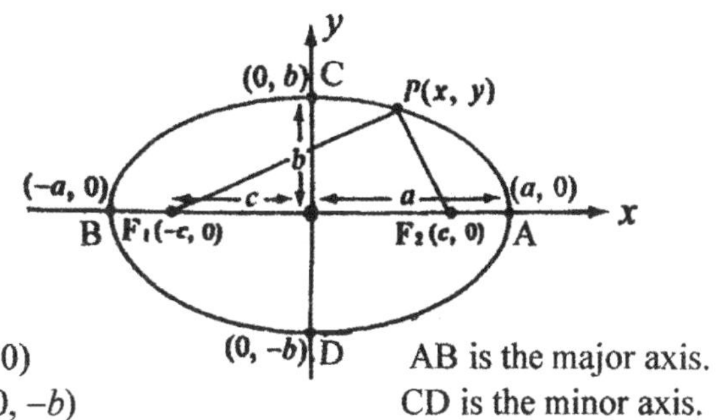

2. $\dfrac{(x-h)^2}{a^2}+\dfrac{(y-k)^2}{b^2}=1$

Center (h, k)

$c^2=a^2-b^2$

3. $\dfrac{x^2}{b^2}+\dfrac{y^2}{a^2}=1$

Center (0, 0)

$c^2=a^2-b^2$

4. $\dfrac{(x-h)^2}{b^2}+\dfrac{(y-k)^2}{a^2}=1$

Center (h, k)

$c^2=a^2-b^2$

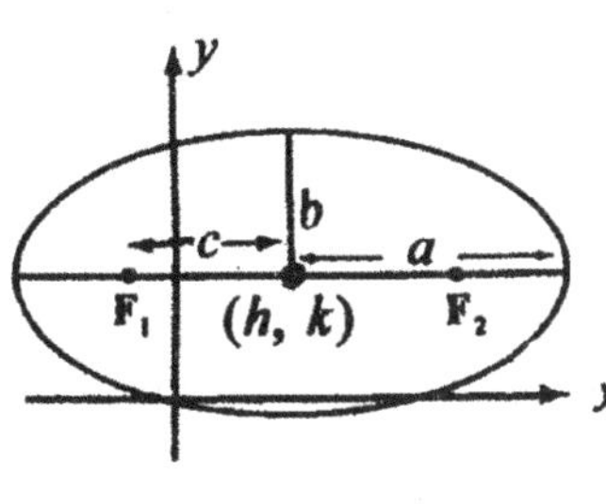

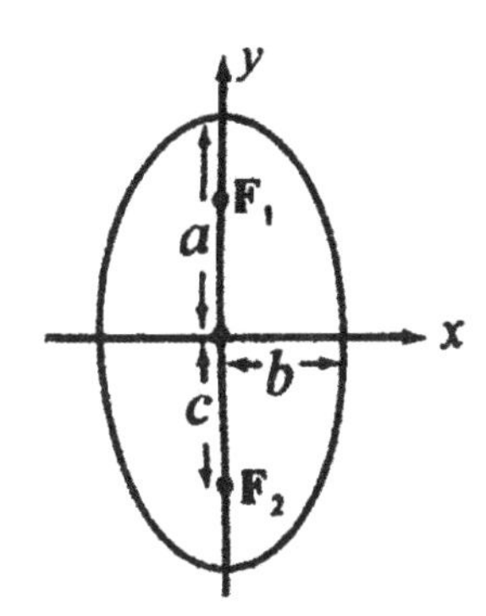

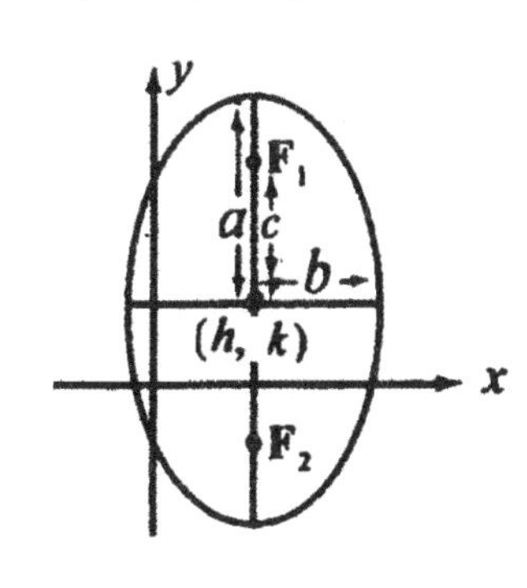

Example

1. Find the equation of an ellipse having foci (0, 3) and (0, –3) and the sum of its focal radii is 10.

   Solution:

$$PF_1+PF_2=10$$

$$\sqrt{(x-0)^2+(y-3)^2}+\sqrt{(x-0)^2+(y+3)^2}=10$$

$$\sqrt{x^2+(y-3)^2}=10-\sqrt{x^2+(y+3)^2}$$

$$\cancel{x^2}+(y-3)^2=100-20\sqrt{x^2+(y+3)^2}+\cancel{x^2}+(y+3)^2$$

$$\cancel{y^2}-6y+\cancel{9}=100-20\sqrt{x^2+(y+3)^2}+\cancel{y^2}+6y+\cancel{9}$$

$$12y+100=20\sqrt{x^2+(y+3)^2}$$

$$3y+25=5\sqrt{x^2+y^2+6y+9}$$

$$9y^2+150y+625=25(x^2+y^2+6y+9)$$

$$9y^2+\cancel{150y}+625=25x^2+25y^2+\cancel{150y}+225$$

$$25x^2+16y^2=400$$

$$\frac{25x^2}{400}+\frac{16y^2}{400}=1 \quad \therefore \frac{x^2}{16}+\frac{y^2}{25}=1$$

Examples

2. Graph $9x^2+25y^2=225$.

Solution:

$9x^2+25y^2=225$

Divided both sides by 225:

$$\frac{9x^2}{225}+\frac{25y^2}{225}=1$$

$$\frac{x^2}{25}+\frac{y^2}{9}=1$$

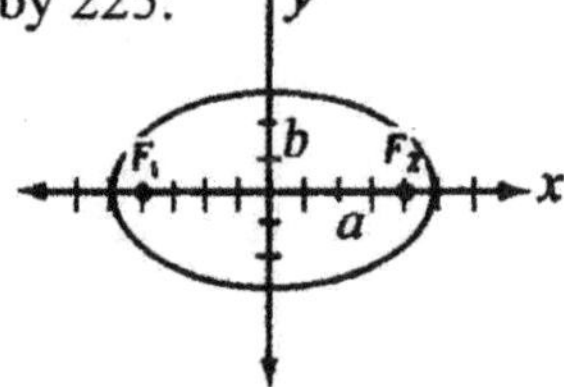

The major axis is horizontal.

Center(0, 0), $a=5$, $b=3$

$c^2=a^2-b^2=25-9=16$

$\therefore c=4$

3. Graph $25x^2+9y^2=225$.

Solution:

$25x^2+9y^2=225$

Divided both sides by 225:

$$\frac{25x^2}{225}+\frac{9y^2}{225}=1$$

$$\frac{x^2}{9}+\frac{y^2}{25}=1$$

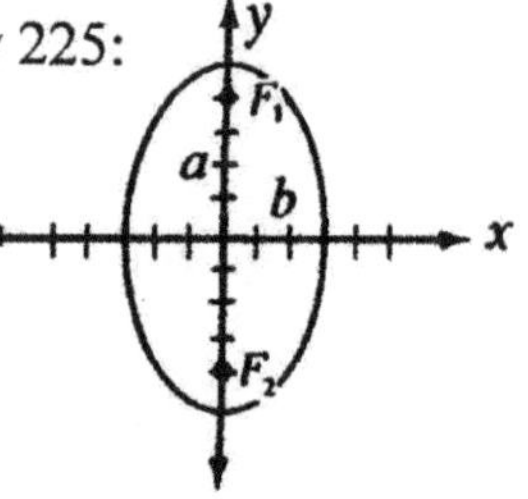

The major axis is vertical.

Center(0, 0), $a=5$, $b=3$

$c^2=a^2-b^2=25-9=16$

$\therefore c=4$

4. Find the equation of an ellipse having foci (0, 3) and (0, –3) and the sum of its focal radii is 10.

Solution:

Center (0, 0) and $c=3$

$2a=10 \quad \therefore a=5$

$c^2=a^2-b^2$

$3^2=5^2-b^2$

$b^2=16 \quad \therefore b=4$

The major axis is vertical:

$$\frac{x^2}{b^2}+\frac{y^2}{a^2}=1 \quad \therefore \frac{x^2}{16}+\frac{y^2}{25}=1. \text{ Ans.}$$

(See graph on Example 1)

5. Find the equation of an ellipse having vertices at (0, 5) and (0, –5) and foci at (0, 3) and (0, –3).

Solution:

Center (0, 0) and $a=5$, $c=3$

We find b:

$c^2=a^2-b^2$

$3^2=5^2-b^2$

$b^2=16 \quad \therefore b=4$

The major axis is vertical:

$$\frac{x^2}{b^2}+\frac{y^2}{a^2}=1 \quad \therefore \frac{x^2}{16}+\frac{y^2}{25}=1$$

(See graph on Example 1)

6. Graph $4x^2+y^2-16x+10y+25=0$

Solution:

Completing the square:

$4(x^2-4x)+(y^2+10y)+25=0$

$4[(x^2-4x+4)-4]+(y^2+10y+25)-25+25=0$

$4[(x-2)^2-4]+(y+5)^2=0$

$4(x-2)^2+(y+5)^2=16$

Divided both sides by 16:

$\frac{(x-2)^2}{4}+\frac{(y+5)^2}{16}=1$, The major axis is vertical.

Center $(2,-5)$; $a=4$; $b=2$

$c^2=a^2-b^2=16-4=12$

$\therefore c=\sqrt{12}=3.36$

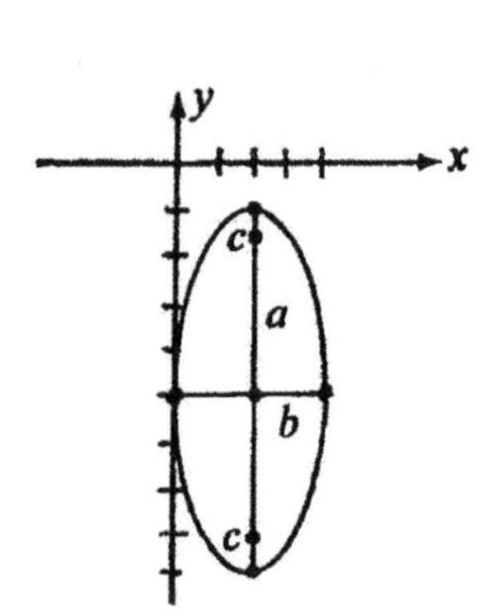

Examples

7. Find the equation of an ellipse passing through the point (3, 5) and having the foci at (3, 2) and (−1, 2).

Solution:

We graph and find: Center (1, 2) and $c = 2$

The major axis is horizontal.

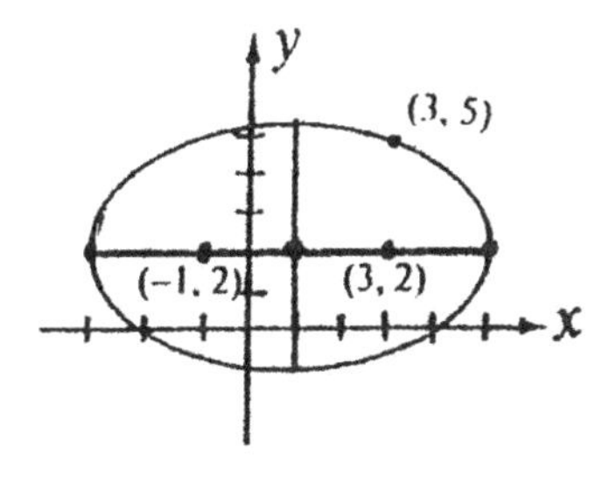

$$\frac{(x-1)^2}{a^2}+\frac{(y-2)^2}{b^2}=1$$

Passing the point (3, 5), we apply the definition of an ellipse:

$$\sqrt{(5-2)^2+(3-3)^2}+\sqrt{(5-2)^2+(3+1)^2}=2a$$

$$\sqrt{9}+\sqrt{25}=2a$$

$$3+5=2a \quad \therefore a=4$$

Since $c^2=a^2-b^2$, $c=2$ we have $4=16-b^2$ $\therefore b=\sqrt{12}$

$$\therefore \frac{(x-1)^2}{16}+\frac{(y-2)^2}{12}=1. \text{ Ans.}$$

8. The earth's orbit around the sun is shaped as an ellipse in which the sun is located and fixed at one of its two foci. At its closest distance (called the perihelion), the earth is 91.45 million miles from the center of the sun. At its farthest distance (called the aphelion), the earth is 94.55 million miles from the center of the sun. Write the equation of the orbit.

Solution:

$2a = 91.45 + 94.55 = 186$ million miles

$\therefore a = 93$ million miles

We have: $c = 93 - 91.45 = 1.55$ million miles

$$c^2=a^2-b^2$$

$$(1{,}55)^2=(93)^2-b^2$$

$$b^2=(93)^2-(1.55)^2=8646.60$$

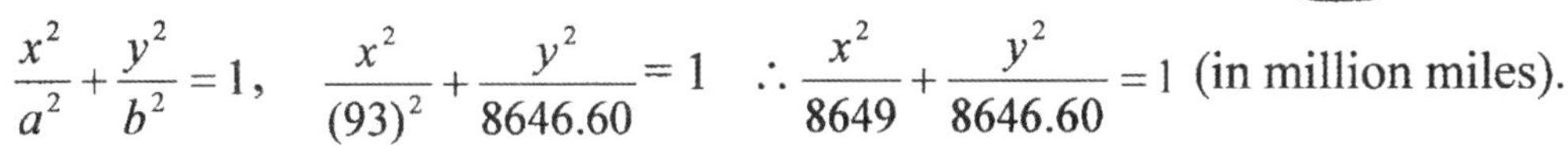

$$\frac{x^2}{a^2}+\frac{y^2}{b^2}=1, \quad \frac{x^2}{(93)^2}+\frac{y^2}{8646.60}=1 \quad \therefore \frac{x^2}{8649}+\frac{y^2}{8646.60}=1 \text{ (in million miles).}$$

It is nearly circular.

On graduation day, a high school graduate said to his math teacher.

Student: Thank you very much for giving me a "D" in your class.
For your kindness, do you want me to do anything for you ?

Teacher: Yes, Please do me a favor.
Just don't tell anyone that I was your math teacher.

Notes

3-23 Equation of a Hyperbola

Definition of a hyperbola: A hyperbola is the set of all points in a plane such that for each point, the difference of the distances (called the focal radii) from two fixed points (called the foci) is a constant.

Standard Forms of Hyperbolas

1. $\dfrac{x^2}{a^2}-\dfrac{y^2}{b^2}=1$

Center $(0, 0)$, $c^2=a^2+b^2$

Focal radii: $|PF_1-PF_2|=2a$

Two asymptotes are: $y=\pm\frac{b}{a}x$

The length of transverse axis $=2a$

The length of conjugate axis $=2b$

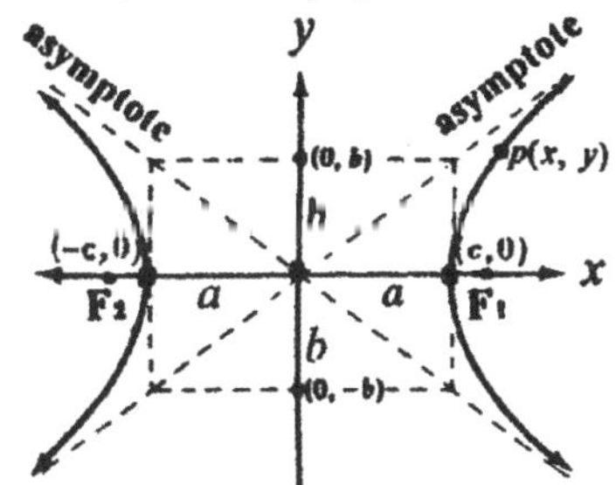

2. $\dfrac{(x-h)^2}{a^2}-\dfrac{(y-k)^2}{b^2}=1$

Center (h, k), $c^2=a^2+b^2$

Two asymptotes are: $y-k=\pm\frac{b}{a}(x-h)$

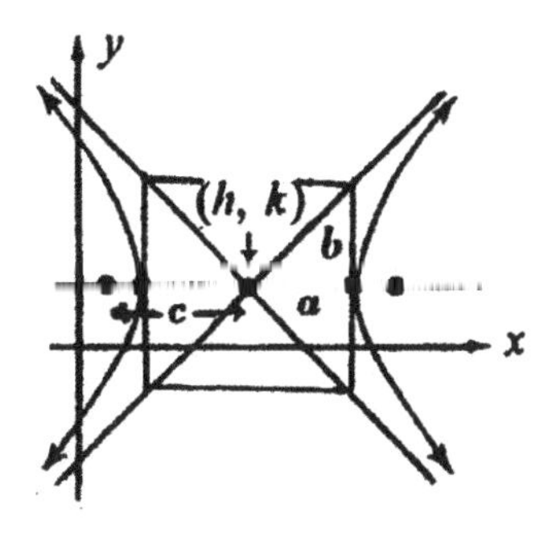

3. $\dfrac{y^2}{a^2}-\dfrac{x^2}{b^2}=1$

Center $(0, 0)$

$c^2=a^2+b^2$

Two asymptotes are:

$y=\frac{a}{b}x$ and $y=-\frac{a}{b}x$

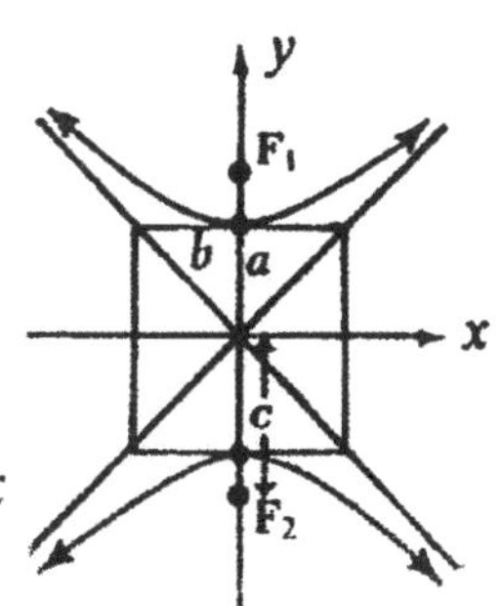

4. $\dfrac{(y-k)^2}{a^2}-\dfrac{(x-h)^2}{b^2}=1$

Center (h, k)

$c^2=a^2+b^2$

Two asymptotes are:

$y-k=\pm\frac{a}{b}(x-h)$

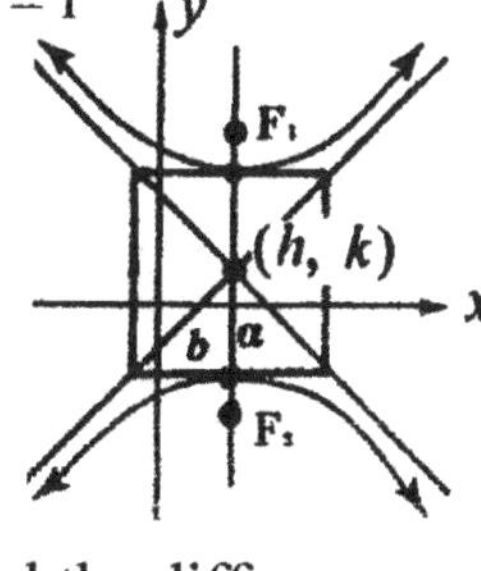

Example

1. Find the equation of a hyperbola having foci $(0, 3)$ and $(0, -3)$ and the difference of its focal radii is 4.

Solution: $|PF_1=PF_2|=4$

$$\left|\sqrt{(x-0)^2+(y-3)^2}-\sqrt{(x-0)^2+(y+3)^2}\right|=4$$

$$\sqrt{x^2+(y-3)^2}-\sqrt{x^2+(y+3)^2}=\pm 4$$

$$\sqrt{x^2+(y-3)^2}=\sqrt{x^2+(y+3)^2}\pm 4$$

$$x^2+(y-3)^2=x^2+(y+3)^2\pm 8\sqrt{x^2+(y+3)^2}+16$$

$$y^2-6y+9=y^2+6y+9\pm 8\sqrt{x^2+(y+3)^2}+16$$

$$12y+16=\pm 8\sqrt{x^2+(y+3)^2},\quad 3y+4=\pm 2\sqrt{x^2+(y+3)^2}$$

$$9y^2+24y+16=4(x^2+y^2+6y+9)$$

$$5y^2-4x^2=20,\quad \frac{5y^2}{20}-\frac{4x^2}{20}=1 \quad \therefore \frac{y^2}{4}-\frac{x^2}{5}=1.\ \text{Ans.}$$

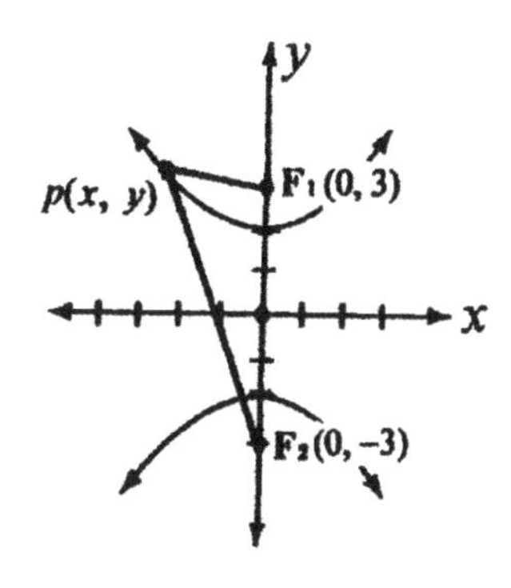

Examples

2. Graph $16x^2 - 9y^2 = 144$.

Solution:

$16x^2 - 9y^2 = 144$

Divided both sides by 144:

$\frac{16x^2}{144} - \frac{9y^2}{144} = 1$

$\frac{x^2}{9} - \frac{y^2}{16} = 1$

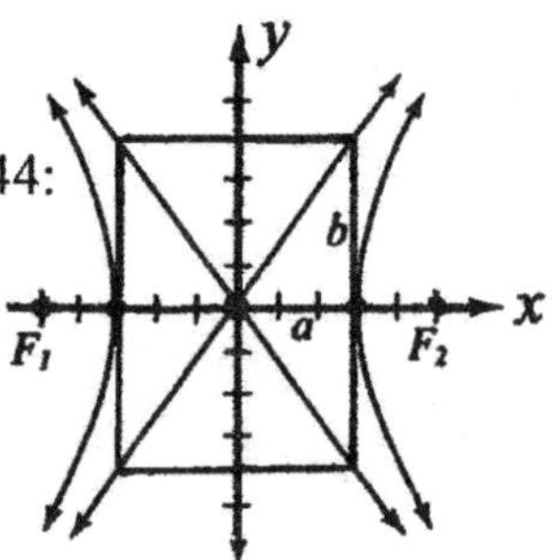

The transverse axis is horizontal (x – axis).

Center (0, 0), $a = 3$, $b = 4$

$c^2 = a^2 + b^2 = 9 + 16 = 25 \quad \therefore c = 5$

3. Graph $4y^2 - 9x^2 = 36$.

Solution:

$4y^2 - 9x^2 = 36$

Divided both sides by 36:

$\frac{4y^2}{36} - \frac{9x^2}{36} = 1$

$\frac{y^2}{9} - \frac{x^2}{4} = 1$

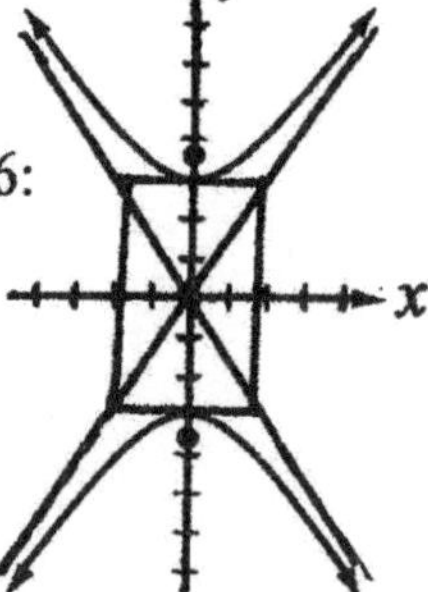

The transverse axis is vertical (y – axis).

Center (0, 0), $a = 3$, $b = 2$

$c^2 = a^2 + b^2 = 9 + 4 = 13 \quad \therefore c = \sqrt{13}$

4. Find the equation of a hyperbola having foci (0, 3) and (0, –3) and the difference of its focal radii is 4.

Solution:

Center (0, 0), $c = 3$

$2a = 4 \quad \therefore a = 2$

We find b: $c^2 = a^2 + b^2$

$3^2 = 2^2 + b^2$

$b^2 = 5 \quad \therefore b = \sqrt{5}$

The transverse axis is vertical (y – axis).

$\frac{y^2}{a^2} - \frac{x^2}{b^2} = 1 \quad \therefore \frac{y^2}{4} - \frac{x^2}{5} = 1$. Ans.

(See graph on Example 1)

5. Find the equation of a hyperbola having vertices at (0, 2) and (0, –2) and foci (0. 3) and (0, –3).

Solution:

Center (0, 0) and $a = 2$, $c = 3$

We find b: $c^2 = a^2 + b^2$

$3^2 = 2^2 + b^2$

$b^2 = 5$

$\therefore b = \sqrt{5}$

The transverse axis is vertical ((y – axis).

$\frac{y^2}{a^2} - \frac{x^2}{b^2} = 1 \quad \therefore \frac{y^2}{4} - \frac{x^2}{5} = 1$. Ans.

(See graph on Example 1)

6. Graph $y^2 - 2x^2 + 6y + 8x - 3 = 0$

Solution:

Completing the square:

$(y^2 + 6y) - 2(x^2 - 4x) - 3 = 0$

$(y^2 + 6y + 9) - 9 - 2[(x^2 - 4x + 4) - 4] - 3 = 0$

$(y+3)^2 - 9 - 2(x-2)^2 + 8 - 3 = 0$

$(y+3)^2 - 2(x-2)^2 = 4$

Divided both sides by 4:

$\frac{(y+3)^2}{4} - \frac{(x-2)^2}{2} = 1$

The transverse axis is vertical ($x = 2$).

Center (2, –3), $a = 2$, $b = \sqrt{2}$

$c^2 = a^2 + b^2 = 4 + 2 = 6 \quad \therefore c = \sqrt{6}$

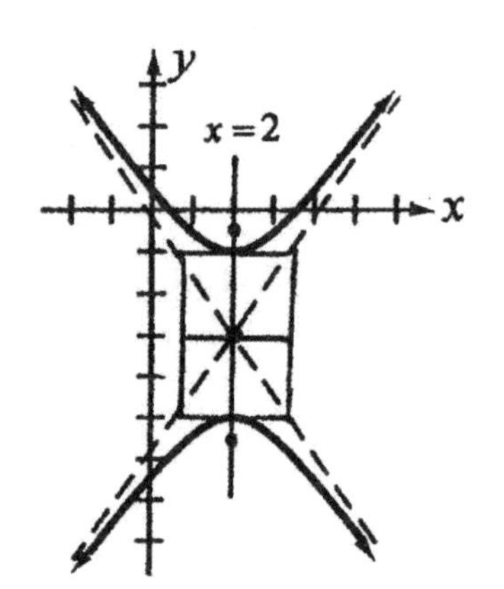

Examples

7. Find the equation of a hyperbola passing through the point (1, –5) and having the foci at (–2, –1) and (–2, –5).

Solution:

We graph and find: Center (–2, –3) and $c = 2$

The transverse axis is vertical.

We have the equation: $\frac{(y+3)^2}{a^2} - \frac{(x+2)^2}{b^2} = 1$

Passing the point (1, –5), we apply the definition of a hyperbola:

$$\left|\sqrt{(1+2)^2+(-5+1)^2} - \sqrt{(1+2)^2+(-5+5)^2}\right| = 2a$$

$$\left|\sqrt{25} - \sqrt{9}\right| = 2a$$

$$|5-3| = 2a \quad \therefore a = 1$$

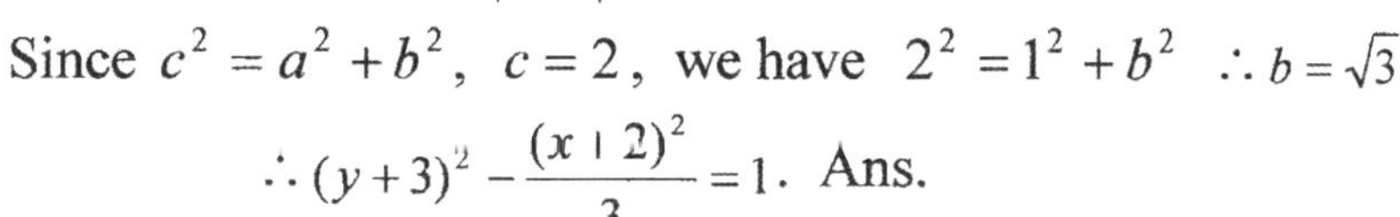

Since $c^2 = a^2 + b^2$, $c = 2$, we have $2^2 = 1^2 + b^2 \quad \therefore b = \sqrt{3}$

$$\therefore (y+3)^2 - \frac{(x+2)^2}{3} = 1. \text{ Ans.}$$

8. Find the equations of the asymptotes of the hyperbola $x^2 - 4y^2 = 4$.

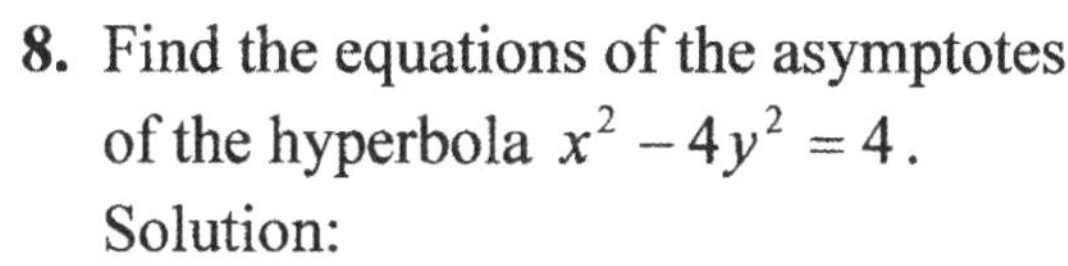

Solution:

$$x^2 - 4y^2 = 4$$

Divided both sides by 4:

$$\frac{x^2}{4} - \frac{y^2}{1} = 1 \quad \therefore a = 2,\ b = 1$$

The equations of the asymptotes are:

$$y = \pm\tfrac{b}{a}x$$

$$\therefore y = \pm\tfrac{1}{2}x. \text{ Ans.}$$

9. Find the equations of the asymptotes of the hyperbola $3y^2 - 4x^2 = 36$.

Solution:

$$3y^2 - 4x^2 = 36$$

Divided both sides by 36:

$$\frac{y^2}{12} - \frac{x^2}{9} = 1 \quad \therefore a = \sqrt{12},\ b = 3$$

The equations of the asymptotes are:

$$y = \pm\tfrac{a}{b}x$$

$$\therefore y = \pm\tfrac{2\sqrt{3}}{3}x. \text{ Ans.}$$

10. Find the equations of the asymptotes of the hyperbola $9x^2 - 4y^2 - 36x - 24y - 36 = 0$.

Solution:

$$9x^2 - 4y^2 - 36x - 24y - 36 = 0$$

Completing the square, we have:

$$\frac{(x-2)^2}{4} - \frac{(y+3)^2}{9} = 1 \quad \therefore a = 2,\ b = 3$$

The equations of the asymptotes are:

$$y - k = \pm\tfrac{b}{a}(x-h)$$

$$\therefore y + 3 = \pm\tfrac{3}{2}(x-2). \text{ Ans.}$$

11. Find the equations of the asymptotes of the hyperbola $y^2 - x^2 - 6y - 4x + 1 = 0$.

Solution:

$$y^2 - x^2 - 6y - 4x + 1 = 0$$

Completing the square, we have:

$$\frac{(y-3)^2}{4} - \frac{(x+2)^2}{4} = 1 \quad \therefore a = b = 2$$

(Hint: It is an **equilateral hyperbola**)

The equations of the asymptotes are:

$$y - k = \pm\tfrac{a}{b}(x-h)$$

$$\therefore y - 3 = \pm(x+2). \text{ Ans.}$$

Notes

3-24 Polar Coordinates and Graphs

A point $p(x, y)$ in the rectangular coordinates x and y can be represented in the **polar coordinates** $p(r, \theta)$**:** (Hint: Read Chapter 4-2 if needed.)

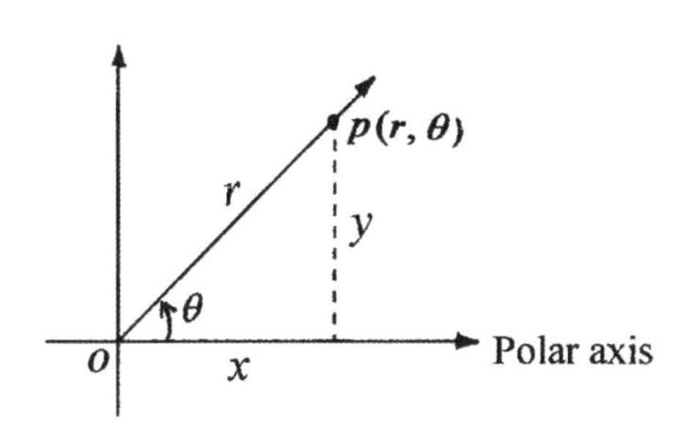

o: the **pole** or **origin**.
r: directed distance from o to p.
θ : directed angle, counterclockwise from polar axis to $\overline{op}$.

The relationship between polar and rectangular coordinates are:

$$x = r\cos\theta, \quad y = r\sin\theta$$

$$\tan\theta = \frac{y}{x}, \quad r = \sqrt{x^2 + y^2}$$

In the polar coordinates, the coordinates of the pole is $p(0, \theta)$, where θ can be any angle.

In the polar coordinates, a point $p(r, \theta)$ has infinitely many representations.

The point $p(r, \theta)$ can be represented as $p(r, \theta \pm 2n\pi)$ or $p(-r, \theta \pm (2n\pi + \pi))$, where n is an integer.

Example: $(5, \frac{\pi}{6}) = (5, \frac{\pi}{6} + 2\pi) = (5, \frac{13\pi}{6})$

$(5, \frac{\pi}{6}) = (5, \frac{\pi}{6} - 2\pi) = (5, -\frac{11\pi}{6})$

$(5, \frac{\pi}{6}) = (-5, \frac{\pi}{6} + \pi) = (-5, \frac{7\pi}{6})$

$(5, \frac{\pi}{6}) = (-5, \frac{\pi}{6} - \pi) = (-5, -\frac{5\pi}{6})$

To locate the point $(-5, \frac{7\pi}{6})$, we use the ray in the direction opposite the terminal side of $\frac{7\pi}{6}$ at a distance 5 units from the pole.

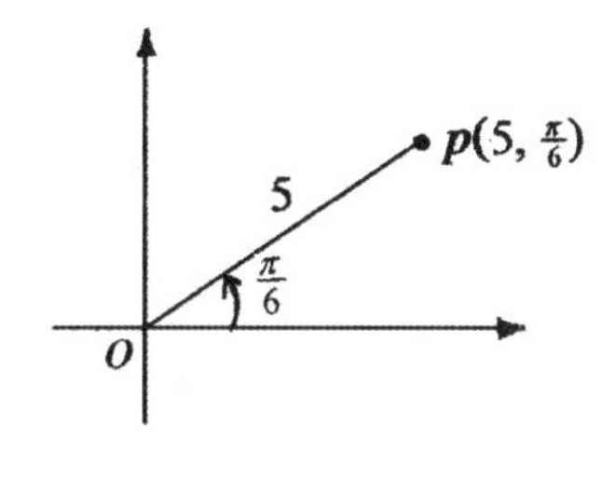

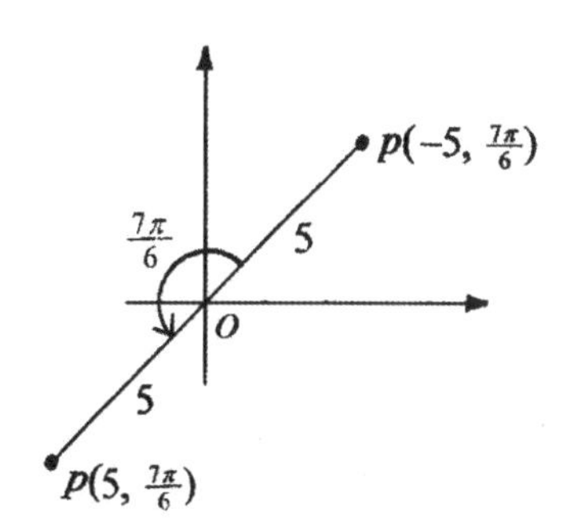

Examples

1. Express the point $(\sqrt{3}, 1)$ in terms of polar coordinates.
   Solution:

   $r=\sqrt{x^2+y^2}=\sqrt{(\sqrt{3})^2+1^2}=2$, $\tan\theta=\frac{y}{x}=\frac{1}{\sqrt{3}}$ $\therefore\theta=\frac{\pi}{6}$.

   The polar coordinates of p are $(2, \frac{\pi}{6})$. Ans.

2. Find the polar coordinates of the point $(0, -5)$.
   Solution:

   The point lies on the negative $y-axis$.

   $r=5$, $\theta=\frac{3\pi}{2}$. The polar coordinates of p are $(5, \frac{3\pi}{2})$. Ans.

3. Find the polar coordinates of $p(-1, -\sqrt{3})$.
   Solution:

   $r=\sqrt{x^2+y^2}=\sqrt{(-1)^2+(-\sqrt{3})^2}=2$. $\tan\alpha=\frac{-\sqrt{3}}{-1}=\sqrt{3}$, $\alpha=\frac{\pi}{3}$.

   The point lies on the 3rd quadrant. $\therefore\theta=\frac{\pi}{3}+\pi=\frac{4\pi}{3}$.

   The polar coordinates of p are $(2, \frac{4\pi}{3})$. Ans.

4. Find the rectangular coordinates of the point $(2, \frac{\pi}{6})$.
   Solution:

   $x=r\cos\theta=2\cos\frac{\pi}{6}=2(\frac{\sqrt{3}}{2})=\sqrt{3}$

   $y=r\sin\theta=2\sin\frac{\pi}{6}=2(\frac{1}{2})=1$

   The rectangular coordinates of the point are $(\sqrt{3}, 1)$. Ans.

5. Transform the equation $xy=2$ to polar form.
   Solution:

   $xy=2$, $(r\cos\theta)(r\sin\theta)=2$, $r^2\cos\theta\sin\theta=2$, $r^2(\frac{1}{2}\sin 2\theta)=2$

   The polar form of the equation is $r^2\sin 2\theta=4$. Ans.

6. Transform the equation $r^2\sin 2\theta=4$ to rectangular coordinates.
   Solution:

   $r^2\sin 2\theta=4$, $r^2(2\sin\theta\cos\theta)=4$, $r^2\sin\theta\cos\theta=2$

   $(r\cos\theta)(r\sin\theta)=2$

   The rectangular coordinates of the equation is $xy=2$. Ans.

Graphing Polar Equations

One method to graph a polar equation is to convert the equation to rectangular form. However, it is not always easy or helpful to graph a polar equation by converting it to rectangular form. Usually, we graph a polar equation by **checking for symmetry and plotting points**.

Checking for symmetry: (See Patterns of Special Polar Graphs on Page 303)

1. If we replace (r, θ) by $(r, -\theta)$ and the equation is unchanged, it is symmetry with respect to the polar axis.
2. If we replace (r, θ) by $(-r, \theta)$ and the equation is unchanged, it is symmetry with respect to the pole (the origin).
3. If we replace (r, θ) by $(r, \pi - \theta)$ and the equation is unchanged, it is symmetry with respect to the line $\theta = \frac{\pi}{2}$.

However, the above rules are not necessary conditions for symmetry. It is possible for a graph to have certain symmetry status which the above rules fail to test.

Examples:

1. Graph the polar equation $\theta = \dfrac{3\pi}{4}$.

Solution:

Converting it to rectangular form.

$\tan\theta = \frac{y}{x}$, $\tan\frac{3\pi}{4} = -\tan 45° = -1$, $\frac{y}{x} = -1$

$\therefore x + y = 0$

It is a line passing through the pole at an angle of $135°$ with the polar axis (regardless of the value of r.)

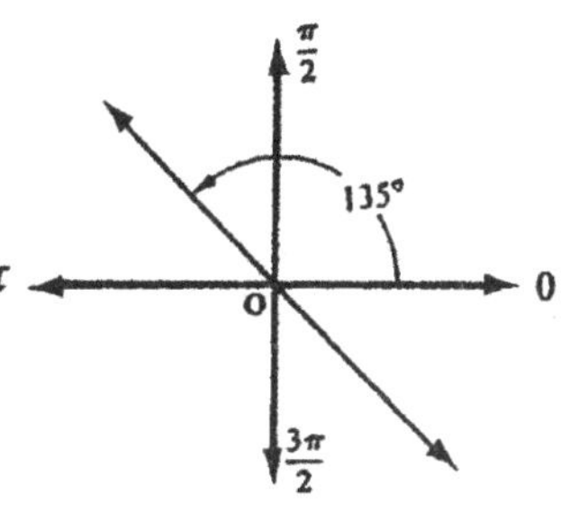

2. Graph the polar equation $r = -4$.

Solution:

Converting it to rectangular form.

$r = \sqrt{x^2 + y^2} = -4$, $\therefore x^2 + y^2 = 16$

It is a circle with center (0, 0) and radius 4.

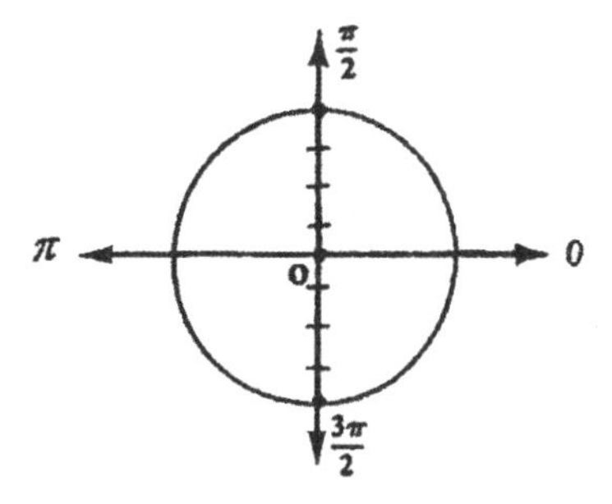

3. Graph the polar equation $r = -4\cos\theta$.

Solution:

Converting it to rectangular form.

$r = -4\cos\theta$, $r^2 = -4r\cos\theta$, $x^2 + y^2 = -4x$

$x^2 + y^2 + 4x = 0$

Completing the square: $(x+2)^2 + y^2 = 4$.

It is a circle with center (–2, 0) and radius 2.

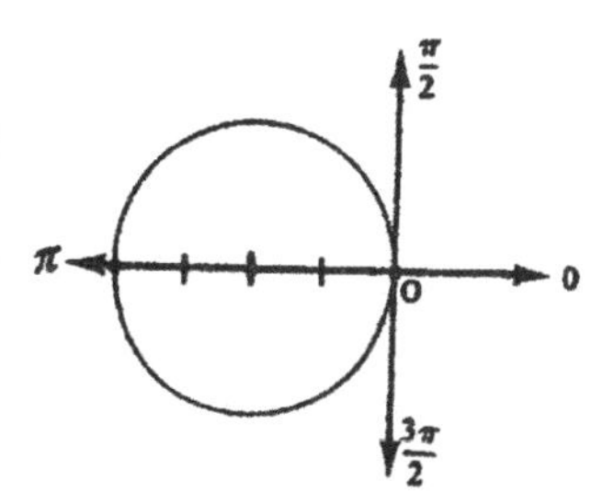

Examples

4. Graph the polar equation $r = 1 - \cos\theta$.

Solution:

Replacing θ by $(-\theta)$, $r = 1 - \cos(-\theta) = 1 - \cos\theta$.
The graph is symmetric with respect to the polar axis.
Plotting the points, we only need to select values of θ from 0 to π.

θ	0	$\frac{\pi}{6}$	$\frac{\pi}{3}$	$\frac{\pi}{2}$	$\frac{2\pi}{3}$	$\frac{5\pi}{6}$	π
$\cos\theta$	1	$\frac{\sqrt{3}}{2}$	$\frac{1}{2}$	0	$-\frac{1}{2}$	$-\frac{\sqrt{3}}{2}$	-1
$r = 1 - \cos\theta$	0	0.13	0.5	1	1.5	1.87	2

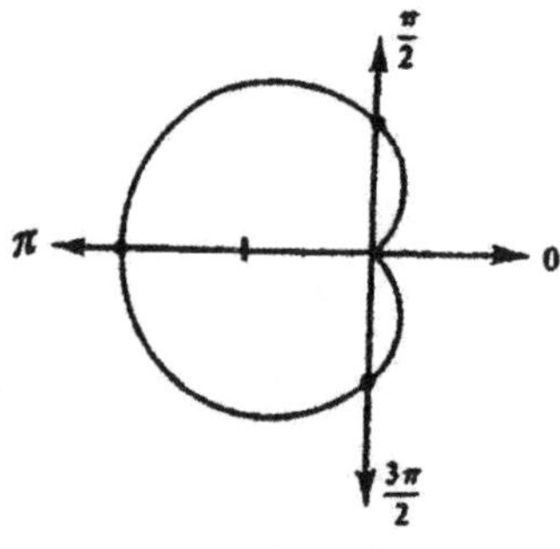

It is a **Cardioid.**
(A heart-shaped curve).

5. Graph the polar equation $r = 3 + 2\sin\theta$.

Solution:

Replacing θ by $(\pi - \theta)$, $r = 3 + 2\sin(\pi - \theta) = 3 + 2\sin\theta$.
The graph is symmetric with respect to the line $\theta = \frac{\pi}{2}$.
Plotting the points, we only need to select values of θ from $-\frac{\pi}{2}$ to $\frac{\pi}{2}$.

θ	$-\frac{\pi}{2}$	$-\frac{\pi}{3}$	$-\frac{\pi}{6}$	0	$\frac{\pi}{6}$	$\frac{\pi}{3}$	$\frac{\pi}{2}$
$\sin\theta$	-1	$-\frac{\sqrt{3}}{2}$	$-\frac{1}{2}$	0	$\frac{1}{2}$	$\frac{\sqrt{3}}{2}$	1
$r = 3 + 2\sin\theta$	1	1.27	2	3	4	4.73	5

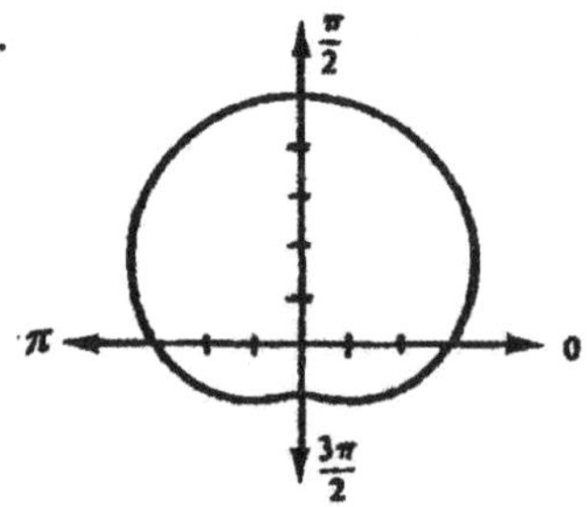

It is a **Limaçon.**

6. Graph the polar equation $r = 4\cos 2\theta$.

Solution:

Replacing θ by $(-\theta)$, $r = 4\cos 2(-\theta) = 4\cos 2\theta$.
Replacing θ by $(\pi - \theta)$,

$r = 4\cos 2(\pi - \theta) = 4\cos(2\pi - 2\theta) = 4\cos 2\theta$.

The graph is symmetric with respect to the polar axis and the line $\theta = \frac{\pi}{2}$.
Plotting the points, we only need to select values of θ from 0 to $\frac{\pi}{2}$.

θ	0	$\frac{\pi}{6}$	$\frac{\pi}{4}$	$\frac{\pi}{3}$	$\frac{\pi}{2}$
$\cos 2\theta$	1	$\frac{1}{2}$	0	$-\frac{1}{2}$	-1
$r = 4\cos 2\theta$	4	2	0	-2	-4

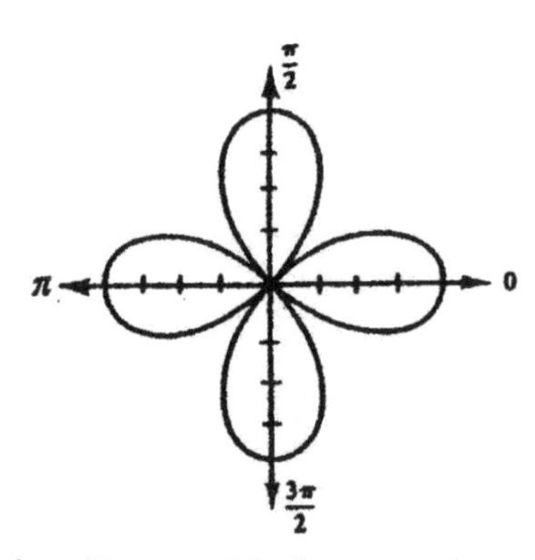

It is a **Rose** with four petals.

Example 7.

Graph the polar equation $r = 1 + 2\sin\theta$.

Solution: Replacing θ by $(\pi - \theta)$: $r = 1 + 2\sin(\pi - \theta) = 1 + 2\sin\theta$.

The graph is symmetric with respect to the line $\theta = \frac{\pi}{2}$.

Plotting the points, we only need to select values of θ from $-\frac{\pi}{2}$ to $\frac{\pi}{2}$.

θ	$-\frac{\pi}{2}$	$-\frac{\pi}{3}$	$-\frac{\pi}{6}$	0	$\frac{\pi}{6}$	$\frac{\pi}{3}$	$\frac{\pi}{2}$
$\sin\theta$	-1	$-\frac{\sqrt{3}}{2}$	$-\frac{1}{2}$	0	$\frac{1}{2}$	$\frac{\sqrt{3}}{2}$	1
$r = 1 + 2\sin\theta$	-1	-0.73	0	1	2	2.73	3

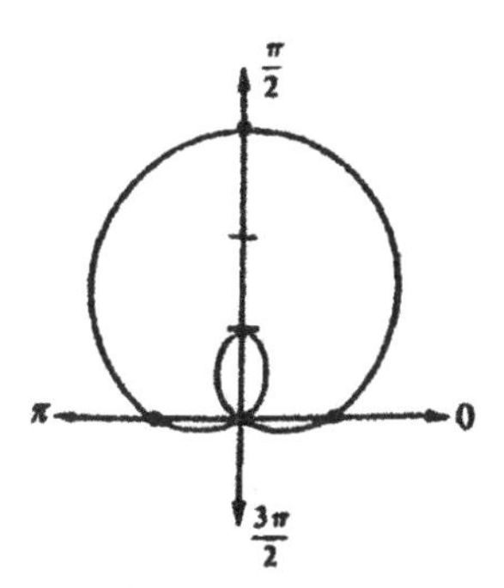

It is a **Limaçon** with an inner loop.

Polar Forms of Conics: In order to find a convenient form to represent a conic, we locate a focus at the pole and apply the definition of a conic.

Definition of a conic: A conic is the set of all points such that the distance from a fixed point (called the focus) is in a constant ratio to its distance from a fixed line (called the directrix). The constant ratio is the eccentricity of the conic and is denoted by e.

$$\frac{\overline{PF}}{\overline{PD}} = e \quad \therefore \overline{PF} = e \cdot \overline{PD}$$

$$r = e(p + r\cos\theta)$$
$$r = ep + er\cos\theta$$
$$r - er\cos\theta = ep$$
$$r(1 - e\cos\theta) = ep$$

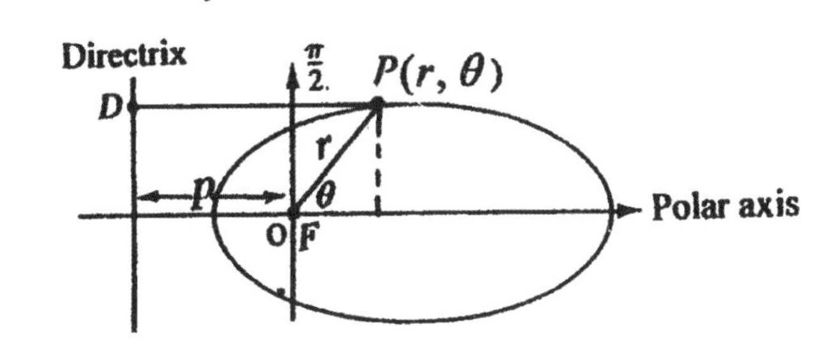

$$\therefore r = \frac{ep}{1 - e\cos\theta}$$ (Directrix is located at the left of the pole.)

Similarly, we can obtain the following polar forms of conics:

$$r = \frac{ep}{1 + e\cos\theta}$$ (Directrix is located at the right of the pole.)

$$r = \frac{ep}{1 - e\sin\theta}$$ (Directrix is located below the pole.)

$$r = \frac{ep}{1 + e\sin\theta}$$ (Directrix is located above the pole.)

Hint: $e = 1$ (parabola), $e < 1$ (ellipse), $e > 1$ (hyperbola)

Example 8.

Identify and graph the equation $r = \dfrac{9}{3 - 3\sin\theta}$.

Solution: $r = \frac{3}{1 - \sin\theta}$ and $e = 1$

It is a parabola with focus at the pole.

$ep = 3 \quad \therefore p = 3$

Directrix is 3 units below the pole.

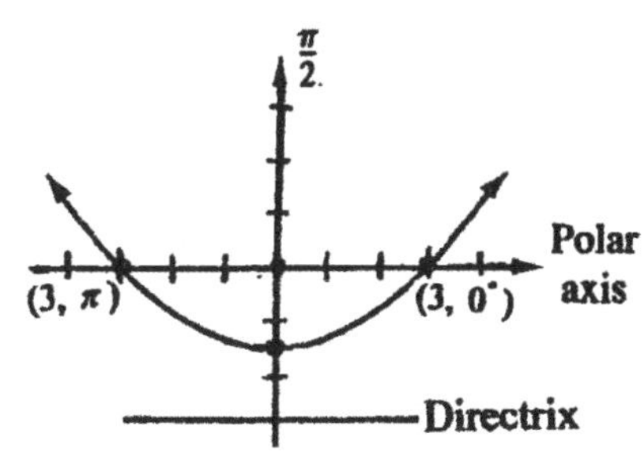

We plot two addition points $(3,\ 0^\circ)$ and $(3,\ \pi)$ to assist in graphing.

Polar Coordinates involving Complex Numbers

A **complex number** $z = a + b\,i$ is formed when we solve an algebraic equation having an imaginary root. A complex number $z = a + b\,i$ can be interpreted geometrically as the corresponding point (*a*, *b*) in the rectangular plane. The polar coordinates used to represent the complex number is called the **complex plane.**

In a complex plane, the horizontal axis is called the **real axis**, and the vertical axis is called the **imaginary axis.**

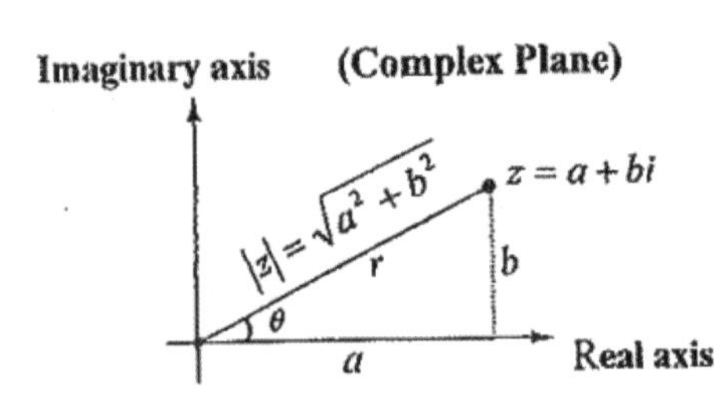

$z = a + b\,i$ and $|z| = \sqrt{a^2 + b^2}$

$a = r\cos\theta$

$b = r\sin\theta$

$r = \sqrt{a^2 + b^2}$, $\tan\theta = \frac{b}{a}$

$a + b\,i = r\cos\theta + (r\sin\theta)\,i$

$\therefore\ z = r(\cos\theta + i\sin\theta)$ and $|z| = r$

The number r is called the **modulus** (or **amplitude**) of z. The angle θ is called an **argument** of z.

The **absolute value** of a complex number $z = a + b\,i$ is defined as $|z| = \sqrt{a^2 + b^2}$. It is the distance between z and the origin in the complex plane.

Since there are infinitely many choices for θ, the polar form of a complex number is not unique. We choice the values of θ for $0 \le \theta < 2\pi$, or $\theta < 0$.

The complex number i is represented by the point (0, 1) and $i = 1(\cos 90^\circ + i\sin 90^\circ)$.

A complex number represented in polar form will give us an efficient and easy way to do some applications and computations (add, subtract, multiply, and divide).

To add or subtract two complex numbers, we use the **parallelogram rule.**

Example: Find the sum and difference of $z_1 = 2 + 6i$ and $z_2 = 4 - 2i$ by using the polar polar coordinates.

Solution:

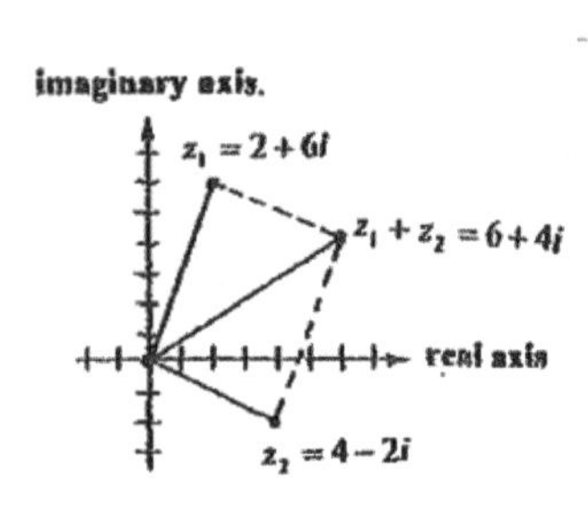

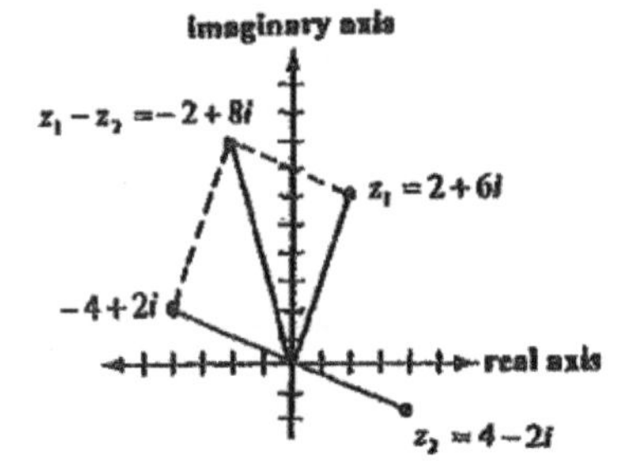

Check:

$z_1 + z_2 = (2 + 6i) + (4 - 2i) = 6 + 4i$

$z_1 - z_2 = (2 + 6i) - (4 - 2i)$

$= (2 + 6i) + (-4 + 2i)$

$= -2 + 8i$

To multiply or divide two complex numbers in polar forms, we use the following two formulas: $z_1 = r_1(\cos\theta_1 + i\sin\theta_1)$, $z_2 = r_2(\cos\theta_2 + i\sin\theta_2)$

1) $z_1 \cdot z_2 = r_1 r_2[\cos(\theta_1 + \theta_2) + i\sin(\theta_1 + \theta_2)]$

2) $\dfrac{z_1}{z_2} = \dfrac{r_1}{r_2}[\cos(\theta_1 - \theta_2) + i\sin(\theta_1 - \theta_2)]$, $z_2 \neq 0$

3-25 Coordinate Geometry in Three-Dimensions

In Chapter 2, we have learned how to plot a point in a **plane** in the two-dimensional coordinate system. The plane is called the Cartesian plane determined by two perpendicular number lines. The lines are called the x-**axis** and the y-**axis**. A point in the plane is determined by an **ordered pair** (x, y).

To plot a point in a **space**, we add a third dimension by a z-**axis** perpendicular to both the x-axis and the y-axis at the origin. In a three-dimensional coordinate system, the axes determine three coordinate planes, the xy-**plane**, the xz-**plane**, and the yz-**plane**. A point in the space is determined by an **ordered triple** (x, y, z). The three axes divide the space into eight **octants**. In the first octant, all three coordinates are positive.

The geometry determined by the coordinate system is called the **Coordinate Geometry**, or **Analytic Geometry**. The geometry determined by the three-dimensional coordinate system is also called the **Solid Analytic Geometry**.

Coordinate System in Three-Dimensions

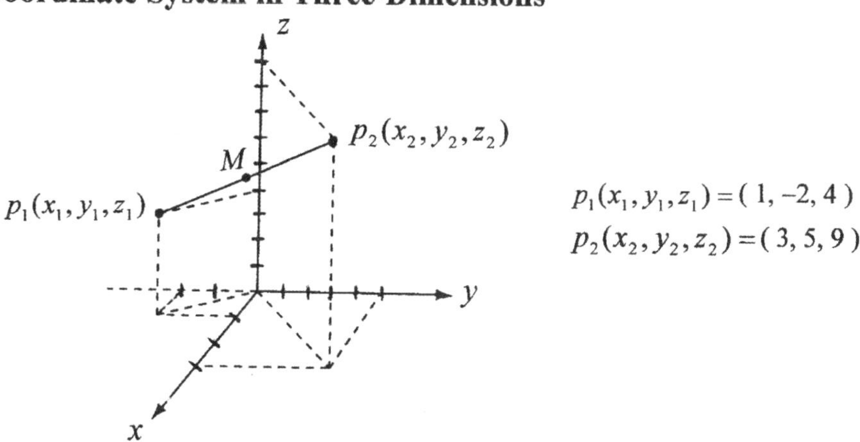

In algebra and geometry, many formulas in three-dimensional coordinate system are similar to the formulas in two-dimensional coordinate system (see Chapter 3-12).

Distance Formula: The distance between two points p_1 and p_2 is

$$d = \overline{p_1 p_2} = \sqrt{(x_2 - x_1)^2 + (y_2 - y_1)^2 + (z_2 - z_1)^2}$$

Midpoint Formula: The midpoint between two points p_1 and p_2 is

$$M = \left(\frac{x_1 + x_2}{2}, \frac{y_1 + y_2}{2}, \frac{z_1 + z_2}{2}\right)$$

Example

1. Find the distance and midpoint between (1, –2, 4) and (3, 5, 9).
   Solution:

$$d = \sqrt{(3-1)^2 + (5+2)^2 + ((9-4)^2} = \sqrt{4+49+25} = \sqrt{78} \approx 8.83 \text{ Ans.}$$

$$M = \left(\frac{1+3}{2}, \frac{-2+5}{2}, \frac{4+9}{2}\right) = (2,\ 1.5,\ 6.5) \text{ Ans.}$$

Linear Equations in Three-Dimensions

In Chapter 2-5, we have learned that the linear equation $ax+by=c$ is a straight line in two-dimensional system. If we write the equation $ax+by=c$ by solving for y in terms of x, we have a linear function $y=-\frac{a}{b}x+\frac{c}{b}$, or $f(x)=-\frac{a}{b}x+\frac{c}{b}$.

The linear equation $ax+by+cz=d$ is a plane in three-dimensional system. If we write the equation $ax+by+cz=d$ by solving for z in terms of x and y, we have a linear function $z=-\frac{a}{c}x-\frac{b}{c}y+\frac{d}{c}$, or $f(x, y)=-\frac{a}{c}x-\frac{b}{c}y+\frac{d}{c}$.

To graph a linear equation in three-dimensions, we find the points at which the graph intersects the three axes. Then, we connect these points of intercepts with lines to form a triangular plane that lies in the first octant and extends to all other octants.

Examples

1. Write the linear equation $2x+3y+4z=12$ as a function of x and y.

   Solution:

   $2x+3y+4z=12$

   Solving for z:

   $4z=12-2x-3y$

   $z=\frac{1}{4}(12-2x-3y)$. Ans.

   Or: $f(x, y)=\frac{1}{4}(12-2x-3y)$. Ans.

2. Graph the linear equation $2x+3y+4z=12$ in the first octant.

   Solution:

   Find the intercepts:

   Let $x=0$, $y=0$, z-intercept $=3$.

   Let $x=0$, $z=0$, y-intercept $=4$.

   Let $y=0$, $z=0$, x-intercept $=6$.

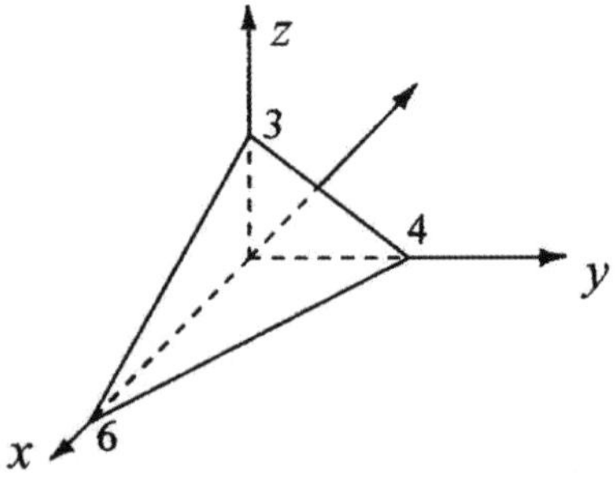

3. Find the volume of the rectangular solid with a given point (2, 6, 5).

   Solution:

   Length: $\ell=6$
   Width: $w=2$
   Height: $h=5$

   Volume: $V=\ell\times w\times h=6\times2\times5=60$. Ans.

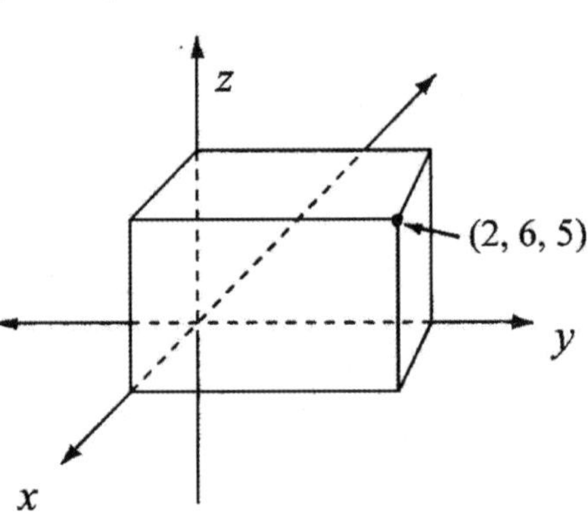

The Equation of a Sphere in Three-Dimensions

In Chapter 3-15, we have learned that, in two-dimensions, the equation $x^2 + y^2 = r^2$ is a circle whose center is the origin (0, 0) and whose radius is r. The equation $(x-h)^2 + (y-k)^2 = r^2$ is a circle whose center is the point (h, k) and whose radius is r.

In three-dimensions, the equation of a sphere is:

$$x^2 + y^2 + z^2 = r^2$$ **(Standard Form)**

Its center is the origin (0, 0, 0) and radius is r.

In three-dimensions, the equation of a sphere is:

$$(x-h)^2 + (y-k)^2 + (z-j)^2 = r^2$$ **(Standard Form)**

Its center is the point (h, k, j) and radius is r.

In three-dimensions, the equation $x^2 + y^2 + z^2 + ax + by + cz = 0$ is the **general form** of a sphere. To find its center and radius, we rewrite it to match the standard form by the method of **completing the square** (see Page 88).

Examples

1. Find the equation of the sphere whose center is (2, 5, 3) and whose radius is 3.
   Solution:

$$(x-2)^2 + (y-5)^2 + (z-3)^2 = 3^2$$

$$(x-2)^2 + (y-5)^2 + (z-3)^2 = 9. \text{ Ans.}$$

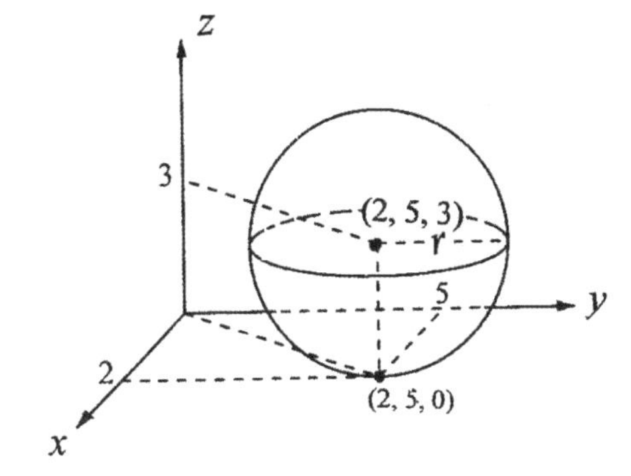

2. Find the center and radius of the sphere:

$$x^2 + y^2 + z^2 - 4x - 10y - 6z + 29 = 0$$

   Solution:

$$x^2 + y^2 + z^2 - 4x - 10y - 6z + 29 = 0$$

Completing the square:

$$(x^2 - 4x) + (y^2 - 10y) + (z^2 - 6z) + 29 = 0$$

$$(x^2 - 4x + 4) - \cancel{4} + (y^2 - 10y + 25) - \cancel{25} + (z^2 - 6z + 9) - 9 + \cancel{29} = 0$$

$$(x^2 - 4x + 4) + (y^2 - 10y + 25) + (z^2 - 6z + 9) = 9$$

$$(x-2)^2 + (y-5)^2 + (z-3)^2 = 3^2$$

$\therefore$ Center (2, 5, 3) and radius = 3. Ans.

(See the graph in Problem 1.)

Distance Formulas between Lines and Planes

1. The distance between a point (x_1, y_1) and a straight line $ax+by+c=0$ is given by the formula:

$$d=\overline{PQ}=\frac{|ax_1+by_1+c|}{\sqrt{a^2+b^2}}$$

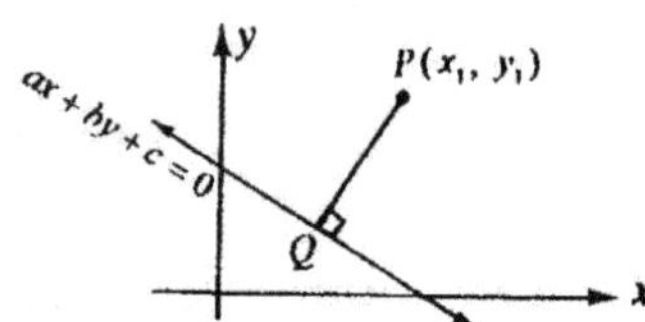

2. The distance between two parallel lines $ax+by+c_1=0$ and $ax+by+c_2=0$ is given by the formula:

$$d=\overline{PQ}=\frac{|c_1-c_2|}{\sqrt{a^2+b^2}}$$

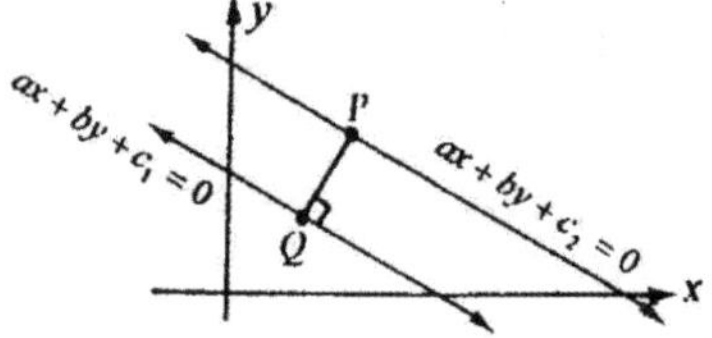

3. The distance between a point (x_1, y_1, z_1) and a plane $ax+by+cz+d=0$ is given by the formula:

$$d=\overline{PQ}=\frac{|ax_1+by_1+cz_1+d|}{\sqrt{a^2+b^2+c^2}}$$

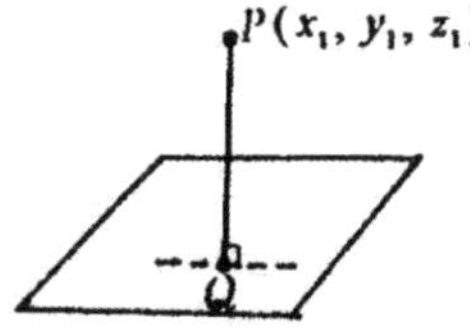

Examples

1. Find the distance between the line $2x+4y-5=0$ and the point (3, -1).
   Solution:

$$d=\frac{|ax_1+by_1+c|}{\sqrt{a^2+b^2}}=\frac{|2\cdot3+4(-1)+(-5)|}{\sqrt{2^2+4^2}}=\frac{|-3|}{\sqrt{20}}=\frac{3}{4.472}=0.67. \text{ Ans.}$$

2. Find the distance between the two parallel lines $x-2y+3=0$ and $x-2y-5=0$.
   Solution:

$$d=\frac{|c_1-c_2|}{\sqrt{a^2+b^2}}=\frac{|3-(-5)|}{\sqrt{1^2+(-2)^2}}=\frac{8}{\sqrt{5}}=\frac{8}{2.236}=3.58. \text{ Ans.}$$

3. Find the distance between the point (1, 2, 1) and the plane $2x+y-z-1=0$.
   Solution:

$$d=\frac{|ax_1+by_1+cz_1+d|}{\sqrt{a^2+b^2+c^2}}=\frac{|2\cdot1+1\cdot2-1\cdot1-1|}{\sqrt{2^2+1^2+(-1)^2}}=\frac{2}{\sqrt{6}}=\frac{2}{2.449}=0.82. \text{ Ans.}$$

4. Two parallel planes are $2x-3y+z-1=0$ and $2x-3y+z-5=0$. Find the distance between the planes.
   Solution:

   Find one point on the first plane (0, 0, 1)

$$d=\frac{|ax_1+by_1+cz_1+d|}{\sqrt{a^2+b^2+c^2}}=\frac{|0+0+1-5|}{\sqrt{4+9+1}}=\frac{4}{\sqrt{14}}=\frac{4}{3.742}=1.07. \text{ Ans.}$$

Examples

5. Find the distance between the origin and the point (2, 5, 3)

Solution:

$$d = \sqrt{2^2 + 5^2 + 3^2} = \sqrt{38} = 6.16 \text{. Ans.}$$

6. Determine whether the point (3, 4, –2) is outside or inside the sphere $x^2 + y^2 + z^2 = 9$.

Solution:

The center of the sphere is the origin (0, 0, 0) and the radius is 3.
A point is inside the sphere if the distance between it and the origin is smaller than the radius.

$$d = \sqrt{3^2 + 4^2 + (-2)^2} = \sqrt{9 + 16 + 4} = \sqrt{29} \approx 5.34 > 3$$

The point (3, 4, -2) is not inside the sphere. Ans.

In a toy store, a five-year old girl wants her father to buy her an expensive toy.

Father: No, it is too expensive.
Girl: Dad, I like it. Honey and Sweetheart.
Father: Why are you calling me "Honey and Sweetheart" ?
Girl: Because whenever mom wants to buy a new dress, she calls you "Honey" or "Sweetheart".

Notes

SAMPLES QUESTIONS

CHAPTER 3: GEOMETRY and MEASUREMENTS

1. How many lines can be determined by two distinct points ?
Solution:
One line. Ans.

2. How many planes can be determined by three distinct points ?
Solution:
One plane. Ans.

3. What could be their intersection if two distinct lines intersect ?
Solution:
A point. Ans.

4. What could be their intersection if two planes intersect ?
Solution:

The empty set,	a plane,	or a line
(parallel)	(concide)	(distinct)

5. What could be their intersection if two distinct planes intersect ?
Solution:
A line. Ans.

6. How many lines can be determined by three noncollinear points ?
Solution:
3 lines. Ans.

7. How many lines can be determined by four noncollinear points ?
Solution:
6 lines. Ans.

8. How many lines can be determined by eight noncollinear points ?
Solution: (Apply sequence)

Points	2	3	4	5	6	7	8
Lines	1	3	6	10	15	21	**28**

Ans. 28 lines

9. Six times the measure of the complement of an angle is equal to 5° less than the measure of the supplement of that angle. What is the measure of the angle ?
Solution:
Let x = the measure of the angle
The equation is :

$$6(90-x)=(180-x)-5$$
$$540-6x=175-x$$
$$365=5x \qquad \therefore x=73^\circ. \text{ Ans.}$$

10. In the figure, what is the value of x ?

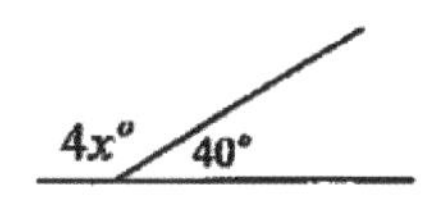

Solution:
$4x=140 \quad \therefore x=35$. Ans.

SAMPLES QUESTIONS

CHAPTER 3: GEOMETRY and MEASUREMENTS

1. In the figure, what is the value of x ?
Solution:

$$m < OBC + m < OCB = 180 - 150 = 30$$

$$\therefore x = 180 - (50 + 40 + 30) = 60. \text{ Ans.}$$

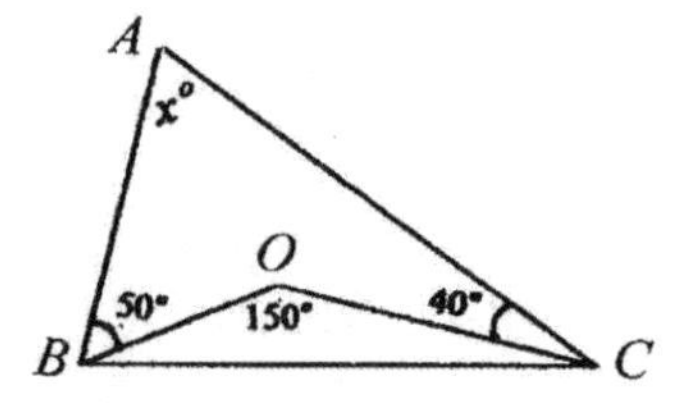

2. The figure is a right circular cone. What is the lateral surface area ?
Solution:

$$(5\sqrt{2})^2 + r^2 = (5\sqrt{3})^2$$

$$50 + r^2 = 75$$

$$r^2 = 25, \ r = 5$$

$$\therefore LA = \pi rs = \pi \cdot 5 \cdot 5\sqrt{3} = 25\sqrt{3}\,\pi. \text{ Ans.}$$

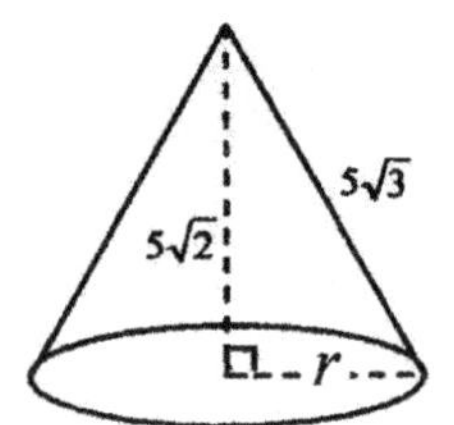

3. In the figure, $\overline{AB}$ and $\overline{AC}$ are tangent to the circle O at B and C. If $OA = 13$ and $OB = 4$, what is the value of $AB + AC$?
Solution:

$$< ABO = 90^o, \ < ACO = 90^o, \ \overline{AB} = \overline{AC}$$

$$(AB)^2 + 4^2 = 13^2$$

$$(AB)^2 = 169 - 16 = 153$$

$$AB = 12.37$$

$$\therefore AB + AC = 12.37 + 12.37 = 24.74. \text{ Ans.}$$

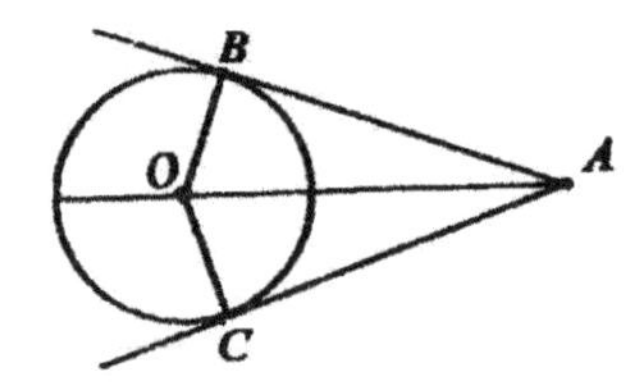

4. In the triangle, $\overline{DE} // \overline{BC}$, what is the measure of x ?
Solution:

$$\Delta ADE \sim \Delta ABC$$

$$\frac{x}{4.5} = \frac{4}{7}$$

$$\therefore x = \frac{4(4.5)}{7} = 2\frac{4}{7}. \text{ Ans.}$$

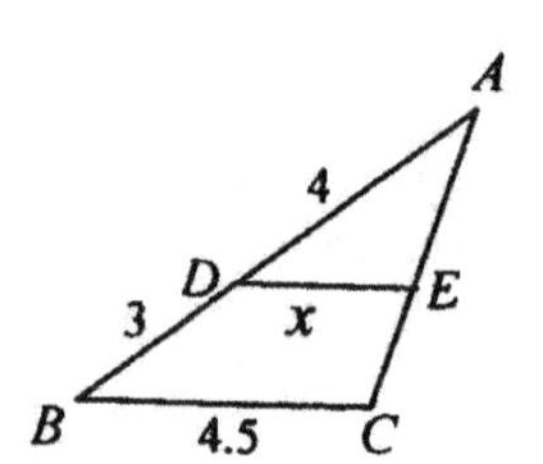

5. In the triangle, what is the measure of x ?
Solution:

$$18^2 + (\tfrac{1}{2}x)^2 = x^2, \ 324 + \tfrac{1}{4}x^2 = x^2$$

$$324 = \tfrac{3}{4}x^2$$

$$3x^2 = 1296$$

$$x^2 = 432 \ \therefore x = \sqrt{432} = 12\sqrt{3}. \text{ Ans.}$$

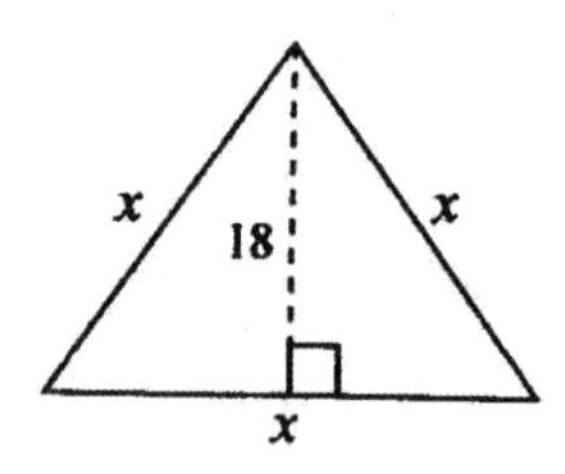

SAMPLES QUESTIONS

CHAPTER 3: GEOMETRY and MEASUREMENTS

1. The triangle is rotated about the side AB. What is the volume of the resulting solid ?
Solution:

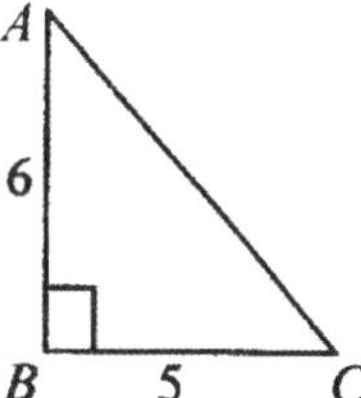

$$V = \tfrac{1}{3}\pi\, r^2 \cdot h$$
$$= \tfrac{1}{3}\pi\,(5^2)(6) = 50\,\pi. \text{ Ans.}$$

2. The line segment joining the points (–5, 4) and (3, 6) is rotated by 90° (counterclockwise) with the origin as the center of rotation. What is the corresponding line segment after the rotation ?
Solution:

Using the Formula for Rotation of a Point:

$$\begin{bmatrix} x' \\ y' \end{bmatrix} = \begin{bmatrix} \cos\theta & -\sin\theta \\ \sin\theta & \cos\theta \end{bmatrix} \cdot \begin{bmatrix} x \\ y \end{bmatrix}$$

$$\begin{bmatrix} \cos 90^o & -\sin 90^o \\ \sin 90^o & \cos 90^o \end{bmatrix} \cdot \begin{bmatrix} -5 \\ 4 \end{bmatrix} = \begin{bmatrix} -4 \\ -5 \end{bmatrix}$$

$$\begin{bmatrix} \cos 90^o & -\sin 90^o \\ \sin 90^o & \cos 90^o \end{bmatrix} \cdot \begin{bmatrix} 3 \\ 6 \end{bmatrix} = \begin{bmatrix} -6 \\ 3 \end{bmatrix}$$

Ans: The corresponding line segment joining the points (–4, –5) and (–6, 3).

3. The radius of the circle is 12. The measure of the minor arc is 45°. What is the area of the shaded region ?
Solution:

Area of the shaded region
$$= \pi(12^2) - 144\,\pi\,(\tfrac{45}{360})$$
$$= 144\pi - 18\pi$$
$$= 126\,\pi. \text{ Ans.}$$

4. In the triangle, what is the length of side AC ?
Solution:

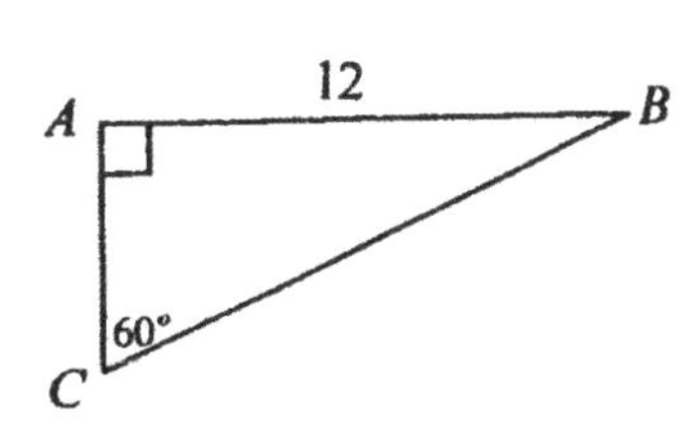

$$AB = \sqrt{3}(AC)$$
$$\therefore AC = \frac{AB}{\sqrt{3}} = \frac{12}{\sqrt{3}} = 4\sqrt{3}. \text{ Ans.}$$

SAMPLES QUESTIONS

CHAPTER 3: GEOMETRY and MEASUREMENTS

1. In the graph, what is the perimeter of the regular hexagon (6 sides) if its apothem $OC = 12\sqrt{3}$?

Solution:

ΔOAC is a 30-60-90 triangle.

$OC = \sqrt{3}(AC)$

$AC = \frac{OC}{\sqrt{3}} = \frac{12\sqrt{3}}{\sqrt{3}} = 12$

$AB = 2 \times (AC) = 2 \times 12 = 24$

$\therefore$ Perimeter $= 24 \times 6 = 144$. Ans.

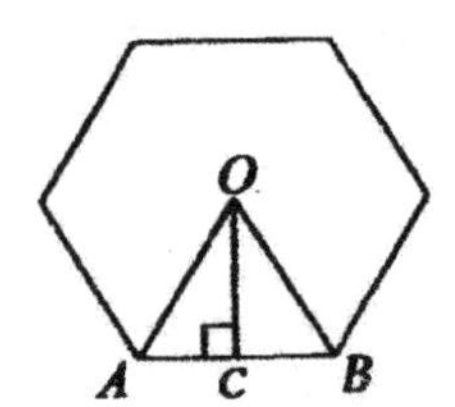

2. In the graph graph, $\Delta ABC \sim \Delta CDE$, what is the value $\frac{x}{y}$?

Solution:

$\frac{x}{1.8} = \frac{y}{2.3}$

$\therefore \frac{x}{y} = \frac{1.8}{2.3} = 0.78$. Ans.

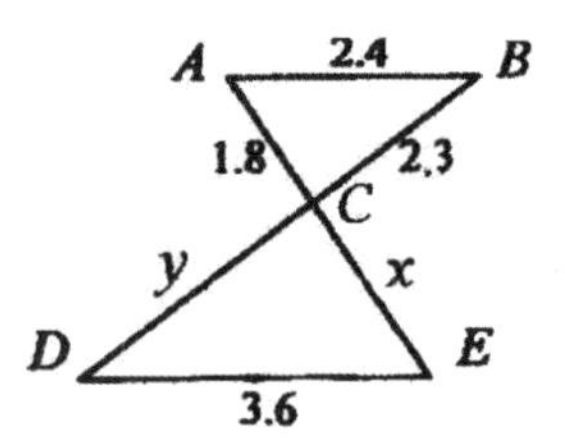

3. In the triangle, what is the length of side BC ?

Solution:

$(x-1)^2 + (x+1)^2 = (5\sqrt{2})^2$

$x^2 - 2x + 1 + x^2 + 2x + 1 = 50$

$2x^2 + 2 = 50$

$2x^2 = 48$

$x^2 = 24$

$x = \sqrt{24} = 2\sqrt{6}$

$\therefore BC = x + 1 = 2\sqrt{6} + 1 = 5.90$. Ans.

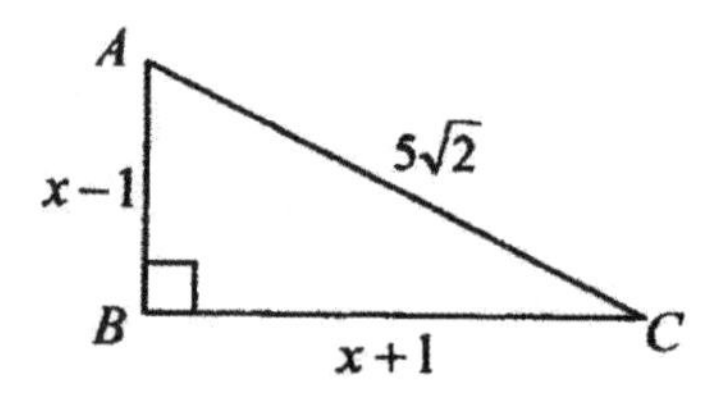

4. In the figure, $BC = 12$ and $AB = 18$. $\overline{AB}$ is a tangent line to the circle O at A. What is the length of the diameter of the circle ?

Solution:

$< OAB = 90^o$

$r^2 + 18^2 = (r+12)^2$

$r^2 + 324 = r^2 + 24r + 144$

$24r = 180, \quad r = 7.5$

$\therefore$ Diameter $= 2r = 2(7.5) = 15$. Ans.

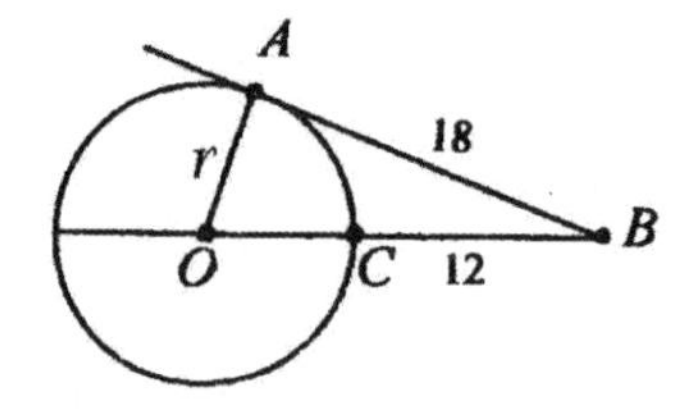

SAMPLES QUESTIONS

CHAPTER 3: GEOMETRY and MEASUREMENTS

1. In the triangle, $DE // BC$, what is the value of $x + y$?

Solution:

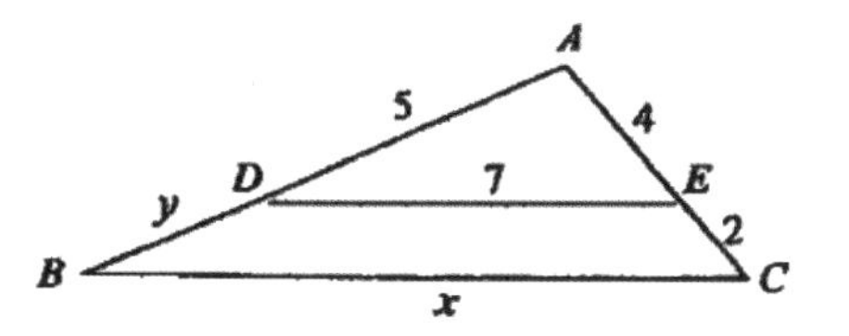

$\Delta ABC \sim \Delta ADE$

$\frac{4}{6} = \frac{7}{x}$, $4x = 42$, $x = 10.5$

$\frac{4}{6} = \frac{5}{5+y}$, $20 + 4y = 30$, $4y = 10$, $y = 2.5$

$\therefore x + y = 10.5 + 2.5 = 13$. Ans.

2. In the figure, the arc AB of the semicircle has length 4π and the arc AC of the semicircle has length 3π. What is the area of ΔABC ?

Solution:

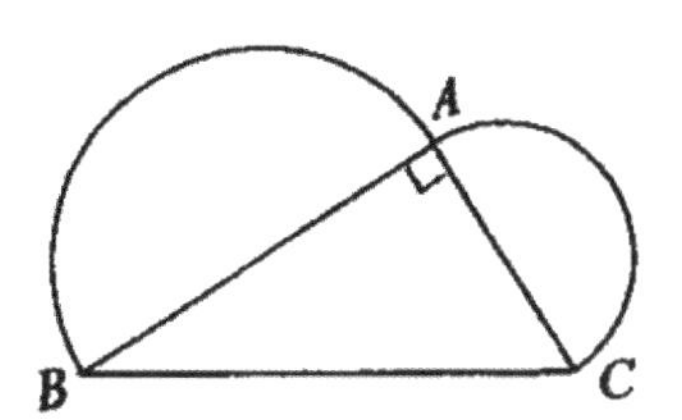

Circumference $C = \pi D$

$2(4\pi) = \pi(AB)$, $AB = 8$

$2(3\pi) = \pi(AC)$, $AC = 6$

$\therefore$ Area of $\Delta ABC = \frac{1}{2}(8)(6) = 24$. Ans.

3. In the figure, what is the area of ΔABC ?

Solution:

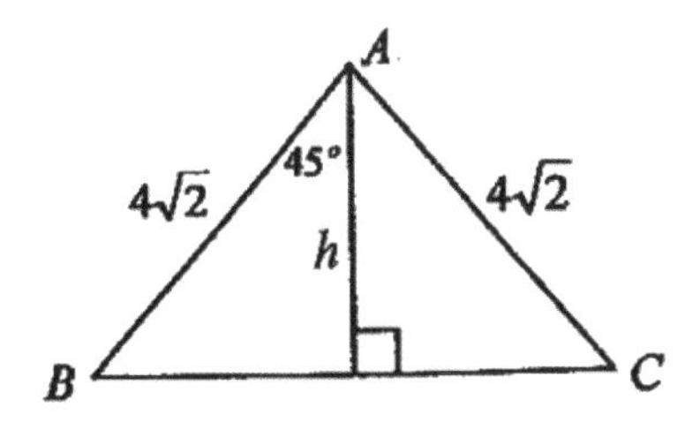

$(BC)^2 = (4\sqrt{2})^2 + (4\sqrt{2})^2 = 32 + 32 = 64$

$BC = 8$

$h^2 + 4^2 = (4\sqrt{2})^2$

$h^2 + 16 = 32$, $h^2 = 16$, $h = 4$

Area of $\Delta ABC = \frac{1}{2}(BC)(h) = \frac{1}{2}(8)(4) = 16$. Ans.

4. In the figure, a large pulley is belted to a small pulley. If the larger pulley runs 80 revolutions per minutes, what is the speed of the small pulley, in revolutions per minute ?

Solution:

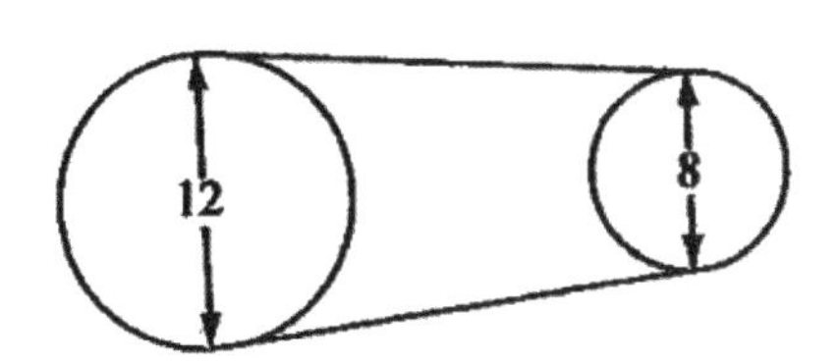

Speed of the small pulley

$= \frac{\pi(12)(80)}{\pi(8)}$

$= \frac{960}{8}$

$= 120$ revolutions/minute. Ans.

Notes

CHAPTER FOUR

TRIGONOMETRY

4-1 Degree Measure of Angles

In geometry, the degree measures of an angle is between 0° and 180°.
In trigonometry, we consider an angle as being formed by the rotation of its **initial ray** about its **vertex** to its **terminal ray**.
An angle is often denoted by θ (theta), ϕ (phi), α (alpha), β (beta), or γ (gamma).

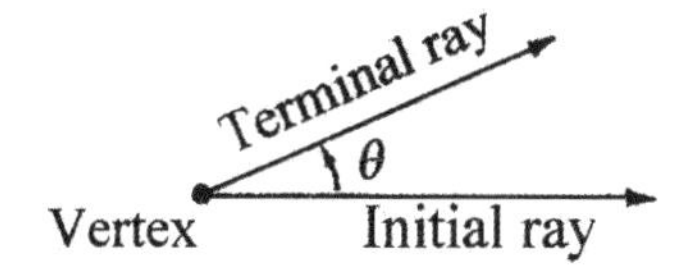

An angle is **positive** if it rotates counterclockwise. An angle is **negative** if it rotates clockwise. An angle is said to measure 360° if it is formed by rotating the initial side exactly one revolution in the counterclockwise direction until it meets with itself. A **right angle** is an angle that measures 90°. A **straight angle** is an angle that measures 180°.

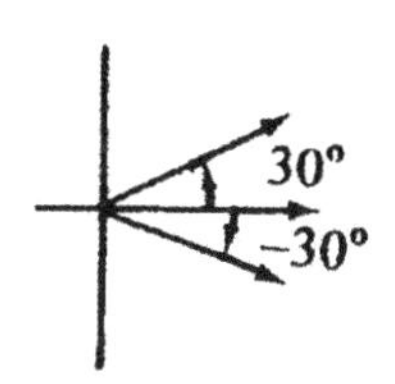

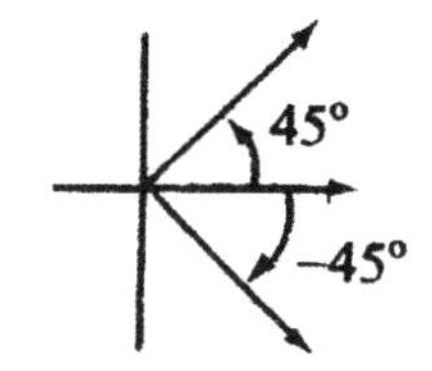

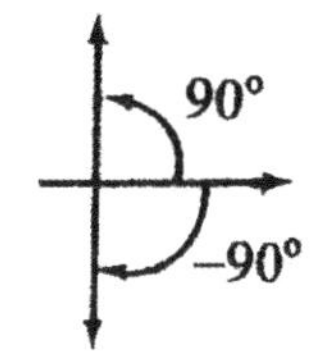

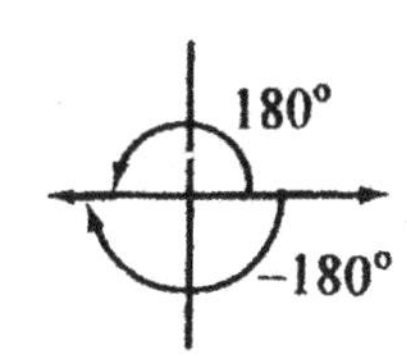

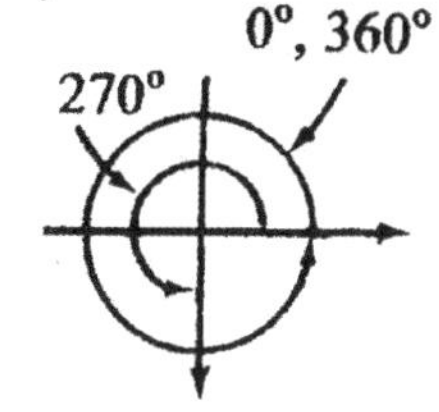

Coterminal Angles are angles which have the same initial and terminal rays. There are infinite number of angles coterminal with an angle θ.
The coterminal angles of θ are: $\pm 360^o n + \theta$, $n = 0, 1, 2, 3, \cdots$.
If $\theta = 30^o$, then the coterminal angles are: $\cdots, -690^o, -330^o, 30^o, 390^o, 750^o, \cdots$

There are two different ways to express the degree measure of an angle.

1. Decimal degrees: 7.5944^o

2. Degrees, minutes, seconds: $7^o\ 35'\ 40''$ = 7 degree, 35 minutes, 40 seconds
(1 degree = 60 minutes, 1 minute = 60 seconds, 1 degree = 3600 seconds)

Examples

1. Express 7.5944^o in degree, minutes, seconds.
Solution:

$0.5944^o = 0.5944 \times 60 = 35.664$ minutes

0.664 minutes $= 0.664 \times 60 \approx 40$ seconds

$7.5944^o = 7^o\ 35'\ 40''$. Ans.

2. Express $7^o\ 35'\ 40''$ in decimal degrees.
Solution:

$7^o\ 35'\ 40'' = 7^o + (\frac{35}{60})^o + (\frac{40}{3600})^o$

$= 7^o + 0.5833^o + 0.0111^o$

$= 7.5944^o$. Ans.

Examples

3. Express 6.30^o in degrees, minutes, seconds.
Solution:
$6.30^o = 6^o + (0.30 \times 60)' = 6^o18'$. Ans.

4. Express -60.50^o in degrees, minutes, seconds.
Solution:
$-60.50^o = -60^o - (0.50 \times 60)' = -60^o - 30'$
$= -60^o30'$. Ans.

5. Express 60.5030^o in degrees, minutes, seconds.
Solution:
$60.5030^o = 60^o + (0.5030 \times 60)' = 60^o + 30.18' = 60^o + 30' + (0.18 \times 60)''$
$= 60^o + 30' + 0.108'' \approx 60^o30'11''$. Ans.

6. Express -120.3553^o in degrees, minutes, seconds.
Solution:
$-120.3553^o = -120^o - (0.3553 \times 60)' = -120^o - 21.3180'$
$= -120^o - 21' - (0.3180 \times 60)'' = -120^o - 21' - 19.08''$
$\approx -120^o21'19''$. Ans.

7. Express $6^o30'$ in decimal degrees.
Solution:
$6^o30' = 6^o + (\frac{30}{60})^o = 6^o + 0.5^o = 6.5^o$. Ans.

8. Express $6^o50'30''$ in decimal degrees.
Solution:
$6^o50'30'' = 6^o + (\frac{50}{60})^o + (\frac{30}{3600})^o$
$= 6^o + 0.833^o + 0.0083^o$
$= 6.8416^o$. Ans.

9. Express $-120^o35'53''$ in decimal degrees.
Solution:
$-120^o35'53'' = -120^o - (\frac{35}{60})^o - (\frac{53}{3600})^o = -120^o - 0.5833^o - 0.0147^o$
$= 120.598^o$. Ans.

A middleaged single lady told her friend.
Lady: I prefer to marry an archaeologist rather than a doctor or a lawyer.
Friend: Why do you want to marry an archaeologist ?
Lady: If my husband is an archaeologist, the more I get older, the more he is interested in me.

4-2 Six Trigonometric Functions in a Triangle

In trigonometry, we start to find six ratios of an acute angle θ in a right triangle. Each ratio is a function of the angle. We call these six ratios " **Six Trigonometric Functions** ". If θ is an acute angle in a right (90°) triangle, the definitions of these six trigonometric functions are:

1. The **sine** of θ $= \dfrac{opposite \quad leg}{hypotenuse} = \dfrac{y}{r}$

2. The **cosine** of θ $= \dfrac{adjacent \quad leg}{hypotenuse} = \dfrac{x}{r}$

3. The **tangent** of θ $= \dfrac{opposite \quad leg}{adjacent \quad leg} = \dfrac{y}{x}$

4. The **cotangent** of θ $= \dfrac{adjacent \quad leg}{opposite \quad leg} = \dfrac{x}{y}$

5. The **secant** of θ $= \dfrac{hypotenuse}{adjacent \quad leg} = \dfrac{r}{x}$

6. The **cosecant** of θ $= \dfrac{hypotenuse}{opposite \quad leg} = \dfrac{r}{y}$

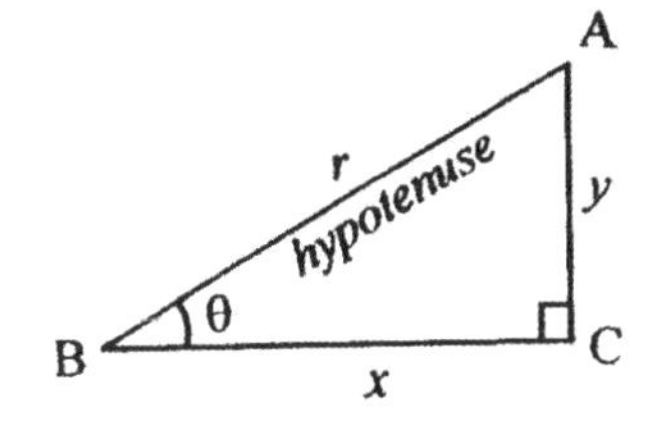

These trigonometry functions can be written in abbreviated forms as:

$$\sin\theta = \frac{y}{r}, \quad \cos\theta = \frac{x}{r}, \quad \tan\theta = \frac{y}{x}, \quad \cot\theta = \frac{x}{y}, \quad \sec\theta = \frac{r}{x}, \quad \csc\theta = \frac{r}{y}$$

Examples

1. If the terminal side of an angle θ passes through (3, 2), find all six basic trigonometric functions.

   Solution:

$$r^2 = 3^2 + 2^2 = 13 \quad \therefore r = \sqrt{13}$$

$$\sin\theta = \frac{y}{r} = \frac{2}{\sqrt{13}} = \frac{2\sqrt{13}}{13}$$

$$\cos\theta = \frac{x}{r} = \frac{3}{\sqrt{13}} = \frac{3\sqrt{13}}{13}$$

$$\tan\theta = \frac{y}{x} = \frac{2}{3}, \quad \cot\theta = \frac{x}{y} = \frac{3}{2}$$

$$\sec\theta = \frac{r}{x} = \frac{\sqrt{13}}{3}, \quad \csc\theta = \frac{r}{y} = \frac{\sqrt{13}}{2}$$

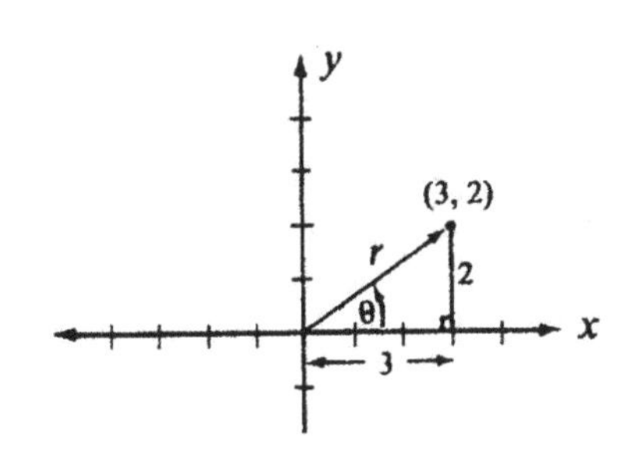

Examples

2. If the terminal side of an angle θ passes through (–1, 3), find all six basic trigonometric functions.

   Solution:

$$r^2 = (-1)^2 + 3^2 = 10 \quad \therefore r = \sqrt{10}$$

$$\sin\theta = \frac{y}{r} = \frac{3}{\sqrt{10}} = \frac{3\sqrt{10}}{10}$$

$$\cos\theta = \frac{x}{r} = \frac{-1}{\sqrt{10}} = -\frac{\sqrt{10}}{10}$$

$$\tan\theta = \frac{y}{x} = \frac{3}{-1} = -3$$

$$\cot\theta = \frac{x}{y} = \frac{-1}{3} = -\frac{1}{3}$$

$$\sec\theta = \frac{r}{x} = \frac{\sqrt{10}}{-1} = -\sqrt{10}$$

$$\csc\theta = \frac{r}{y} = \frac{\sqrt{10}}{3}$$

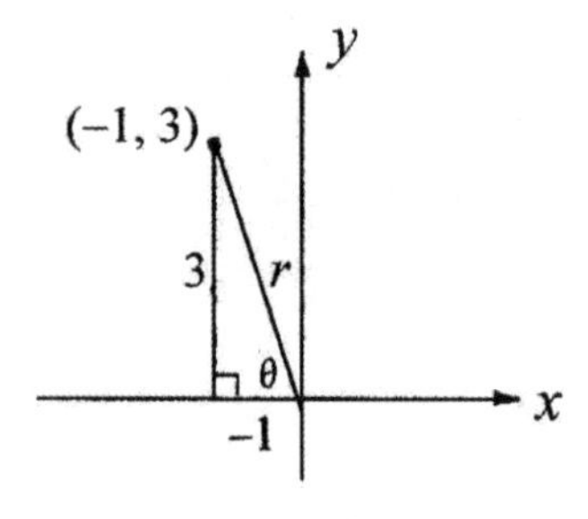

3. If the terminal side of an angle θ passes through (1, –3), find all six basic trigonometric functions

   Solution:

$$r^2 = 1^2 + (-3)^2 = 10 \quad \therefore r = \sqrt{10}$$

$$\sin\theta = \frac{y}{r} = \frac{-3}{\sqrt{10}} = -\frac{3\sqrt{10}}{10}$$

$$\cos\theta = \frac{x}{r} = \frac{1}{\sqrt{10}} = \frac{\sqrt{10}}{10}$$

$$\tan\theta = \frac{y}{x} = \frac{-3}{1} = -3$$

$$\cot\theta = \frac{x}{y} = \frac{1}{-3} = -\frac{1}{3}$$

$$\sec\theta = \frac{r}{x} = \frac{\sqrt{10}}{1} = \sqrt{10}$$

$$\csc\theta = \frac{r}{y} = \frac{\sqrt{10}}{-3} = -\frac{\sqrt{10}}{3}$$

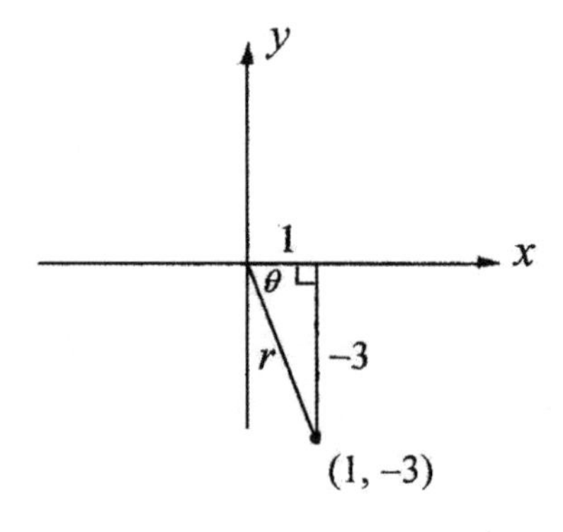

4. If $\theta < 90^o$ and $\sin\theta = \frac{1}{2}$, find $\cos\theta$.

   Solution:

$$x^2 + 1^2 = 2^2, \quad \therefore x = \sqrt{3}$$

$$\cos\theta = \frac{x}{r} = \frac{\sqrt{3}}{2}. \text{ Ans.}$$

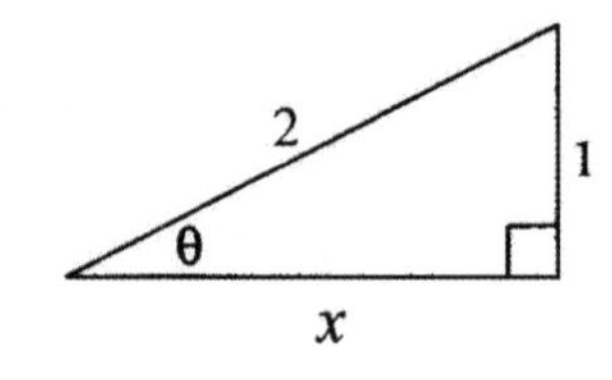

4-3 Solving a Right Triangle

We can use the basic six trigonometric functions to find the value of a specific side or angle of a right triangle if certain information of the triangle are given.

$$\sin\theta = \frac{y}{r}\ ,\ \cos\theta = \frac{x}{r}\ ,\ \tan\theta = \frac{y}{x}\ ,\ \cot\theta = \frac{x}{y},\ \sec\theta = \frac{r}{x}\ ,\ \csc\theta = \frac{r}{y}$$

Examples

1. Find a and b.

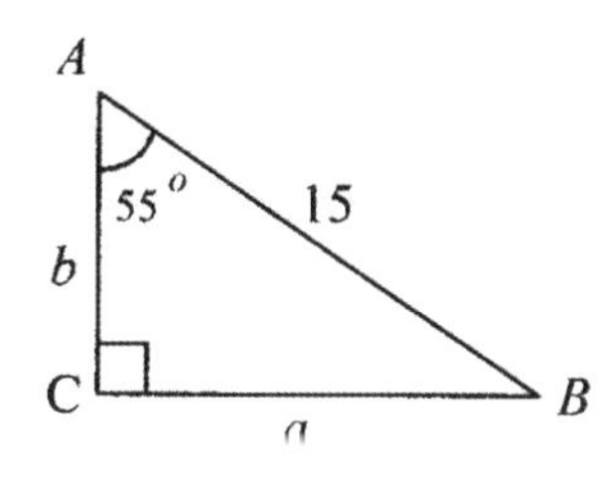

Solution:

$$\sin 55^o = \frac{a}{15}$$

$$\therefore a = 15\sin 55^o = 15\ (0.8192) = 12.3$$

$$\cos 55^o = \frac{b}{15}$$

$$\therefore b = 15\cos 55^o = 15\ (0.5736) = 8.6.\ \text{Ans.}$$

2. Find a.

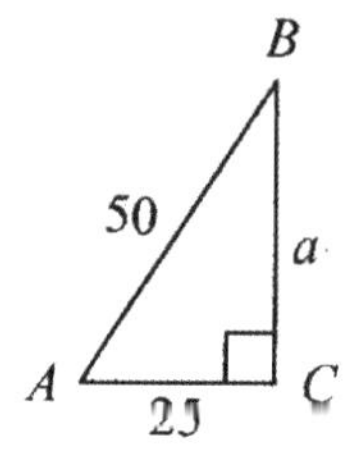

Solution:

$$\cos A = \frac{25}{50} = 0.5$$

$$\therefore A = 60^o$$

$$\tan A = \frac{a}{25}$$

$$\therefore a = 25\tan A = 25\tan 60^o = 25\ (1.732) = 43.3.\ \text{Ans.}$$

3. A ladder of length 20 feet leans against the side of a building by an angle of 25° with the building. Find the distance from the base of the ladder to the bottom of the building.

Solution:

$$\sin 25^o = \frac{x}{20}$$

$$x = 20(\sin 25^o) = 20(0.4226) \approx 8.45\ \text{feet. Ans.}$$

Ladder 20 ft.
25°
Building
x

Examples

4. A pilot is required to approach the airport at a 15° angle of descent toward the runway. If the plane is at an altitude of 10,000 feet, at what distance measured on the ground from the airport should the decent begin ?
   Solution:

$$\tan 15^\circ = \frac{10000}{x}$$

$$x = \frac{10000}{\tan 15^\circ} = \frac{10000}{0.2679} \approx 37{,}327 \text{ feet. Ans.}$$

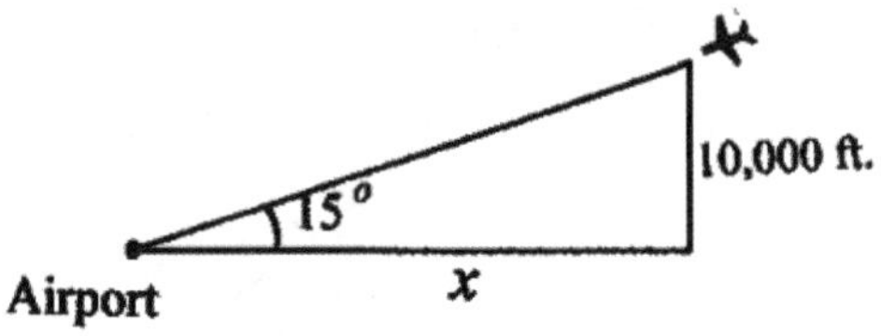

10,000 ft.
15°
Airport
x

In the court room, three defendants are waiting for the judge to sentence them for the time they need to stay in jail for the crimes they have committed.

Judge: Do you have any good reason why you think I should sentence you lesser time in jail ?
Defendant A: I was your high school classmate.
Judge: 1 week.
Defendant B: I built your house.
Judge: 1 month.
Defendant C: I introduced a girl friend to you 20 years ago. She is your wife now.
Judge: 20 years.

4-4 Trigonometric Identities

A **trigonometric identity** is an equation that is true for all values of the variable for which both sides of the equation are defined. We may simply say that trigonometric identities mean **trigonometric formulas** or **trigonometric relationships**.

Using trigonometric identities, we can simplify trigonometric expressions or solve trigonometric equations.

From the definition of the six trigonometric functions, we have developed the following trigonometric identities which are useful in the study of trigonometry.

1) Reciprocal Relationships

$$\sin\theta = \frac{1}{\csc\theta}, \quad \cos\theta = \frac{1}{\sec\theta}, \quad \tan\theta = \frac{1}{\cot\theta}$$

$$\csc\theta = \frac{1}{\sin\theta}, \quad \sec\theta = \frac{1}{\cos\theta}, \quad \cot\theta = \frac{1}{\tan\theta}$$

2. Basic Relationships

$$\tan\theta = \frac{\sin\theta}{\cos\theta}$$

$$\cot\theta = \frac{\cos\theta}{\sin\theta}$$

3) Cofunction Identities

$$\sin\theta = \cos(90^o - \theta)$$
$$\cos\theta = \sin(90^o - \theta)$$
$$\tan\theta = \cot(90^o - \theta)$$
$$\cot\theta = \tan(90^o - \theta)$$
$$\sec\theta = \csc(90^o - \theta)$$
$$\csc\theta = \sec(90^o - \theta)$$

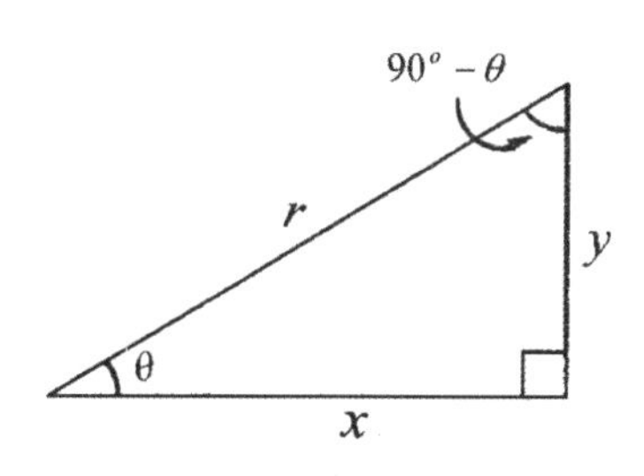

4) Pythagorean Identities

$$\sin^2\theta + \cos^2\theta = 1 \;\rightarrow\; \sin^2\theta \text{ means } (\sin\theta)^2$$
$$1 + \tan^2\theta = \sec^2\theta$$
$$1 + \cot^2\theta = \csc^2\theta$$

Examples

1. Simplify $\cos^2\theta(1+\tan^2\theta)$.

Solution:

$$\cos^2\theta(1+\tan^2\theta)$$
$$= \cos^2\theta(\sec^2\theta)$$
$$= \frac{1}{\sec^2\theta}\cdot\sec^2\theta = 1. \text{ Ans.}$$

2. Simplify $\sin x + \cos x \cot x$.

Solution:

$$\sin x + \cos x \cot x$$
$$= \sin x + \cos x \cdot \frac{\cos x}{\sin x}$$
$$= \sin x + \frac{\cos^2 x}{\sin x}$$
$$= \frac{\sin^2 x + \cos^2 x}{\sin x} = \frac{1}{\sin x} = \csc x. \text{ Ans.}$$

Examples

3. Simplify $\csc\theta - \cos\theta\cot\theta$.

Solution:

$$\csc\theta - \cos\theta\cot\theta = \frac{1}{\sin\theta} - \cos\theta\cdot\frac{\cos\theta}{\sin\theta} = \frac{1}{\sin\theta} - \frac{\cos^2\theta}{\sin\theta} = \frac{1-\cos^2\theta}{\sin\theta}$$
$$= \frac{\sin^2\theta}{\sin\theta} = \sin\theta. \text{ Ans.}$$

4. Simplify $\dfrac{1}{\cot(90^\circ - \theta)}$.

Solution:

$$\frac{1}{\cot(90^\circ - \theta)} = \frac{1}{\tan\theta} = \cot\theta. \text{ Ans.}$$

5. Prove $\dfrac{\tan\theta(1+\cot^2\theta)}{\cot\theta} = \sec^2\theta$.

Proof:

$$\frac{\tan\theta(1+\cot^2\theta)}{\cot\theta} = \frac{\tan\theta\csc^2\theta}{\cot\theta} = \frac{\tan\theta}{1/\tan\theta} = \tan^2\theta\csc^2\theta = \frac{\cancel{\sin^2\theta}}{\cos^2\theta}\cdot\frac{1}{\cancel{\sin^2\theta}}$$
$$= \frac{1}{\cos^2\theta} = \sec^2\theta. \text{ Ans.}$$

6. Simplify $\dfrac{\sin^2 x}{1+\cos x}$.

Solution:

$$\frac{\sin^2 x}{1+\cos x} = \frac{\sin^2 x}{1+\cos x}\cdot\frac{1-\cos x}{1-\cos x} = \frac{\sin^2 x(1-\cos x)}{1-\cos^2 x} = \frac{\cancel{\sin^2 x}(1-\cos x)}{\cancel{\sin^2 x}} = 1-\cos x. \text{ Ans.}$$

7. Simplify $\sin^2 x + \cos^2 x - \csc^2 x + \cot^2 x$.

Solution:

$$\sin^2 x + \cos^2 x - \csc^2 x + \cot^2 x = 1 + \cot^2 x - \csc^2 x = \csc^2 x - \csc^2 x = 0. \text{ Ans.}$$

8. Given $\cos x = 0.866$, find the value of $(\sin x)(\cos x)(\cot x)$.

Solution:

$$(\sin x)(\cos x)(\cot x) = (\cancel{\sin x})(\cos x)\left(\frac{\cos x}{\cancel{\sin x}}\right) = \cos^2 x = (0.866)^2 \approx 0.75. \text{ Ans.}$$

9. Solve $\sin x = \cos 20^\circ$.

Solution:

$$\sin x = \cos(90^\circ - x)$$
$$\sin x = \cos 20^\circ \text{(given)}$$
$$\therefore \cos(90^\circ - x) = \cos 20^\circ$$
$$90^\circ - x = 20^\circ$$
$$\therefore x = 70^\circ. \text{ Ans.}$$

10. Solve $\csc x = \sec 50^\circ$.

Solution:

$$\csc x = \sec(90^\circ - x)$$
$$\csc x = \sec 50^\circ \text{(given)}$$
$$\therefore \sec(90^\circ - x) = \sec 50^\circ$$
$$90^\circ - x = 50^\circ$$
$$\therefore x = 40^\circ. \text{ Ans.}$$

4-5 Trigonometric Functions of Special Angles

To memorize the values of the trigonometric functions of special angles 30°, 45°, 60° that appear often in the study of trigonometry, we use the following two special triangles:

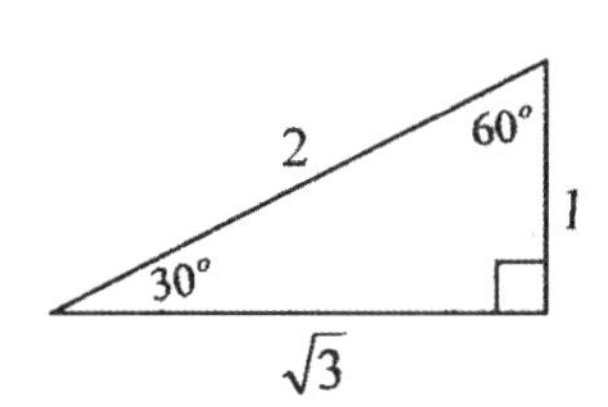

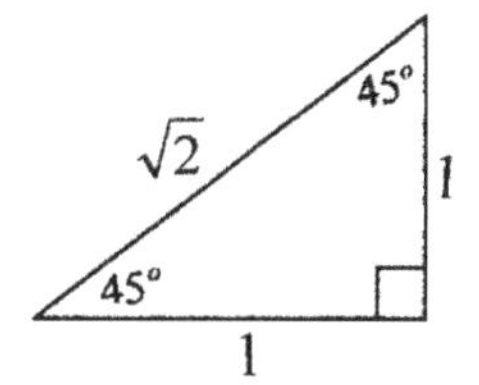

$\pi = 180^\circ$

θ	$\sin\theta$	$\cos\theta$	$\tan\theta$	$\cot\theta$	$\sec\theta$	$\csc\theta$
$30^\circ, \frac{\pi}{6}$	$\frac{1}{2}$	$\frac{\sqrt{3}}{2}$	$\frac{\sqrt{3}}{3}$	$\sqrt{3}$	$\frac{2\sqrt{3}}{3}$	2
$45^\circ, \frac{\pi}{4}$	$\frac{\sqrt{2}}{2}$	$\frac{\sqrt{2}}{2}$	1	1	$\sqrt{2}$	$\sqrt{2}$
$60^\circ, \frac{\pi}{3}$	$\frac{\sqrt{3}}{2}$	$\frac{1}{2}$	$\sqrt{3}$	$\frac{\sqrt{3}}{3}$	2	$\frac{2\sqrt{3}}{3}$

Examples

1. Find the value of $\sin 30^\circ \cos 60^\circ$.
Solution:

$$\sin 30^\circ \cos 60^\circ = \frac{1}{2}\cdot\frac{1}{2} = \frac{1}{4}. \text{ Ans.}$$

2. Find the value of $\tan 60^\circ \sec 45^\circ$.
Solution:

$$\tan 60^\circ \sec 45^\circ = \sqrt{3}\cdot\sqrt{2} = \sqrt{6} \approx 2.45. \text{ Ans.}$$

3. Find the value of $\sin\frac{\pi}{4}\cos\frac{\pi}{6}$.

Solution:

$$\sin\frac{\pi}{4}\cos\frac{\pi}{6} = \frac{\sqrt{2}}{2}\cdot\frac{\sqrt{3}}{2} = \frac{\sqrt{6}}{4} \approx 0.612. \text{ Ans.}$$

Examples

4. Find the value of $\tan\frac{\pi}{4} - \cos\frac{\pi}{6}$.

Solution:

$$\tan\frac{\pi}{4} - \cos\frac{\pi}{6} = 1 - \frac{\sqrt{3}}{2} \approx 1 - 0.866 \approx 0.134. \text{ Ans.}$$

In the court room, three defendants are waiting for the judge to sentence them for the time they need to stay in jail for the crimes they have committed.

Judge: Do you have any good reason why you think I should sentence you lesser time in jail ?
Defendant A: I was your high school classmate.
Judge: 1 week.
Defendant B: I built your house.
Judge: 1 month.
Defendant C: I introduced a girl friend to you 20 years ago. She is your wife now.
Judge: 20 years.

4-6 Radian Measure of Angles

We have learned the degree measure of an angle. The other way for measuring an angle is the **radian measure**.

1 radian (1^R): 1^R of an angle is the measure of the angle with counterclockwise rotation corresponding to an arc length equal to the **radius *r*** of the circle.

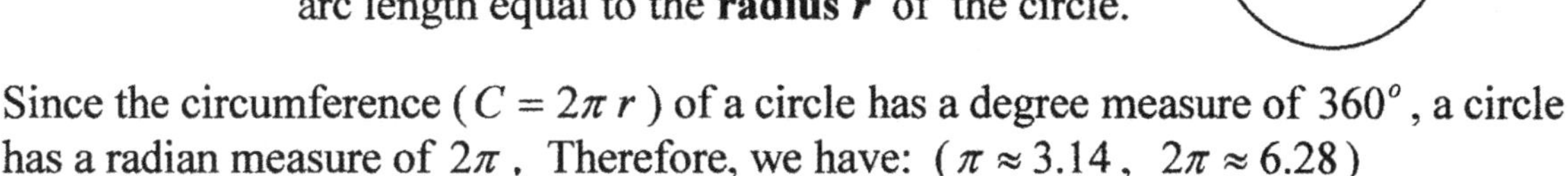

Since the circumference ($C = 2\pi r$) of a circle has a degree measure of 360°, a circle has a radian measure of 2π. Therefore, we have: ($\pi \approx 3.14$, $2\pi \approx 6.28$)

$$360^o = (2\pi)^R \approx 6.28 \text{ radians} \quad ; \quad 180^o = \pi^R \approx 3.14 \text{ radians}$$

We may simply say: $2\pi = 360^o$; $\pi = 180^o$.

Formulas: **1.** $1^o = \dfrac{\pi}{180} \approx 0.0175$ radians **2.** $1^R = \dfrac{180}{\pi} \approx 57.3^o$

To change the measure of an angle from degrees to radians, or from radians to degrees, we use the following formulas:

Formulas: **3.** $\theta^R = \dfrac{\pi}{180} \cdot \theta^o$ **4.** $\theta^o = \dfrac{180}{\pi} \cdot \theta^R$

For a circle of radius r and a central angle of θ in radians, the arc length s is:

Formula: **5.** $s = r\theta$

Examples

1. Express 72° in radians.
Solution:

$$\theta^R = \frac{\pi}{180} \cdot \theta^o = \frac{\pi}{180} \cdot 72$$
$$= 0.4\pi \approx 1.26 \text{ radians. Ans.}$$

2. Express -75° in radians.
Solution:

$$\theta^R = \frac{\pi}{180} \cdot \theta^o = \frac{\pi}{180}(-75)$$
$$= -(\tfrac{5}{12}\pi) \approx -1.31 \text{ radians. Ans.}$$

3. Express 1.26 radians in degree.
Solution:

$$\theta^o = \frac{180}{\pi} \cdot \theta^R = \frac{180}{\pi} \cdot (1.26)$$
$$\approx 180(0.40) \approx 72^o. \text{ Ans.}$$

4. Express -1.31 radians in degrees.
Solution:

$$\theta^o = \frac{180}{\pi} \cdot \theta^R = \frac{180}{\pi} \cdot (-1.31)$$
$$\approx 180(-0.417) \approx 75^o. \text{ Ans.}$$

5. Express 0.4π radians in degrees.
Solution:

$$\theta^o = \frac{180}{\pi} \cdot \theta^o = \frac{180}{\pi}(0.4\pi)$$
$$= 180(0.4) = 72^o. \text{ Ans.}$$

Quick Method:

$$\theta^o = 0.4\pi = 0.4(180) = 72^o. \text{ Ans.}$$

Examples

6. Express $-\frac{5}{12}\pi$ radians in degrees.

Solution:

$$\theta^o = -\tfrac{5}{12}\pi = -\tfrac{5}{12}(180) = -75^o. \text{ Ans.}$$

7. Express 90° in radians and express it in terms of π.

Solution:

$$\theta^R = \frac{\pi}{180}\cdot 90 = \frac{\pi}{2} \text{ radians. Ans.}$$

8. Find the arc length of a circle of radius 4 meters and subtended by a central angle of 0.45 radian.

Solution:

$$s = r\theta = 4(0.45) = 1.8 \text{ meters. Ans.}$$

$$\text{Or: } C = 2\pi r, \quad s = 2\pi(4)\cdot\frac{0.45}{2\pi} = 1.8 \text{ meters. Ans.}$$

9. Find the arc length of a circle of radius 5 feet and subtended by a central angle of 60°.

Solution:

$$s = r\theta = 5\cdot\frac{\pi}{180}(60) = 5(\frac{\pi}{3}) \approx 5.23 \text{ feet. Ans.}$$

$$\text{Or: } C = 2\pi r, \quad s = 2\pi(5)\cdot\frac{60}{360} = \frac{10\pi}{6} \approx 5.23 \text{ feet. Ans.}$$

A lady said to her friend.

Lady: I always keep quiet whenever my husband argues with me.

Friend: How do you control your anger ?

Lady: I clean the toilet.

Friend: How does that help ?

Lady: I use his toothbrush.

4-7 Trigonometric Functions of General Angles

In the previous sections, we have learned to find the values of trigonometric functions for acute angles ($\theta < 90^o$).
Since the values of x and y have different signs (+ or –) in the various quadrants, the signs of the trigonometric function of an angle θ for $\theta > 90^o$ can be determined by the quadrant of the angle.
The following figure gives the terminal ray of an angle θ which lies in a particular quadrant, and a point (a, b) on the terminal side.

$$\sin\theta = \frac{y}{r}, \quad \cos\theta = \frac{x}{r}, \quad \tan\theta = \frac{y}{x}, \quad \cot\theta = \frac{x}{y}, \quad \sec\theta = \frac{r}{x}, \quad \csc\theta = \frac{r}{y}$$

The signs (+ or –) depend on the quadrant where the angle θ lies.

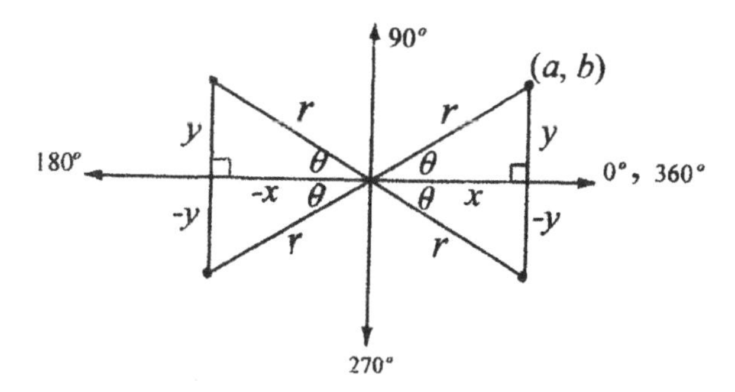

1. The value of each of the trigonometric functions at the **quadrantal angles**:

(Hint: $0^o = 0$, $90^o = \frac{\pi}{2}$, $180^o = \pi$, $270^o = \frac{3\pi}{2}$, $360^o = 2\pi$)

$\sin 0^o = 0$	$\cos 0^o = 1$	$\tan 0^o = 0$	$\cot 0^o =$ undefined	$\sec 0^o = 1$	$\csc 0^o =$ undefined
$\sin 90^o = 1$	$\cos 90^o = 0$	$\tan 90^o =$ undefined	$\cot 90^o = 0$	$\sec 90^o =$ undefined	$\csc 90^o = 1$
$\sin 180^o = 0$	$\cos 180^o = -1$	$\tan 180^o = 0$	$\cot 180^o =$ undefined	$\sec 180^o = -1$	$\csc 180^o =$ undefined
$\sin 270^o = -1$	$\cos 270^o = 0$	$\tan 270^o =$ undefined	$\cot 270^o = 0$	$\sec 270^o =$ undefined	$\csc 270^o = -1$
$\sin 360^o = 0$	$\cos 360^o = 1$	$\tan 360^o = 0$	$\cot 360^o =$ undefined	$\sec 360^0 = 1$	$\csc 360^o =$ undefined

2. The value of each of the trigonometric functions at the **coterminal angles**:
Coterminal angles are angles which have the same initial and terminal rays.
The coterminal angles of θ are: $\pm 360^o n + \theta$ or $\pm 2\pi n + \theta$, where $n = 0, 1, 2, 3, \cdots$.

$\sin(\pm 360^o n + \theta) = \sin\theta$
$\cos(\pm 360^o n + \theta) = \cos\theta$
$\tan(\pm 360^o n + \theta) = \tan\theta$
$\cot(\pm 360^o n + \theta) = \cot\theta$
$\sec(\pm 360^o n + \theta) = \sec\theta$
$\csc(\pm 360^o n + \theta) = \csc\theta$

Examples

1. $\sin 390^o = \sin(360^o + 30^o) = \sin 30^o = \frac{1}{2}$. Ans.

2. $\cos 390^o = \cos(360^o + 30^o) = \cos 30^o = \frac{\sqrt{3}}{2}$. Ans.

3. $\tan 405^o = \tan(360^o + 45^o) = \tan 45^o = 1$. Ans.

4. $\sec\frac{5\pi}{2} = \sec(2\pi + \frac{\pi}{2}) = \sec\frac{\pi}{2} =$ undefined. Ans.

5. $\sin 3\pi = \sin(2\pi + \pi) = \sin\pi = 0$. Ans.

3. Since the six basic trigonometric functions have the reciprocal relationships and only the signs (+ or –) of x and y determine the signs of the functions in the various quadrants, we say that $\sin\theta$ and $\csc\theta$ have the same sign in each quadrant, $\cos\theta$ and $\sec\theta$ have the same sign in each quadrant, and $\tan\theta$ and $\cot\theta$ have the same sign in each quadrant.

Functions	**Quadrants** 1st	2nd	3rd	4th
$\sin\theta$ and $\csc\theta$	+	+	–	–
$\cos\theta$ and $\sec\theta$	+	–	–	+
$\tan\theta$ and $\cot\theta$	+	–	+	–

To find the value of the trigonometric functions of any angle θ when $\theta > 90^o$, we use the **reference angle** of θ. A reference angle is always an acute angle between 0° and 90°. The value of each of the trigonometric functions of an angle θ equals the value of the trigonometric functions of its reference angle, except for the correct sign (+ or –). The sign depends on the quadrant in which the angle θ lies.
In the figures, $\alpha = 20^o$ is called the **reference angle** of the angles $\theta = 160^o$, 200°, or 340°. We can find the value of each of the trigonometric functions of the angle θ by multiplying the same value of its reference angle by the signs ($+1$ or -1) according to the quadrant in which the angle θ lies.

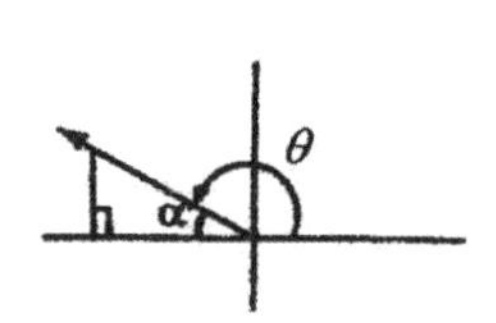

$\theta = 160^o$

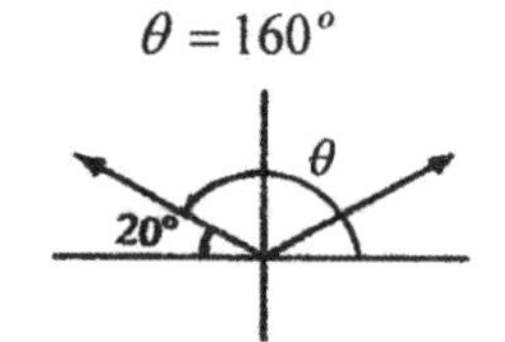

$\theta = 200^o$

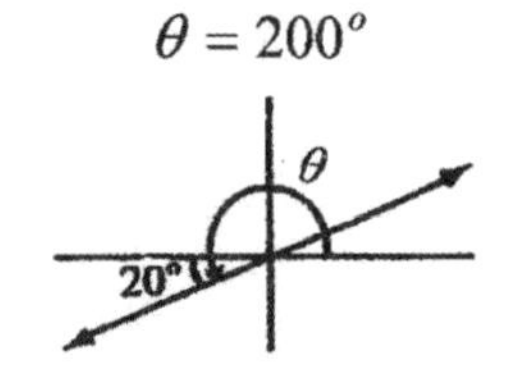

$\theta = 340^o$

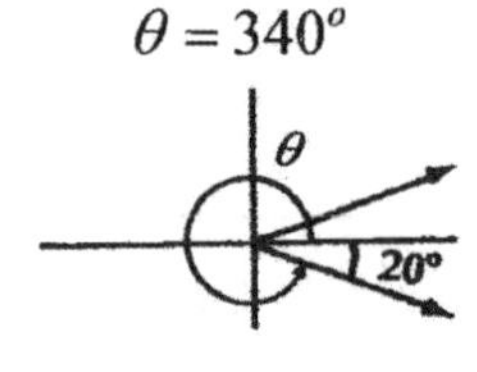

Given $\sin 20^o = 0.3420$ and $\cos 20^o = 0.9397$, we have:

$\sin 160^o = 0.3420$ (2nd qdt.) $\qquad \cos 160^o = -0.9397$ (2nd qdt.)
$\sin 200^o = -0.3420$ (3rd qdt.) $\qquad \cos 200^o = -0.9397$ (3rd qdt.)
$\sin 340^o = -0.3420$ (4th qdt.) $\qquad \cos 340^o = 0.9397$ (4th qdt)

We use the **coterminal angle** or the **reference angle** to find the value of a trigonometric function.

Examples

1. Find $\tan 160^o$.
Solution:

$$\tan 160^o = -\tan 20^o = -0.3640 \text{ (2nd qdt.). Ans.}$$

2. Find $\sin 4.35^R$.
Solution:

$$\text{reference angle} = 4.35 - 3.14 = 1.21$$
$$\sin 4.35^R = -\sin 1.21 = -0.9385 \text{ (3rd qdt.). Ans.}$$

Examples

3. Find $\cos 850^o 20'$.

Solution:

$$\cos 850^o 20' = \cos(720^o + 130^o 20') = \cos 130^o 20' = -\cos 49^o 40' = -\cos 49.67^o$$
$$= -0.647 \text{ (2nd qdt). Ans.}$$

4. Find $\tan 120^o$.

Solution:

$$\tan 120^o = -\tan 60^o = -\sqrt{3} \text{ (2nd qdt.). Ans.}$$

5. Find $\cot 315^o$.

Solution:

$$\cot 315^o = \cot(360^o - 45^o) = -\cot 45^o = -1 \text{ (4th qdt.). Ans.}$$

6. Find $\tan(-30^o)$.

Solution:

$$\tan(-30^o) = -\tan 30^o = -\tfrac{\sqrt{3}}{3} \text{ (4th qdt.). Ans.}$$

7. Find $\sec(-225^o)$.

Solution:

$$\sec(-225^o) = \sec(-360^o + 135^o) = \sec 135^o = -\sec 45^o = -\sqrt{2} \text{ (2nd qdt). Ans.}$$

8. Find $\csc(-315^o)$.

Solution:

$$\csc(-315^o) = \csc(-360^o + 45^o) = \csc 45^o = \sqrt{2} \text{ (1st qdt.). Ans.}$$

9. Find $\cos \dfrac{5\pi}{6}$.

Solution:

$$\cos\frac{5\pi}{6} = -\cos\frac{\pi}{6} = -\frac{\sqrt{3}}{2} \text{ (2nd qdt.). Ans.}$$

10. Find $\tan\left(-\dfrac{\pi}{3}\right)$.

Solution:

$$\tan\left(-\frac{\pi}{3}\right) = -\tan\frac{\pi}{3} = -\sqrt{3} \text{ (4th qdt.) Ans.}$$

11. Find $\sec\left(-\dfrac{7\pi}{4}\right)$.

Solution:

$$\sec\left(-\frac{7\pi}{4}\right) = \sec\left(-2\pi + \frac{\pi}{4}\right) = \sec\frac{\pi}{4} = \sqrt{2} \text{ (1st qdt). Ans.}$$

Examples

12. Find θ if $\csc\theta = \frac{2\sqrt{3}}{3}$ for $0^\circ \le \theta \le 360^\circ$.

Solution:

$\csc\theta$ is positive in 1st and 2nd quadrants.

$\therefore \theta = 60^\circ$ and 120°. Ans.

13. Find $\tan\theta$ if $\sec\theta = -\frac{\sqrt{5}}{2}$ and θ is in the 3rd quadrant.

Solution:

$\sec\theta = \frac{r}{x} = \frac{\sqrt{5}}{-2}$

$y^2 + (-2)^2 = (\sqrt{5})^2$

$y^2 + 4 = 5 \qquad \therefore y = \pm 1$

x and y are negative in 3rd quadrant.

$\therefore \tan\theta = \frac{y}{x} = \frac{-1}{-2} = \frac{1}{2}$. Ans.

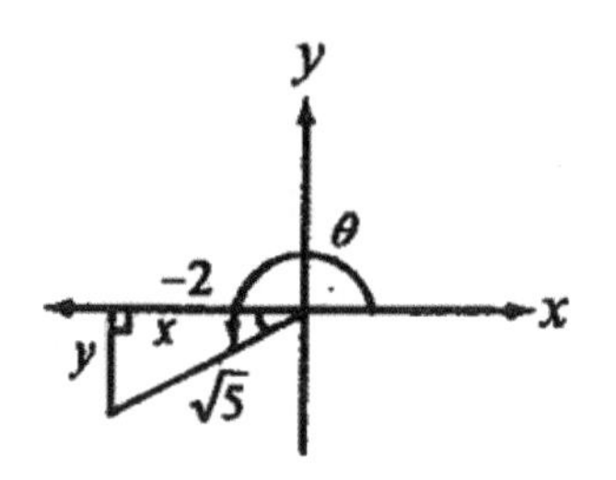

A young girl walks around in a shopping mall. Her hair and eyes are painted with multiple colors. An old man looks at her frequently.

Girl: What is wrong with you, old man ? You never got wild when you were young ?

Man: I am wondering either your father or mother is a peacock. They got wild when they were young.

4-8 The Law of Sines

The **Law of Sines** states that for any triangle, the ratio of the sine of each angle to the length of the opposite side is a constant (proportion). We can use the Law of Sines to find the unknown sides if :

1. two sides and one of their opposite angle are given.
2. two angles and one side are given.

The Law of Sines: $\frac{\sin A}{a} = \frac{\sin B}{b} = \frac{\sin C}{c}$

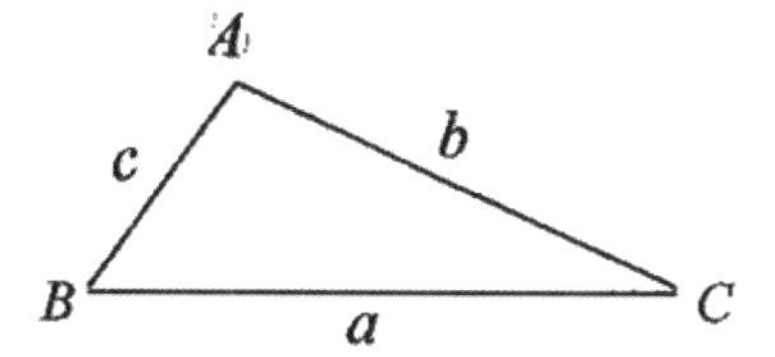

Examples

1. In ΔABC, find $< B$.

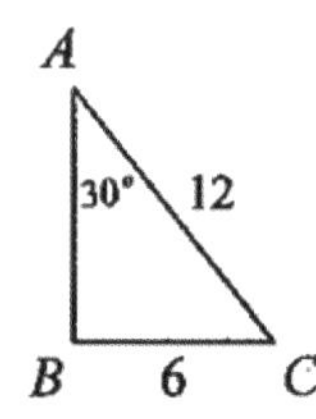

Solution:

$$\frac{\sin A}{a} = \frac{\sin B}{b}, \quad \frac{\sin 30^o}{6} = \frac{\sin B}{12}$$

$$\sin B = \frac{12 \sin 30^o}{6} = 2 \sin 30^o = 2(\tfrac{1}{2}) = 1$$

$\sin 90^o = 1 \quad \therefore < B = 90^o$. Ans.

2. In ΔABC, find the length of c.

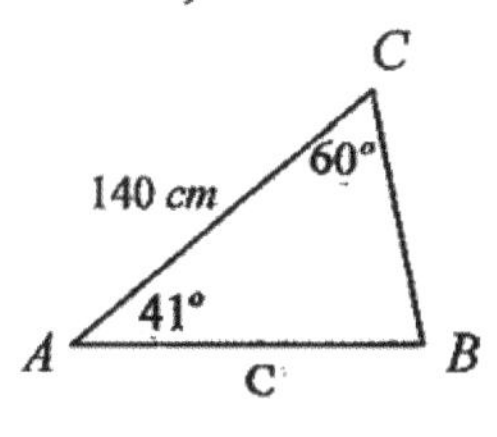

Solution:

$$< B = 180^o - 60^o - 41^o = 79^o$$

$$\frac{\sin B}{b} = \frac{\sin C}{c}, \quad \frac{\sin 79^o}{140} = \frac{\sin 60^o}{c}$$

$$\therefore \ c = \frac{140 \sin 60^o}{\sin 79^o} = \frac{140(0.866)}{0.982} \approx 123.46\,cm. \text{ Ans.}$$

3. In ΔABC, $a = 12$, $b = 8$, $< A = 98^o$, find $< B$, $< C$, and c.

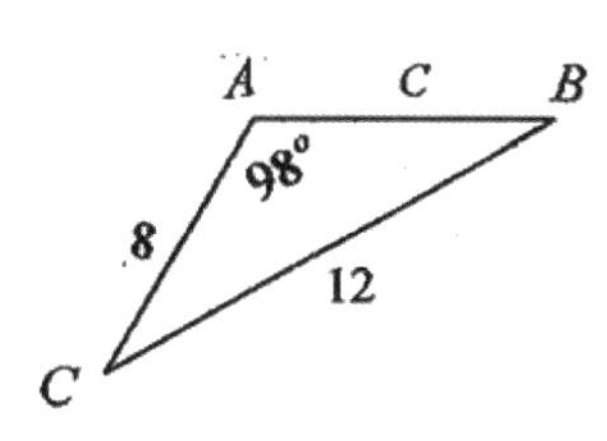

Solution:

$$\frac{\sin 98^o}{12} = \frac{\sin B}{8} = \frac{\sin C}{c}$$

$$\sin B = \frac{8(\sin 98^o)}{12} = \frac{8(0.99)}{12} = 0.66$$

$\therefore < B \approx 41.3^o$. Ans.

$< C = 180^o - 98^o - 41.3^o \approx 40.7^o$. Ans.

$$c = \frac{8(\sin C)}{\sin B} = \frac{8(\sin 40.7^o)}{\sin 41.3^o} = \frac{8(0.65)}{0.66} \approx 7.88. \text{ Ans.}$$

Examples

4. In a triangle, the lengths of two sides are 6 and 3. The opposite angle of side of length 3 is 70°. Find the opposite angle of side of length 6.

Solution:

$$\frac{\sin 70^0}{3} = \frac{\sin\theta}{6}$$

$$\sin\theta = \frac{6(\sin 70^o)}{3} = 2(\sin 70^o) \approx 2(0.94) \approx 1.88$$

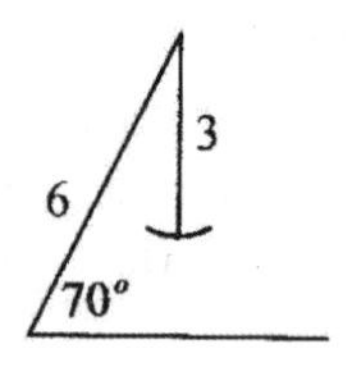

Since there is no angle measure of A for $\sin A > 1$, there is no triangle with the given measurements.

Ans: No solution. (It is an ambiguous case.)

5. In the figure, find the opposite angles of side of length 7.

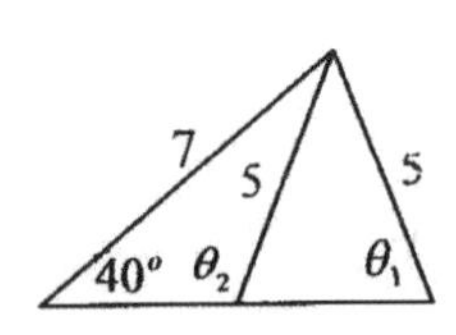

Solution:

$$\frac{\sin 40^o}{5} = \frac{\sin\theta}{7}$$

$$\sin\theta = \frac{7(\sin 40^o)}{5} \approx \frac{7(0.643)}{5} \approx 0.9$$

$\theta_1 \approx 64.16^o$ or $\theta_2 = 180^o - 64.16 = 115.84^o$. Ans.

(There are two answers. It is an ambiguous case.)

6. An airplane sends the radar signals to two ground Stations A and B which are 10 miles away. The signal indicates that the bearing (direction) of the airplane from Station A is $E\,40^o N$ (40° north of east). The signal indicates that the bearing (direction) of the airplane from Station B is $W\,60^0 N$ (60° north of west). How far is the airplane from each station ?

Solution:

The measure of the third angle $= 180^o - 40^o - 60^o = 80^o$

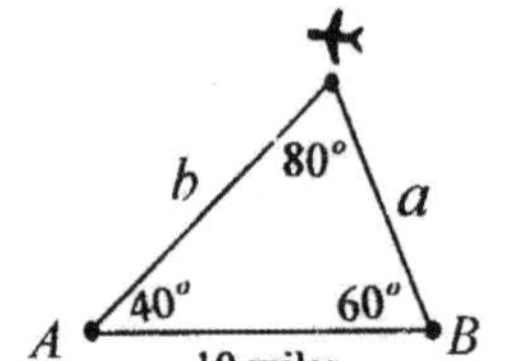

$$\frac{\sin 40^o}{a} = \frac{\sin 80^o}{10} \quad \therefore a = \frac{10(\sin 40^o)}{\sin 80^o} \approx \frac{10(0.643)}{0.985} \approx 6.53 \text{ miles. Ans.}$$

$$\frac{\sin 60^o}{b} = \frac{\sin 80^o}{10} \quad \therefore b = \frac{10(\sin 60^o)}{\sin 80^o} \approx \frac{10(0.866)}{0.985} \approx 8.79 \text{ miles. Ans.}$$

On graduation day, a high school graduate said to his math teacher.
Student: Thank you very much for giving me a "D" in your class. For your kindness, do you want me to do anything for you ?
Teacher: Yes. Please do me a favor. Just don't tell anyone that I was your math teacher.

4-9 The Law of Cosines

The **Law of Cosines** states that for any triangle, the square of one side equals the sum of the squares of the other two sides minus twice their product times the cosine of their included angle.

We can use the Law of Cosines to find unknown dimensions of any triangle if:

1. Two sides and the included angle are given.

2. Three sides are given.

The law of Cosines:
1. $a^2 = b^2 + c^2 - 2bc\cos A$
2. $b^2 = a^2 + c^2 - 2ac\cos B$
3. $c^2 = a^2 + b^2 - 2ab\cos C$

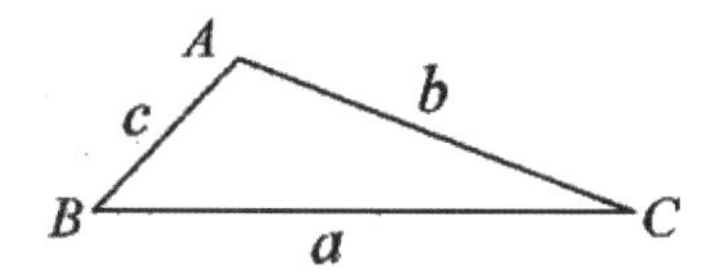

Examples

1. In ΔABC, find b.

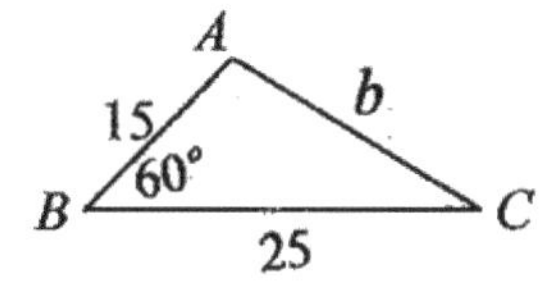

Solution:

$$\begin{aligned} b^2 &= a^2 + c^2 - 2ac\cos 60^\circ \\ &= 25^2 + 15^2 - 2(25)(15)(\tfrac{1}{2}) \\ &= 625 + 225 - 375 \\ &= 475 \end{aligned}$$

$\therefore b = \sqrt{475} \approx 21.8$. Ans.

2. In ΔABC, find $< A$.

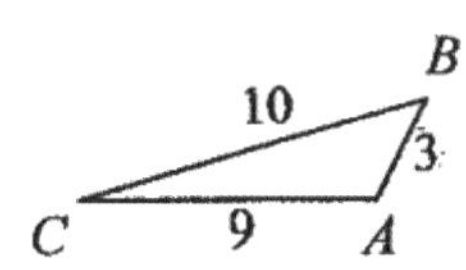

Solution:

$$\begin{aligned} a^2 &= b^2 + c^2 - 2bc\cos A \\ 10^2 &= 9^2 + 3^2 - 2(9)(3)\cos A \\ 100 &= 81 + 9 - 54\cos A \\ 10 &= -54\cos A \end{aligned}$$

$\cos A = -\dfrac{10}{54} \approx -0.1852 \quad \therefore < A \approx 100.67^\circ$. Ans.

3. In the triangle, find x.

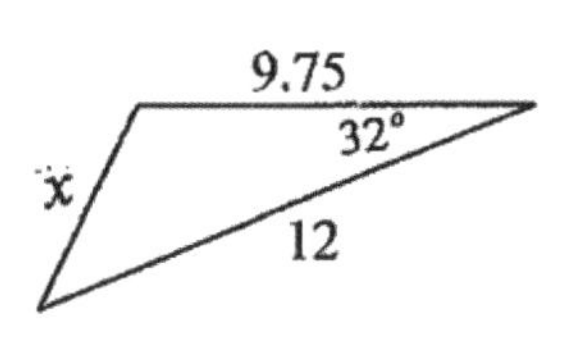

Solution:

$$\begin{aligned} x^2 &= 12^2 + (9.75)^2 - 2(12)(9.75)\cos 32^\circ \\ &= 144 + 95.1 - 234(0.8480) \\ &= 239.1 - 198.43 \\ &= 40.67 \end{aligned}$$

$x = \sqrt{40.67} \approx 6.38$. Ans.

Examples

4. Solve the triangle.

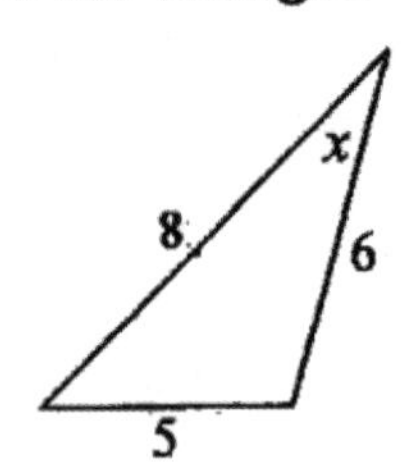

Solution:

$$5^2 = 8^2 + 6^2 - 2 \cdot 8 \cdot 6 \cos x$$
$$25 = 64 + 36 - 96 \cos x$$
$$25 = 100 - 96 \cos x$$
$$\cos x = \frac{100 - 25}{96} \approx 0.781$$
$\therefore x \approx 38.65^\circ$. Ans.

5. In Δ ABC, if $a = 3$, $b = 10$, $c = 9$, find $\angle$ B.

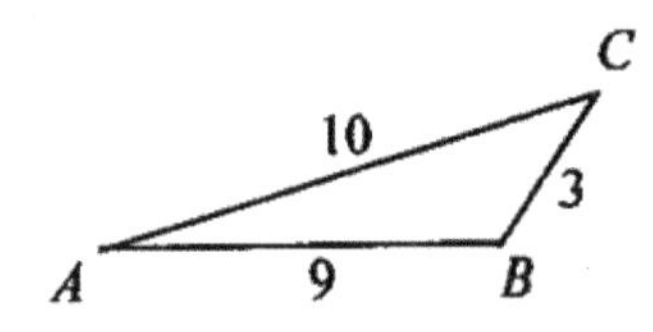

Solution:

$$b^2 = a^2 + c^2 - 2ac \cos B$$
$$10^2 = 3^2 + 9^2 - 2 \cdot 3 \cdot 9 \cos B$$
$$100 = 9 + 81 - 54 \cos B$$
$$\cos B = \frac{10}{-54} \approx -0.1852$$
$\therefore < B \approx 100.67^\circ$. Ans.

In a toy store, a five-year old girl wants her father to buy her an expensive toy.
Father: No, it is too expensive.
Girl: Dad, I like it. Honey and Sweetheart.
Father: Why are you calling me "Honey and Sweetheart" ?
Girl: Because whenever mom wants to buy a new dress, she calls you Honey, or Sweetheart.

4-10 Double-Angle Formulas

We use double-Angle formulas to find the **exact values** of trigonometric function whose sine, cosine, or tangent are known exactly.

1. $\sin 2A = 2\sin A\cos A$

2. $\cos 2A = \cos^2 A - \sin^2 A$

$= 1 - 2\sin^2 A \quad \rightarrow \quad$ **4.** $\sin^2\theta = \dfrac{1-\cos 2\theta}{2}$

$= 2\cos^2 A - 1 \quad \rightarrow \quad$ **5.** $\cos^2\theta = \dfrac{1+\cos 2\theta}{2}$

3. $\tan 2A = \dfrac{2\tan A}{1-\tan^2 A}$ **6.** $\tan^2\theta = \dfrac{1-\cos 2\theta}{1+\cos 2\theta}$

Evaluate each expression by its exact value (in fraction, finite decimal, radical, and π)

1. $2\sin 15^o\cos 15^o = \sin 2(15^o) = \sin 30^o = \dfrac{1}{2}$

2. $\cos^2 15^o - \sin^2 15^o = \cos 2(15^o) = \cos 30^o = \dfrac{\sqrt{3}}{2}$

3. $1 - 2\sin^2 105^o = \cos 2(105^o) = \cos 210^o = -\cos 30^o = -\dfrac{\sqrt{3}}{2}$

4. $2\cos^2 120^o - 1 = \cos 2(120^o) = \cos 240^o = -\cos 60^o = -\dfrac{1}{2}$

5. $\dfrac{2\tan 22.5^o}{1-\tan^2 22.5^o} = \tan 2(22.5^o) = \tan 45^o = 1$

Examples

1. If $\sin\theta = \frac{3}{5}$, find the exact value of $\sin 2\theta$ if $0^o < \theta < 90^o$.

Solution:

Find $\cos\theta$: $\sin\theta = \frac{3}{5} = \frac{y}{r}$, $y = 3$, $r = 5$, $\theta \approx 37^o$

$x^2 + 3^2 = 5^2$, $x^2 = 16$, $x = 4$ $\therefore \cos\theta = \frac{x}{r} = \frac{4}{5}$

$\sin 2\theta = 2\sin\theta\cos\theta = 2(\frac{3}{5})(\frac{4}{5}) = \frac{24}{25}$ (2θ is in 1st qdt.). Ans.

2. If $\cos\theta = \frac{3}{5}$, find the exact value of $\cos 2\theta$ if $0^o < \theta < 90^o$

Solution:

Find $\sin\theta$: $\cos\theta = \frac{3}{5} = \frac{x}{r}$, $x = 3$, $r = 5$, $\theta \approx 53^o$

$3^2 + y^2 = 5^2$, $y^2 = 16$, $y = 4$ $\therefore \sin\theta = \frac{y}{r} = \frac{4}{5}$

$\cos 2\theta = 1 - 2\sin^2\theta = 1 - 2(\frac{4}{5})^2 = 1 - \frac{32}{25} = -\frac{7}{25}$ (2θ is in 2nd qdt.). Ans.

Examples

3. If $\sin\theta = \frac{2}{5}$, find the exact value of $\cos 2\theta$ if $90^\circ < \theta < 180^\circ$

   Solution:

   $\sin\theta = \frac{2}{5}$, $\theta \approx 42^\circ$ (reference angle), $\theta \approx 138^\circ$

   2θ lies in 4th quadrant.

   $\cos 2\theta = 1 - 2\sin^2\theta = 1 - 2(\frac{2}{5})^2 = \frac{17}{25}$. Ans.

4. If $\tan\theta = \frac{1}{2}$, find the exact value of $\sin 2\theta$ if $\pi < \theta < \dfrac{3\pi}{2}$.

   Solution:

   $\tan\theta = \frac{1}{2}$, $\theta \approx 27^\circ$ (reference angle), $\theta \approx 207^\circ$

   $\tan\theta = \frac{y}{x}$, $y = 1$, $x = 2$

   $2^2 + 1^2 = r^2$, $r = \sqrt{5}$ $\therefore \sin\theta = \frac{y}{r} = \frac{1}{\sqrt{5}} = \frac{\sqrt{5}}{5}$, $\cos\theta = \frac{x}{r} = \frac{2}{\sqrt{5}} = \frac{2\sqrt{5}}{5}$

   2θ lies in 1st quadrant.

   $\sin 2\theta = 2\sin\theta\cos\theta = 2(\frac{\sqrt{5}}{5})(\frac{2\sqrt{5}}{5}) = \frac{4}{5}$. Ans.

Simplify each expression

5. $\cos^2 2\theta - \sin^2 2\theta = \cos 2(2\theta) = \cos 4\theta$

6. $\cos^4\theta - \sin^4\theta = (\cos^2\theta + \sin^2\theta)(\cos^2\theta - \sin^2\theta) = \cos^2\theta - \sin^2\theta = \cos 2\theta$

A high school boy is playing the violin.
His girlfriend sits next to him.
Boy: Do you like the song I played ?
Girl: Yes. It makes me miss my father.
Boy: Was your father a musician ?
Girl: No. He was a sawmill worker.

4-11 Half-Angle Formulas

We use half-angle formulas to find the **exact values** of trigonometric functions whose sine or cosine are known exactly.
Using the formulas in Chapter 4-10, we obtain the following half-angle formulas:

1. $\sin\frac{A}{2} = \pm\sqrt{\frac{1-\cos A}{2}}$

2. $\cos\frac{A}{2} = \pm\sqrt{\frac{1+\cos A}{2}}$

3. $\tan\frac{A}{2} = \frac{\sin A}{1+\cos A} = \frac{1-\cos A}{\sin A}$

4. $\tan\frac{A}{2} = \pm\sqrt{\frac{1-\cos A}{1+\cos A}}$

The signs (+ or −) is determined by the quadrant of the angle $\frac{A}{2}$.

Evaluate each expression by its exact value (in fraction, finite decimal, radical, and π)

1. $\cos 22.5^\circ = \cos\frac{45^\circ}{2} = \sqrt{\frac{1+\cos 45^\circ}{2}} = \sqrt{\frac{1+\sqrt{2}/2}{2}} = \sqrt{\frac{2+\sqrt{2}}{4}} = \frac{\sqrt{2+\sqrt{2}}}{2}$ (1st qdt.)

2. $\sin(-15^\circ) = -\sin 15^\circ$ (4th qdt.)

$= -\sin\frac{30^\circ}{2} = -\sqrt{\frac{1-\cos 30^\circ}{2}} = -\sqrt{\frac{1-\sqrt{3}/2}{2}} = -\sqrt{\frac{2-\sqrt{3}}{4}} = -\frac{\sqrt{2-\sqrt{3}}}{2}$

3. $\sin(-67.5^\circ) = -\sin 67.5^\circ = -\sin\frac{135^\circ}{2}$ (4th qdt.)

$= -\sqrt{\frac{1-\cos 135^\circ}{2}} = -\sqrt{\frac{1-(-\cos 45^\circ)}{2}} = -\sqrt{\frac{1+\sqrt{2}/2}{2}} = -\sqrt{\frac{2+\sqrt{2}}{4}} = -\frac{\sqrt{2+\sqrt{2}}}{2}$

4. $\tan 157.5^\circ = -\tan 22.5^\circ = -\tan\frac{45^\circ}{2}$ (2nd qdt.)

$= -\frac{\sin 45^\circ}{1+\cos 45^\circ} = -\frac{\sqrt{2}/2}{1+\sqrt{2}/2} = -\frac{\sqrt{2}}{2+\sqrt{2}}\cdot\frac{2-\sqrt{2}}{2-\sqrt{2}} = -\frac{2\sqrt{2}-2}{2} = -(\sqrt{2}-1) = 1-\sqrt{2}$

5. $\tan\frac{9\pi}{8} = \tan\frac{\pi}{8} = \tan 22.5^\circ$ (3rd qdt.)

$= \frac{\sin 45^\circ}{1+\cos 45^\circ} = \frac{\sqrt{2}/2}{1+\sqrt{2}/2} = \frac{\sqrt{2}}{2+\sqrt{2}}\cdot\frac{2-\sqrt{2}}{2-\sqrt{2}} = \frac{2\sqrt{2}-2}{2} = \sqrt{2}-1$

A student said to a priest.
Student: I don't believe that heaven is as good as you said. You have never been there.
Priest: Have you ever heard of anyone who came back here because he didn't like it there. ?

Examples

1. If $\sin\theta = \frac{3}{5}$ and $0^o < \theta < 90^o$, find the exact value of **a.** $\sin\frac{\theta}{2}$ **b.** $\cos\frac{\theta}{2}$ **c.** $\tan\frac{\theta}{2}$.

Solution:

Find $\cos\theta$: $\sin\theta = \frac{y}{r}$, $y = 3$, $r = 5$, $\theta \approx 37^o$

$x^2 + 3^2 = 5^2$, $x^2 = 16$, $x = 4$ $\therefore \cos\theta = \frac{x}{r} = \frac{4}{5}$

$\theta/2$ lies in 1st quadrant.

a. $\sin\frac{\theta}{2} = \sqrt{\frac{1-\cos\theta}{2}} = \sqrt{\frac{1-4/5}{2}} = \sqrt{\frac{1/5}{2}} = \sqrt{\frac{1}{10}} = \frac{\sqrt{10}}{10}$. Ans.

b. $\cos\frac{\theta}{2} = \sqrt{\frac{1+\cos\theta}{2}} = \sqrt{\frac{1+4/5}{2}} = \sqrt{\frac{9/5}{2}} = \sqrt{\frac{9}{10}} = \frac{3\sqrt{10}}{10}$. Ans.

c. $\tan\frac{\theta}{2} = \sqrt{\frac{1-\cos\theta}{1+\cos\theta}} = \sqrt{\frac{1-4/5}{1+4/5}} = \sqrt{\frac{1/5}{9/5}} = \sqrt{\frac{1}{9}} = \frac{1}{3}$. Ans.

Or: $\tan\frac{\theta}{2} = \frac{\sin\theta/2}{\cos\theta/2} = \frac{\sqrt{10}/10}{3\sqrt{10}/10} = \frac{1}{3}$. Ans.

2. If $\cos\theta = -\frac{2}{5}$ and $\pi < \theta < 3\pi/2$, find the exact value of **a.** $\sin\frac{\theta}{2}$ **b.** $\cos\frac{\theta}{2}$ **c.** $\tan\frac{\theta}{2}$.

Solution:

$\cos\theta = -\frac{2}{5}$, $\theta \approx 66^o$ (reference angle), $\theta \approx 246^o$

$\theta/2$ lies in 2nd quadrant.

a. $\sin\frac{\theta}{2} = \sqrt{\frac{1-\cos\theta}{2}} = \sqrt{\frac{1-(-2/5)}{2}} = \sqrt{\frac{7/5}{2}} = \sqrt{\frac{7}{10}} = \frac{\sqrt{70}}{10}$. Ans.

b. $\cos\frac{\theta}{2} = -\sqrt{\frac{1+\cos\theta}{2}} = -\sqrt{\frac{1+(-2/5)}{2}} = -\sqrt{\frac{3/5}{2}} = -\sqrt{\frac{3}{10}} = -\frac{\sqrt{30}}{10}$. Ans.

c. $\tan\frac{\theta}{2} = -\sqrt{\frac{1-\cos\theta}{1+\cos\theta}} = -\sqrt{\frac{1-(-2/5)}{1+(-2/5)}} = -\sqrt{\frac{7/5}{3/5}} = -\frac{\sqrt{21}}{3}$. Ans.

In a medical school, a professor asks a student about his major.
Professor: Why do you think a dentist can make more money than a heart surgeon ?
Student: A patient has 32 teeth, but only one heart.

4-12 Sum and Difference Formulas

We use sum and difference formulas to find the **exact values** of trigonometric functions whose sine, cosine, or tangent are known exactly.

1. $\sin(A+B)=\sin A\cos B+\cos A\sin B$
2. $\sin(A-B)=\sin A\cos B-\cos A\sin B$
3. $\cos(A+B)=\cos A\cos B-\sin A\sin B$
4. $\cos(A-B)=\cos A\cos B+\sin A\sin B$
5. $\tan(A+B)=\dfrac{\tan A+\tan B}{1-\tan A\tan B}$
6. $\tan(A-B)=\dfrac{\tan A-\tan B}{1+\tan A\tan B}$

Evaluate each expression by its exact value (in fraction, finite decimal, radical, and π)

1. $\sin 75^\circ=\sin(45^\circ+\sin 30^\circ)=\sin 45^\circ\cos 30^\circ+\cos 45^\circ\sin 30^\circ=\frac{\sqrt{2}}{2}\cdot\frac{\sqrt{3}}{2}+\frac{\sqrt{2}}{2}\cdot\frac{1}{2}=\frac{\sqrt{6}+\sqrt{2}}{4}$

2. $\sin 15^\circ=\sin(45^\circ-30^\circ)=\sin 45^\circ\cos 30^\circ-\cos 45^\circ\sin 30^\circ=\frac{\sqrt{2}}{2}\cdot\frac{\sqrt{3}}{2}-\frac{\sqrt{2}}{2}\cdot\frac{1}{2}=\frac{\sqrt{6}-\sqrt{2}}{4}$

3. $\cos 120^\circ=\cos(90^\circ+30^\circ)=\cos 90^\circ\cos 30^\circ-\sin 90^\circ\sin 30^\circ=0\cdot\frac{\sqrt{3}}{2}-1\cdot\frac{1}{2}=-\frac{1}{2}$

4. $\cos 150^\circ=\cos(180^\circ-30^\circ)=\cos 180^\circ\cos 30^\circ+\sin 180^\circ\sin 30^\circ=-1\cdot\frac{\sqrt{3}}{2}+0\cdot\frac{1}{2}=-\frac{\sqrt{3}}{2}$

5. $\sin 20^\circ\cos 25^\circ+\cos 20^\circ\sin 25^\circ=\sin(20^\circ+25^\circ)=\sin 45^\circ=\frac{\sqrt{2}}{2}$

6. $\sin 75^\circ\cos 15^\circ-\cos 75^\circ\sin 15^\circ=\sin(75^\circ-15^\circ)=\sin 60^\circ=\frac{\sqrt{3}}{2}$

7. $\sin\frac{\pi}{4}\cos\frac{\pi}{12}-\cos\frac{\pi}{4}\sin\frac{\pi}{12}=\sin(\frac{\pi}{4}-\frac{\pi}{12})=\sin\frac{\pi}{6}=\frac{1}{2}$

8. $\cos 85^\circ\cos 40^\circ+\sin 85^\circ\sin 40^\circ=\cos(85^\circ-40^\circ)=\cos 45^\circ=\frac{\sqrt{2}}{2}$

9. $\tan 105^\circ=\tan(60^\circ+45^\circ)=\frac{\tan 60^\circ+\tan 45^\circ}{1-\tan 60^\circ\tan 45^\circ}=\frac{\sqrt{3}+1}{1-\sqrt{3}\cdot 1}=\frac{1+\sqrt{3}}{1-\sqrt{3}}=\frac{1+\sqrt{3}}{1-\sqrt{3}}\cdot\frac{1+\sqrt{3}}{1+\sqrt{3}}$
$=\frac{1+2\sqrt{3}+3}{1-3}=\frac{4+2\sqrt{3}}{-2}=-2-\sqrt{3}$.

Evaluate each expression by its exact value

10. $\tan 15^\circ = \tan(45^\circ - 30^\circ) = \dfrac{\tan 45^\circ - \tan 30^\circ}{1 + \tan 45^\circ \tan 30^\circ} = \dfrac{1 - \frac{\sqrt{3}}{3}}{1 + 1\cdot\frac{\sqrt{3}}{3}} = \dfrac{3-\sqrt{3}/3}{3+\sqrt{3}/3} = \dfrac{3-\sqrt{3}}{3+\sqrt{3}} \cdot \dfrac{3-\sqrt{3}}{3-\sqrt{3}}$

$= \dfrac{9 - 6\sqrt{3} + 3}{9 - 3} = \dfrac{12 - 6\sqrt{3}}{6} = 2 - \sqrt{3}$

Simplify each expression

11. $\cos(x-y)\cos y + \sin(x-y)\sin y = \cos[(x-y)-y] = \cos(x-2y)$.

12. $\sin(x-y)\cos y + \cos(x-y)\sin y = \sin[(x-y)+y] = \sin x$.

13. $\dfrac{\tan 70^\circ - \tan 25^\circ}{1 + \tan 70^0 \tan 25^\circ} = \tan(70^\circ - 25^\circ) = \tan 45^\circ = 1$.

14. $\dfrac{\tan x + \tan 2x}{1 - \tan x \tan 2x} = \tan(x + 2x) = \tan 3x$.

15. $\dfrac{\tan\frac{\pi}{4} - \tan\frac{\pi}{12}}{1 + \tan\frac{\pi}{4}\tan\frac{\pi}{12}} = \tan(\dfrac{\pi}{4} - \dfrac{\pi}{12}) = \tan\dfrac{\pi}{6} = \dfrac{\sqrt{3}}{3}$.

16. $\sin(\dfrac{\pi}{2} + \theta) = \sin\dfrac{\pi}{2}\cos\theta + \cos\dfrac{\pi}{2}\sin\theta = 1\cdot\cos\theta + 0\cdot\sin\theta = \cos\theta$

17. $\tan(\pi - \theta) = \dfrac{\tan\pi - \tan\theta}{1 + \tan\pi\tan\theta} = \dfrac{0 - \tan\theta}{1 + 0\cdot\tan\theta} = -\tan\theta$

18. Prove $\sin 3x = 3\sin x - 4\sin^3 x$.

Proof:

$\sin 3x = \sin(2x + x) = \sin 2x\cos x + \cos 2x\sin x = 2\sin x\cos x\cdot\cos x + (1 - 2\sin^2 x)\sin x$

$= 2\sin x\cos^2 x + \sin x - 2\sin^3 x = 2\sin x(1 - \sin^2 x) + \sin x - 2\sin^3 x$

$= 2\sin x - 2\sin^3 x + \sin x - 2\sin^3 x = 3\sin x - 4\sin^3 x$. Ans.

19. Prove $\cos 3x = 4\cos^3 x - 3\cos x$.

Proof:

$\cos 3x = \cos(2x + x) = \cos 2x\cos x - \sin 2x\sin x = (2\cos^2 x - 1)\cos x - (2\sin x\cos x)\sin x$

$= 2\cos^3 x - \cos x - 2\sin^2 x\cos x = 2\cos^3 x - \cos x - 2(1 - \cos^2 x)\cos x$

$= 2\cos^3 x - \cos x - 2\cos x + 2\cos^3 x = 4\cos^3 x - 3\cos x$. Ans.

4-13 The Area of a Triangle

The area of a triangle can be obtained by the following formulas if two sides and the included angle are known.

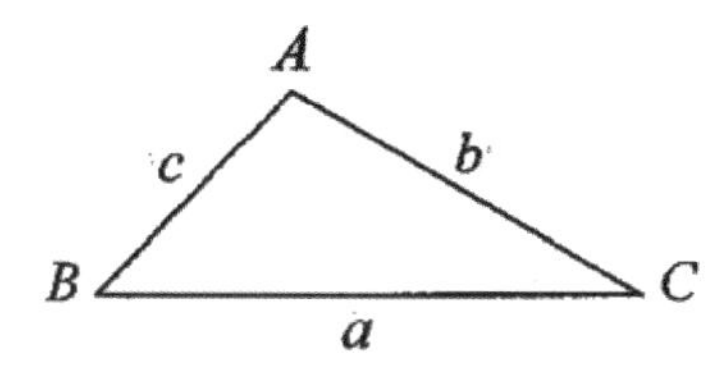

1. $Area = \frac{1}{2}ab\sin C$ 2. $Area = \frac{1}{2}bc\sin A$

3. $Area = \frac{1}{2}ac\sin B$

4. $Area = \sqrt{s(s-a)(s-b)(s-c)}$ where $s = \dfrac{a+b+c}{2}$

(It is called **Heron's Formula.**)

Examples

1. Find the area of ΔABC.

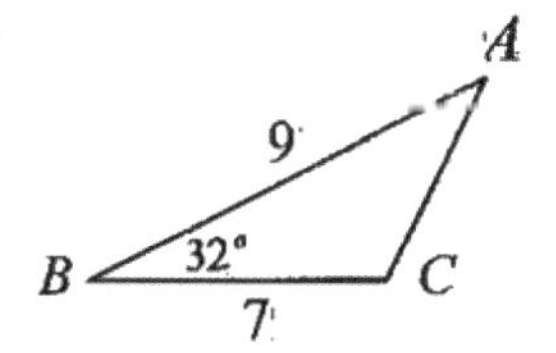

Solution:

$$Area = \tfrac{1}{2}ac\sin 32^o$$
$$= \tfrac{1}{2}(7)(9)(\sin 32^o)$$
$$\approx \tfrac{1}{2}(7)(9)(0.5299) \approx 16.69. \text{ Ans.}$$

2. Find the area of the triangle.

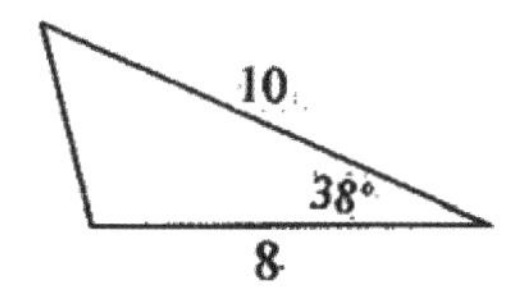

Solution:

$$Area = \tfrac{1}{2}(10)(8)(\sin 38^o)$$
$$\approx \tfrac{1}{2}(10)(8)(0.6157) \approx 24.63. \text{ Ans.}$$

3. Find the area of the triangle.

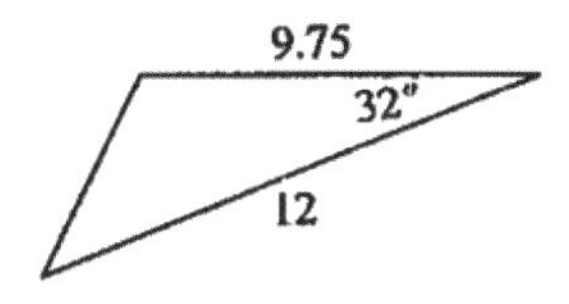

Solution:

$$Area = \tfrac{1}{2}(12)(9.75)(\sin 32^o)$$
$$\approx \tfrac{1}{2}(12)(9.75)(0.5299) \approx 31. \text{ Ans.}$$

4. Find the area of a triangle with the lengths of three sides 12, 6.38, and 9.75.
Solution:

$$s = \frac{12 + 6.38 + 9.75}{2} = 14.07$$
$$Area = \sqrt{s(s-a)(s-b)(s-c)}$$
$$= \sqrt{14.07(14.07-12)(14.07-6.38)(14.07-9.75)}$$
$$= \sqrt{14.07(2.07)(7.69)(4.32)} = \sqrt{967.55} \approx 31. \text{ Ans.}$$

Examples

5. Find the area of the triangle for which $a = 4$, $b = 3$, and the included angle $\alpha = 30^o$

   Solution:

$$Area = \frac{1}{2}ab\sin\alpha = \frac{1}{2}(4)(3)(\sin 30^o) = \frac{1}{2}(4)(3)(\frac{1}{2}) = 3.$$ Ans.

6. Find the area of the shaded region in the circle.

   Solution:

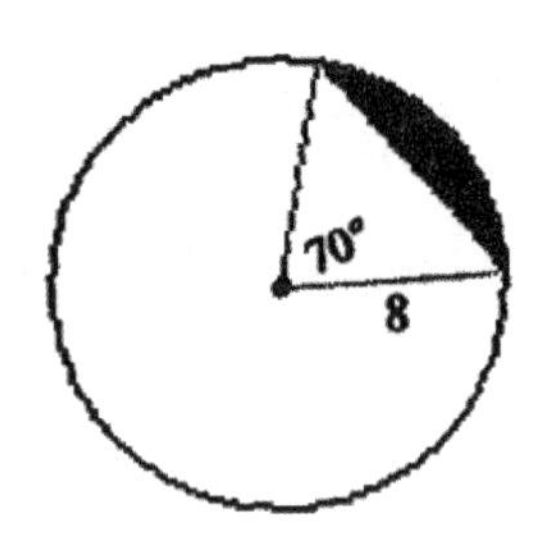

Area of the sector $= \pi\ (8^2)(\frac{70}{360}) \approx 39.095$

Area of the triangle $= \frac{1}{2}(8)(8)(\sin 70^o) \approx 30.07$

Area of the shaded region $\approx 39.095 - 30.07 \approx 9.025$. Ans.

Using a phone, a four-year old boy calls his mother who is working in a foreign country.

Son: Mom, are you coming home today ?

Mother: No, I am very far away at the other side of the earth.

Son: Where is the other side of the earth ?

Mother: Earth is like a basketball. You are on top of the ball. I am at the bottom.

Son: It must be very hard for you with your head upside-down when you drink water.

4-14 Periodic Functions and Graphs

Trigonometric functions are called **periodic functions**. The values of a periodic funtion repeat on a regular interval. This regular interval is called the period of the funtion. The period of a trigonometric function contains a full cycle of the function.
The period of sine or cosine function is 2π radians or 360°. It means that the values of sine or cosine function repeat themselves every cycle (360°). The period of tangent or cotangent function is π radians or 180°. The period of secant or cosecant function is 2π or 360°.
Each of the six trigonometric functions can be graphed on a coordinate plane. The x-axis represents the value of the angle. The y-axis represents the value of the function of each angle.

Graphs of $y = \sin x$ and $y = \cos x$

To draw the graphs of $y = \sin x$ (or $y = \cos x$) in a coordinate plane, we begin by choosing values of x as an angle in radians in the interval $0 \le x \le 2\pi$. The values of $\sin x$ (or $\cos x$) repeat every 2π units with a smooth, continuous curve. These graphs are called **sine waves** and **cosine waves**. The values of $\sin x$ or $\cos x$ vary from -1 to 1. We say that both $\sin x$ and $\cos x$ have a **period** of 2π and an **amplitude** of 1.

A) Sine wave $\rightarrow$ $y = \sin x$; $-1 \le y \le 1$, $|\sin x| \le 1$

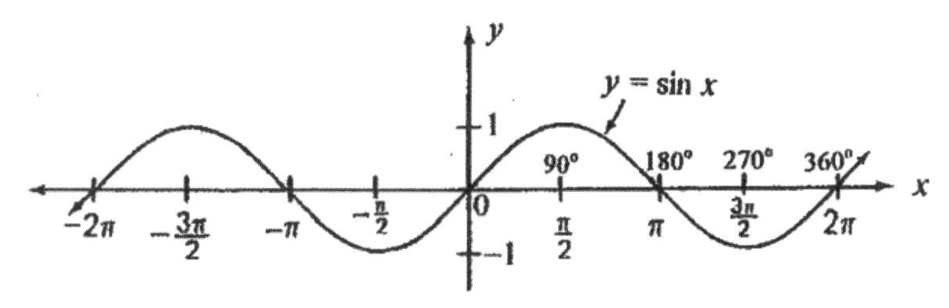

B) Cosine wave $\rightarrow$ $y = \cos x$; $-1 \le y \le 1$, $|\cos x| \le 1$

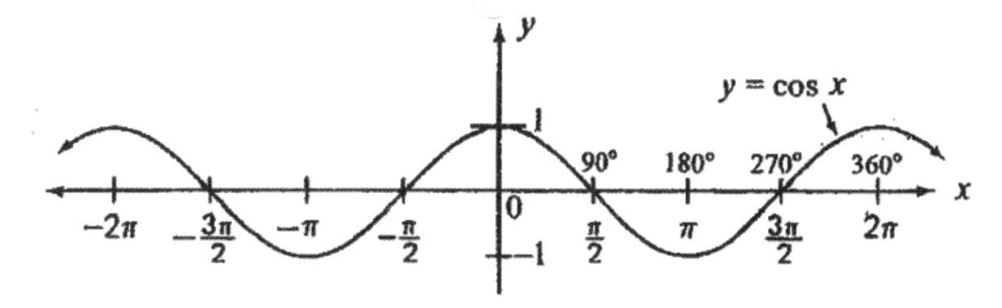

Since the sine and cosine functions have a period of 2π, we have the following rules:

$$\sin(2\pi + x) = \sin x$$
$$\cos(2\pi + x) = \cos x$$

The **amplitudes** and **periods** of sine and cosine functions can be determined by the following ways:

$$y = A \sin kx \quad \text{or} \quad y = A \cos kx$$

$$\textbf{amplitude} = A \ , \ \textbf{period} = \frac{2\pi}{k} \ .$$

Graphing Variations of Trigonometric Functions using Transformations

The graphing transformations introduced in chapter 2-21 can be used to graph trigonometric functions that are variations of the functions. Example: $y = a\sin(kx \pm c)$

1. $y = -\sin x$ It reflects the graph over the x-axis.
2. $y = \sin x \pm c$ It moves (shifts) the graph by c units upward (downward).
3. $y = \sin(x \pm c)$ It moves (shifts) the graph by c units to the left (right).
4. $y = a\sin x$ It stretches the graph vertically if $a > 1$ and compresses the graph vertically if $0 < a < 1$.
5. $y = \sin ax$ It compresses the graph horizontally if $a > 1$ and stretches the graph horizontally if $0 < a < 1$.

Examples

1. Find the amplitude and period of the function $y = 2\sin x$.

   Solution:

   $A = 2$ (amplitude) ; The function varies from –2 to 2.

   $k = 1$ $\therefore \text{period} = \frac{2\pi}{k} = \frac{2\pi}{1} = 2\pi$. Ans.

2. Find the amplitude and period of the function $y = \frac{2}{3}\cos 2x$.

   Solution:

   $A = \frac{2}{3}$ (amplitude) ; The function varies from $-\frac{2}{3}$ to $\frac{2}{3}$.

   $k = 2$ $\therefore \text{period} = \frac{2\pi}{k} = \frac{2\pi}{2} = \pi$. Ans.

3. Graph $y = \sin 2x$.

   Solution: $A = 1$ (amplitude), $\text{period} = \frac{2\pi}{k} = \frac{2\pi}{2} = \pi$.

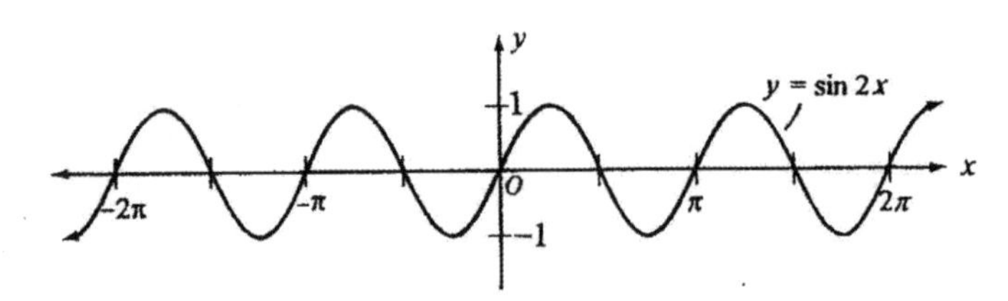

4. Graph $y = 2\cos\frac{1}{2}x$.

   Solution: $A = 2$ (amplitude) ; $\text{period} = \frac{2\pi}{k} = \frac{2\pi}{\frac{1}{2}} = 4\pi$.

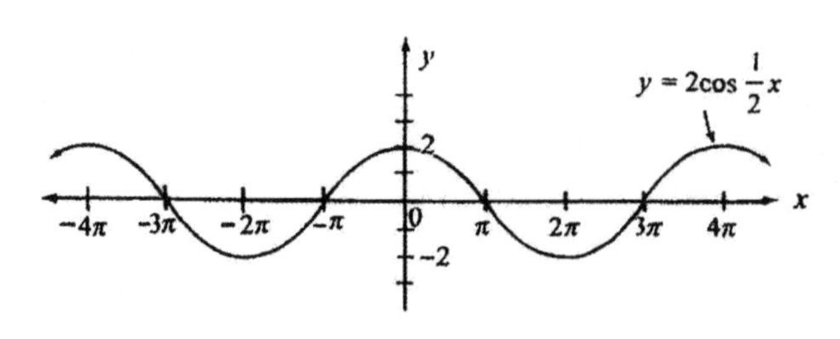

Graphs of $y = \tan x$ and $y = \cot x$

To draw the graphs of $y = \tan x$ (or $y = \cot x$) in a coordinate plane, we begin by choosing values of x as an angle in radians. The graphs of tangent and cotangent functions have no amplitude. Their graphs are drawn between successive vertical asymptotes and repeat the pattern to the right and to the left.

A) $y = \tan x$

x	$-\frac{\pi}{2}$	$-\frac{\pi}{3}$	$-\frac{\pi}{4}$	$-\frac{\pi}{6}$	0	$\frac{\pi}{6}$	$\frac{\pi}{4}$	$\frac{\pi}{3}$	$\frac{\pi}{2}$
y	undefined	$-\sqrt{3}$	-1	$-\frac{\sqrt{3}}{3}$	0	$\frac{\sqrt{3}}{3}$	1	$\sqrt{3}$	undefined

1. y is undefined when $x = n\pi + \frac{\pi}{2}$, where n is an integer.
2. y increases as x increases.
3. The equations of the asymptotes are: $x = n\pi + \frac{\pi}{2}$, n is an integer.
4. A period of π

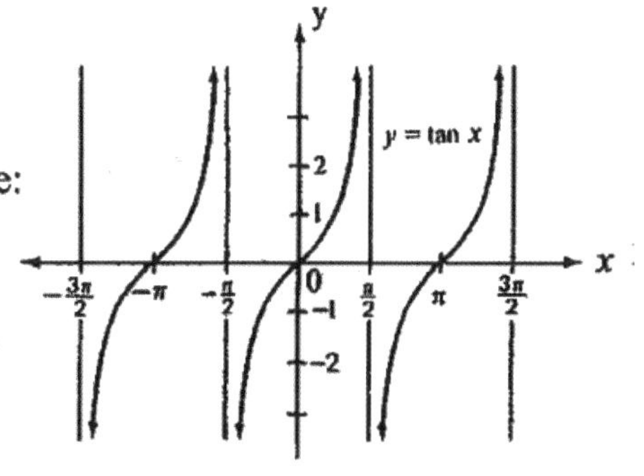

B) $y = \cot x$

x	0	$\frac{\pi}{6}$	$\frac{\pi}{4}$	$\frac{\pi}{3}$	$\frac{\pi}{2}$	$\frac{2\pi}{3}$	$\frac{3\pi}{4}$	$\frac{5\pi}{6}$	π
y	undefined	$\sqrt{3}$	1	$\frac{\sqrt{3}}{3}$	0	$-\frac{\sqrt{3}}{3}$	-1	$-\sqrt{3}$	undefined

1. y is undefined when $x = n\pi$, where n is an integer.
2. y decreases as x increases.
3. The equations of asymptotes are: $x = n\pi$, n is an integer.
4. A period of π

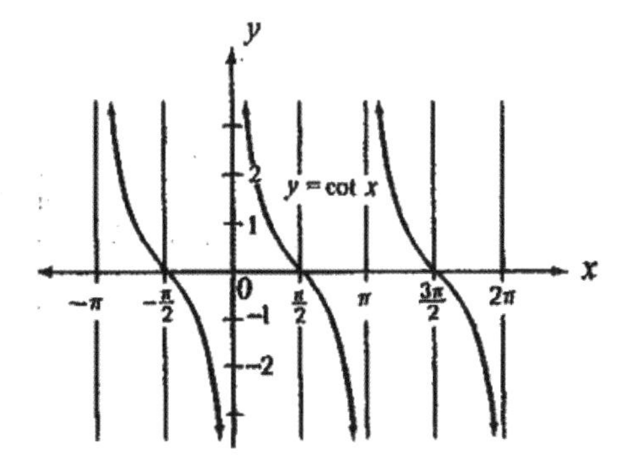

The periods of tangent and cotangent functions can be determined by the following way:

$$\text{If } y = \tan kx \text{ or } y = \cot kx, \text{ then } \textbf{period} = \frac{\pi}{k}.$$

The asymptotes occur every π/k units along the x-axis.

Examples

1. Graph the function $y = 3\tan 2x$.

Solution:

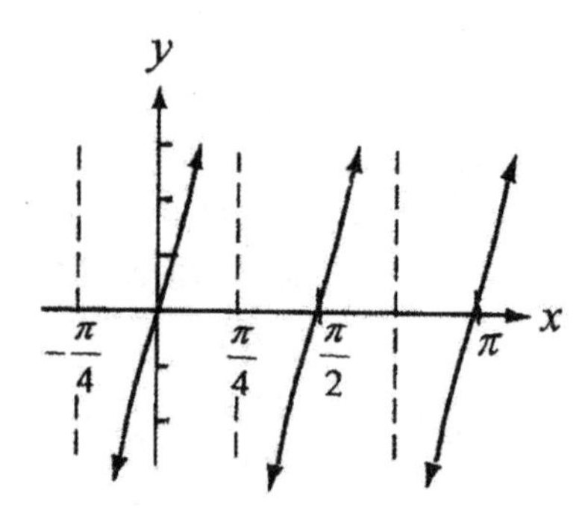

$$k = 2, \text{ period } = \frac{\pi}{k} = \frac{\pi}{2}$$

Find successive vertical asymptotes near the origin:

$\tan 2x = \text{undefined}, \quad 2x = -\frac{\pi}{2}$ and $2x = \frac{\pi}{2}$

Vertical asymptotes are $x = -\frac{\pi}{4}$ and $x = \frac{\pi}{4}$

Graph the function between these asymptotes. The same pattern occurs to the right and to the left.

2. Graph the function $y = \cot 2x$.

Solution:

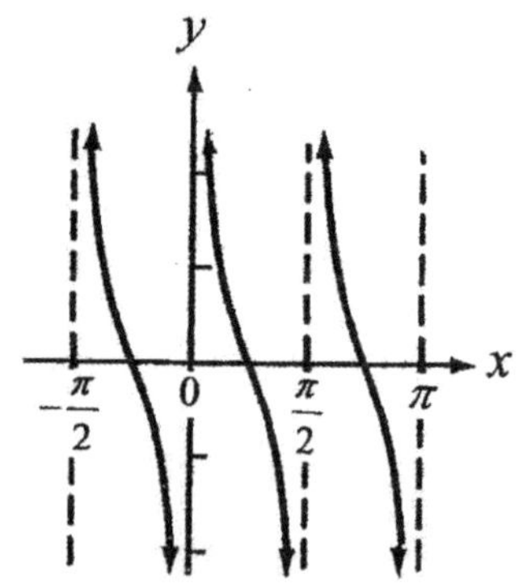

$$k = 2, \text{ period } = \frac{\pi}{k} = \frac{\pi}{2}$$

Find successive vertical asymptotes near the origin:

$\cot 2x = \text{undefined}, \quad 2x = -\pi, \ 2x = 0$ and $2x = \pi$

Vertical asymptotes are $x = -\frac{\pi}{2}$, $x = 0$ and $x = \frac{\pi}{2}$.

Graph the function between these asymptotes. The same pattern occurs to the right and to the left.

3. Find the period of $y = \tan(x + \frac{\pi}{4})$ and two successive vertical asymptotes near the origin, one to the left, one to the right.

Solution:

$$k = 1, \text{ period } = \frac{\pi}{k} = \frac{\pi}{1} = \pi. \text{ Ans.}$$

$\tan(x + \frac{\pi}{4}) = \text{undefined}, \quad x + \frac{\pi}{4} = -\frac{\pi}{2}$ and $x + \frac{\pi}{4} = \frac{\pi}{2}$

The two successive vertical asymptotes near the origin are:

$$x = -\frac{3\pi}{4} \text{ and } x = \frac{\pi}{4}. \text{ Ans.}$$

4. Find the period of $y = \cot(2x - \frac{\pi}{2})$ and two successive vertical asymptotes to the right of the origin.

Solution: $\quad k = 2, \text{ period } = \frac{\pi}{k} = \frac{\pi}{2}. \text{ Ans.}$

$\cot(2x - \frac{\pi}{2}) = \text{undefined}, \quad 2x - \frac{\pi}{2} = 0$ and $2x - \frac{\pi}{2} = \pi$

The two successive vertical asymptotes near the origin are:

$$x = \frac{\pi}{4} \text{ and } x = \frac{3\pi}{4}. \text{ Ans.}$$

Graphs of $y = \sec x$ and $y = \csc x$

To draw the graphs of $y = \sec x$ (or $y = \csc x$) in a coordinate plane, we begin by choosing values of x as an angle in radians. The graphs of secant and cosecant functions have no amplitude.

Their graphs of secant and cosecant functions can be drawn by making use of the reciprocals of corresponding cosine and sine functions: $\sec x = \dfrac{1}{\cos x}$ and $\csc x = \dfrac{1}{\sin x}$

For example, if $x = 0$, $\sin 0 = 0$, then $\csc = \frac{1}{0} =$ undefined.

Their graphs can also be drawn between successive vertical asymptotes and repeat the pattern to the right and to the left.

A) $y = \sec x$ $\quad y \le -1$ or $y \ge 1$, $|\sec x| \ge 1$

x	0	$\frac{\pi}{4}$	$\frac{\pi}{2}$	$\frac{3\pi}{4}$	π	$\frac{5\pi}{4}$	$\frac{3\pi}{2}$	$\frac{7\pi}{4}$	2π
y	1	$\sqrt{2}$	undefined	$-\sqrt{2}$	-1	$-\sqrt{2}$	undefined	$\sqrt{2}$	1

1. y is undefined when $x = n\pi + \frac{\pi}{2}$, where n is an integer.
2. Intersects with y-axis at (0, 1)
3. No graph between $y = 1$ and $y = -1$
4. Graph discontinues at the asymptotes: $x = n\pi + \frac{\pi}{2}$, n is an integer.
5. Symmetric about y-axis: $\sec(-x) = \sec x$
6. A period of 2π

y
y = sec x
1
x
$-\frac{\pi}{2}$ 0 $\frac{\pi}{2}$ π $\frac{3\pi}{2}$ 2π $\frac{5\pi}{2}$
-1

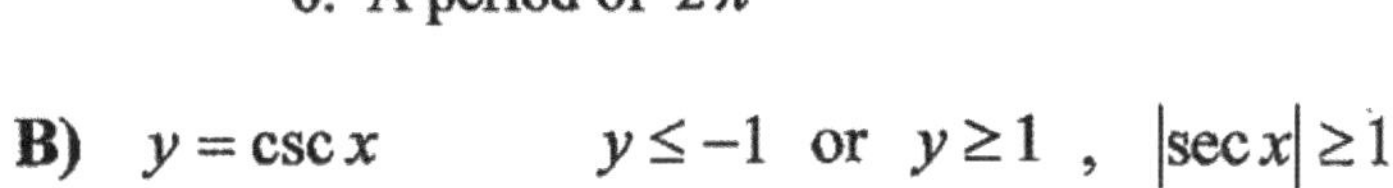

B) $y = \csc x$ $\quad y \le -1$ or $y \ge 1$, $|\sec x| \ge 1$

x	0	$\frac{\pi}{4}$	$\frac{\pi}{2}$	$\frac{3\pi}{4}$	π	$\frac{5\pi}{4}$	$\frac{3\pi}{2}$	$\frac{7\pi}{4}$	2π
y	undefined	$\sqrt{2}$	1	$\sqrt{2}$	undefined	$-\sqrt{2}$	-1	$-\sqrt{2}$	undefined

1. y is undefined when $x = n\pi$, where n is an integer.
2. No intersection with y-axis
3. No graph between $y = 1$ and $y = -1$
4. Graph discontinues at the asymptotes: $x = n\pi$, n is an integer.
5. Symmetric through origin: $\csc(-x) = -\csc x$
6. A period of 2π

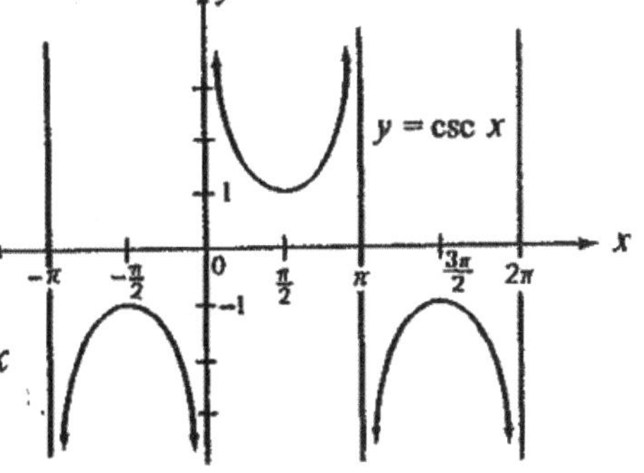

The periods of secant and cosecant functions can be determined by the following way:

If $y = \sec kx$ or $y = \csc kx$, then **period** $= \dfrac{2\pi}{k}$

Examples

1. Graph the function $y = \sec 2x$.

   Solution:

   $k = 2$, period $= \frac{2\pi}{2} = \pi$

   Find successive vertical asymptotes near the origin:

   $\sec 2x =$ undefined, $2x = -\frac{\pi}{2}$, $2x = \frac{\pi}{2}$, and $2x = \frac{3\pi}{2}$

   Vertical asymptotes are $x = -\frac{\pi}{4}$, $x = \frac{\pi}{4}$, and $\frac{3\pi}{4}$

   Graph the function between these asymptotes.

   The same pattern occurs to the right and to the left.

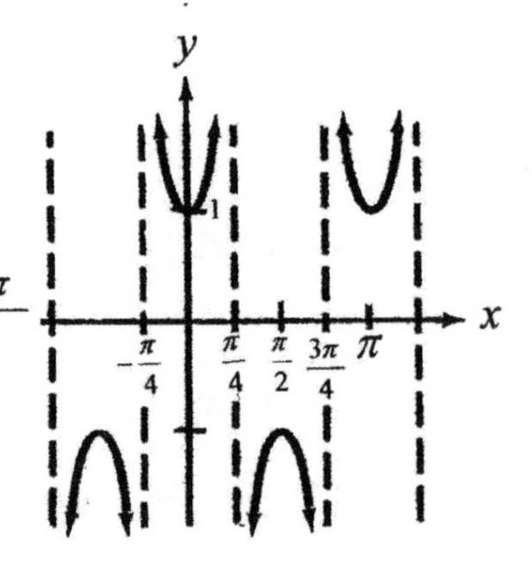

2. Graph the function $y = 2\csc x$.

   Solution:

   $k = 1$, period $= \frac{2\pi}{k} = \frac{2\pi}{1} = 2\pi$

   Find successive vertical asymptotes near the origin:

   $\csc x =$ undefined, $x = -\pi$, $x = 0$, and $x = \pi$

   Graph the function between these asymptotes.

   The same pattern occurs to the right and to the left.

   The graph is vertical stretched by multiplying 2 units.

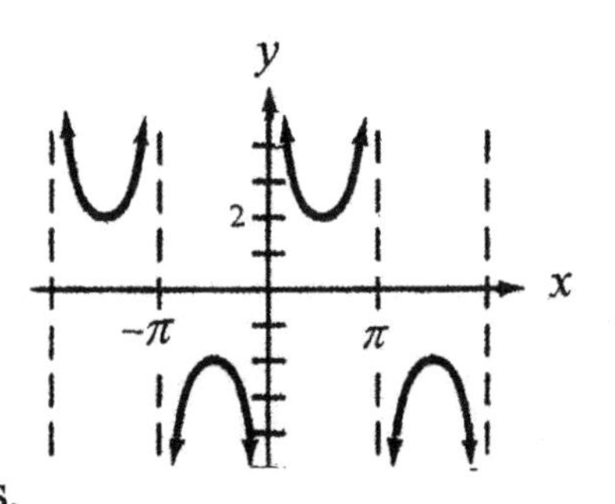

3. Find the period of $y = 2\sec(x - \frac{\pi}{4})$ and three successive vertical asymptotes near the origin, one to the left, two to the right.

   Solution:

   $$k = 1, \text{ period} = \frac{2\pi}{k} = \frac{2\pi}{1} = 2\pi. \text{ Ans.}$$

   $\sec(x - \frac{\pi}{4}) =$ undefined, $x - \frac{\pi}{4} = -\frac{\pi}{2}$, $x - \frac{\pi}{4} = \frac{\pi}{2}$ and $x - \frac{\pi}{4} = \frac{3\pi}{2}$

   The three successive vertical asymptotes near the origin are:

   $$x = -\frac{\pi}{4},\ x = \frac{3\pi}{4}, \text{ and } x = \frac{7\pi}{4}. \text{ Ans.}$$

   Hint: It shifts the graph of $y = 2\sec x$ to the right by $\frac{\pi}{4}$ units.

> A kid said to his mother.
> Kid: Did you wear contact lenses this morning ?
> Mother: No. Why do you ask ?
> Kid: I hear you yell to father this morning. You said "After 20 years of marriage, I have finally seen what kind of man you are ! ".

4-15 Trigonometric Equations

A **trigonometric equation** is an equation involving trigonometric functions. We solve a trigonometric equation just as we solve any algebraic equation. To solve trigonometric equations, we use algebraic transformations and trigonometric formulas.

Examples

1. Determine whether $\theta = 30^o$ is a solution of the equation $\sin\theta = \frac{1}{2}$.

Solution:

$$\sin 30^o = \tfrac{1}{2} \quad \therefore \theta = 30^o \text{ is a solution of the equation. Ans.}$$

2. Determine whether $\theta = \pi/4$ is a solution of the equation $\cos\theta = \frac{\sqrt{3}}{2}$.

Solution:

$$\cos \pi/4 = \tfrac{\sqrt{2}}{2} \neq \tfrac{\sqrt{3}}{2} \quad \therefore \theta = \pi/4 \text{ is not a solution of the equation. Ans.}$$

3. Determine whether $x = 150^o$ is a solution of the equation $\sin x = \frac{1}{2}$.

Solution:

$$\sin 150^o = \sin 30^o = \tfrac{1}{2} \text{ (2nd qdt)}$$

$$\therefore x = 150^o \text{ is a solution of the equation. Ans.}$$

4. Solve $2\sin x = 1$ for $0 \le x \le 360^o$.

Solution:

$$2\sin x = 1, \quad \sin x = \tfrac{1}{2}$$

$$\therefore x = 30^o \text{ and } 150^o. \quad \text{or } x = \pi/6 \text{ and } 5\pi/6. \text{ Ans.}$$

5. Solve $2\sin^2 x - \sin x = 1$ for $0 \le x \le 2\pi$.

Solution:

$$2\sin^2 x - \sin x = 1$$

$$2\sin^2 x - \sin x - 1 = 0$$

$$(2\sin x + 1)(\sin x - 1) = 0$$

$2\sin x + 1 = 0$	$\sin x - 1 = 0$
$\sin x = -\frac{1}{2}$	$\sin x = 1$
$x = 210^o,\ 330^o$	$x = 90^o$

Ans: $x = 90^o, 210^o, 330^o$.

or $x = \pi/2,\ 7\pi/6,\ 11\pi/6$

Examples

6. Solve $\cos x + 2\sin^2 x = 1$ for $0 \le x \le 360^o$.

Solution:

$$\cos x + 2\sin^2 x = 1$$
$$\cos x + 2(1 - \cos^2 x) = 1$$
$$\cos x + 2 - 2\cos^2 x = 1$$
$$-2\cos^2 x + \cos x + 1 = 0$$
$$2\cos^2 x - \cos x - 1 = 0$$
$$(2\cos x + 1)(\cos x - 1) = 0$$

$2\cos x + 1 = 0$	$\cos x - 1 = 0$
$\cos x = -\frac{1}{2}$	$\cos x = 1$
$x = 120^o,\ 240^o$	$x = 0^o,\ 360^o$

Ans: $x = 0^o,\ 120^o,\ 240^o,\ 360^o$
or $x = 0,\ \frac{2\pi}{3},\ \frac{4\pi}{3},\ 2\pi$

7. Solve $\tan\theta = \cot\theta$ for $0 \le \theta \le 360^o$.

Solution:

$$\tan\theta = \cot\theta,\ \tan\theta = \frac{1}{\tan\theta},\ \tan^2\theta = 1,\ \therefore \tan\theta = \pm 1$$

$$\tan\theta = 1,\quad \theta = 45^o,\ 225^o$$
$$\tan\theta = -1,\quad \theta = 135^o,\ 315^o$$

Ans: $\theta = 45^o,\ 135^o,\ 225^o,\ 315^o$
or $\theta = \frac{\pi}{4},\ \frac{3\pi}{4},\ \frac{5\pi}{4},\ \frac{7\pi}{4}$

8. Find the general solution of $\cos 2x = 5\cos x + 2$.

Solution:

$$\cos 2x = 5\cos x + 2$$
$$2\cos^2 x - 1 = 5\cos x + 2$$
$$2\cos^2 x - 5\cos x - 3 = 0$$
$$(2\cos x + 1)(\cos x - 3) = 0$$

$2\cos x + 1 = 0$	$\cos x - 3 = 0$
$\cos x = -\frac{1}{2}$	$\cos x = 3$
$x = \frac{2\pi}{3},\ \frac{4\pi}{3}$	$\lvert\cos\rvert \le 1$, no solution

The cosine function has a period of 2π.

$\therefore$ The general solution is $x = \frac{2\pi}{3} + 2\pi \cdot n$ and $x = \frac{4\pi}{3} + 2\pi \cdot n$,
where $n = 1 \cdot 2 \cdot 3 \cdots\cdots$. Ans.

4-16 Inverse Trigonometric Functions

We have defined the trigonometric functions for an acute angle.

Function: $y = \sin x$

Example: $y = \sin 30^o = \frac{1}{2}$

In order to represent an angle, we restrict the domain of $y = \sin x$ to the interval of $-\frac{\pi}{2} \le x \le \frac{\pi}{2}$ to define a one-to-one inverse function $x = \sin y$ by interchanging x and y and using the symbol of **inverse trigonometric function.**

Function: $y = \sin x$ where $-1 \le y \le 1$ and $-\frac{\pi}{2} \le x \le \frac{\pi}{2}$

Inverse Function: $y = \sin^{-1} x$ where $-1 \le x \le 1$ and $-\frac{\pi}{2} \le y \le \frac{\pi}{2}$

Example: $y = \sin^{-1}(\frac{1}{2}) = 30^o$

$\sin^{-1} x$ (or $\arcsin x$) is called " inverse sine of x " or " $\arcsin x$ ".

$y = \sin^{-1} x$ and $x = \sin y$ are equivalent and read " y is the angle whose sine equals x.".

$y = \sin^{-1} x$ and $y = \sin x$ are **inverses.** We say that $y = \sin^{-1} x$ means " y is the inverse sine of x "

The graphs of $y = \sin x$ and its inverse $y = \sin^{-1} x$ are reflection of each other about the line $y = x$.

$y = \sin x$ and $-1 \le y \le 1,\ -\frac{\pi}{2} \le x \le \frac{\pi}{2}$

$y = \sin^{-1} x$ and $-1 \le x \le 1,\ -\frac{\pi}{2} \le y \le \frac{\pi}{2}$

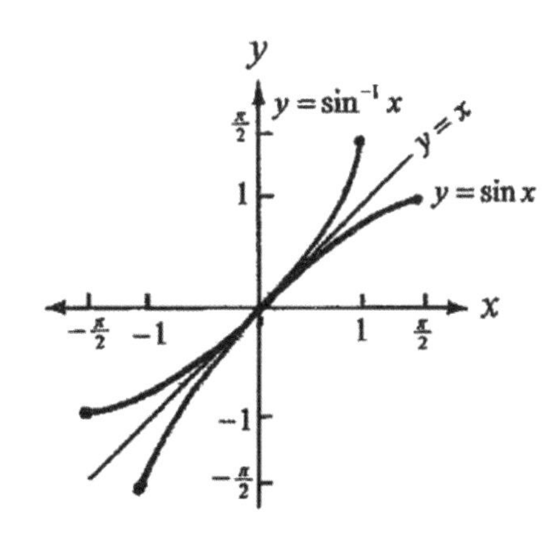

Notice that: $y = \sin^{-1} x$ and $x = \sin y$ are not inverses each other. They are only one same function in different forms. They are **equivalent.**

$y = \sin x$ and $y = \sin^{-1} x$ are inverses each other.

Similarly, an inverse function can be defined for each of the other five trigonometric functions.

$y = \cos x$ and $-1 \le y \le 1$, $0 \le x \le \pi$

$y = \cos^{-1} x$ and $-1 \le x \le 1$, $0 \le y \le \pi$

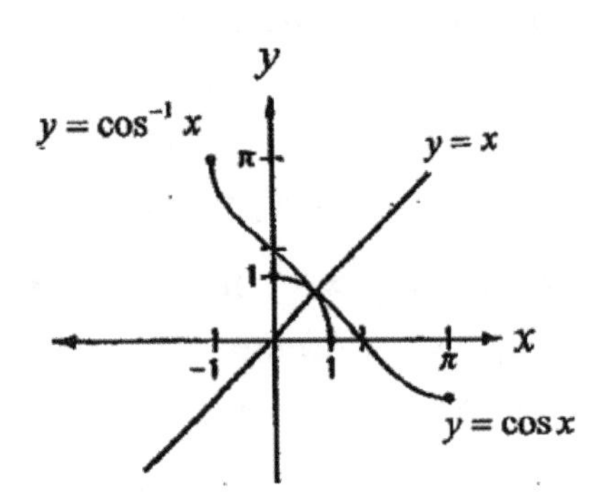

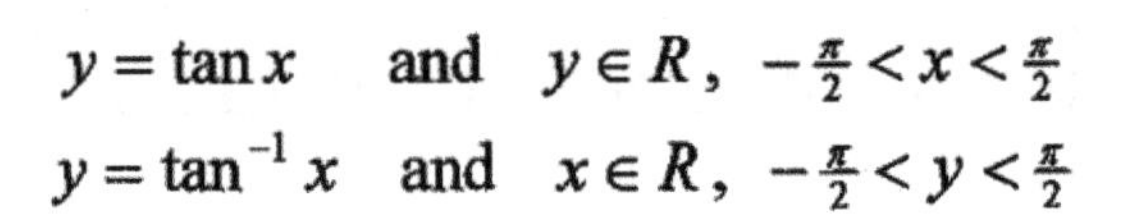

$y = \tan x$ and $y \in R$, $-\frac{\pi}{2} < x < \frac{\pi}{2}$

$y = \tan^{-1} x$ and $x \in R$, $-\frac{\pi}{2} < y < \frac{\pi}{2}$

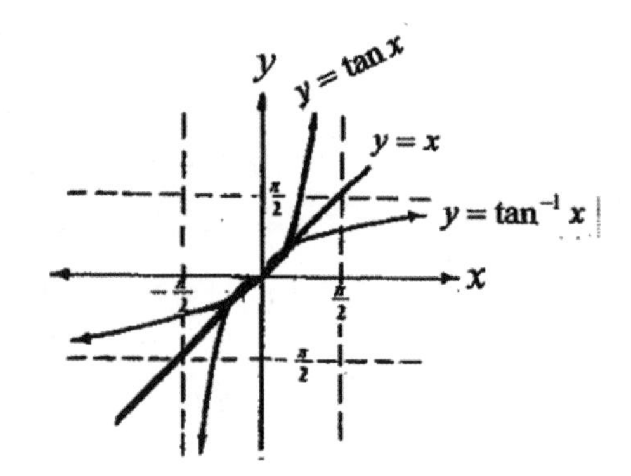

$y = \cot x$ and $y \in R$, $0 < x < \pi$

$y = \cot^{-1} x$ and $x \in R$, $0 < y < \pi$

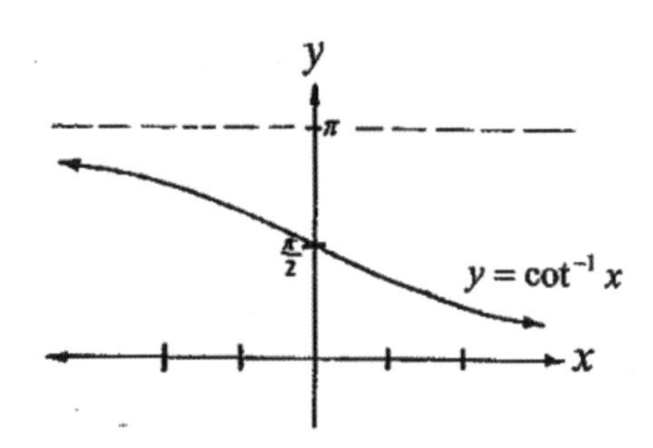

$y = \sec x$ and $|y| \ge 1$, $0 \le x \le \pi$

$y = \sec^{-1} x$ and $|x| \ge 1$, $0 \le y \le \pi$

$\sec^{-1} x \ne \frac{\pi}{2}$

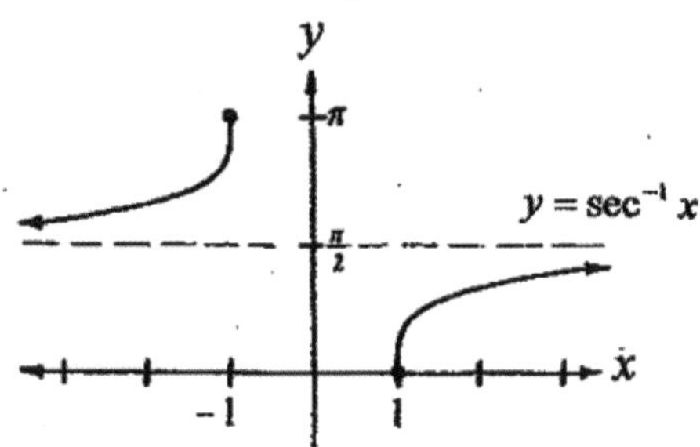

$y = \csc x$ and $|y| \ge 1$, $-\frac{\pi}{2} \le x \le \frac{\pi}{2}$

$y = \csc^{-1} x$ and $|x| \ge 1$, $-\frac{\pi}{2} \le y \le \frac{\pi}{2}$

$\csc^{-1} x \ne 0$

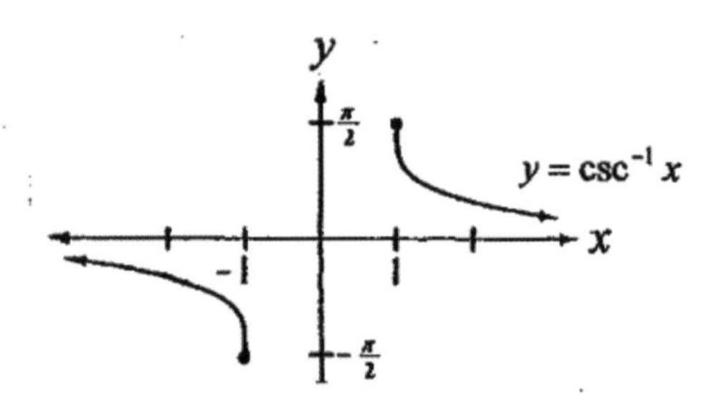

Notice that: 1. $\cot^{-1} x = \tan^{-1}\frac{1}{x}$, $\sec^{-1} x = \cos^{-1}\frac{1}{x}$, $\csc^{-1} x = \sin^{-1}\frac{1}{x}$

2. Since $f^{-1}(f(x)) = x$ and $f(f^{-1}(x)) = x$,

we have: $\sin^{-1}(\sin x) = x$ and $\sin(\sin^{-1}(x)) = x$

$\cos^{-1}(\cos x) = x$ and $\cos(\cos^{-1} x) = x$

$\tan^{-1}(\tan x) = x$ and $\tan(\tan^{-1} x) = x$

The above rules apply to other 3 trigonometric functions as well.

Examples

Find the exact value of each expression

(Hint: Exact value means value with fraction, finite decimal, radical, and π.)

1. $\sin^{-1}(\frac{1}{2})$ **2.** $\sin^{-1}(-\frac{1}{2})$ **3.** $\cos^{-1}(-\frac{\sqrt{2}}{2})$ **4.** $\tan^{-1}(-1)$

5. $\sec^{-1}2$ **6.** $\csc^{-1}(-2)$

Solution:

1. Let $y = \sin^{-1}(\frac{1}{2})$, we have

$\sin y = \frac{1}{2}$ and $-\frac{\pi}{2} \le y \le \frac{\pi}{2}$

$y = \frac{\pi}{6}$ is the only angle in the interval.

$\therefore \sin^{-1}(\frac{1}{2}) = \frac{\pi}{6}$. Ans.

2. Let $y = \sin^{-1}(-\frac{1}{2})$, we have

$\sin y = -\frac{1}{2}$ and $-\frac{\pi}{2} \le y \le \frac{\pi}{2}$

$y = -\frac{\pi}{6}$ is the only angle in the interval.

$\therefore \sin^{-1}(-\frac{1}{2}) = -\frac{\pi}{6}$. Ans.

3. Let $y = \cos^{-1}(-\frac{\sqrt{2}}{2})$, we have

$\cos y = -\frac{\sqrt{2}}{2}$ and $0 \le y \le \pi$

$y = \frac{3\pi}{4}$ is the only angle in the interval.

$\therefore \cos^{-1}(-\frac{\sqrt{2}}{2}) = \frac{3\pi}{4}$. Ans.

4. Let $y = \tan^{-1}(-1)$, we have

$\tan y = -1$ and $-\frac{\pi}{2} < y < \frac{\pi}{2}$

$y = -\frac{\pi}{4}$ is the only angle in the interval.

$\therefore \tan^{-1}(-1) = -\frac{\pi}{4}$. Ans.

5. Let $y = \sec^{-1}2$, we have

$\sec y = 2$ and $0 \le y \le \pi$

$y = \frac{\pi}{3}$ is the only angle in the interval.

$\therefore \sec^{-1}2 = \frac{\pi}{3}$. Ans.

6. Let $y = \csc^{-1}(-2)$, we have

$\csc y = -2$ and $-\frac{\pi}{2} \le y \le \frac{\pi}{2}$

$y = -\frac{\pi}{6}$ is the only angle in the interval.

$\therefore \csc^{-1}(-2) = -\frac{\pi}{6}$. Ans.

7. Find the exact value of $\cos(\sin^{-1}(\frac{3}{5}))$.

Solution:

Let $\theta = \sin^{-1}(\frac{3}{5})$, we have

$\sin\theta = \frac{3}{5}$ and $-\frac{\pi}{2} \le \theta \le \frac{\pi}{2}$

$\theta \approx 37^o$ is the only angle in the interval.

By Pythagorean Theorem, we have

$x^2 + y^2 = r^2$

$x^2 + 3^2 = 5^2$

$x^2 = 16$

$\therefore x = 4$

(Right triangle with angle θ, hypotenuse r = 5, opposite side y = 3, adjacent side x = 4.)

$\therefore \cos(\sin^{-1}(\frac{3}{5})) = \cos\theta = \frac{4}{5}$.

8. Find the exact value of $\sin(\sin^{-1}\frac{1}{2})$.

Solution:

$\sin(\sin^{-1}\frac{1}{2}) = \sin\frac{\pi}{6} = \frac{1}{2}$.

9. Find the exact value of $\sin^{-1}(\sin\frac{7\pi}{4})$.

Solution:

$\sin^{-1}(\sin\frac{7\pi}{4}) = \sin^{-1}(\sin(2\pi - \frac{\pi}{4}))$

$= \sin^{-1}(-\sin\frac{\pi}{4}) = -\frac{\pi}{4}$.

10. Find the exact value of $\cos(\cos^{-1}\pi)$.

Solution: $y = \cos^{-1}x$, $-1 \le x \le 1$

$\cos^{-1}\pi$ is not defined.

Therefore, $\cos(\cos^{-1}\pi)$ is not defined.

Examples (compare with Problems 1~6)

11. Find the approximate value of $\sin^{-1}(\frac{1}{2})$.

Solution:

Using a calculator: $\sin^{-1}(\frac{1}{2}) = 30^o \approx 0.52$ radians. Ans.

12. Find the approximate value of $\sin^{-1}(-\frac{1}{2})$.

Solution:

Using a calculator: $\sin^{-1}(-\frac{1}{2}) = -30^o \approx -0.52$ radians. Ans.

13. Find the approximate value of $\cos^{-1}(-\frac{\sqrt{2}}{2})$.

Solution:

Using a calculator: $\cos^{-1}(-\frac{\sqrt{2}}{2}) = 135^o \approx 2.36$ radians. Ans.

14. Find the approximate value of $\tan^{-1}(-1)$.

Solution:

Using a calculator: $\tan^{-1}(-1) = -45^o \approx -0.79$ radians. Ans.

15. Find the approximate value of $\sec^{-1} 2$.

Solution:

Using a calculator: $\sec^{-1} 2 = \cos^{-1}(\frac{1}{2}) = 60^o \approx 1.05$ radians. Ans.

16. Find the approximate value of $\csc^{-1}(-2)$.

Solution:

Using a calculator: $\csc^{-1}(-2) = \sin^{-1}(-\frac{1}{2}) = -30^o \approx -0.52$ radians. Ans.

17. Write the trigonometric expression as an algebraic expression.

$$\sin(\sin^{-1} x + \cos^{-1} x)$$

Solution:

The formula: $\sin(u+v) = \sin u \cos v + \cos u \sin v$

Let $u = \sin^{-1} x$ and $v = \cos^{-1} x$

$\sin u = x$ $\cos v = x$

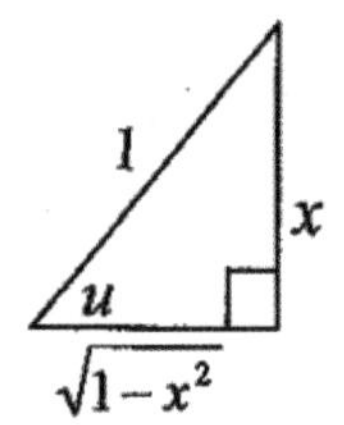

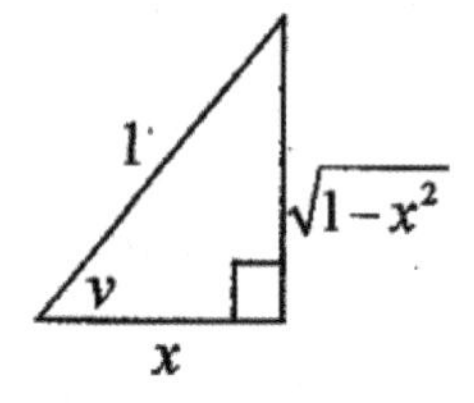

$$\sin(\sin^{-1} x + \cos^{-1} x) = \sin(\sin^{-1} x)\cos(\cos^{-1} x) + \cos(\sin^{-1} x)\sin(\cos^{-1} x)$$

$$= x \cdot x + \sqrt{1-x^2} \cdot \sqrt{1-x^2} = x^2 + (1 - x^2)$$

$$= 1.$$

SAMPLES QUESTIONS

CHAPTER 4: TRIGONOMETRY

1. $\sin x + \cos x \cot x = ?$

Solution:

$$\sin x + \cos x \cot x = \sin x + \cos x \cdot \frac{\cos x}{\sin x}$$
$$= \frac{\sin^2 x + \cos^2 x}{\sin x} = \frac{1}{\sin x} = \csc x. \text{ Ans.}$$

2. $(\sin x + \cos x)^2 + (\sin x - \cos x)^2 = ?$

Solution:

$$(\sin x + \cos x)^2 + (\sin x - \cos x)^2$$
$$= \sin^2 x + 2\sin x \cos x + \cos^2 x + \sin^2 x - 2\sin x \cos x + \cos^2 x$$
$$= (\sin^2 x + \cos^2 x) + (\sin^2 x + \cos^2 x)$$
$$= 1 + 1 = 2. \text{ Ans.}$$

3. If $\sin a^o = 0.573$, then $\sin(a^o - 13^o) = ?$

Solution:

$$\sin a^o = 0.573$$
$$a^o = \sin^{-1}(0.573) = 34.960^o$$
$$\therefore \sin(a^o - 13^o) = \sin(34.960^o - 13^o) = 0.371. \text{ Ans.}$$

4. If $m = \frac{1 + \tan x}{\sec x}$ and $n = \frac{1 - \tan x}{\sec x}$, then $m^2 + n^2 = ?$

Solution:

$$m^2 + n^2 = \frac{(1 + \tan x)^2}{\sec^2 x} + \frac{(1 - \tan x)^2}{\sec^2 x}$$
$$= \frac{1 + 2\tan x + \tan^2 x + 1 - 2\tan x + \tan^2 x}{\sec^2 x}$$

$$= \frac{2 + 2\tan^2 x}{\sec^2 x} = \frac{2(1 + \tan^2 x)}{\sec^2 x} = \frac{2\sec^2 x}{\sec^2 x} = 2. \text{ Ans.}$$

SAMPLES QUESTIONS

CHAPTER 4: TRIGONOMETRY

1. Solve $\sin^2 x + 2\sin x - 1 = 0$.

Solution:

Using Quadratic Formula:

$$\sin x = \frac{-2 \pm \sqrt{4+4}}{2} = \frac{-2 \pm \sqrt{8}}{2} = \frac{-2 \pm 2\sqrt{2}}{2} = -1 \pm \sqrt{2}$$

$\sin x = -1 + \sqrt{2}$ or $\sin x = -1 - \sqrt{2}$

$= 0.414$	Since $\lvert \sin x \rvert \le 1$,
$x = \sin^{-1}(0.414)$	$(-1 - \sqrt{2})$ is not permissible.
$= 0.427$. Ans.	No solution.

2. If $\cos(\sin^{-1} x) = \frac{4}{5}$ and $0 \le x \le \frac{\pi}{2}$, what is the value of x ?

Solution:

$$\cos(\sin^{-1} x) = \tfrac{4}{5}$$

Let $\sin^{-1} x = \theta$

$$\cos\theta = \tfrac{4}{5}$$

$$\theta = \cos^{-1}(\tfrac{4}{5}) = 0.6435^R \text{ (radian)}$$

$$\therefore \sin^{-1} x = 0.6435$$

$$x = \sin(0.6435^R) = 0.60. \text{ Ans.}$$

3. What is the graph of the parametric equation represented By $x = 2\cos\theta$ and $y = 4\sin\theta$?

Solution:

$$\frac{x}{2} = \cos\theta, \quad \frac{y}{4} = \sin\theta$$

$$\left(\frac{x}{2}\right)^2 + \left(\frac{y}{4}\right)^2 = \cos^2\theta + \sin^2\theta = 1$$

$$\frac{x^2}{4} + \frac{y^2}{16} = 1$$

It is an ellipse. Ans.

4. What is the rectangular form of the polar equation $r = 4\sin\theta$?

Solution:

$$r = 4\sin\theta$$

$$r^2 = 4r\sin\theta \text{ where } r^2 = x^2 + y^2, \quad y = r\sin\theta$$

$$x^2 + y^2 = 4y, \quad x^2 + y^2 - 4y = 0. \text{ Ans.}$$

SAMPLES QUESTIONS

CHAPTER 4: TRIGONOMETRY

1. In the triangle,

a. what is the length of side AC ?

b. what is the length of side BC ?

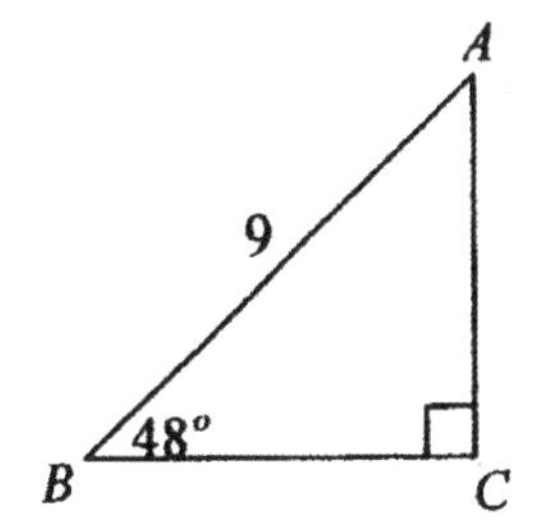

Solution:

a. $AC = 9\sin 48^o = 9(0.7431) = 6.69$. Ans.

b. $AC = 9\cos 48^o = 9(0.6691) = 6.02$. Ans.

2. In the figure, what is the measure of x ?

Solution:

Using The Law of Cosine:

$6^2 = 8^2 + 5^2 - 2\cdot 8\cdot 5\cos x$

$36 = 64 + 25 - 80\cos x$

$36 = 89 - 80\cos x$

$80\cos x = 53$

$\therefore x = \cos^{-1}(\frac{53}{68}) = 48.5^o$. Ans.

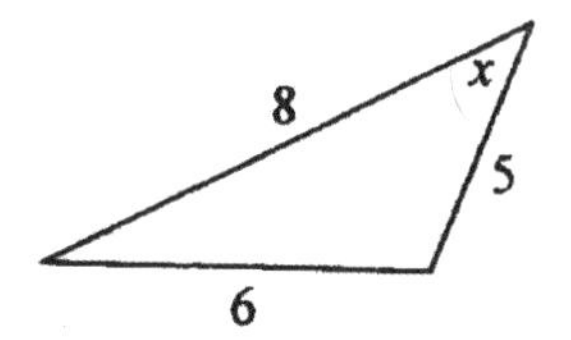

3. In the figure, what is the measure of $< C$?

Solution:

Using The Law of Sine:

$$\frac{\sin C}{11.4} = \frac{\sin 32^o}{8.2}$$

$$\sin C = \frac{11.4\sin 32^o}{8.2} = 0.7367$$

$$\therefore C = \sin^{-1}(0.7367) = 47.45^o$$

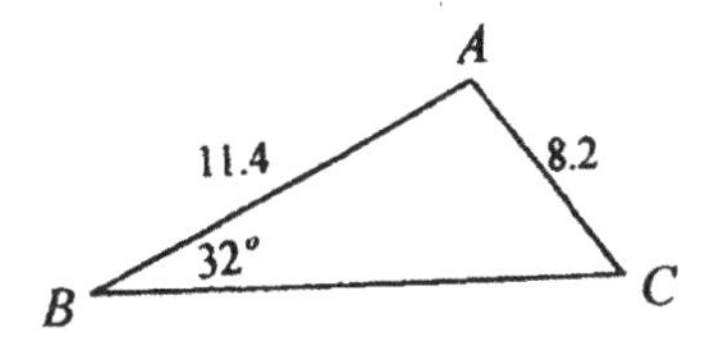

4. In the figure, what is the area ?

Solution:

Using The Area Formula:

Area $= \frac{1}{2}(14)(18)\sin 42^o = 84.3$. Ans

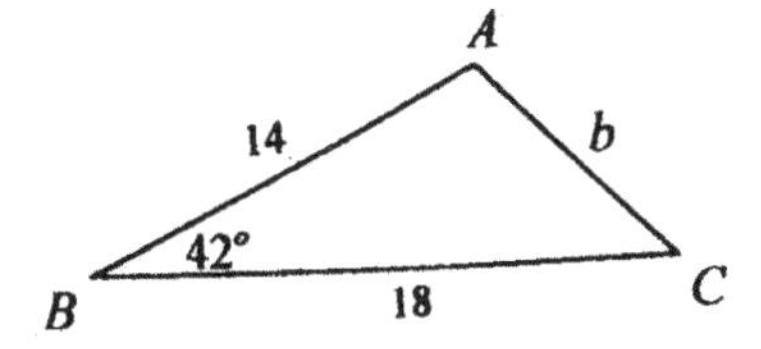

SAMPLES QUESTIONS

CHAPTER 4: TRIGONOMETRY

1. If $f(x) = \sin x$, then $f^{-1}(0.74) = ?$

Solution:

$$f(x) = \sin x$$

Let $y = f(x)$

$$y = \sin x$$

Exchange x and y:

$$x = \sin y$$

$$y = \sin^{-1} x,\quad f^{-1}(x) = \sin^{-1} x$$

$\therefore f^{-1}(0.74) = \sin^{-1}(0.74) = 0.83$ (radian). Ans.

2. If $\tan x = 100$, then $\log \sin x - \log \cos x = ?$

Solution:

$$\log \sin x - \log \cos x = \log \frac{\sin x}{\cos x} = \log \tan x = \log(100) = 2.$$ Ans.

3. The expression $z = 4(\cos 120^o + i \sin 120^o)$ is the polar form of a complex number. What is the rectangular form ?

Solution:

$$z = 4(\cos 120^o + i \sin 120^o)$$

$$= 4(-\tfrac{1}{2} + i \cdot \tfrac{\sqrt{3}}{2})$$

$$= -2 + 2\sqrt{3}\, i.$$ Ans.

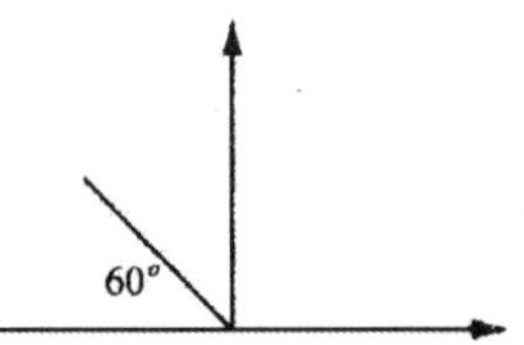

4. What is the rectangular form of the parametric equation represented by $x = 1 - \cot^2 \theta$ and $y = 2 + \csc^2 \theta$?

Solution:

$$x + y = 3 + (\csc^2 \theta - \cot^2 \theta)$$

$$= 3 + (1 + \cot^2 \theta - \cot^2 \theta)$$

$\therefore x + y = 4$. It is a straight line. Ans.

SAMPLES QUESTIONS

CHAPTER 4: TRIGONOMETRY

The Graph of $y = a\sin x + b\cos x$

Formula: $y = a\sin x + b\cos x = \sqrt{a^2 + b^2}\sin(x+\theta)$

Proof:

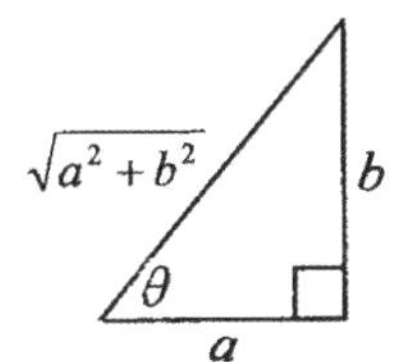

$\sin\theta = \dfrac{b}{\sqrt{a^2+b^2}}$ and $\cos\theta = \dfrac{a}{\sqrt{a^2+b^2}}$

$$y = a\sin x + b\cos x = \sqrt{a^2+b^2}\left(\frac{a}{\sqrt{a^2+b^2}}\sin x + \frac{b}{\sqrt{a^2+b^2}}\cos x\right)$$
$$= \sqrt{a^2+b^2}\left(\cos\theta\sin x + \sin\theta\cos x\right)$$
$$= \sqrt{a^2+b^2}\sin(x+\theta)$$

Examples

1. What is the amplitude of the graph of $y = \sin x + \cos x$?

   Solution:

   Amplitude: $A = \sqrt{a^2+b^2} = \sqrt{1^2+1^2} = \sqrt{2}$. Ans.

   $\sin\theta = \dfrac{1}{\sqrt{2}} = \dfrac{\sqrt{2}}{2}$, $\theta = \dfrac{\pi}{4}$

   $y = \sin x + \cos x = \sqrt{2}\sin(x + \dfrac{\pi}{4})$

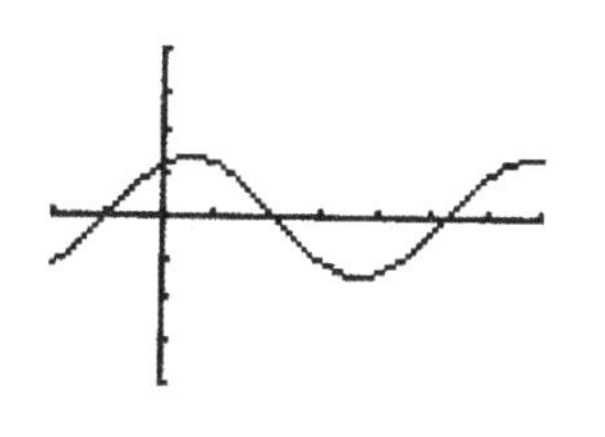

2. What is the amplitude of the graph of $y = 3\sin x + 4\cos x$?

   Solution:

   Amplitude: $A = \sqrt{a^2+b^2} = \sqrt{3^2+4^2} = 5$. Ans.

   $\sin\theta = \dfrac{4}{5}$, $\theta = 53^o$

   $y = 3\sin x + 4\cos x = 5\sin(x + 53^o)$

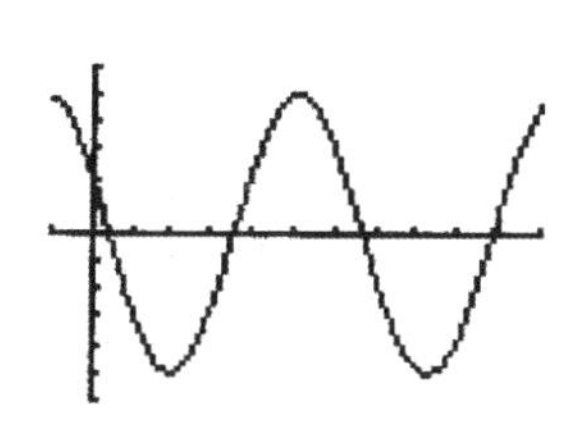

CHAPTER FIVE

Data Analysis, Statistics, and Probability

5-1 Mean, Median, Mode, and Range

In order to find a convenient way to describe and compare the information of the entire collection of data, we need to organize the data and find a single number that is most significant to represent the data. Then we can analyze and make conclusion about the data. **Statistics** is the study of collecting, organizing, and analyzing numerical data.

The **mean**, **median**, **mode**, and **range** are important statistical numbers often used in analyzing numerical data. The mean, median, and mode are three kinds of averages.

Mean: It is the sum of the numbers divided by the number of items in the data. It is the arithmetic average and is most commonly used as an average.
Median: It is the middle number in the data when the data are arranged in an increasing order (from least to greatest). For an even number of items, the average of the two numbers in the middle is the median.
Mode: It is the number that appears most often in the data. There may be no mode or more than one mode.
Range: It is the difference between the highest and the lowest numbers.

The mean, median, or mode provides the information about the **central tendency** of the data. Which of these three measures is the best to represent the central tendency depends on the way in which you need to use the data. If the mean is inflated by a few very larger or very small numbers in the data,, the median can be used as an average. Mode can be used as an average if we want the most frequent number in the data.
The range provides the information about the **spread** or **dispersion** which shows how the numbers in the data are scattered about the mean.

Examples

1. The heartbeats per minute (pulse rates) for 9 students are listed below.

66, 71, 72, 69, 70, 71, 72, 85, 72

Find: **a)** the mean **b)** the median **c)** the mode **d)** the range.
Solution:

a) Divide the sum of the numbers by the number of students.
The mean $= \frac{66+71+72+69+70+71+72+85+72}{9} = \frac{648}{9} = 72$. Ans.

b) Arrange the numbers from least to greatest.
66, 69, 70, 71, **71**, 72, 72, 72, 85. The median = 71. Ans.

c) 72 appears most often.
The mode = 72. Ans.

d) $85 - 66 = 19$
The range = 19. Ans.

Examples

2. Find the median of the data.

15, 19, 6, 17, 12, 12, 17, 11

Solution:

Arrange the number from least to greatest: 6, 11, 12, **12**, **15**, 17, 17, 19

The median $= \frac{12+15}{2} = 13.5$. Ans.

3. John's mean scores in 5 math tests is 77. His four test scores are 72, 65, 84, 79. What is his score of fifth test ?

Solution:

Total scores $= 77 \times 5 = 385$

The score of fifth test $= 385 - (72 + 65 + 84 + 79) = 385 - 300 = 85$. Ans.

4. The mean of a and b is 18. The mean of a, b, and c is 15. What is the value of c ?

Solution:

$a + b = 18 \times 2 = 36$

$a + b + c = 15 \times 3 = 45$

$c = 45 - 36 = 9$. Ans.

A lady said to her friend.

Lady: I always keep quiet whenever my husband argues with me.

Friend: How do you control your anger ?

Lady: I clean the toilet.

Friend: How does that help ?

Lady: I use his toothbrush.

5-2 Quartiles and Interquartile Ranges

Quartiles are used to measure the dispersion of the data. The quartiles are the values in the data that separate the data into four sections, each containing 25% (or $\frac{1}{4}$) of the data. To find the quartiles of the data, we use the **median** to divide the data into **two halves** between the lower extreme (the minimum) and the upper extreme (the maximum). The two extreme values are called the **outliers**.

The **lower (first) quartile** (Q_1) is the median of the **lower half**. The **second quartile** (Q_2) is the median of the **entire data**. The **upper (third) quartile** (Q_3) is the median of the **upper half**.

The **interquartile range** is the difference between Q_1 and Q_3. It represents the range containing the middle 50% (or $\frac{1}{2}$) of the data between Q_1 and Q_3.

Example:

1. Find Q_1, Q_2, Q_3, the range, and the interquartile range of the data as follows.

75, 86, 65, 95, 85, 68, 92, 90, 85, 75, 82, 80

Solution: Arrange the numbers from least to greatest.

65, 68, **75, 75**, 80, **82, 85**, 85, **86, 90**, 92, 95

$Q_2 =$ The median $= \frac{82+85}{2} = 83.5$. Ans.

The median of the numbers below 83.5 is Q_1.

$Q_1 = \frac{75+75}{2} = 75$. Ans.

The median of the numbers above 83.5 is Q_3.

$Q_3 = \frac{86+90}{2} = 88$. Ans.

The range $= 95 - 65 = 30$. Ans.

The interquartile range $= Q_3 - Q_1 = 88 - 75 = 13$. Ans.

2. The test scores of 20 students are listed below.

74 52 70 49 66 63 48 79 81 54
89 94 68 75 78 81 79 90 76 82

a. What are the quartiles Q_1, Q_2, and Q_3 of the data ?

b. What is the interquartile range ?

Solution: Arrange from least to greatest

48, 49, 52, 54, **63, 66**, 68, 70, 74, **75, 76**, 78, 79, 79, **81, 81**, 82, 89, 90, 94

a. $Q_2 =$ Median $= \frac{75+76}{2} = 75.5$. Ans.

$Q_1 = \frac{63+66}{2} = 64.5$. Ans. $Q_3 = \frac{81+81}{2} = 81$. Ans.

b. Interquartile Range $= Q_3 - Q_1 = 81 - 64.5 = 16.5$. Ans.

Notes

5-3 Graphs and Plots

In statistics, graphs and plots are often used to display and represent numerical data. Data represented in graphs and plots are easily to read and understand.
There are many different kinds of statistical graphs and plots we use in statistics, such as **bar graph, line graph, circle graph, picture graph, Stem-and-Leaf plot, Box-and Whisker plot**, and others. We can count or estimate data from graphs.

Examples

1. The bar graph below shows the population of a town between 1930 and 2000.
   1. What data unit is used on the horizontal axis ?
   2. What data unit is used on the vertical axis ?
   3. In which year did the population increase the most ?
   4. Estimate the approximate population in 1980.
   5. Find the approximate total increase in population during the years from 1930 to 2000.
   6. In which year did the population stay the same as the previous year.

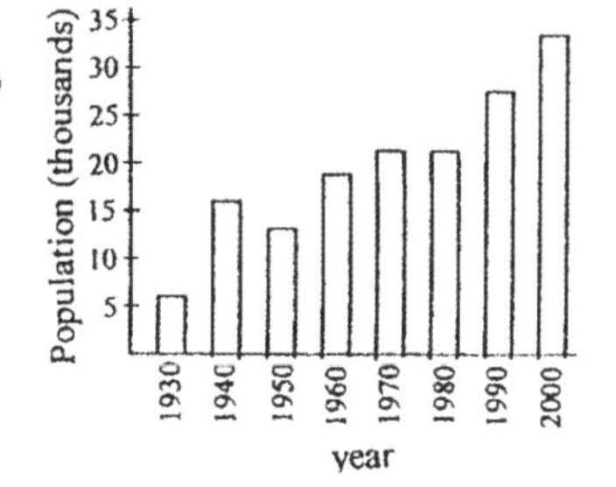

Solution: **1.** The year **2.** Thousands of persons **3.** 1940
4. 22,000 **5.** 34,000 – 6,000 = 28,000 persons **6.** 1980

2. The line graph below shows the given data of population of a town (in Thousands). Find the population of the town from 1930 to 2000.
   Solution:

1930	1940	1950	1960	1970	1980	1990	2000
6	16	13	17	22	22	28	34

(in thousands)

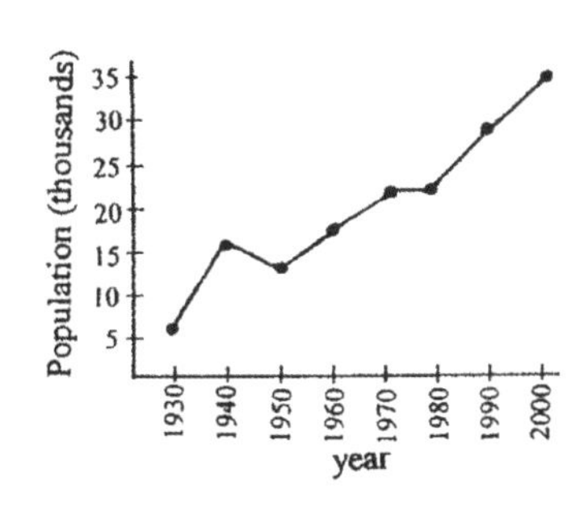

3. The circle graph below shows Ronald's budget to spend his money. His monthly income is $2,700. Find how much money does he plan to spend on food, clothes, housing, girl friends, savings and others.
   Solution:

Food: $\$2,700 \times 25\% = \$675$
Clothes: $\$2,700 \times 8\% = \$216$
Housing: $\$2,700 \times 17\% = \$459$
Girl friends: $\$2,700 \times 3\frac{2}{3}\% = \$99$
Saving: $\$2,700 \times 35\frac{1}{3}\% = \$954$
Others: $\$2,700 \times 11\% = \$297$

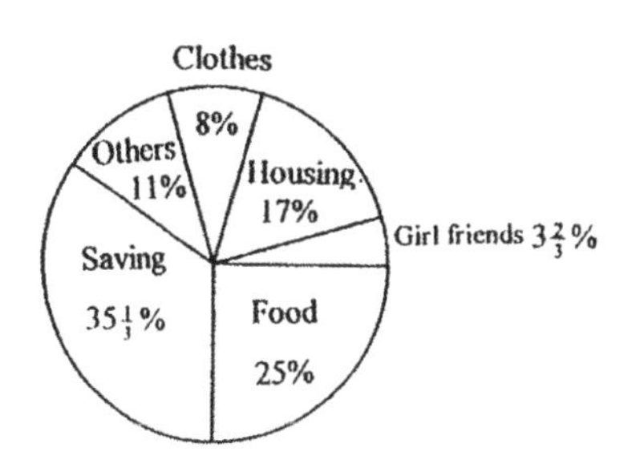

Stem-and-Leaf Plots and Box-and-Whisker Plots

A. Stem-and-Leaf Plot

It is a statistical plot used to organize numerical data. We can use the stem-and-leave plot to convert data into a bar graph or compare two sets of data.

Examples

1. Draw a stem-and-leaf plot of the data.

13, 4, 23, 34, 40, 52, 55, 52, 9, 38
60, 36, 22, 68, 15, 39, 29, 15, 41

Solution:

Identify the lowest number 4.
Identify the highest number 68.
In the plot, use the tens digits as the stems and the ones digits as the leaves.
Draw a vertical line between the stems and the leaves.
Arrange each row of leaves from smallest to largest.
A legend is used if the decimal point is not included in the stems.

Stems	Leaves
0	4, 9
1	3, 5, 5
2	2, 3, 9
3	4, 6, 8, 9
4	0, 1
5	2, 2, 5
6	0, 8

1|3 = 13

2. Draw a stem-and-leaf plot of the data below.

8.9, 4.6, 11.2, 10.2, 8.3, 8.4, 10.4, 8.8
4.2, 8.1, 10.8, 5.7, 10.4, 5.3, 11, 6.4

Solution:

Identify the lowest number 4.2.
Identify the highest number 11.2.
In the plot, use the whole number as the stems and the tenths as the leaves.
Arrange each row of leaves from smallest to largest.

Stems	Leaves
4.	2, 6
5.	3, 7
6.	4
7.	
8.	1, 3, 4, 8, 9
9.	
10.	2, 4, 4, 8
11.	0, 2

B. Box-and-Whisker Plot (See Chapter 5~3)

In a **Box-and-Whisker Plot**, the data are divided into four sections (2 whiskers and 2 boxes) between two outliers (maximum and minimum). The lines to the left and right of the two boxes are the two whickers of the plot. Each whicker and each small box contains 25% (or $\frac{1}{4}$) of the data.

Q_1 is the lower quartile, Q_2 is the second quartile (the median), and Q_3 is the upper quartile. The interquartile range contains the middle 50% of the data between Q_1 and Q_3. The data of Example 1 in Chapter 5~2 can be represented as below.

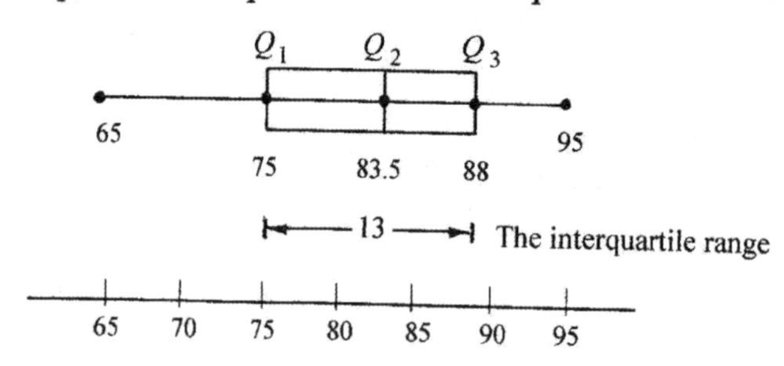

Examples

1. The box-and-whisker plot shows the scores of students on a test.

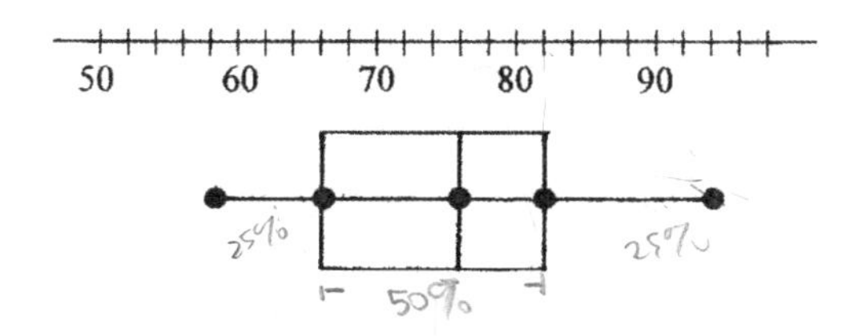

a. What is the median score ?
b. What percent of the scored less than 66 ?
c. What percent of the scored more than 82 ?
d. What is the lower and upper quartiles ?
e. What is the mean score ?
f. What is the interquartile range ?
g. How many students were tested ?

Solution:

a. 76 **b.** 25% **c.** 25%
d. lower quartile = 66, upper quartile = 82
e. The plot does not show the mean.
f. interquartile range $= 82 - 66 = 16$
g. The plot does not show the number of students.

2. The box-and-whicker plot shows the monthly sales in copies of a book.

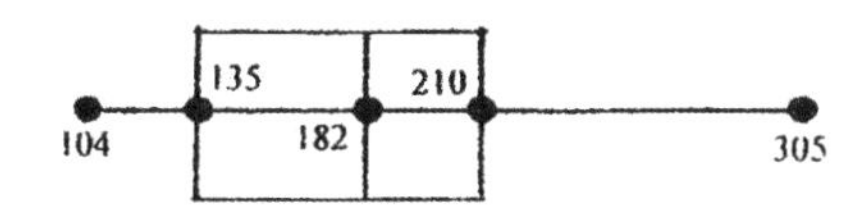

a. What is the median sale ?
b. There are four sections in the plot. Which section contains more data ?
c. Between what two reading is the middle 50% of the copies sold ?
d. What is the lower and upper quartiles from the plot

Solution:

a. 182 copies
b. None. Each section contains the same 25% of the data.
c. 135 and 210 (they are two endpoints of interquartile range.)
d. lower quartile = 135 copies, upper quartile = 210 copies

Notes

5-4 Histogram and Normal Distribution

In statistics, we use a table called **frequency distribution** to show information about the frequency of occurrences of statistical data.

Histogram is a statistical graph to describe a frequency distribution. In a histogram, data are grouped into convenient intervals. A boundary datum in a histogram is included in the interval to its left. If the midpoints at the tops of the bars are connected, a smooth "**bell-shaped**" curve called **normal curve** can result from the plotting of a larger collection of data. The area under the normal curve represents the probability from **normal distribution.** The area under the curve and above the x – axis is 1. The sum of the probabilities of all possible outcomes is 1 (or 100%).

The probability of each region under a normal curve is shown below. We can find the probability of a given region under the normal curve.

Frequency Distribution

Test scores	numbers of students
45 ~ 55	3
55 ~ 65	13
65 ~ 75	21
75 ~ 85	27
85 ~ 95	15
95 ~ 100	7

Histogram

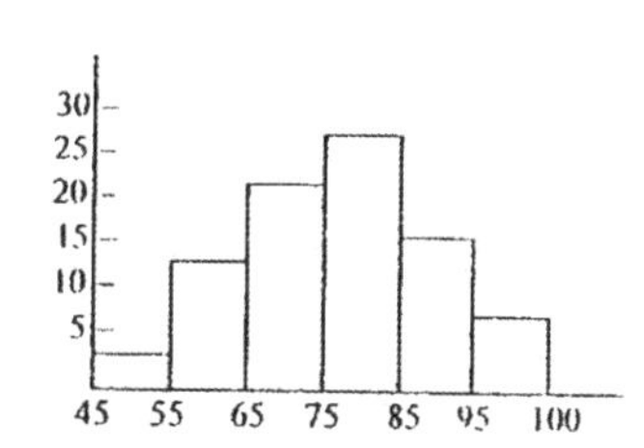

Normal Distribution

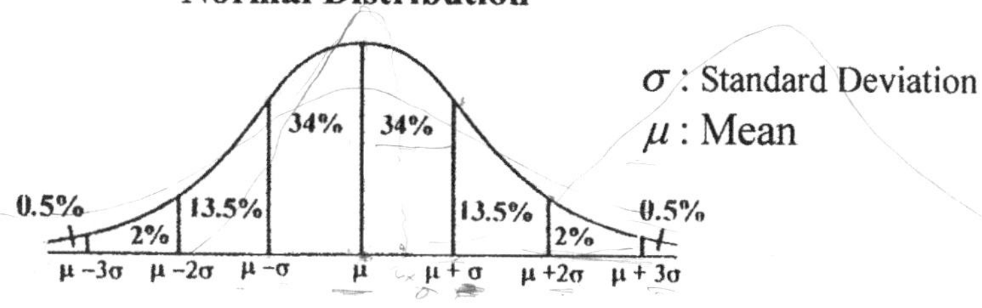

Example

1. The heights of the students in a high school are normally distributed as shown in the graph.

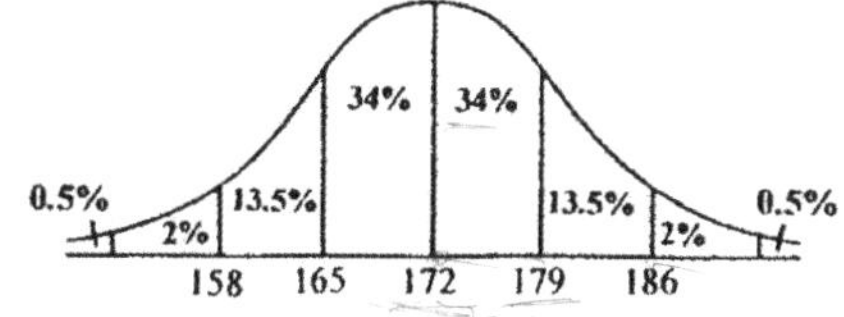

Find the percentage of the students in each the following groups.

a. Height >172 *cm* **b.** Height <172 *cm* **c.** Height >179 *cm* **d.** Height <186 *cm*

Solution:

a. 50% **b.** 50%

c. 13.5% + 2% + 0.5% = 16%

d. 50% + 34% +13.5% = 97.5%

Examples

2. The sales of a math book (in copies) each month are normally distributed as shown in the graph.

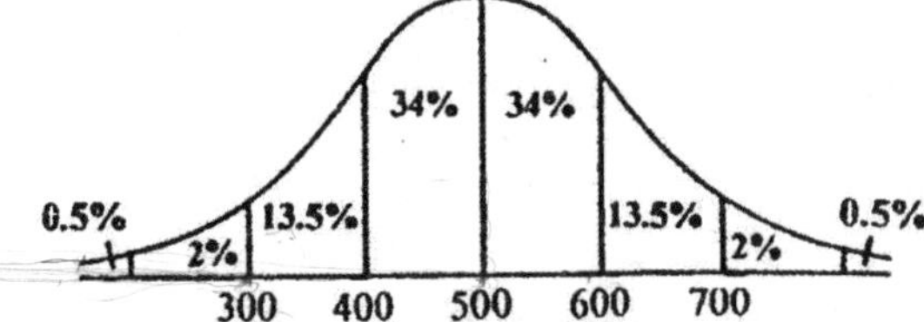

a. What is the probability that more than 600 copies are sold in a month ?
b. What is the probability that less than 300 copies are sold in a month ?

Solution:

a. 13.5% + 2% + 0.5% = 16%. Ans.
b. 2% + 0.5% = 2.5%. Ans.

3. In an experiment, the probability of operating life (in hours) of a new brand battery is normally distributed as shown in the graph.

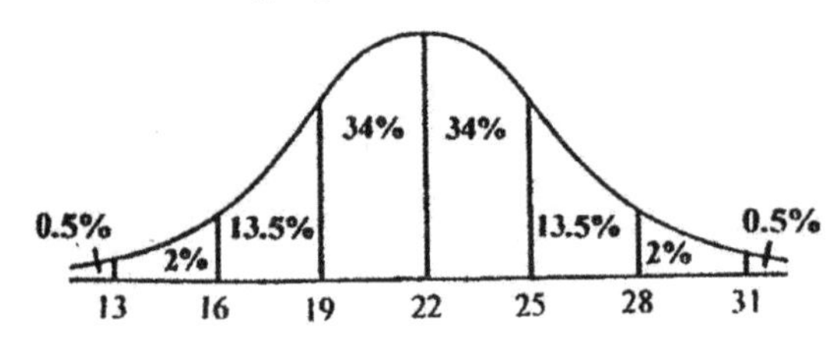

a. What is the probability that a battery will last more than 25 hours ?
b. What is the probability that a battery will last less than 25 hours ?

Solution:

a. 13.5% + 2% + 0.5% = 16%. Ans.
b. 34%+34%+13.5%+2%+0.5% = 84%. Ans.

On the airplane, a three-year old boy asks his mother.
Boy: Where are we going ?
Mother: We are going to grandma's home.
Boy: Why are so many people going to grandma's home ?

5-5 Probability

In mathematics, we want to find the possibilities of uncertainty. For example, we want to know the chance of winning a lottery, or the chance to have a life of more than 90 years. There are countless of uncertainty in our everyday life. To measure the uncertainties, we use the term "the probability". For example, the probability of tossing a fair coin turned up head is one-half, or 50%.

The main use for the theory of probability is to make decision by performing a survey or experiment that will allow us to study, predict, and make conclusions from data. In the study of probability, the various results in a random experiment are called the **outcomes**. The collection of all possible outcomes is called the **sample space**. Any subset of the sample space is called an **event**.

There are two types of probability, **theoretical probability** and **experimental probability**.

Theoretical Probability: (It is simply called the probability.)

If there are n equally likely outcomes and an event A for which there are k of these outcomes, then the probability that the event A will occur is given by:

$$P(A) = \frac{k}{n}$$

The probability of tossing a fair coin turned up head is $P(H) = \frac{1}{2}$.

The probability of tossing a fair coin turned up tail is $P(T) = \frac{1}{2}$.

The probability of rolling a fair die turned up an even number is $P(E) = \frac{3}{6} = \frac{1}{2}$.

The probability of an event must be between 0 and 1 (or 0% and 100%).

The statement $P(A) = 0$ means that event A cannot occur (0%).

The statement $P(A) = 1$ means that event A must occur (100%).

If p is the probability of an event, $1 - p$ is the probability of an event not occurring.

Experimental Probability:

In real-life situations, sometimes it is not possible to find the theoretical probability of an event. Therefore, we need to find the experimental probability by performing an experiment or a survey. We can use the experimental probability to predict future occurrences of events.

If a sufficient number of trials is conducted n times, and an event A occurs e of these times, then the experimental probability that the event A will occur in another trial is given by:

$$P(A) = \frac{e}{n}$$

A school with 2,000 students has 40 students who are left-handed this year. Then, the probability of a student chosen at random will be left-handed is $\frac{40}{2000} = \frac{1}{50}$.

If there is no other information available, we can best estimate that the probability of a student chosen at random next year will be left-handed is $\frac{1}{50}$.

Tree Diagrams

A **tree diagram** is helpful in listing all the possible outcomes (the sample space).
The sample space for tossing two coins is shown below. Use H for heads and T for tails.
There are four possible outcomes (four simple events) {(H, H), (H, T), (T, H), (T, T)}.

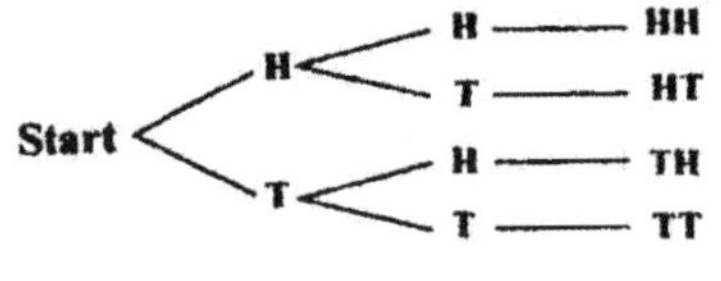

The probability of two heads is $\frac{1}{4}$.
The probability of two tails is $\frac{1}{4}$.
The probability of "one is heads and one is tails" is $\frac{1}{4}+\frac{1}{4}=\frac{1}{2}$.
The probability of "at least one is heads" is $\frac{3}{4}$.

Basic Counting Principal (Principal of Multiplication):

If we choose one from m ways, then we choose another one from n ways. Then, the number of possible choices is: $m \times n$

Examples

1. You have 5 shirts and 6 pants to choose. In how many ways you can choose, one of each kind ?
   Solution: $5 \times 6 = 30$ ways. Ans.

2. In how many different ways can a 5-question true- false test be answered if each question must be answered ?
   Solution: $2 \times 2 \times 2 \times 2 \times 2 = 32$ ways. Ans.

Independent Events: In an experiment, A and B are two independent events if the occurrence of one event does not affect the occurrence of the other.

Dependent Events: In an experiment, A and B are two dependent events if the occurrence of one event affect the occurrence of the other.

Probability of Independent Events:

If A and B are two independent events, then the probability that both A and B occur is
$P(A \cap B) = P(A) \times P(B)$. Or, $P(A \text{ and } B) = P(A) \times P(B)$.

Example 3. A bag contains 5 red balls and 4 white balls. Two balls are drawn at random from the bag. The first ball drawn is put back into the bag before the second ball is drawn. What is the probability that the two balls drawn are both red ?
Solution:
$P\text{ (red)} = \frac{5}{9}$; $P\text{ (red and red)} = \frac{5}{9} \times \frac{5}{9} = \frac{25}{81}$. Ans.

Probability of Dependent Events:

If A and B are two dependent events, then the probability that both A and B occur is $P(A \cap \text{B}) = P(A) \times P(B \mid A)$. Or, $P(A \text{ and } B) = P(A) \times P(B \mid A)$.
$P(B \mid A)$ is called the **conditional probability** of B given that A has occurred.

Example 4. A bag contains 5 red balls and 4 white balls. Two balls are drawn at random from the bag. The first ball drawn is not put back into the bag before the second ball is drawn. What is the probability that the two balls drawn are both red ?
Solution:
$P\text{ (red)} = \frac{5}{9}$; $P(\text{ red} \mid \text{red}) = \frac{4}{8}$; $P(\text{ red and red}) = \frac{5}{9} \times \frac{4}{8} = \frac{20}{72} = \frac{5}{18}$. Ans.

Mutually Exclusive Events

Mutually exclusive events: Two events which have no elements in common.

The probability of the **union** of any two events having elements in common in a sample space is: (not mutually exclusive events)

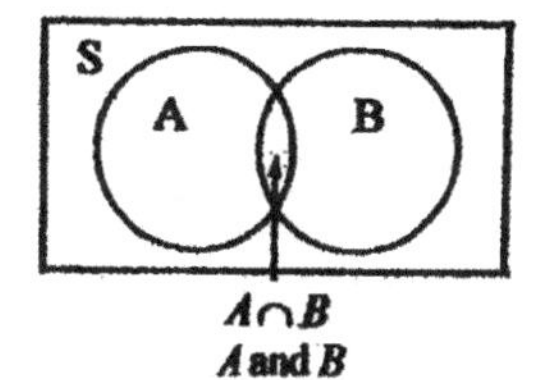

Formula:

$$P(A \cup B) = P(A) + P(B) - P(A \cap B)$$

OR: $P(A \text{ or } B) = P(A) + P(B) - P(A \text{ and } B)$

The probability of the **union of** two **mutually exclusive** events in a sample space is:
(For mutually exclusive events, $P(A \cap B) = 0$)

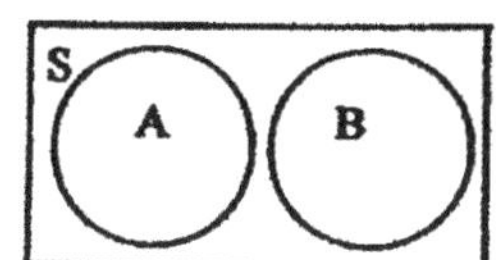

Formula:

$$P(A \cup B) = P(A) + P(B)$$

OR: $P(A \text{ or } B) = P(A) + P(B)$

Examples

1. A number is drawn at random from $\{1, 2, 3, 4, 5, 6\}$. What is the probability that either a number larger than 3 or an even number will occur ?
   Solution:

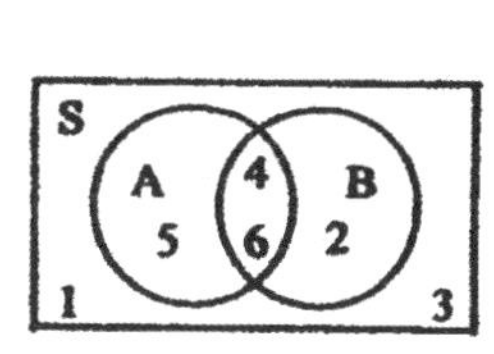

$S = \{1, 2, 3, 4, 5, 6\}$

$A = \{4, 5, 6\}$ $\quad \therefore P(A) = \frac{1}{2}$

$B = \{2, 4, 6\}$ $\quad \therefore P(B) = \frac{1}{2}$

$A \cap B = \{4, 6\}$ $\quad \therefore P(A \cap B) = \frac{2}{6} = \frac{1}{3}$

$A \cup B = \{2, 4, 5, 6\}$

$$P(A \cup B) = P(A) + P(B) - P(A \cap B) = \frac{1}{2} + \frac{1}{2} - \frac{1}{3} = \frac{2}{3}. \quad \text{Ans.}$$

2. A bag contains 5 red, 4 white and 3 black balls. One ball is drawn at random. What is the probability that it is a red or a white ball ?
   Solution: (A and B are mutually exclusive events.)
   Let A $\rightarrow$ the event that the ball is red. B $\rightarrow$ the event that the ball is white.

$$P(A \cup B) = P(A) + P(B) = \frac{5}{12} + \frac{4}{12} = \frac{9}{12} = \frac{3}{4}. \quad \text{Ans.}$$

3. A bag contains 5 red balls and 4 white balls. Two balls are drawn together at random from the bag. What is the probability that the two balls drawn are of the same color ?
   Solution: (A and B are mutually exclusive events.)
   Let A $\rightarrow$ the event that two balls are red. B $\rightarrow$ the event that two balls are white.

$$P(A) = \frac{{}_5C_2}{{}_9C_2} = \frac{10}{36} \; ; \; P(B) = \frac{{}_4C_2}{{}_9C_2} = \frac{6}{36}$$

$$P(A \cup B) = P(A) + P(B) = \frac{10}{36} + \frac{6}{36} = \frac{16}{36} = \frac{4}{9}. \quad \text{Ans.}$$

Examples

1. One card is drawn at random from a 52-card bridge deck. Find the probability of each event.
   a. It is a 10. **b.** It is a heart. **c.** It is the ace of heart. **d.** It is an ace or a face card .
   Solution:
   a. $p=\frac{4}{52}=\frac{1}{13}$. Ans. **b.** $p=\frac{13}{52}=\frac{1}{4}$. Ans. **c.** $p=\frac{1}{52}$. Ans.
   d. $p=\frac{4}{52}+\frac{12}{51}=\frac{16}{52}=\frac{4}{13}$. Ans.

2. A fair die is tossed. Find the probability of each event.
   a. It is an even number. **b.** It is greater than 4. **c.** It is greater than or equal to 2.
   Solution:
   a. $p=\frac{3}{6}=\frac{1}{2}$. Ans. **b.** $p=\frac{2}{6}=\frac{1}{3}$. Ans. **c.** $p=\frac{5}{6}$. Ans.

3. A fair coin is tossed three times, what is the probability that the coined turned up heads exactly twice ?
   Solution:
   Draw a tree diagram. (The tree diagram is left to the students)
   There are 8 possible outcomes. Three of them have heads exactly twice.
   $p=\frac{3}{8}$. Ans.
   Other Method: $p(H)=\frac{1}{2}$, $p(T)=\frac{1}{2}$; outcomes are: *HHT, HTH, THH*
   $P=\frac{1}{2}\times\frac{1}{2}\times\frac{1}{2}+\frac{1}{2}\times\frac{1}{2}\times\frac{1}{2}+\frac{1}{2}\times\frac{1}{2}\times\frac{1}{2}=\frac{1}{8}+\frac{1}{8}+\frac{1}{8}=\frac{3}{8}$. Ans.

4. Three fair coins are tossed, What is the probability that at least one turned up heads ?
   Solution:
   Draw a tree diagram. (The tree diagram is left to the students)
   There are 8 possible outcomes. Seven of them have at least one heads.
   $p=\frac{7}{8}$. Ans.
   Other method: $p(H)=\frac{1}{2}$, $P(T)=\frac{1}{2}$;
   One coin turned up head (HTT, HTH, TTH): $p=\frac{1}{8}+\frac{1}{8}+\frac{1}{8}=\frac{3}{8}$
   Two coins turned up heads (HHT, HTH, THH): $p=\frac{1}{8}+\frac{1}{8}+\frac{1}{8}=\frac{3}{8}$
   Three coins turned up heads (HHH): $p=\frac{1}{8}$

   $\therefore$ At least one turned up heads, $p=\frac{3}{8}+\frac{3}{8}+\frac{1}{8}=\frac{7}{8}$. Ans.

5. Scientists tag 100 salmons in a river. Later, they caught 50 salmons. Of these, 2 have tags. Estimate how many salmons are in the river ?
   Solution:
   $100\div\frac{2}{50}=100\times\frac{50}{2}=\frac{5000}{2}=2{,}500$ salmons.

Examples

6. Spin two spinners as shown below.

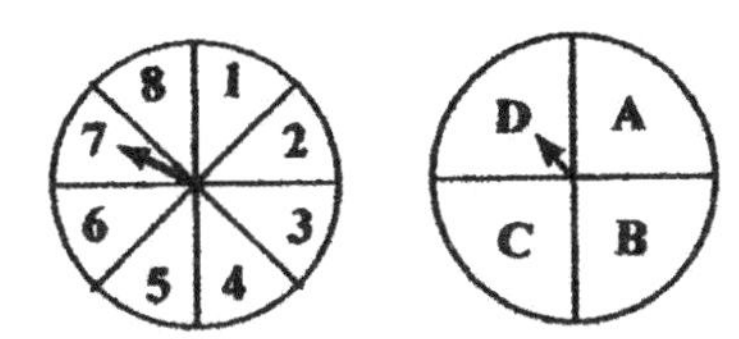

a. What is the probability of getting 7 and D ?
b. What is the probability of getting 4 or 5, followed by A ?
c. What is the probability of getting 2, followed by A and B ?

Solution:

a. $p = \frac{1}{8} \times \frac{1}{4} = \frac{1}{32}$. Ans. **b.** $p = \frac{2}{8} \times \frac{1}{4} = \frac{1}{16}$. Ans. **c.** $p = \frac{1}{8} \times \frac{2}{4} = \frac{1}{16}$. Ans.

7. A bag contains 5 red balls and 4 white balls. Three balls are drawn at random from the bag. The ball drawn is put back into the bag before next ball is drawn. What is the probability that the three balls drawn are all red ?

Solution:

$$p = \frac{5}{9} \times \frac{4}{8} \times \frac{3}{7} = \frac{5}{42}. \text{ Ans.}$$

8. The probability that you will pass the math test is $\frac{9}{15}$. What is the probability that you will not pass the next math test.

Solution:

$$p = 1 - \frac{9}{15} = \frac{6}{15} = \frac{2}{5}. \text{ Ans.}$$

9. The probability of a person being colorblind is $\frac{1}{40}$, and the probability of a person being left-handed is $\frac{1}{30}$. There are 2,400 students in a high school. How many students can we estimate that are both colorblind and left-handed ?

Solution:

$$2400 \times \frac{1}{40} \times \frac{1}{30} = \frac{2400}{1200} = 2 \text{ students. Ans.}$$

10. A and B are two independent events. $p(A) = 0.3$, $p(B) = 0.9$. Find $p(A \text{ and } B)$.

Solution:

$$p(A \text{ and } B) = p(A) \cdot p(B) = 0.3(0.9) = 0.27. \text{ Ans.}$$

11. Tell whether A and B are dependent or independent events.

$$p(A) = 0.24, \quad p(B) = 0.45, \quad p(A \cap B) = 0.036$$

Solution:

$p(A) \cdot p(B) = 0.24(0.45) = 0.036$

Given $p(A \cap B) = 0.036$

$p(A) \cdot p(B) = p(A \cap B)$ They are independent. Ans.

Examples

12. In the basketball game, the probability that you can make the basket on free throws is 0.25 (or $\frac{1}{4}$). How many times can you expect to make the basket if you make 120 free throws ?

Solution: $120 \times 0.25 = 30$ times. Ans.

13. in the basketball game, the probability that you can make the basket on free throws is 0.25 (or $\frac{1}{4}$). What is the probability that you can expect to make at least one basket in next three throws ?

Solution:

In each throw, the probability that you will miss the basket is $\frac{3}{4}$.

In three throws, the probability that you will miss the basket is $\frac{3}{4} \times \frac{3}{4} \times \frac{3}{4} = \frac{27}{64}$.

The probability that you can make at least one basket in three throws is

$$p = 1 - \frac{27}{64} = \frac{37}{64}. \text{ Ans.}$$

Other Method: $p(\text{basket}) = \frac{1}{4}$, $p(\text{missing}) = \frac{3}{4}$

Make one basket (BMM, MBM, MMB): $p = \frac{1}{4} \times \frac{3}{4} \times \frac{3}{4} + \frac{3}{4} \times \frac{1}{4} \times \frac{3}{4} + \frac{3}{4} \times \frac{3}{4} \times \frac{1}{4} = \frac{27}{64}$

Make two baskets (BBM, BMB, MBB): $p = \frac{1}{4} \times \frac{1}{4} \times \frac{3}{4} + \frac{1}{4} \times \frac{3}{4} \times \frac{1}{4} + \frac{3}{4} \times \frac{1}{4} \times \frac{1}{4} = \frac{9}{64}$

Make three baskets (BBB): $p = \frac{1}{4} \times \frac{1}{4} \times \frac{1}{4} = \frac{1}{64}$

$\therefore$ Make at least one baskets in three throws: $p = \frac{27}{64} + \frac{9}{64} + \frac{1}{64} = \frac{37}{64}$. Ans.

Using a phone, a four-year old boy calls his mother who is working in a foreign country.

Son: Mom, are you coming home today ?

Mother: No, I am very far away at the other side of the earth.

Son: Where is the other side of the earth ?

Mother: Earth is like a basketball. You are on top of the ball. I am at the bottom.

Son: It must be very hard for you with your head upside-down when you drink water.

5-6 Least-Squares Linear Regression

In real life situation, we want to organize and analyze numerical data according to time and make prediction. The data arranged with the time are called a **time series** (or **sequence**). A line of best-fit for measuring trends of data is called a **regression line** (or least-squares line). A curve of best-fit is called a **regression curve** (or least-squares curve).
In this section, we find a linear function $y = ax + b$ (or $f(x) = ax + b$) to be best-fit to represent the data. The best-fit linear function is to be as simple and as accurate as possible.

Suppose we have the time series as the points $\{(x_1, y_1), (x_2, y_2), \cdots, (x_n, y_n)\}$ in the coordinate plane. The least-squares regression line is $y = ax + b$ (or $f(x) = ax + b$).
We need to find formulas for *a* and *b*. Since the data are scattered about the regression line, we add the squares of the differences between the actually y-values and the y-values of the regression line $f(x) = ax + b$ to obtain the sum of the squared errors *S*:

$$S = \sum [y_i - f(x_i)]^2 = \sum (y_i - ax_i - b)^2$$

where $i = 1, 2, \cdots, n$

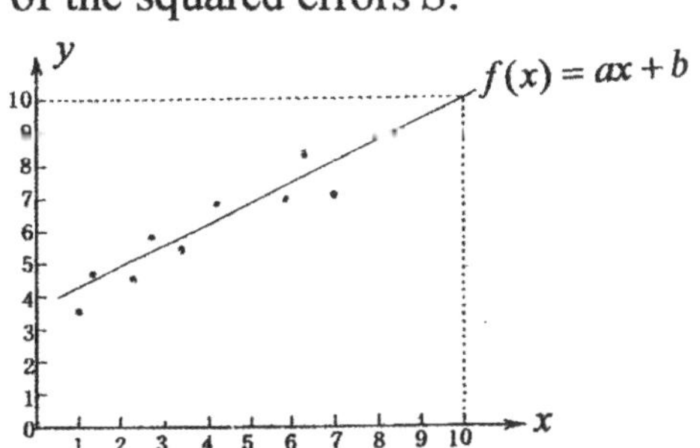

$[y_i - f(x_i)]$ is the vertical distances between the y-values of the regression line and the given points in the plane. If the regression line is perfect, then $S = 0$.
In advanced mathematics, we can prove the following theorem by the method of **partial derivatives in calculus** ($\frac{\partial S}{\partial a} = 0$ and $\frac{\partial S}{\partial b} = 0$) to find *a* and *b* under minimum of S.

Theorem for Least-Squares Regression Line

The least-squares regression line for $\{(x_1, y_1), (x_2, y_2), \cdots, (x_n, y_n)\}$ is given by $f(x) = ax + b$, where

$$a = \frac{n\sum xy - \sum x \sum y}{n\sum x^2 - \left(\sum x\right)^2} \quad \text{and} \quad b = \frac{\sum y \sum x^2 - \sum x \sum xy}{n\sum x^2 - \left(\sum x\right)^2} = \frac{\sum y}{n} - a\left(\frac{\sum x}{n}\right)$$

To simplify the above formula, we could let the median of the x-values be represented by 0. Then $\sum x_i = 0$ and the formula for *a* and *b* simplify to:

$$a = \frac{\sum xy}{\sum x^2} \quad \text{and} \quad b = \frac{\sum y}{n}$$

To learn the proof of the above theorem, read the book " Mathematics for Management, written by Rong Yang, 1986 ".
Fortunately, we have already developed the computer program based on the above formulas to find the regression line directly and simple with a graphing calculator without the tedious process by using the formulas.

Example

The profits of a company are shown in the table below for the selected years. Assume the least-squares linear regression is used to represent the data (in millions of dollars),

a. find the least-squares linear regression for the data.

b. find the profit of the company at 15th year.

Years	1	2	3	4	5	6	7
Profit	−4	−2	1	2	5	8	9

(in millions of dollars)

Solution:

Let $x =$ the years, $y =$ the profits

a. Find the least-squares linear regression for the data by a graphing calculator. (TI-83 or TI-84)

Enter the date: Press **STAT → 1:EDIT → ENTER**

The screen shows:

L1	L2
1	−4
2	−2
3	1
4	2
5	5
6	8
7	9

Find the linear regression:

Press: **STAT → CALC → LinReg(** $ax+b$ **) →ENTER**

The screen shows: **LinReg (** $ax+b$ **)**

Select L1 as x-list and L2 as y-list:

Press: **2nd → L1 → , → 2nd → L2 → ,**

The screen shows: **LinReg (** $ax+b$ **) L1, L2,**

Press: **VARS → Y-VARS → ENTER → ENTER**

The screen shows: **LinReg L1, L2, Y1**

Press: **ENTER**

The screen shows the linear regression equation:

LinReg

$y = ax + b$

$a = 2.25$

$b = -6.28571486$

b. Evaluate y-value at $x = 15$.

Press: **2nd → CALC → 1:Value → ENTER**

Adjust the viewing window if screen shows **ERR: INVALID**:

$x_{min} = 0$, $x_{max} = 20$, $x_{scl} = 1$

$y_{min} = -5$, $y_{max} = 30$, $y_{scl} = 1$

Type: $x = 15$ → **ENTER**

The screen shows:

$y = 27.464286$

(millions of dollars)

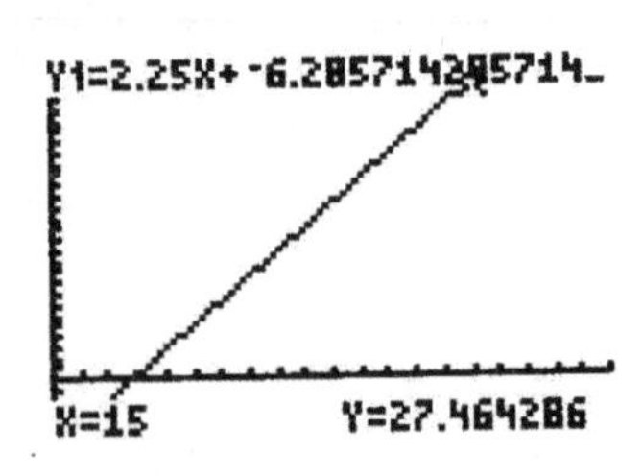

5-7 Standard Deviation

In real life situation, we want to organize and analyze numerical data and make prediction. In statistics, we use statistical (numerical) values to summarize and compare data. The **mean**, **median**, or **mode** provides the measures of **central tendency** of the data. The **range** provides the measure of **spread** or **dispersion** of the data.

Another measure of spread or dispersion is the **standard deviation** of the data.

To find the standard deviation of the data, we need the measures of mean, deviation, and variation.

Mean is the arithmetic average of the data. It is denoted by M, μ, or $\bar{x}$.
Deviation is the difference of each datum from the mean.
Variation is the dispersion of the data. It is denoted by v.

The variation shows how the data are scattered about the mean. It is computed by squaring each deviation from the mean, adding those squares, and dividing their sum by the number of the entries.

Variation: $v = \dfrac{\sum (x_i - \bar{x})^2}{n}$ where $i = 1, 2, \cdots, n$

Standard Deviation of a Set of Data (It is denoted by σ)
The standard deviation of a set of data is the square root of the variance:

Standard Deviation: $\sigma = \sqrt{\dfrac{\sum (x_i - \bar{x})^2}{n}} = \sqrt{\dfrac{(x_1 - \bar{x})^2 + (x_2 - \bar{x})^2 + \cdots + (x_n - \bar{x})^2}{n}}$

where x_1, x_2, $\cdots$, x_n are each term of the data, and $\bar{x}$ is the mean.

To compare two sets of data, the data set with greater standard deviation are more spread out about the mean.

Example

The scores of 5 games of a high school sport teams are 8, 15, 15, 31, 41.
Find the standard deviation.
Solution:

The mean $\bar{x} = \dfrac{8 + 15 + 15 + 31 + 41}{5} = 22$

The standard deviation $\sigma = \sqrt{\dfrac{(8-22)^2 + (15-22)^2 + (15-22)^2 + (31-22)^2 + (41-22)^2}{5}}$

$= \sqrt{\dfrac{736}{5}} = \sqrt{147.2} \approx 12.13$. Ans.

The Properties of Normal Distribution

In Chapter 5-4, we have learned the concepts of **histogram** and **normal distribution.** The graph of normal distribution is a smooth bell-shaped curve from the plotting of a large collecting of data. The shape of the curve indicates that the frequency of occurrences of data are concentrated around the center (the mean).

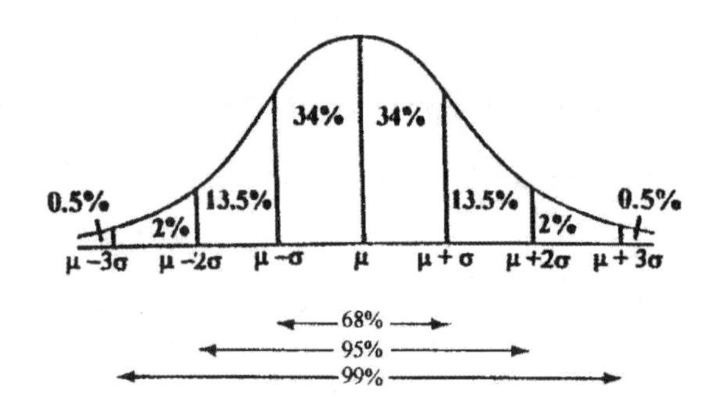

$$y = \frac{1}{\sigma\sqrt{2\pi}} e^{-(x-\mu)^2/2\sigma^2}$$

Standard deviation: σ
Mean: μ
$e = 2.71828\cdots\cdots$
Area = 1

Properties of a Normal Distribution

1. The data are symmetrical about the mean.
2. The mean and median are about equal.
3. About **68%** of the values are within one standard deviation from the mean.
4. About **95%** of the values are within two standard deviations from the mean.
5. About **99%** of the values are within three standard deviations from the mean.

Example

The mean of the heights of the 3,000 students in a high school was found to be 172 *cm* with a standard deviation of 7 *cm*.

1. Draw a normal curve showing the heights at one, two, and three standard deviations from the mean.
2. Find the percentage of students whose heights are between 179 and 193 *cm*.
3. How many students whose heights are between 165 and 179 *cm* ?
4. How many students whose heights are less than 158 *cm* ?

Solution:

1.

$\mu \pm \sigma = 172 \pm 7 \quad = 179$ and 165

$\mu \pm 2\sigma = 172 \pm 2(7) = 186$ and 158

$\mu \pm 3\sigma = 172 \pm 3(7) = 193$ and 151

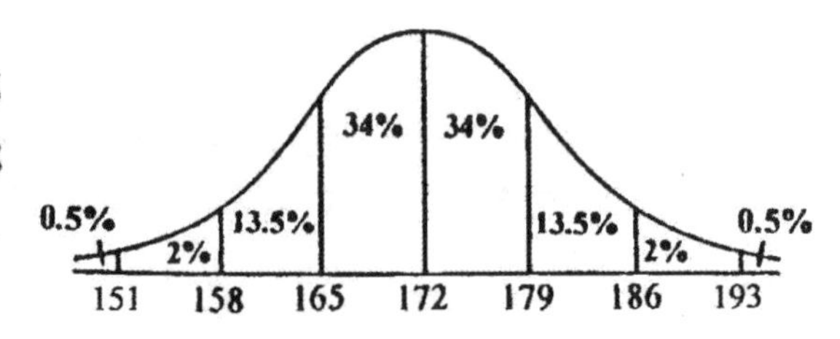

2. $13.5\% + 2\% = 15.5\%$. Ans.
3. $3{,}000 \times 68\% = 2040$ students. Ans.
4. $3{,}000 \times 2.5\% = 75$ students. Ans.

5-8 Least-Squares Quadratic Regression

In Chapter 5-6, we have learned to find a best-fit linear function to represent numerical data.

In this section, we will find a **least-squares quadratic regression** $f(x) = ax^2 + bx + c$.

In advanced mathematics, we can find the following formulas for *a*, *b*, and *c* by taking the partial derivatives in calculus.

Least-squares quadratic regression: $f(x) = ax^2 + bx + c$

$$a = \frac{n\sum x^2 y - \sum y \sum x^2}{n\sum x^4 - \left(\sum x^2\right)^2}, \quad b = \frac{\sum xy}{\sum x^2}, \quad c = \frac{\sum x^4 \sum y - \sum x^2 \sum x^2 y}{n\sum x^4 - \left(\sum x^2\right)^2}$$

To learn the proof of the above formulas, read the book "Mathematics for Management, written by Rong Yang, 1986".

Fortunately, we have already developed the computer program based on the above formulas to find the quadratic regression directly and simple with a graphing calculator without the tedious process by using the formulas.

Example: Use a graphing calculator to find the least-squares quadratic regression of the following data and evaluate $f(x)$ at $x = 11$.

x	1	2	3	4	5	6	7
$f(x)$	7	2	−1	−2	−1	2	7

Solution: **a.** Find the least-squares quadratic regression for the data by a graphing calculator. (TI-83 or TI-84)

Enter the data: Press **STAT → 1:EDIT → ENTER**

The screen shows:

L1	L2
1	7
2	2
3	−1
4	−2
5	−1
6	2
7	7

Find the quadratic regression:
Press: **STAT → CALC → 5:QuadReg → ENTER**
The screen shows: **QuadReg**
Select L1 as x list and L2 as y list:
Press: **2nd → L1 → , → 2nd → L2 → ,**
The screen shows: **QuadReg L1 , L2 ,**
Press: **VARS → Y-VARS → ENTER → ENTER**
The screen shows: **Quadreg L1 , L2 , Y1**
Press: **ENTER**

The screen shows the exponential regression equation:

QuadReg $y = ax^2 + bx + c \quad a = 1 \quad b = -8 \quad c = 14$

b. Evaluate $f(x)$ at $x = 11$.

Press: **2nd → CALC → 1:value → ENTER**

Adjust the viewing window if screen shows **ERR:INVALID**:

$x_{min} = -5, \quad x_{max} = 15, \quad x_{scl} = 1$

$y_{min} = -5, \quad y_{max} = 50, \quad x_{scl} = 1$

Type: $x = 11$ → **ENTER** The screen shows: $y = 47$

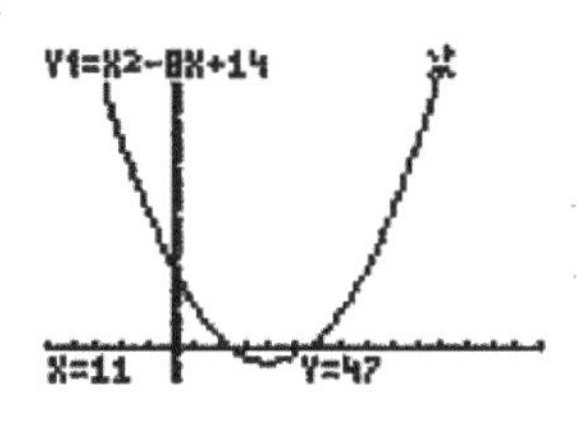

NOTES

5-9 Least-Squares Exponential Regression

In Chapter 5-6 and 5~8, we have learned to find a best-fit linear and exponential functions to represent numerical data.

In this section, we will find a **least-squares exponential regression** $f(x) = ab^x$.

Example

The table below shows the raw materials sold in tons of a company for selected years. Assume the least-squares exponential regression is used to represent the data.

a. Find the least-squares exponential regression for the data.

b. Predict the raw materials sold at 13rd year.

Years	1	2	3	4	5	6	7	8	9
Materials	12.6	10.6	8.9	7.5	6.3	5.3	4.4	3.7	3.1

(in tons)

Solution:

Let $x =$ the years, $y =$ the materials

a. Find the least-squares exponential regression for the data by a graphing calculator. (TI-83 or TI-84)

Enter the data: Press **STAT → 1:EDIT → ENTER**

The screen shows:

L1	L2
1	12.6
2	10.6
3	8.9
4	7.5
5	6.3
6	5.3
7	4.4
8	3.7
9	3.1

Find the exponential regression:

Press: **STAT → CALC → 0:ExReg →ENTER**

The screen shows: **ExpReg**

Select L1 as x list and L2 as y list:

Press: **2nd → L1 → , → 2nd → L2 → ,**

The screen shows: **ExpReg L1 , L2 ,**

Press: **VARS → Y-VARS → ENTER → ENTER**

The screen shows: **ExpReg L1 , L2 , Y1**

Press: **ENTER**

The screen shows the exponential regression equation:

ExpReg

$y = a \times b^\wedge x$ $a = 15.074808$ $b = 0.8391376398$

b. Evaluate y-value at $x = 13$.

Press: **2nd → CALC → 1:value → ENTER**

Adjust the viewing window if screen shows **ERR:INVALID**:

$x_{\min} = 0$, $x_{\max} = 20$, $x_{scl} = 1$

$y_{\min} = -5$, $y_{\max} = 20$, $x_{scl} = 1$

Type: $x = 13$ **→ ENTER**

The screen shows:

$y = 1.5419966$ (tons)

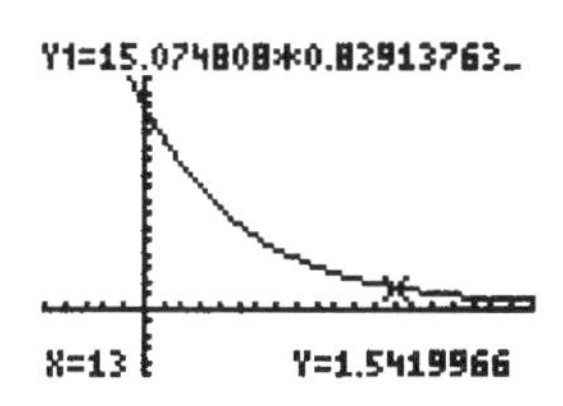

Example

The table below shows the revenue of a company for the selected years. Assume the least-square exponential regression is used to represent the data. Predict the revenue for the company at 17 years.

Years	0	2	5	8
Revenues	5	6	13	24

(in millions of dollars)

Solution:

Let $x =$ the years, $y =$ the revenue

a. Find the least-squares exponential regression for the data by a graphing calculator. (TI-83 or TI-84)

Enter the data: Press **STAT → 1:EDIT → ENTER**

The screen shows:

L1	L2
0	5
2	6
5	13
8	24

Find the exponential regression:
Press: **STAT → CALC → 0:ExReg →ENTER**
The screen shows: **ExpReg**
Select L1 as x list and L2 as y list:
Press: **2nd → L1 → , → 2nd → L2 → ,**
The screen shows: **ExpReg L1 , L2 ,**
Press: **VARS → Y-VARS → ENTER → ENTER**
The screen shows: **ExpReg L1 , L2 , Y1**
Press: **ENTER**

The screen shows the exponential regression equation:

ExpReg

$y = a \times b\^{}x \qquad a = 4.55608634 \qquad b = 1.2277988$

b. Evaluate y-value at $x = 17$.

Press: **2nd → CALC → 1:value → ENTER**

Adjust the viewing window if screen shows **ERR:INVALID**:

$x_{min} = 0, \quad x_{max} = 20, \quad x_{scl} = 1$

$y_{min} = -5, \quad y_{max} = 200, \quad x_{scl} = 1$

Type: $x = 17$ → **ENTER**

The screen shows:

$y = 148.19505 \approx 148$ (millions of dollars)

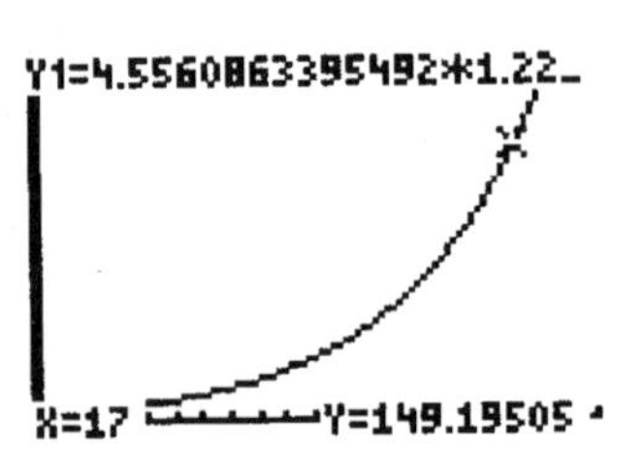

SAMPLES QUESTIONS

CHAPTER 5: DATA ANALYSIS, STATISTICS, and PROBABILITY

1. The average height of 12 students is 172 *cm*. The average height of 4 other students is 164 *cm*. What is the average height in *cm* of all 16 students ?
Solution:

$$\text{Average} = \frac{172 \times 12 + 164 \times 4}{16} = \frac{2064 + 656}{16} = 170. \text{ Ans.}$$

2. If the average (arithmetic mean) of 4, 8, 12, 20, and a is 18, what is the value of a ?
Solution:

$$\frac{4 + 8 + 12 + 20 + a}{5} = 18$$

$$\frac{44 + a}{5} = 18$$

$$44 + a = 90$$

$$\therefore a = 90 - 44 = 46. \text{ Ans.}$$

3.. The mean of 7 consecutive even integers is n. What is the median of these 7 integers ?
Solution:

a, $a + 2$, $a + 4$, $a + 6$, $a + 8$, $a + 10$, $a + 12$
The median is $(a + 6)$.
Write the median in terms of n :

$$\text{median } n = \frac{7a + 42}{7} = a + 6. \text{ Ans.}$$

4. The average of Roger's scores (arithmetic mean) for 8 tests is 76. What does the average score of the next 2 tests have to be if the average of the entire 10 tests equals 80 ?
Solution:

Let x = the average score of the next 2 tests
The average score of the entire 10 tests is:

$$\frac{76 \times 8 + (x \times 2)}{10} = 80$$

$$608 + 2x = 800$$

$$2x = 192$$

$$x = 96. \text{ Ans.}$$

SAMPLES QUESTIONS

CHAPTER 5: DATA ANALYSIS, STATISTICS, and PROBABILITY

1. A school with 3,500 students has 70 students who are left-handed this year. What is the probability of a student chosen at random will be left-handed ?
Solution:

$$p=\frac{70}{3500}=\frac{1}{50}. \text{ Ans.}$$

2. In the figure, a small triangle is placed in the center of a large rectangular. The small triangle has a base 3 and height 2. The large rectangle has a length 8 and width 5. If you throw a dart at random to hit the large rectangle, what is the probability that the dart will hit the small triangle ?
Solution:

$$p=\frac{\frac{1}{2}\times 3\times 2}{5\times 8}=\frac{3}{40}. \text{ Ans.}$$

8
5
2
3

3. A bag contains 3 red balls and 12 white balls. Two balls are drawn at random from the bag with replacement. What is the probability that the two balls drawn are both red ?
Solution:

$$p=\frac{3}{15}\times\frac{3}{15}=\frac{1}{25}. \text{ Ans.}$$

4. A bag contains 3 red balls and 12 white balls. Two balls are drawn at random from the bag without replacement. What is the probability that the two balls drawn are both red ?
Solution:

$$p=\frac{3}{15}\times\frac{2}{14}=\frac{1}{35}. \text{ Ans.}$$

5. A box contains 4 red balls, 8 white balls, and 6 green balls. What is the probability of drawing a red ball and then drawing another green ball from remaining balls ? (without replacement)
Solution:

$$p=\frac{4}{18}\times\frac{6}{17}=\frac{4}{51}. \text{ Ans.}$$

6. A box contains 4 red balls, 8 white balls, and 6 green balls. What is the probability of drawing a red ball and then a green ball if the ball is put back after the first draw ? (with replacement)
Solution:

$$p=\frac{4}{18}\times\frac{6}{18}=\frac{2}{27}. \text{ Ans.}$$

SAMPLES QUESTIONS

CHAPTER 5: DATA ANALYSIS, STATISTICS, and PROBABILITY

1. In the data below, what is the interquartile range ?

48, 49, 52, 54, 63, 66, 68, 70, 74, 75

Solution:

The median $= \dfrac{63+66}{2} = 64.5$

$Q_1 = 52$ (first quartile)

$Q_3 = 70$ (third quartile)

The interquartile range $= Q_3 - Q_1 = 70 - 52 = 18$. Ans.

2. In the figure, a small circle is placed in the center of a larger circle. The small circle has a radius 2 and the larger circle has a radius 6. If you throw a dart at random to hit the larger circle, what is the probability that the dart will hit the small circle ?

Solution:

$$p = \frac{\pi \cdot (2)^2}{\pi \cdot (6)^2} = \frac{4}{36} = \frac{1}{9}.$$ Ans.

3. The probability of a person being left-handed is $\frac{1}{30}$. The probability of a person being colorblind is $\frac{1}{50}$. There are 4,500 students in a high school. How many students would probably be both left-handed and colorblind ?

Solution:

$$n = 4{,}500 \times \frac{1}{30} \times \frac{1}{50} = \frac{4{,}500}{1{,}500} = 3$$ students. Ans.

4. A player spins the spinner below twice. What is the probability of getting a sum of 5 ?

Solution:

Using a tree diagram (see Page 292), we have the following result of getting a sum of 5:

1st spin	2nd spin
1	4
2	3
3	2
4	1

$$p = \frac{4}{5 \times 5} = \frac{4}{25}$$ or 0.16 . Ans.

SAMPLES QUESTIONS

CHAPTER 5: DATA ANALYSIS, STATISTICS, and PROBABILITY

1. A player spins the spinner below twice. What is the probability of getting a sum of 5 and 10 ?

Solution:

Using a tree diagram (Read Page 292), we have the following result of getting a sum of 5 or 10:

1st spin	2nd spin
1	4
2	3
3	2
4	1
5	5

$p = \dfrac{5}{5 \times 5} = \dfrac{1}{5}$. Ans.

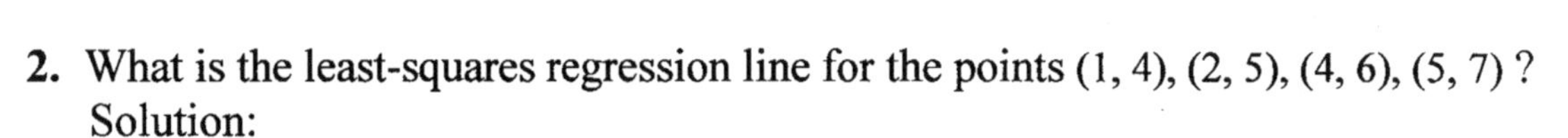

2. What is the least-squares regression line for the points (1, 4), (2, 5), (4, 6), (5, 7) ?

Solution:

Method 1: Using the formula (Read Page 297)

	x	y	xy	x^2
	1	4	4	1
	2	5	10	4
	4	6	24	16
	5	7	35	25
Σ	12	22	73	46

$$a = \frac{n\sum xy - \left(\sum x\right)\left(\sum y\right)}{n\sum x^2 - \left(\sum x\right)^2} = \frac{4(73) - (12)(22)}{4(46) - (12)^2} = \frac{292 - 264}{184 - 144} = \frac{28}{40} = 0.7$$

$$b = \frac{\sum y}{n} - a\left(\frac{\sum x}{n}\right) = \frac{22}{4} - (0.7) \cdot \frac{12}{4} = 5.5 - 2.1 = 3.4$$

The least-square regression line for the given points is $y = 0.7x + 3.4$. Ans.

Method 2: Using a graphing calculator

(Read the Example on Page 298 and try to solve it yourself.)

Notes

Notes

PRACTICE TEST 1
(PSAT / NMSQT)

SECTION 1
Time — 25 minutes
20 Questions

Direction: For this section, solve each problem and decide which is the best of the choices given. Fill in the corresponding circle on the answer sheet. You may use any available space for scratch work.

Notes:
1. The use of a calculator is permitted.
2. All numbers used are real numbers.
3. Figures that accompany problems in this test are intended to provide information useful in solving the problems. They are drawn as accurately as possible EXCEPT when it is stated in a specific problem that the figure is not drawn to scale. All figures lie in a plane unless otherwise indicated.
4. Unless otherwise specified, the domain of any function f is to be the set of all real number x for which $f(x)$ is a real number.

Reference Information

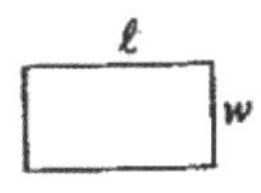

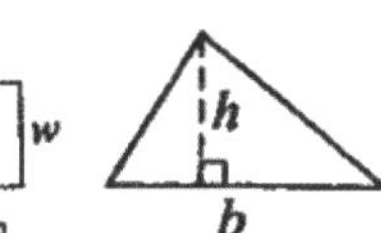

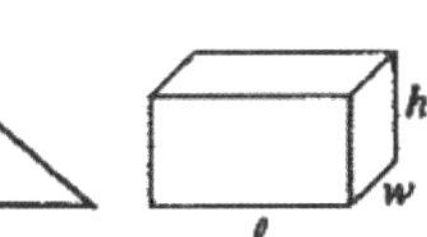

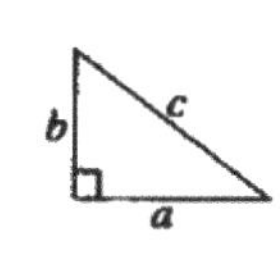

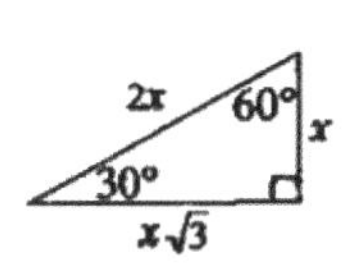

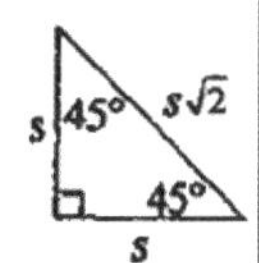

$A = \pi r^2$ $\quad C = 2\pi r$ $\quad A = \ell w$ $\quad A = \frac{1}{2}bh$ $\quad V = \ell wh$ $\quad V = \pi r^2 h$ $\quad c^2 = a^2 + b^2$ $\quad$ Special Right Triangles

The number of degrees of arc in a circle is 360.
The sum of the measures in degrees of the angles of a triangle is 180.

1. If $2x + 3y = 12$, then $8x + 12y - 3 =$

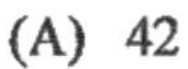

(A) 42
(B) 43
(C) 44
(D) 45
(E) 46

2. If $3^{2x} = 81$, then $x =$

(A) 1
(B) 2
(C) 3
(D) 6
(E) 9

4x° 40°

3. In the figure above, what is the value of x ?

(A) 20
(B) 25
(C) 30
(D) 35
(E) 40

GO ON TO THE NEXT PAGE

4. A 5,000 cubic inches of water is poured into a rectangular container with base 10 inches by 25 inches. What is the height in inches of the water ?

(A) 10
(B) 15
(C) 20
(D) 25
(E) 40

5. In a food fair, John bought 20 pizzas at a price of 2 for $10. He cut each pizza into 8 slices and sold for $1.50 per slice. He sold 16 pizzas. What is his profit ?

(A) $84
(B) $92
(C) $98
(D) $100
(E) $110

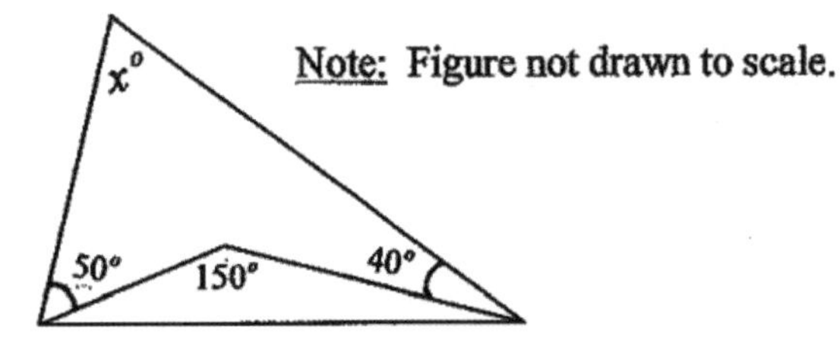

Note: Figure not drawn to scale.

6. In the figure above, what is the value of x ?

(A) 60
(B) 75
(C) 70
(D) 75
(E) 80

7. Each of the 7 basketball teams plays each of the other teams just once. How many games will there be ?

(A) 17
(B) 18
(C) 19
(D) 20
(E) 21

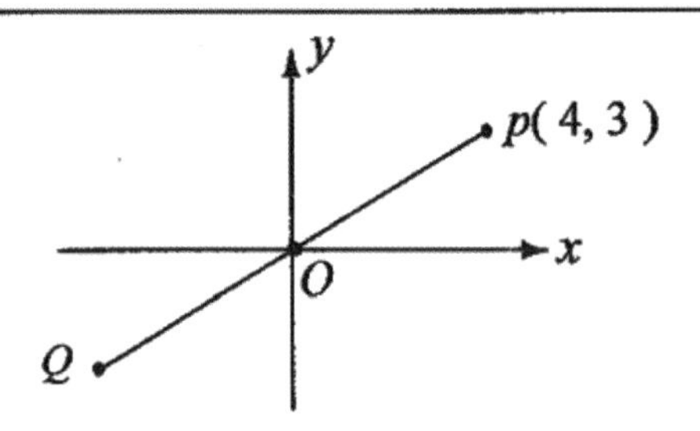

8. In the figure above, PQ is a line segment and $PO = OQ$. What are the coordinates of point Q ?

(A) (3, 4)
(B) (–3, 4)
(C) (–4, 3)
(D) (–4, –3)
(E) (–3, –4)

GO ON TO THE NEXT PAGE

9. A, B, C, and D are points on a number line in that order. $AD = 48$ and $AB = BC = CD$. A point P is on the line at the middle of CD. What does AP equal ?

(A) 42
(B) 40
(C) 38
(D) 36
(E) 34

10. The product of two consecutive integers equals 600. What is the larger number ?

(A) 20
(B) 25
(C) 30
(D) 35
(E) 40

11. The ratio of the numbers of red balls and white balls in a basket is 5 : 9. There are 280 balls in the basket. How many red balls are in the basket ?

(A) 140
(B) 130
(C) 120
(D) 110
(E) 100

12. An alarm system uses a three–letter code and no letter can be used more than once. How many possible codes are there ?

(A) 13,824
(B) 15,000
(C) 15,600
(D) 17,000
(E) 17,576

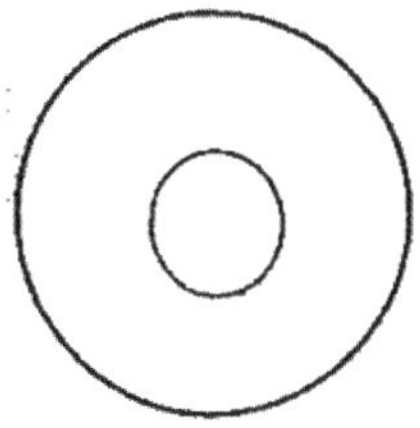

13. In the figure above, a small circle is placed in the center of a large circle. The small circle has a radius 2 and the large circle has a radius 6. If you throw a dart at random to hit the large circle, what is the probability that the dart will hit the small circle ?

(A) $\frac{1}{3}$
(B) $\frac{1}{4}$
(C) $\frac{1}{8}$
(D) $\frac{1}{9}$
(E) $\frac{1}{12}$

GO ON TO THE NEXT PAGE

14. k represents the age of Kevin, and j represents the age of Jack. Kevin is 20 years old. Jack is 10 years old. K is what percent greater than j ?

(A) 200 %
(B) 100 %
(C) 50 %
(D) 10 %
(E) 2 %

15. ΔABC is a triangle and $<A:<B:<C=3:4:5$. What is the measure in degrees of $<C$?

(A) 75
(B) 80
(C) 85
(D) 90
(E) 95

16. The mean of a and b is 21. The mean of a, b, and c is 26. What is the value of c ?

(A) 32
(B) 33
(C) 34
(D) 35
(E) 36

17. Scientists tag 100 salmons in a river. Later, they caught 200 salmons. Of these, 4 have tags. Estimate how many salmons are in the river ?

(A) 2,000
(B) 5,000
(C) 8,000
(D) 10,000
(E) 12,000

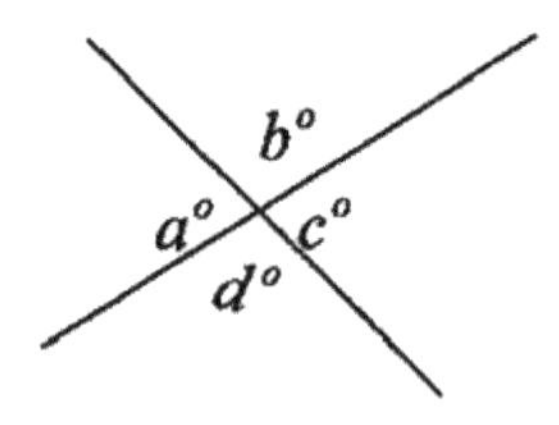

18. Two lines intersects as shown above, which of the following must be true ?

I. $b=d$
II. $a+c=b+d$
III. $a+b=c+d$

(A) I only
(B) II only
(C) I and II only
(D) I and III only
(E) None

GO ON TO THE NEXT PAGE

19. The heartbeats per minute (pulse rates) for 12 students are listed below.

82, 86, 69, 75, 80, 66, 65, 71, 69, 75, 82, 70

What is the median of the heartbeats ?

(A) 70
(B) 71
(C) 72
(D) 73
(E) 74

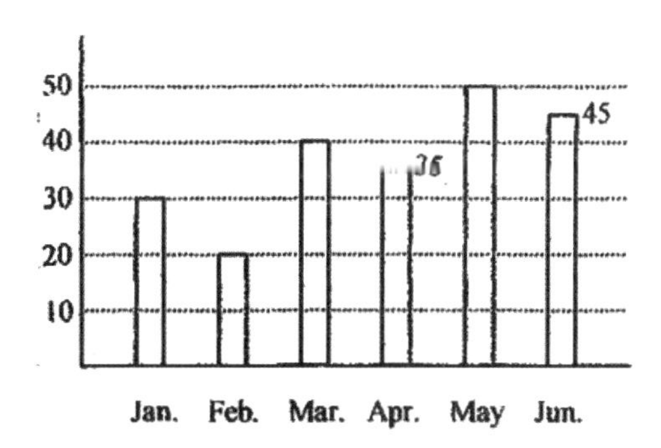

20. The bar graph above shows the number of a product sold during a 6-month period. What percent of the product were sold in March during the 6-month period ? Round the answer to the nearest integer.

(A) 18%
(B) 25%
(C) 40%
(D) 48%
(E) 52%

STOP

If you finish before time is called, you may check your work on this section only.
Do not turn to any other section in this test.

SECTION 2

Time — 25 minutes
18 Questions

Direction: This section contains two types of questions. You have 25 minutes to complete both types. For questions 21-28, solve each problem and decide which is the best of the choices given. Fill in the corresponding circle on the answer sheet. You may use any available space for scratch work.

Notes:

1. The use of a calculator is permitted.
2. All numbers used are real numbers.
3. Figures that accompany problems in this test are intended to provide information useful in solving the problems. They are drawn as accurately as possible EXCEPT when it is stated in a specific problem that the figure is not drawn to scale. All figures lie in a plane unless otherwise indicated.
4. Unless otherwise specified, the domain of any function f is assumed to be the set of all real number x for which $f(x)$ is a real number.

Reference Information

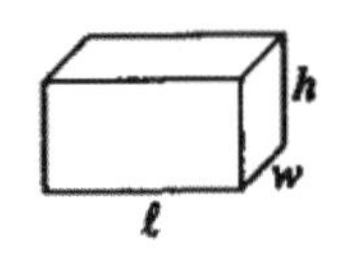

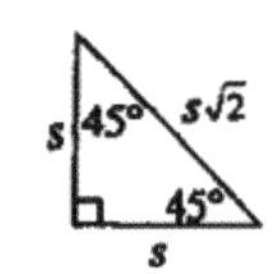

$A = \pi r^2$ $A = \ell w$ $A = \frac{1}{2}bh$ $V = \ell wh$ $V = \pi r^2 h$ $c^2 = a^2 + b^2$ Special Right Triangles

$C = 2\pi r$

The number of degrees of arc in a circle is 360.
The sum of the measures in degrees of the angles of a triangle is 180.

21. If John drives 60 miles per hour, how many miles does he drives in 20 minutes ?

(A) 10
(B) 20
(C) 30
(D) 40
(E) 45

22. What is the value of a if $2(a-4+3a)=12$?

(A) 1
(B) 2
(C) $2\frac{1}{2}$
(D) $3\frac{1}{2}$
(E) 4

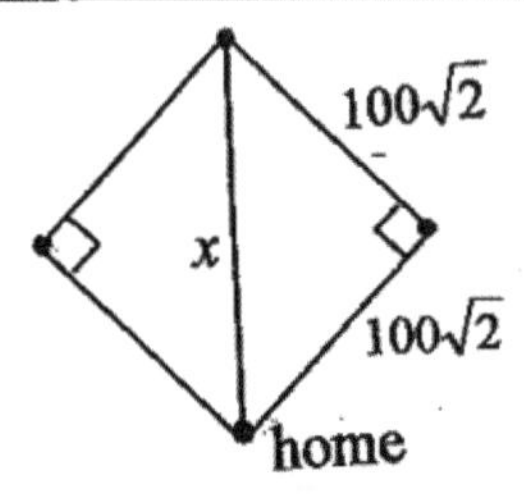

23. A baseball field is shaped as above. Each side is $100\sqrt{2}$ feet. What is the distance in feet from home plate to second base ?

(A) 140
(B) 160
(C) 200
(D) 220
(E) 280

24. What is the slope of the line with the equation $5x + 6y = 15$?

(A) $\frac{6}{5}$
(B) $\frac{5}{6}$
(C) $-\frac{6}{5}$
(D) $-\frac{5}{6}$
(E) $-\frac{1}{3}$

25. Which of the following is the perimeter of the figure below ?

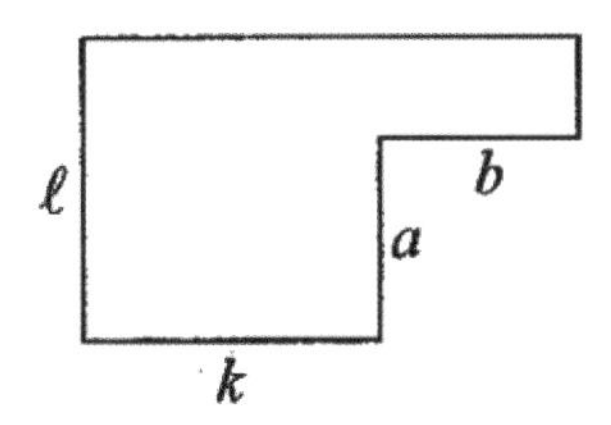

(A) $\ell + k + a + b$
(B) $2(l + k) + a + b$
(C) $2(\ell + k + a) - b$
(D) $2(\ell + k - a) + b$
(E) $2(l + k + b)$

26. If $x = \frac{1}{2}$ and $y = \frac{1}{3}$, what is the value of $\dfrac{x+y}{x-y}$?

(A) -1
(B) 0
(C) 1
(D) 3
(E) 5

27. How many seat arrangements can be made for five students in a row if one of them does not want to take the first seat ?

(A) 120
(B) 96
(C) 85
(D) 80
(E) 78

28. A triangle has lengths of sides 9, 13, and x. What is the possible value of x ?

(A) $4 < x < 22$
(B) $9 < x < 13$
(C) $0 < x < 13$
(D) $13 < x < 22$
(E) $0 < x < 9$

GO ON TO THE NEXT PAGE

Direction: For Student-Produced Response questions 29-38, use the grids at the bottom of the answer sheet page on which you have answered questions 21-28.

Each of the remaining 10 questions requires you to solve the problem and enter your answer by marking the circle in the special grid, as shown in the examples below. You may use any available space for scratch work.

Answer: $7/12$

Write answer in box → 7 / 1 2 ←Fraction line

Grid in → result

Answer: 2.5

2 . 5 ←Decimal point

Answer: 201

Either position is correct.

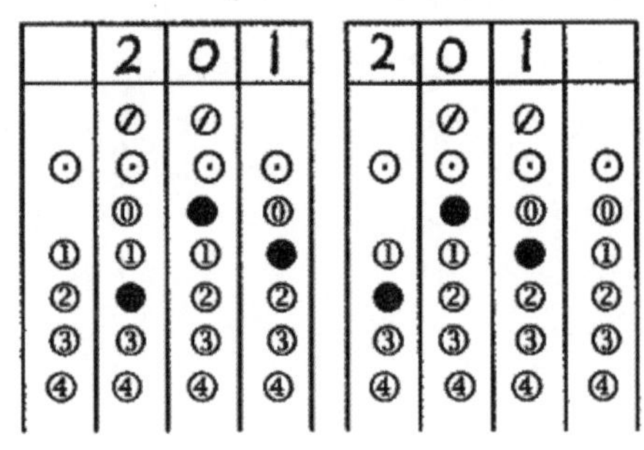

Note: You may start your answers in any column, space permitting. Columns not needed should be left blank.

- Mark no more than one circle in any column.
- Because the answer sheet will be machine-scored, **you will receive credit only if the circles are filled in correctly.**
- Although not required, it is suggested that you write your answers in the boxes at the top of the columns to help you fill in the circles accurately.
- Some problems may have more than one correct answer. In such cases, grid only one answer.
- No question has a negative answer.
- **Mixed numbers such as $3\frac{1}{2}$** must be gridded as 3.5 or 7/2. (If 3 1 / 2 is gridded, it will be interpreted as $\frac{31}{2}$, not $3\frac{1}{2}$
- **Decimal Answers:** If you obtain a decimal answer with more digits than the grid can accommodate, it may be either rounded or truncated, but it must fill the entire grid. For example, if you obtain an answer such as 0.6666…, you should record you result as 0.666 or 0.667. **A less accurate value such as 0.66 or 0.67 will be scored as incorrect.**

Acceptable ways to grid 2/3 are:

2 / 3 . 6 6 6 . 6 6 7

29. If $4x+5y=10$, what is the value of $12x+15y$?

30. The sum of 4 consecutive integers is 1,258. What is the value of the greatest of these integer ?

GO ON TO THE NEXT PAGE

31. A number divided by 3 is equal to the number decreased by 3. Find the number.

32. A 25-gallon salt solution contains 40% pure salt. How much water in gallons should be added to produce the solution to 16% salt ?

33. What is the value of $(\sqrt{5}+\sqrt{2})(\sqrt{5}-\sqrt{2})$?

34. David can paint a warehouse in 10 hours. Maria can paint the same warehouse in 6 hours. If they work together, how many hours will it take them to paint the warehouse ?

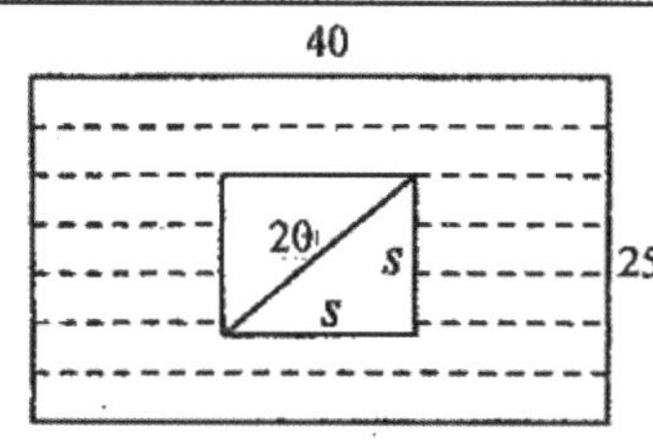

35. In the figure above, a square is cut out from a rectangle of length 40 and width 25. Find the area of the dashed region.

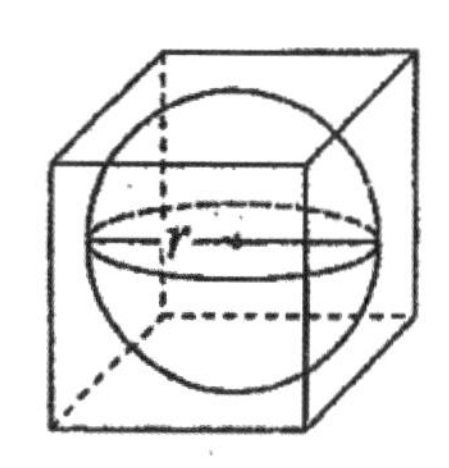

36. In the figure above, a sphere is inscribed in a cube. The ratio of the area of the sphere to the area of the cube is $\pi : n$. What is the value of n ?
(Hint: surface area of a sphere $A = 4\pi r^2$)

37. If $f(x) = x^2 + 18$ and n is a positive number such that $f(2n) = 2f(n)$, find the value of n.

38. The probability of a person being left-handed is $\frac{1}{30}$. The probability of a person being colorblind is $\frac{1}{50}$. There are 4,500 students in a high school. How many students would probably be both left-handed and colorblind ?

STOP

If you finish before time is called, you may check your work on this section only. Do not turn to any other section in the test.

PRACTICE TEST 1
(PSAT/NMSQT Math)

Page 311 ~ 319

Answer Key

Section 1: **1.** D **2.** B **3.** D **4.** C **5.** B **6.** A **7.** E **8.** D **9.** B **10.** B

11. E **12.** C **13.** D **14.** B **15.** A **16.** E **17.** B **18.** D **19.** D **20.** A

Section 2: **21.** B **22.** C **23.** C **24.** D **25.** E **26.** E **27.** B **28.** A

29. 30 **30.** 316 **31.** 4.5 **32.** 37.5 **33.** 3 **34.** 3.75 **35.** 800 **36.** 6

37. 3 **38.** 3

CALCULATE YOUR SCORE

	Correct		Incorrect	
Questions 1 ~ 28	( )		( )	
	+			
Questions 29 ~ 38	( )			
Total Unrounded Raw Score	( )	–	( ) × 0.25	= ( )
Total Rounded Raw Score				( )
Your Math Scaled Score (See Table Below)				( )

PSAT MATH SCORE CONVERSION TABLE

Raw Score	Scaled Score	Raw Score	Scaled Score
–7 to –1	20	19	49
0	21	20	49
1	23	21	51
2	24	22	52
3	27	23	52
4	28	24	53
5	31	25	54
6	33	26	55
7	34	27	56
8	35	28	58
9	38	29	60
10	39	30	62
11	40	31	63
12	41	32	64
13	42	33	67
14	43	34	68
15	44	35	69
16	46	36	72
17	46	37	75
18	47	38	80

PRACTICE TEST 2
(PSAT / NMSQT)

SECTION 1
Time — 25 minutes
20 Questions

Direction: For this section, solve each problem and decide which is the best of the choices given. Fill in the corresponding circle on the answer sheet. You may use any available space for scratch work.

Notes:
1. The use of a calculator is permitted.
2. All numbers used are real numbers.
3. Figures that accompany problems in this test are intended to provide information useful in solving the problems. They are drawn as accurately as possible EXCEPT when it is stated in a specific problem that the figure is not drawn to scale. All figures lie in a plane unless otherwise indicated.
4. Unless otherwise specified, the domain of any function f is to be the set of all real number x for which $f(x)$ is a real number.

Reference Information

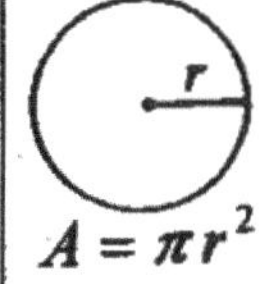

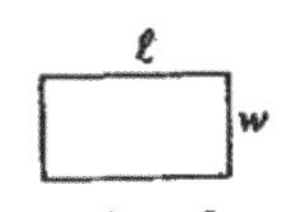

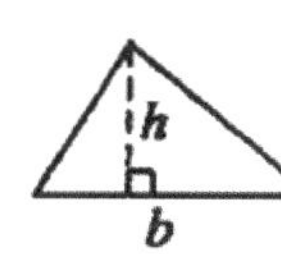

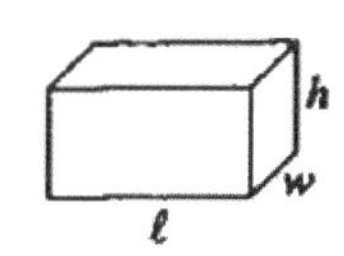

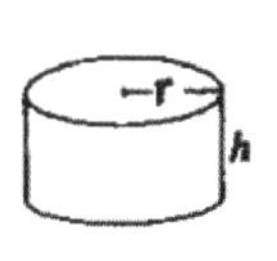

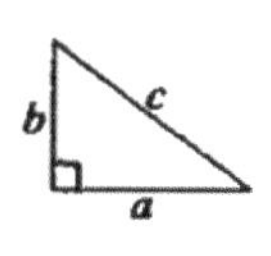

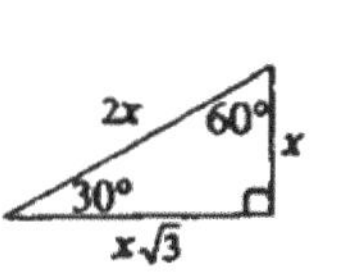

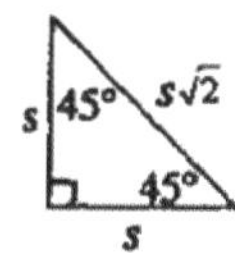

$A = \pi r^2$ $\quad C = 2\pi r$ $\quad A = \ell w$ $\quad A = \frac{1}{2}bh$ $\quad V = \ell wh$ $\quad V = \pi r^2 h$ $\quad c^2 = a^2 + b^2$ $\quad$ Special Right Triangles

The number of degrees of arc in a circle is 360.
The sum of the measures in degrees of the angles of a triangle is 180.

1. What is the value of x for $4^{3x} = 16^{x+2}$?

(A) 0
(B) 2
(C) 3
(D) 4
(E) 5

2. If the average (arithmetic mean) of 4, 8, 12, 20, and a is 18, what is the value of a ?

(A) 30
(B) 37
(C) 43
(D) 46
(E) 50

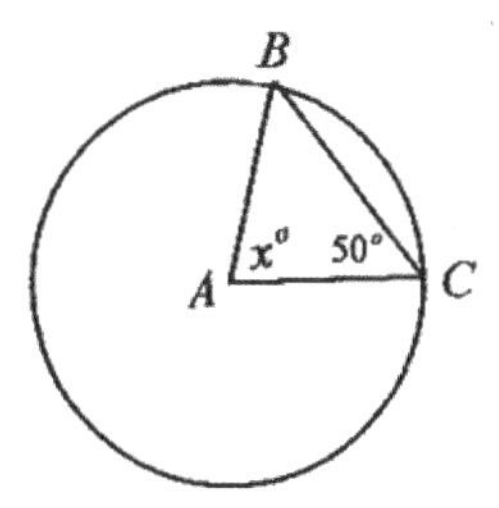

3. In the figure above, A is the center of the circle. What is the value of x ?

(A) 80
(B) 75
(C) 70
(D) 65
(E) 60

GO ON TO THE NEXT PAGE

4. If $a^2 + b^2 = 32$ and $ab = 14$, what is the value of $(a - b)^2$?

(A) 2
(B) 4
(C) 6
(D) 8
(E) 9

5. What is the sum of $1+2+3+\cdots\cdots+50$?

(A) 1450
(B) 1325
(C) 1300
(D) 1295
(E) 1275

6. $\dfrac{10}{x-2} + \dfrac{3}{2-x} =$

(A) $\dfrac{13}{x-2}$

(B) $\dfrac{13}{2-x}$

(C) $\dfrac{7}{x-2}$

(D) $\dfrac{7}{2-x}$

(E) None of the above

7. The population of a city increased from 60,000 to 74,400. What is the percent of increase ?

(A) 20%
(B) 24%
(C) 30%
(D) 32%
(E) 34%

8. If $\dfrac{4}{5}$ of $\dfrac{3}{8} = \dfrac{1}{7}$ of $\dfrac{x}{10}$, what is the value of x ?

(A) 10
(B) 12
(C) 15
(D) 18
(E) 21

GO ON TO THE NEXT PAGE

9. A 30-gallon alcohol-water solution contains 20% alcohol. How much alcohol in gallons should be added to produce the solution to 25% alcohol ?

(A) 2
(B) 3
(C) 4
(D) 5
(E) 6

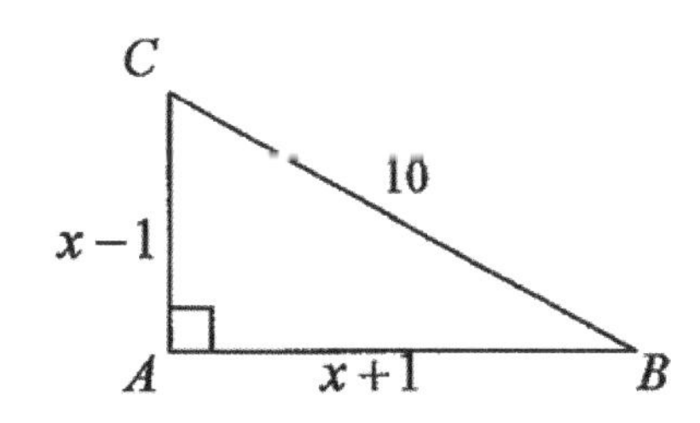

Note: Figure not drawn to scale.

10. In the right triangle above, what is the length of side AC ?

(A) 10
(B) 9
(C) 8
(D) 7
(E) 6

11. If $\frac{a-6}{a}=\frac{2}{5}$, then $a=$

(A) 30
(B) 25
(C) 20
(D) 15
(E) 10

12. If $5x+y=2x-1$, what is y in terms of x ?

(A) $3x-1$
(B) $7x-1$
(C) $3x+1$
(D) $-3x+1$
(E) None of the above

13. What are all values of x for which $|x-4|<7$?

(A) $3<x<11$
(B) $0<x<11$
(C) $-3<x<11$
(D) $-11<x<-3$
(E) $-11<x<3$

14. If $4^8 \times 16^{20} = 4^{2n}$, then $n=$

(A) 12
(B) 14
(C) 18
(D) 24
(E) 28

GO ON TO THE NEXT PAGE

Questions 15 and 16 are based on information in the circle graph below.

Roger's budget to spend his money monthly

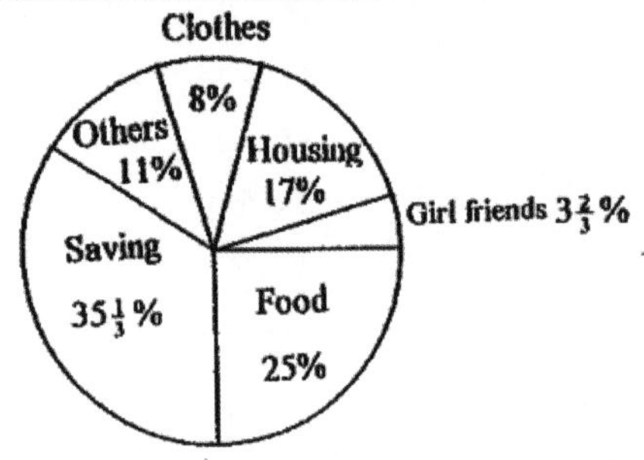

15. His monthly income is $2,400. How much money does he plan to spend on food ?

(A) $600
(B) $700
(C) $800
(D) $900
(E) $990

16. His monthly income is $2,400. How much money does he plan to spend on girl friends ?

(A) $2,400
(B) $142
(C) $88
(D) $20
(E) $0

17. What is the equation of the line having slope –7 and y-intercept 5 ?

(A) $7x-y=5$
(B) $7x+y=5$
(C) $7x+y=-5$
(D) $7x-y=-5$
(E) $x-7y=5$

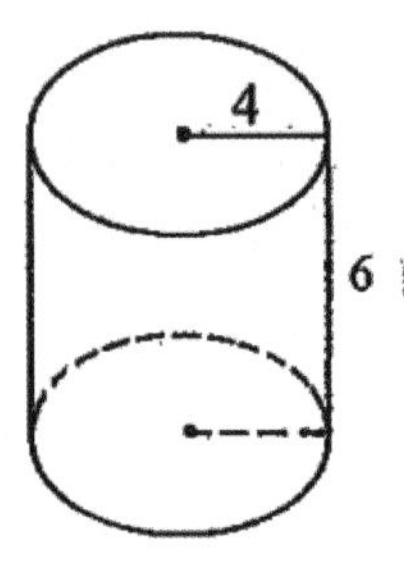

18. Find the volume of the right cylinder shown above.

(A) 24π
(B) 36π
(C) 48π
(D) 96π
(E) 144π

19. Which of the following equations is perpendicular to the line $y=4x-8$?

(A) $y=4x+8$
(B) $y=-4x+8$
(C) $y=\frac{1}{4}x-7$
(D) $y=\frac{1}{4}x+8$
(E) $y=-\frac{1}{4}x-9$

GO ON TO THE NEXT PAGE

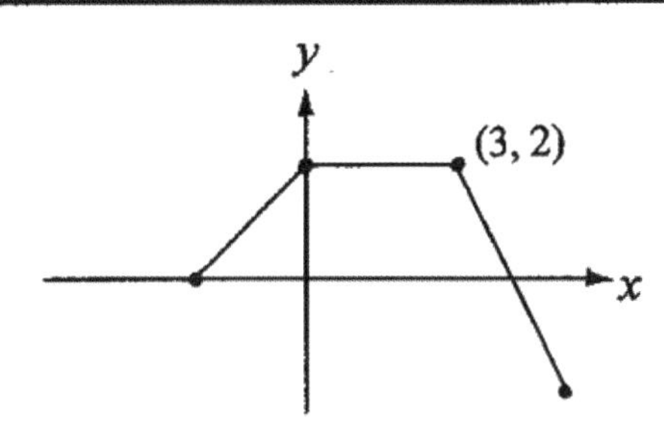

Note: Figure not drawn to scale.

20. The graph of a function $y = f(x)$ is shown above. Which of the following could be the graph of $y = f(x) + 2$?

(A)

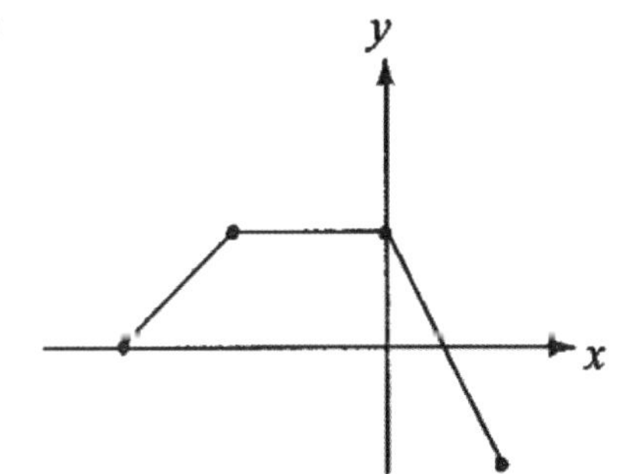

(B)

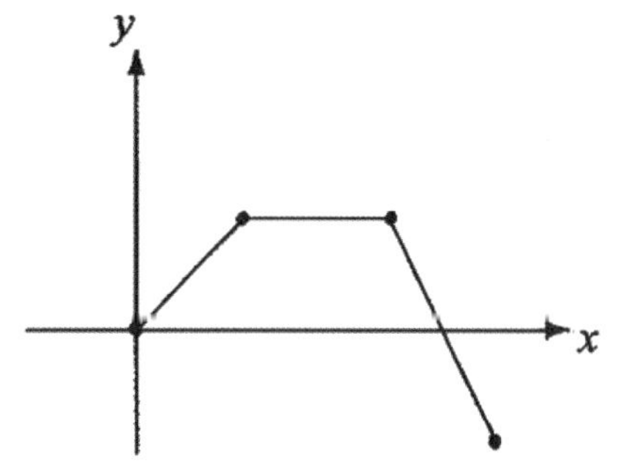

(C)

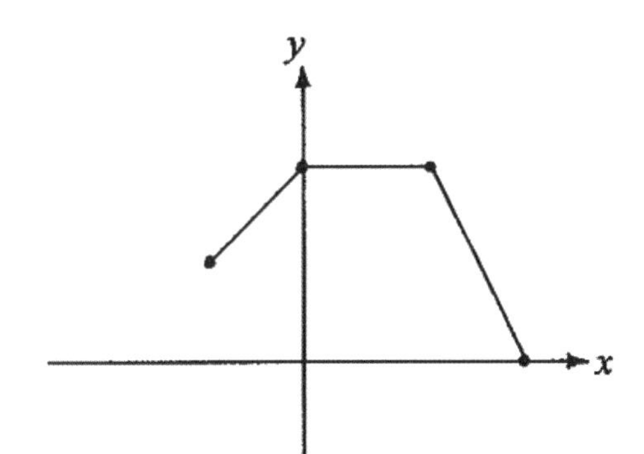

(D)

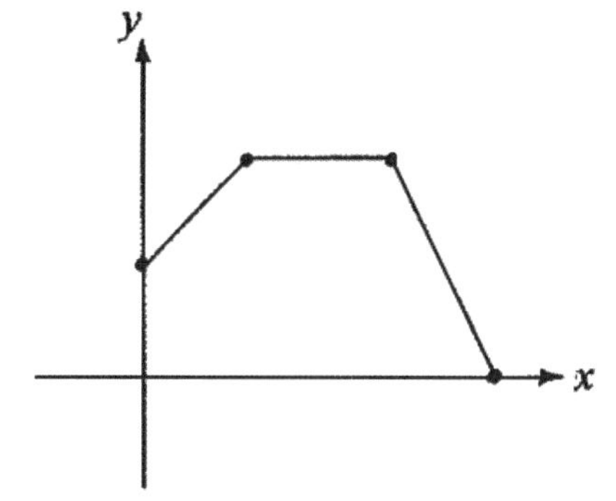

(E)

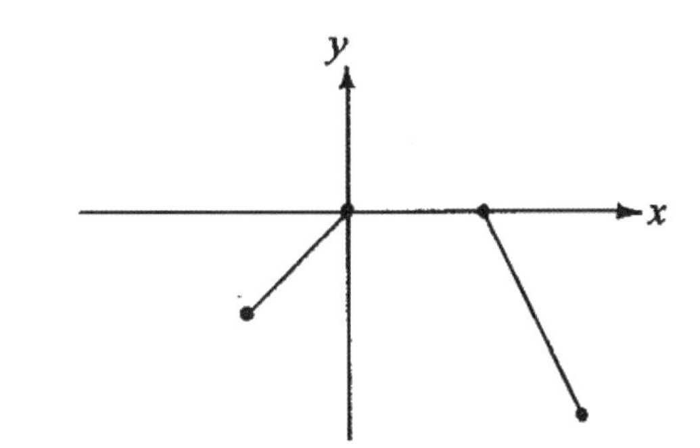

STOP

If you finish before time is called, you may check your work on this section only.
Do not turn to any other section in this test.

SECTION 2

Time — 25 minutes
18 Questions

Direction: This section contains two types of questions. You have 25 minutes to complete both types. For questions 21-28, solve each problem and decide which is the best of the choices given. Fill in the corresponding circle on the answer sheet. You may use any available space for scratch work.

Notes:

1. The use of a calculator is permitted.
2. All numbers used are real numbers.
3. Figures that accompany problems in this test are intended to provide information useful in solving the problems. They are drawn as accurately as possible EXCEPT when it is stated in a specific problem that the figure is not drawn to scale. All figures lie in a plane unless otherwise indicated.
4. Unless otherwise specified, the domain of any function f is assumed to be the set of all real number x for which $f(x)$ is a real number.

Reference Information

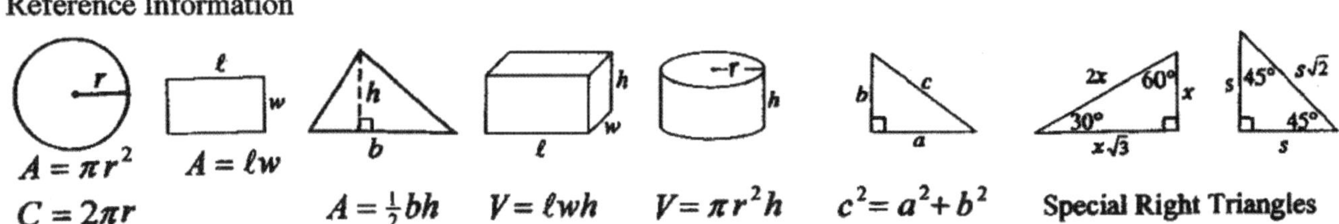

The number of degrees of arc in a circle is 360.
The sum of the measures in degrees of the angles of a triangle is 180.

21. What is the value of a for $10+\sqrt{a}=19$?

(A) 25
(B) 49
(C) 81
(D) 121
(E) 169

22. If a is 20% of 30 and b is 50% of 48, what is the value of $a \times b$?

(A) 144
(B) 121
(C) 100
(D) 78
(E) 50

23. If $f(x)=3x+12$, what is the solution of the equation $f(2x)=-18$?

(A) 9
(B) –5
(C) –12
(D) –15
(E) –20

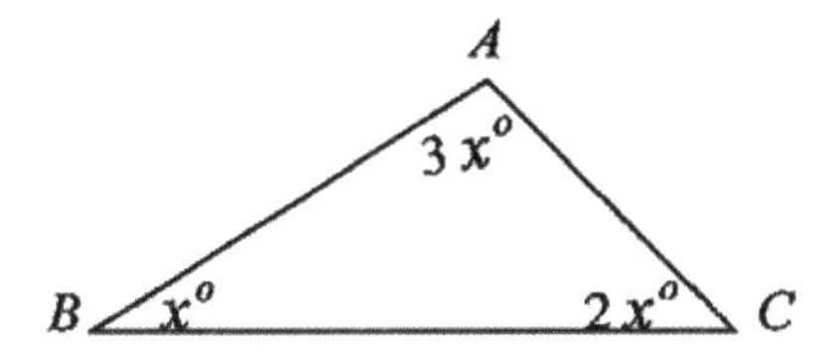

Note: Figure not drawn to scale.

24. In the figure above, what is the measure of angle A ?

(A) 90°
(B) 80°
(C) 60°
(D) 45°
(E) 30°

25. David rented a car. The car rental was \$80 plus \$0.35 per mile. If the total bill was \$157, how many miles did he drive by using this car ?

(A) 150
(B) 180
(C) 200
(D) 220
(E) 240

26. How many degree of arc are there in $\frac{3}{4}$ of a circle ?

(A) 70°
(B) 90°
(C) 120°
(D) 150°
(E) 270°

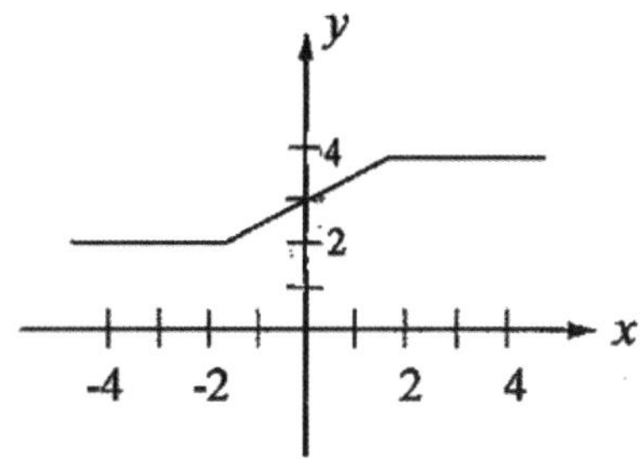

27. The above figure shows the graph of the function $y = f(x)$. If $f(a) = 2$, what is the possible value of a ?

(A) 4
(B) 1
(C) 0
(D) −1
(E) −3

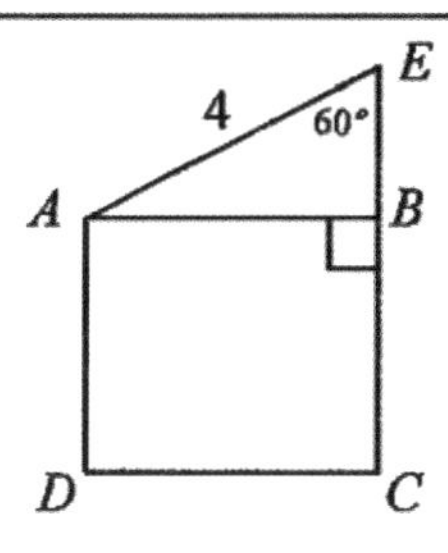

28. In the figure above, $ABCD$ is a square. The length of AE is 4. What is the area of the square ?

(A) 48
(B) 24
(C) 20
(D) 16
(E) 12

GO ON TO THE NEXT PAGE

Direction: For Student-Produced Response questions 29-38, use the grids at the bottom of the answer sheet page on which you have answered questions 21-28.

Each of the remaining 10 questions requires you to solve the problem and enter your answer by marking the circle in the special grid, as shown in the examples below. You may use any available space for scratch work.

Answer: $7/12$ — Write answer in box → 7 / 1 2 ← Fraction line; Grid in → result

Answer: 2.5 — 2 . 5 ← Decimal point

Answer: 201
Either position is correct.

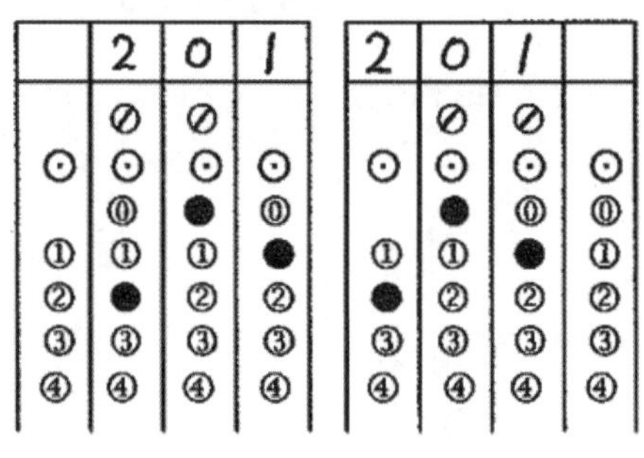

Note: You may start your answers in any column, space permitting. Columns not needed should be left blank.

- Mark no more than one circle in any column.
- Because the answer sheet will be machine-scored, **you will receive credit only if the circles are filled in correctly.**
- Although not required, it is suggested that you write your answers in the boxes at the top of the columns to help you fill in the circles accurately.
- Some problems may have more than one correct answer. In such cases, grid only one answer.
- No question has a negative answer.
- **Mixed numbers such as** $3\frac{1}{2}$ must be gridded as 3.5 or 7/2. (If 3 1 / 2 is gridded, it will be interpreted as $\frac{31}{2}$, not $3\frac{1}{2}$.

- **Decimal Answers:** If you obtain a decimal answer with more digits than the grid can accommodate, it may be either rounded or truncated, but it must fill the entire grid. For example, if you obtain an answer such as 0.6666····, you should record you result as 0.666 or 0.667. **A less accurate value such as 0.66 or 0.67 will be scored as incorrect.**

Acceptable ways to grid 2/3 are:

2 / 3   .666   .667

29. If $4x+5y=16$, what is the value of x-intercept ?

30. In a alcohol-water solution, the ratio by volume of alcohol to water is 2 to 7. How many liters of pure alcohol will there be in 45 liters of the solution ?

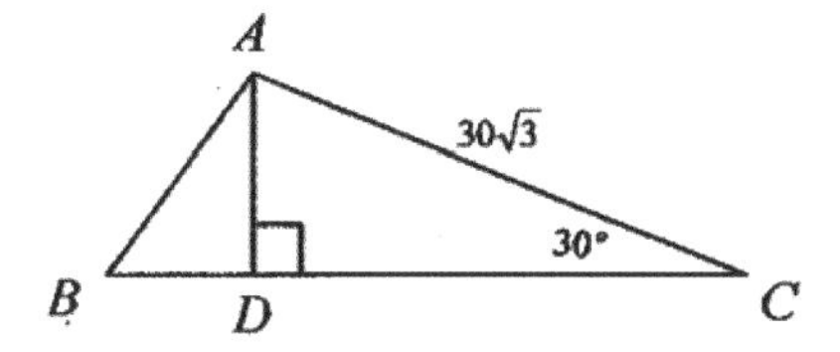

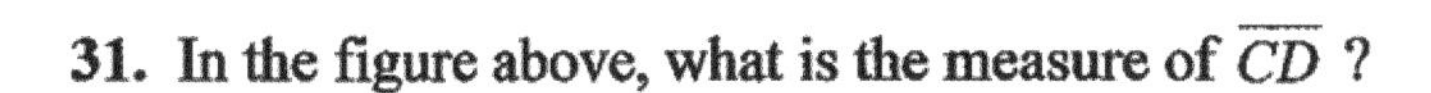

31. In the figure above, what is the measure of $\overline{CD}$?

32. A bag contains 6 red balls and 4 white balls. Two balls are drawn at random from the bag without replacement. What is the probability that the two balls drawn are both white ?

33. A rocket is fired upward with an initial speed of 120 feet per second. The height (h), in feet, of the rocket is given by the formula $h = 120t - 16t^2$, where t is the time in second. What is the height in feet of the rocket in 5 seconds after being fired upward ?

34. If $\frac{x-6}{x} = \frac{7}{13}$, then $x =$

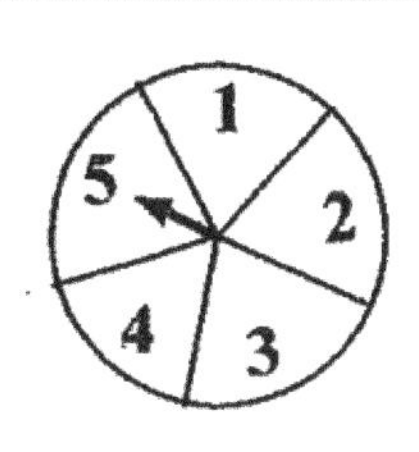

35. A player spins the spinner above twice. What is the probability of getting a sum of 5 ?

36. What is the sum of the measures in degrees of a polygon with 22 sides ?

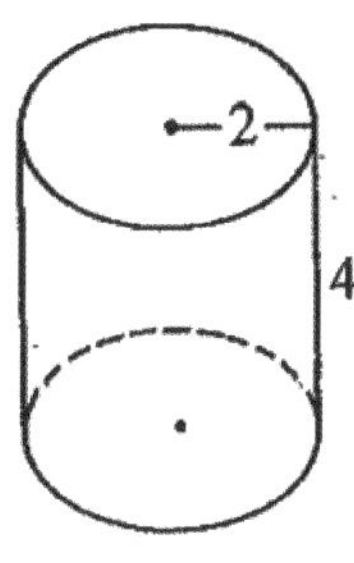

37. In the right cylinder above, what is the total surface area ? ($\pi = 3.14$)

38. If $a \otimes b$ be defined as the product of all integers larger than a and smaller than b. What is the value of $(4 \otimes 8) - (3 \otimes 7)$?

STOP

If you finish before time is called, you may check your work on this section only.
Do not turn to any other section in the test.

PRACTICE TEST 2
(PSAT/NMSQT Math)

Page 321 ~ 329

Answer Key

Section 1: **1.** D **2.** D **3.** A **4.** B **5.** E **6.** C **7.** B **8.** E **9.** A **10.** E

11. E **12.** E **13.** C **14.** D **15.** A **16.** C **17.** B **18.** D **19.** E **20.** C

Section 2: **21.** C **22.** A **23.** B **24.** A **25.** D **26.** E **27.** E **28.** E

29. 4 **30.** 10 **31.** 45 **32.** $\frac{2}{15}$ or .133 **33.** 200 **34.** 13 **35.** $\frac{4}{25}$ or .16

36. 3600 **37.** 75.4 **38.** 90

CALCULATE YOUR SCORE

	Correct		Incorrect	
Questions 1 ~ 28	( )		( )	
	+			
Questions 29 ~ 38	( )			
Total Unrounded Raw Score	( )	−	( ) × 0.25	= ( )
Total Rounded Raw Score				( )
Your Math Scaled Score (See Table Below)				( )

PSAT MATH SCORE CONVERSION TABLE

Raw Score	Scaled Score	Raw Score	Scaled Score
−7 to −1	20	19	49
0	21	20	49
1	23	21	51
2	24	22	52
3	27	23	52
4	28	24	53
5	31	25	54
6	33	26	55
7	34	27	56
8	35	28	58
9	38	29	60
10	39	30	62
11	40	31	63
12	41	32	64
13	42	33	67
14	43	34	68
15	44	35	69
16	46	36	72
17	46	37	75
18	47	38	80

PRACTICE TEST 3
(SAT MATH)

SECTION 1
Time — 25 minutes
20 Questions

Direction: For this section, solve each problem and decide which is the best of the choices given. Fill in the corresponding circle on the answer sheet. You may use any available space for scratch work.

Notes:
1. The use of a calculator is permitted.
2. All numbers used are real numbers.
3. Figures that accompany problems in this test are intended to provide information useful in solving the problems. They are drawn as accurately as possible EXCEPT when it is stated in a specific problem that the figure is not drawn to scale. All figures lie in a plane unless otherwise indicated.
4. Unless otherwise specified, the domain of any function f is to be the set of all real number x for which $f(x)$ is a real number.

Reference Information

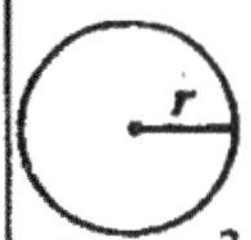

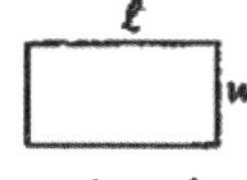

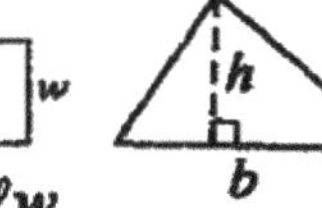

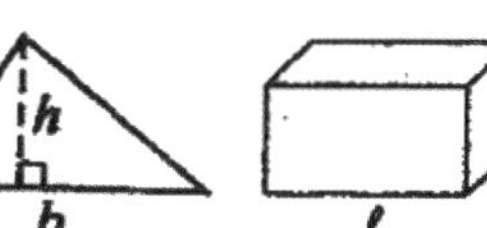

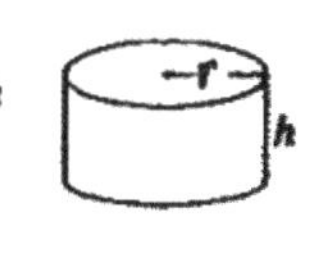

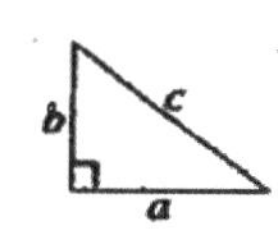

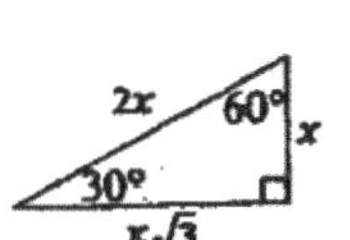

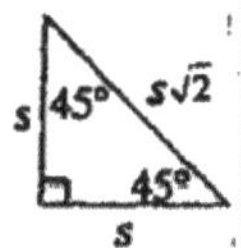

$A = \pi r^2$ $C = 2\pi r$ $A = \ell w$ $A = \frac{1}{2}bh$ $V = \ell wh$ $V = \pi r^2 h$ $c^2 = a^2 + b^2$ Special Right Triangles

The number of degrees of arc in a circle is 360.
The sum of the measures in degrees of the angles of a triangle is 180.

1. In a rectangular a coordinate plane, which of following points must be a point on the line $3x - y = 4$?

(A) $(x,\ 4x-3)$
(B) $(x,\ 4x+3)$
(C) $(x,\ -3x-4)$
(D) $(x,\ 3x-4)$
(E) $(x,\ -3x+4)$

2. If $6(a+3) = 24$, then $a =$

(A) 1
(B) 3
(C) 5
(D) 9
(E) 12

3. All numbers divisible by 3 and 8 are also divisible by which of the following ?

(A) 5
(B) 6
(C) 9
(D) 14
(E) 36

GO ON TO THE NEXT PAGE

4. If the vertices of a rectangle are at (4, 3), (–2, 3), (–2, –1), and (4, –1), what is the area of the rectangle ?

(A) 24
(B) 26
(C) 28
(D) 30
(E) 32

5. If $|x+y| = |x| + |y|$, which of the following could be the values of x and y ?

I. $x = 0$ and $y = 0$
II. $x > 0$ and $y < 0$
III. $x > 0$ and $y > 0$

(A) I only
(B II only
(C) III only
(D) I and III only
(E) I, II, and III

6. What is the 10th term of the following sequence ?
2, 3, 5, 8, 12, 17, ····

(A) 43
(B) 44
(C) 45
(D) 46
(E) 47

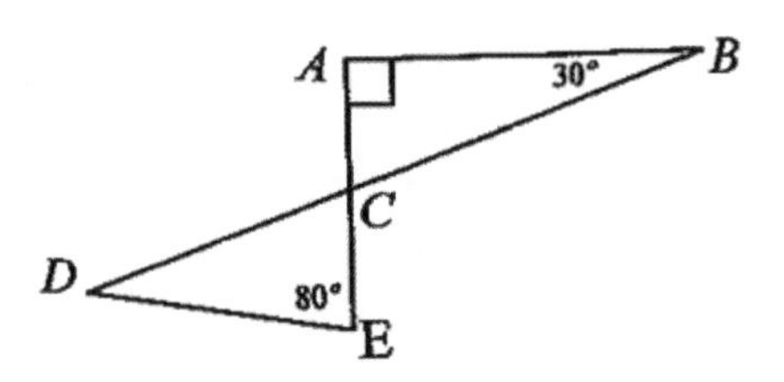

Note: Figure not drawn to scale.

7. In the figure above, what is the measure of $< D$?

(A) 30°
(B) 35°
(C) 40°
(D) 45°
(E) 50°

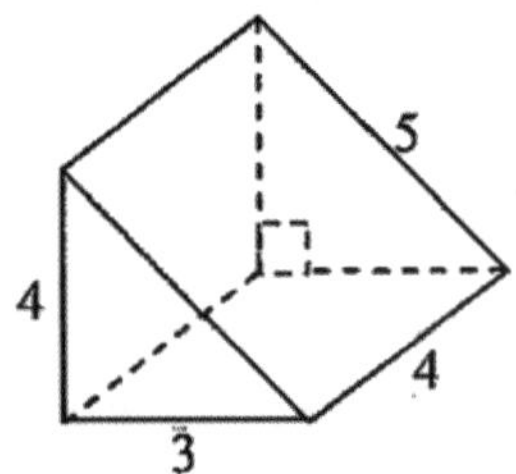

Note: Figure not drawn to scale.

8. In the figure above, what is the surface area of this right triangular prism ?

(A) 30
(B) 40
(C) 50
(D) 60
(E) 70

GO ON TO THE NEXT PAGE

9. You drove at an average speed of 60 miles per hour for the first 2 hours and then at an average speed of 65 miles per hour for the next three hours. What is the average speed for the entire trip ?

(A) 62 miles
(B) 62.5 miles
(C) 63 miles
(D) 63.5 miles
(E) 64 miles

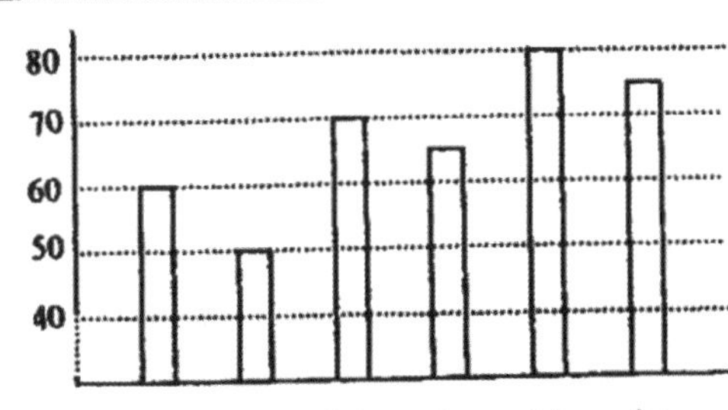

10. The bar graph above shows the number of a product sold during a 6-month period. Which month were the month that sold the product two times as many as February ?

(A) January
(B) March
(C) May
(D) June
(E) None of the above

11. $\frac{5}{a-b}+\frac{2}{b-a}=$

(A) $\frac{3}{a-b}$

(B) $\frac{7}{a-b}$

(C) $\frac{3}{b-a}$

(D) $\frac{7}{b-a}$

(E) $\frac{-3}{a-b}$

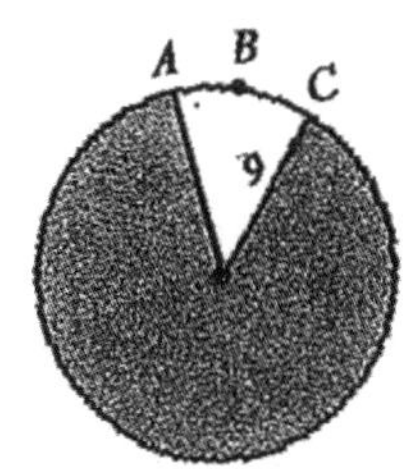

Note: Figure not drawn to scale.

12. The radius of the circle above is 9 . The measure of the arc ABC is 40° . What is the area of the shaded region ?

(A) 70π
(B) 71π
(C) 72π
(D) 73π
(E) 74π

1.6 meters

1.2 meters

13. The graph above shows the length and the width of a rectangular field on a map. If the actual width is 312 meters, what is the actual length ?

(A) 315 meters
(B) 416 meters
(C) 486 meters
(D) 514 meters
(E) 528 meters

GO ON TO THE NEXT PAGE

14. If you deposited \$7,500 in a saving account at an annual simple interest rate 4%, how much will be in the account after one year ?

(A) \$7,300
(B) \$7,400
(C) \$7,600
(D) \$7,800
(E) \$8,200

15. The third term and each term thereafter of the following sequence is the average of the two preceding terms. What is the value of the first term in the sequence that is not an integer ?

18, 14, ……

(A) 13.5
(B) 14.5
(C) 15.5
(D) 16.5
(E) 17.5

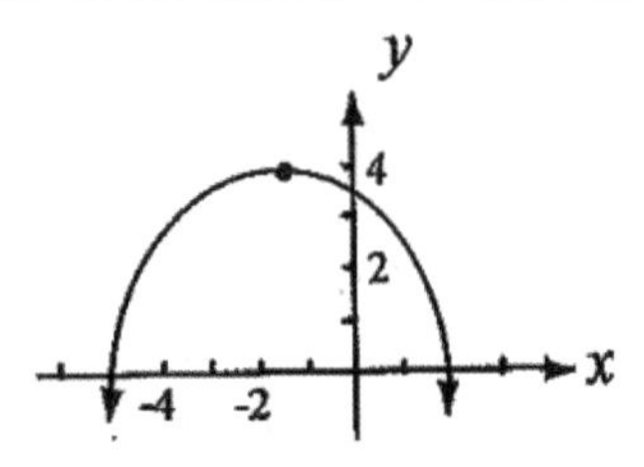

16. The figure above shows the graph of a quadratic function $y = g(x)$. The maximum value of the function is $g(-1.5)$. If $g(a) = 0$, which of the following could be the value of a ?

(A) -5
(B) -2
(C) 0
(D) 1
(E) 4

17. If $3a + b + c = 10$ and $3a + c = b$, what is the value of b ?

(A) 2
(B) 3
(C) 5
(D) 8
(E) 10

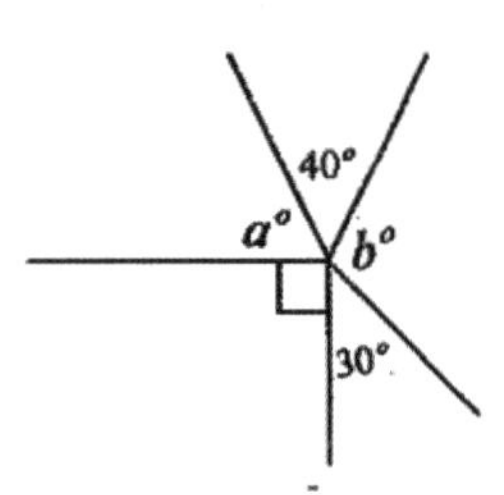

Note: Figure not drawn to scale.

18. In the figure above, what is the value of a in terms of b ?

(A) $200 - b$
(B) $220 - b$
(C) $240 - b$
(D) $260 - b$
(E) $280 - b$

GO ON TO THE NEXT PAGE

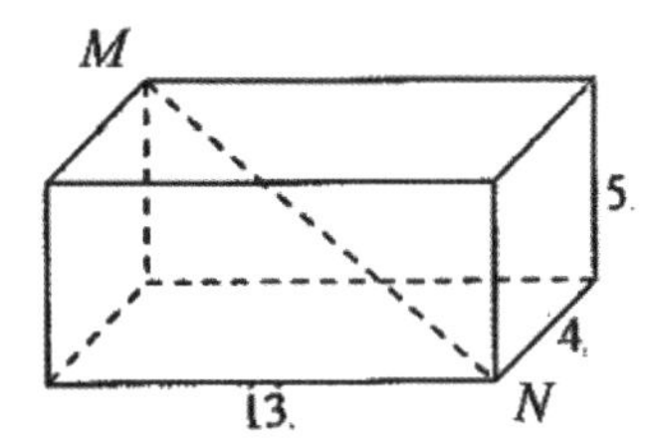

Note: Figure not drawn to scale.

19. In the figure above, what is the length of diagonal $\overline{MN}$ in the rectangular solid ?

(A) 10.25
(B) 14.49
(C) 14.98
(D) 13.40
(E) 13.81

---

20. If $x^2 + ax + 12 = (x-3)(x+b)$, where a and b are constants, what is the value of a ?

(A) 15
(B) 12
(C) 8
(D) –5
(E) –7

STOP

If you finish before time is called, you may check your work on this section only.
Do not turn to any other section in this test.

SECTION 2

Time — 25 minutes
18 Questions

Direction: This section contains two types of questions. You have 25 minutes to complete both types. For questions 1-8, solve each problem and decide which is the best of the choices given. Fill in the corresponding circle on the answer sheet. You may use any available space for scratch work.

Notes:

1. The use of a calculator is permitted.
2. All numbers used are real numbers.
3. Figures that accompany problems in this test are intended to provide information useful in solving the problems. They are drawn as accurately as possible EXCEPT when it is stated in a specific problem that the figure is not drawn to scale. All figures lie in a plane unless otherwise indicated.
4. Unless otherwise specified, the domain of any function f is assumed to be the set of all real number x for which $f(x)$ is a real number.

Reference Information

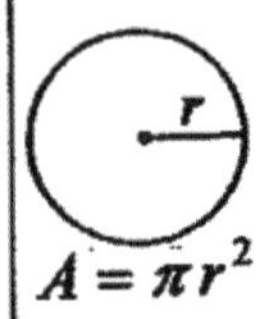

$A = \pi r^2$
$C = 2\pi r$

$A = \ell w$

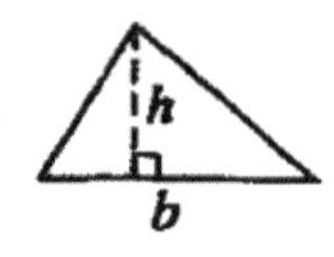

$A = \frac{1}{2}bh$

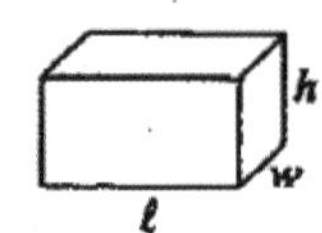

$V = \ell wh$

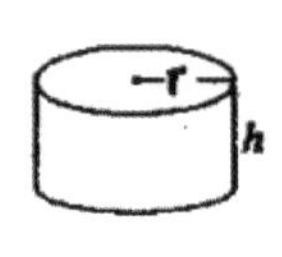

$V = \pi r^2 h$

$c^2 = a^2 + b^2$

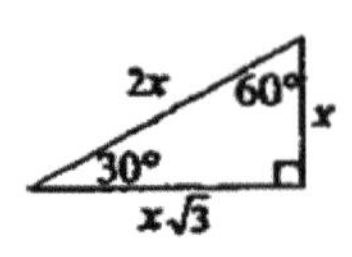

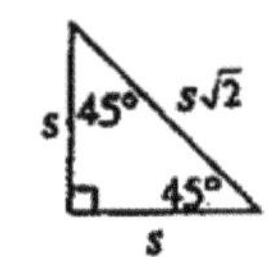

Special Right Triangles

The number of degrees of arc in a circle is 360.
The sum of the measures in degrees of the angles of a triangle is 180.

1. What is the value of n for $(n^2)^3 = 64$?

(A) 10
(B) 7
(C) 6
(D) 2
(E) 1

2. A car uses 5 gallons of gasoline to travel 110 miles. How many gallons can the car use to travel 242 miles ?

(A) 10
(B) 11
(C) 12
(D) 13
(E) 15

3. Roger picked up two-thirds of the books in the shelf. There are 15 books left in the shelf. How many books were in the shelf to begin with ?

(A) 63
(B) 60
(C) 51
(D) 45
(E) 42

GO ON TO THE NEXT PAGE

4. Roger has three times as much money as Carol. Together they have \$480. What is the fraction of the total money Roger has ?

(A) $\frac{1}{3}$
(B) $\frac{2}{3}$
(C) $\frac{3}{4}$
(D) $\frac{1}{5}$
(E) $\frac{2}{5}$

5. What are the possible values of a for $-8 \leq 2a - 4 \leq 14$?

(A) $-2 \leq a \leq 9$
(B) $-4 \leq a \leq 7$
(C) $-8 \leq a \leq 10$
(D) $4 \leq a \leq 10$
(E) $8 \leq a \leq 18$

6. A function be defined by $f(x) = 3x - 2$. If $\frac{1}{2}f(\sqrt{k}) = 3$, what is the value of k ?

(A) $\frac{7}{4}$
(B) $\frac{39}{5}$
(C) $\frac{25}{3}$
(D) $\frac{62}{7}$
(E) $\frac{64}{9}$

7. The population of males of a city increased 12 % and females decreased 20% this year. The ratio of males to females at the end of this year was how many times the ratio at the beginning of this year ?

(A) $\frac{4}{5}$
(B) $\frac{7}{5}$
(C) $\frac{7}{4}$
(D) $\frac{5}{7}$
(E) $\frac{4}{7}$

8

8. In the figure above, the diameters of the two semicircles are the legs of the right triangle. What is the sum of the areas of of the semicircles ?

(A) 24π
(B) 20π
(C) 16π
(D) 12π
(E) 8π

GO ON TO THE NEXT PAGE

Direction: For Student-Produced Response questions 9-18, use the grids at the bottom of the answer sheet page on which you have answered questions 1-8.

Each of the remaining 10 questions requires you to solve the problem and enter your answer by marking the circle in the special grid, as shown in the examples below. You may use any available space for scratch work.

Answer: $\frac{7}{12}$

Write answer in box → | 7 | / | 1 | 2 |

Grid in → result

←Fraction line

Answer: 2.5

| | 2 | . | 5 |

←Decimal point

Answer: 201

Either position is correct.

| | 2 | 0 | 1 |

| 2 | 0 | 1 | |

Note: You may start your answers in any column, space permitting. Columns not needed should be left blank.

- Mark no more than one circle in any column.
- Because the answer sheet will be machine-scored, **you will receive credit only if the circles are filled in correctly.**
- Although not required, it is suggested that you write your answers in the boxes at the top of the columns to help you fill in the circles accurately.
- Some problems may have more than one correct answer. In such cases, grid only one answer.
- No question has a negative answer.
- **Mixed numbers such as $3\frac{1}{2}$** must be gridded as 3.5 or 7/2. (If | 3 | 1 | / | 2 | is gridded, it will be interpreted as $\frac{31}{2}$, not $3\frac{1}{2}$.)
- **Decimal Answers:** If you obtain a decimal answer with more digits than the grid can accommodate, it may be either rounded or truncated, but it must fill the entire grid. For example, if you obtain an answer such as 0.6666····, you should record you result as 0.666 or 0.667. **A less accurate value such as 0.66 or 0.67 will be scored as incorrect.** Acceptable ways to grid 2/3 are:

| | 2 | / | 3 |

| . | 6 | 6 | 6 |

| . | 6 | 6 | 7 |

9. $(\sqrt{5}+\sqrt{2})(\sqrt{5}-\sqrt{2})=$

10. A number divided by 5 is equal to the number decreased by 5. What is the number ?

GO ON TO THE NEXT PAGE

11. What is the value of a for $7(a-1)=5(a+3)$?

12. What is the value of n for $3^{n+2}=243$?

13. If a varies directly as b, and $a=24$ when $b=5$, what is the value of a when $b=8$?

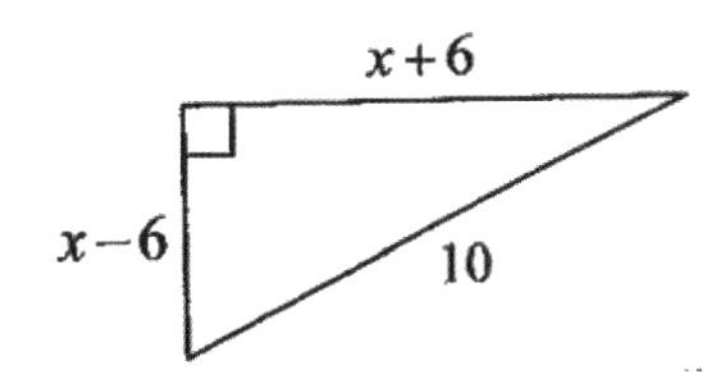

Note: Figure not drawn to scale.

14. In the figure above, what is the value of x ?

15. If $a \neq 0$ and $a^n \cdot a^5 = a^{20}$, what is the value of n ?

Note: Figure not drawn to scale.

16. In the figure above, what is the value of x ?

17. Three points $(-2,-7)$, $(0,-2)$, and $(k, 3)$ lie on the same line. What is the value of k ?

January	✈✈✈✈✈✈
February	✈✈✈✈
March	✈✈✈
April	✈✈✈✈✈
May	✈✈
June	✈✈✈✈

✈: 10 flights canceled

18. The graph above shows an Airline's flights canceled during six months. What is the mean (average) of flights canceled per month ?

STOP

If you finish before time is called, you may check your work on this section only. Do not turn to any other section in the test.

SECTION 3

Time — 20 minutes
16 Questions

Direction: For this section, solve each problem and decide which is the best of the choices given. Fill in the corresponding circle on the answer sheet. You may use any available space for scratch work.

Notes:
1. The use of a calculator is permitted.
2. All numbers used are real numbers.
3. Figures that accompany problems in this test are intended to provide information useful in solving the problems. They are drawn as accurately as possible EXCEPT when it is stated in a specific problem that the figure is not drawn to scale. All figures lie in a plane unless otherwise indicated.
4. Unless otherwise specified, the domain of any function f is to be the set of all real number x for which $f(x)$ is a real number.

Reference Information

$A = \pi r^2$
$C = 2\pi r$

$A = \ell w$

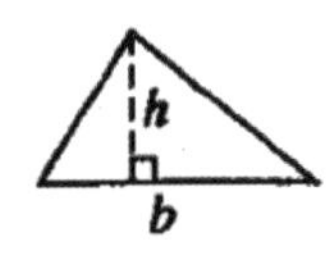

$A = \frac{1}{2}bh$

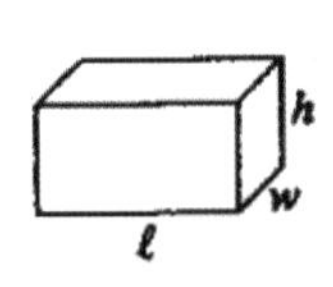

$V = \ell wh$

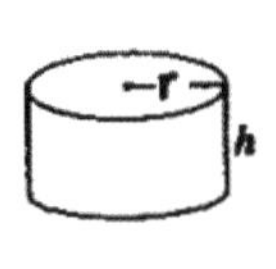

$V = \pi r^2 h$

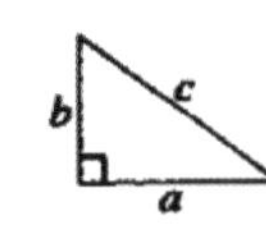

$c^2 = a^2 + b^2$

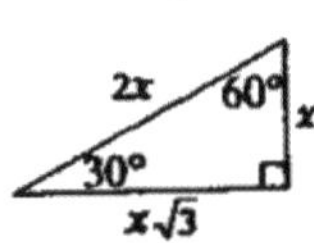

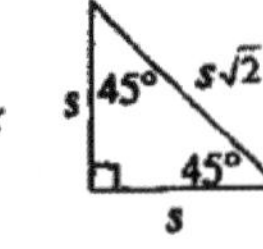

Special Right Triangles

The number of degrees of arc in a circle is 360.
The sum of the measures in degrees of the angles of a triangle is 180.

1. What percent of 8 is 14 ?

(A) 80%
(B) 95%
(C) 120%
(D) 175%
(E) 180%

2. In a mixture of alcohol and water, the ratio of alcohol to water is 2 to 5. How many liters of alcohol will there be in 21 liters of this mixture ?

(A) 6 liters
(B) 7 liters
(C) 9 liters
(D) 11 liters
(E) 12 liters

3. John paid \$8.20 for 3 pounds of beef and 2 pounds of pork. Maria paid \$5.40 for 2 pounds of beef and 1 pound of pork. What is the cost of 1 pound of beef ?

(A) \$3.10
(B) \$2.40
(C) \$2.60
(D) \$1.80
(E) \$1.60

GO ON TO THE NEXT PAGE

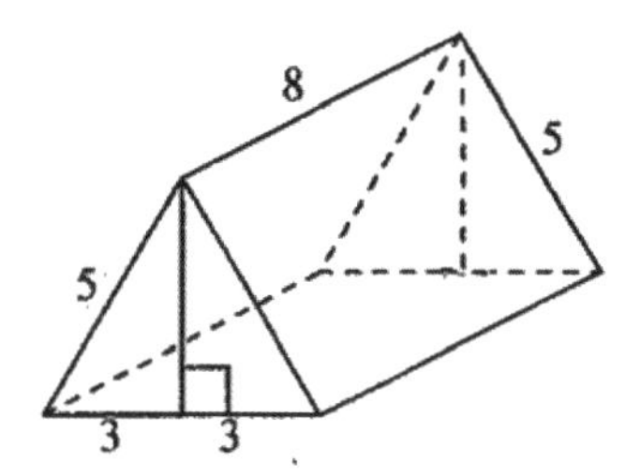

Note: Figure not drawn to scale.

4. In the figure above, what is the surface area of this triangular prism ?

(A) 122
(B) 144
(C) 152
(D) 162
(E) 168

5. If $x+2y=21$ and $3x-y=7$, what is the value of $x+y$?

(A) 10
(B) 13
(C) 15
(D) 18
(E) 20

6. There are one male and 7 females in a room. The male's weight is 2 times the average weight of the female. The male's weight is what fraction of the total weight of the 8 people in the room ?

(A) $\frac{3}{7}$
(B) $\frac{2}{7}$
(C) $\frac{1}{9}$
(D) $\frac{2}{9}$
(E) $\frac{1}{8}$

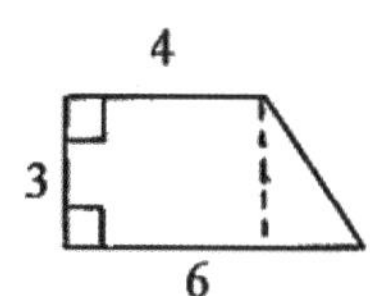

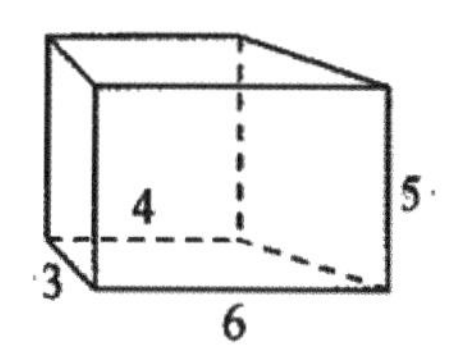

Note: Figure not drawn to scale.

7. In the figures above, the base of the solid is shown on the left. What is the volume of the solid ?

(A) 50
(B) 55
(C) 65
(D) 70
(E) 75

8. What is the maximum number of 3-digit codes can be made from 0 through 9 if no 0 is used for the first digit in the 3-digit code ?

(A) 500
(B) 600
(C) 800
(D) 900
(E) 1000

GO ON TO THE NEXT PAGE

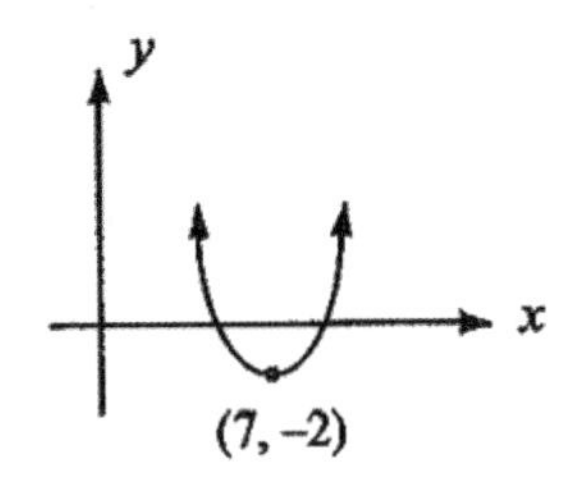

9. The graph above is a parabola. Which of the following could be an equation of the parabola ?

(A) $y=(x+7)^2-2$
(B) $y=(x-7)^2-2$
(C) $y=-(x+7)^2+2$
(D) $y=(x-7)^2+2$
(E) $y=-(x-7)^2-2$

10. A wire of x feet in length is cut into exactly 8 pieces, each 3 feet 2 inches in length. What is the value of x ? (Hint: 1 foot = 12 inches)

(A) $25\frac{1}{3}$
(B) $25\frac{1}{2}$
(C) $27\frac{1}{2}$
(D) 27
(E) 28

11. If $x+y=14$ and $6<x<12$, x and y are positive integers, what is the greatest possible value of $x-y$?

(A) 10
(B) 9
(C) 8
(D) 7
(E) 6

12. The mean of 7 consecutive even integers is n. What is the median of these 7 integers ?

(A) $n-4$
(B) $n-2$
(C) n
(D) $n+2$
(E) $n+4$

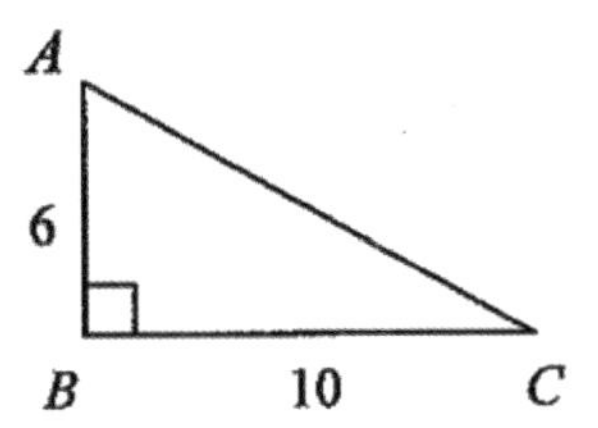

13. The right triangle ABC is revolved about $\overline{AB}$ as an axis generating a right cone. What is the volume of the cone ?

(A) 120π
(B) 150π
(C) 200π
(D) 320π
(E) 360π

GO ON TO THE NEXT PAGE

14. If you take a 5-question true-false test by just guessing at each answer, what is the probability that your score will be 100% ?

(A) $\frac{1}{5}$
(B) $\frac{1}{10}$
(C) $\frac{1}{25}$
(D) $\frac{1}{30}$
(E) $\frac{1}{32}$

Day	Number of computers sold
Monday	💻💻💻💻💻💻💻💻💻
Tuesday	💻💻💻💻💻💻💻
Wednesday	💻💻💻💻💻
Thursday	💻💻💻💻
Friday	💻💻💻💻💻💻
Saturday	💻💻💻💻
Sunday	💻💻💻💻💻💻💻

💻= 10 computers

15. The pictograph above shows the data of computer sold within a week. What is the median of the number of computers sold during the week ?

(A) 6
(B) 18
(C) 35
(D) 60
(E) 64

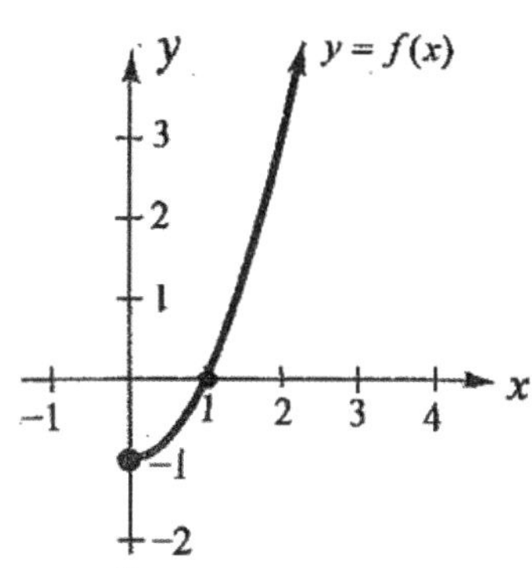

16. The graph of the function $y = f(x)$ is shown above. If the function $g(x)$ is defined by $g(x) = f(x) - 1$, what is the y-intercept of the function $g(x)$?

(A) −4
(B) −2
(C) 0
(D) 2
(E) 4

STOP

If you finish before time is called, you may check your work on this section only. Do not turn to any other section in the test.

PRACTICE TEST 3
(SAT Math)

Page 331 ~ 343

Answer Key

Section 1: **1.** D **2.** A **3.** B **4.** A **5.** D **6.** E **7.** C **8.** D **9.** C **10.** E

11. A **12.** C **13.** B **14.** D **15.** C **16.** A **17.** C **18.** A **19.** B **20.** E

Section 2: **1.** D **2.** B **3.** D **4.** C **5.** A **6.** E **7.** B **8.** E

9. 3 **10.** 6.25 **11.** 11 **12.** 3 **13.** 38.4 **14.** 3.74 **15.** 15 **16.** 55 **17.** 2 **18.** 40

Section 3: **1.** D **2.** A **3.** C **4.** C **5.** B **6.** D **7.** E **8.** D **9.** B **10.** A

11. C **12.** C **13.** C **14.** E **15.** D **16.** B

CALCULATE YOUR SCORE

Questions that you omit are not counted as correct answers or incorrect answers.

	Correct	Incorrect
Section 1		
Questions 1 ~ 20	( )	( )
Section 2	+	+
Questions 1 ~ 8	( )	( )
	+	
Questions 9 ~ 18	( )	
Section 3	+	+
Questions 1 ~ 16	( )	( )

Total Unrounded Raw Score ( ) − ( ) × 0.25 = ( )

Total Rounded Raw Score ( )

Your Math Scaled Score (See Table Below) ( )

SAT MATH SCORE CONVERSION TABLE

Raw Score	Scaled Score	Raw Score	Scaled Score	Raw Score	Scaled Score	Raw Score	Scaled Score
−11 to 0	200	16	400-460	32	520-600	48	670-730
1	220-310	17	410-470	33	520-610	49	680-750
2	240-330	18	420-470	34	530-610	50	690-770
3	250-350	19	430-480	35	540-620	51	730-780
4	270-370	20	430-490	36	550-620	52	760-800
5	280-370	21	440-510	37	560-630	53	780-800
6	290-400	22	450-510	38	560-640	54	800
7	300-410	23	460-520	39	570-650		
8	310-410	24	460-520	40	580-650		
9	320-420	25	470-540	41	600-660		
10	340-430	26	480-540	42	610-670		
11	340-430	27	490-550	43	620-670		
12	360-440	28	490-560	44	620-680		
13	380-440	29	500-560	45	630-690		
14	390-450	30	510-570	46	640-700		
15	400-460	31	510-590	47	660-710		

PRACTICE TEST 4
(SAT MATH)

SECTION 1
Time — 25 minutes
20 Questions

Direction: For this section, solve each problem and decide which is the best of the choices given. Fill in the corresponding circle on the answer sheet. You may use any available space for scratch work.

Notes:
1. The use of a calculator is permitted.
2. All numbers used are real numbers.
3. Figures that accompany problems in this test are intended to provide information useful in solving the problems. They are drawn as accurately as possible EXCEPT when it is stated in a specific problem that the figure is not drawn to scale. All figures lie in a plane unless otherwise indicated.
4. Unless otherwise specified, the domain of any function f is to be the set of all real number x for which $f(x)$ is a real number.

Reference Information

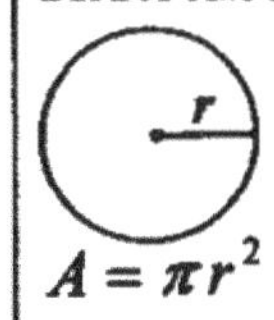

$A = \pi r^2$
$C = 2\pi r$

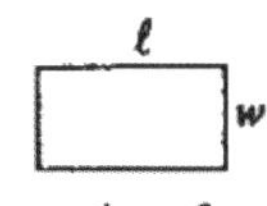

$A = \ell w$

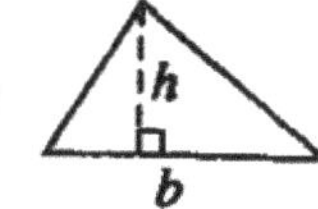

$A = \frac{1}{2}bh$

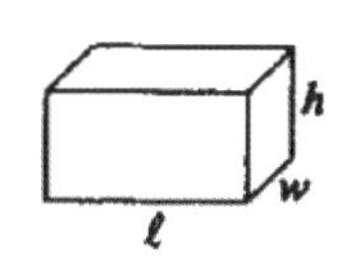

$V = \ell wh$

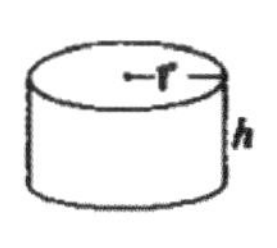

$V = \pi r^2 h$

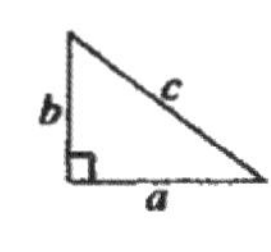

$c^2 = a^2 + b^2$

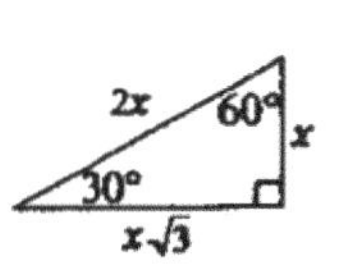

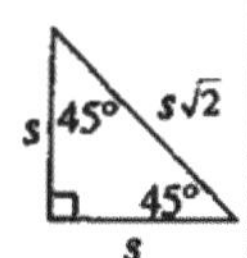

Special Right Triangles

The number of degrees of arc in a circle is 360.
The sum of the measures in degrees of the angles of a triangle is 180.

1. The average (arithmetic mean) of a, b, c, and d is 12. The average of a and b is 14. What is the average of c and d?

(A) 15
(B) 10
(C) 8
(D) 7
(E) 5

2. If $a+b=4$ and $a-b=2$, what is the value of a^2-b^2 ?

(A) 8
(B) 15
(C) 20
(D) 24
(E) 80

3. John can finish $5/7$ of the job in 15 hours. How many hours does he need to finish the whole Job ?

(A) 28
(B) 25
(C) 24
(D) 21
(E) 20

GO ON TO THE NEXT PAGE

4. Which of the following is the graph of a linear function with a negative slope and a negative y-intercept ?

(A)

(B)

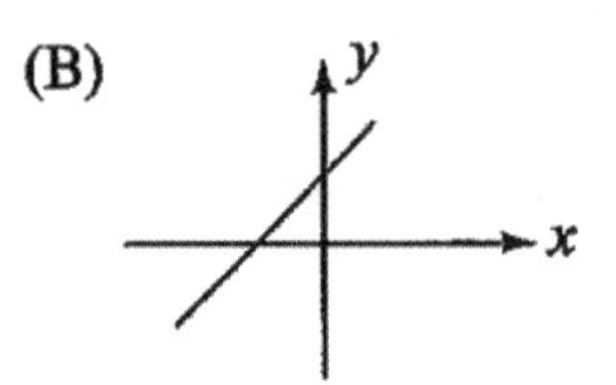

(C)

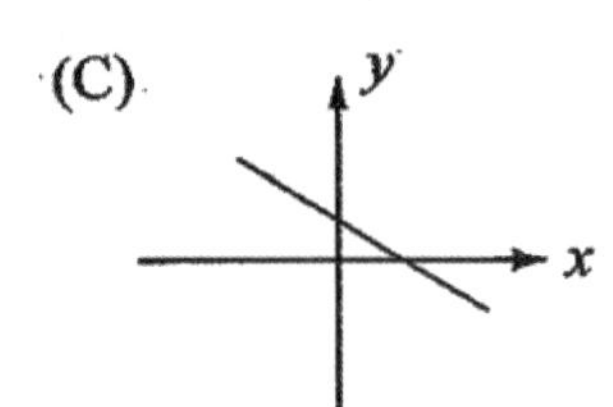

(D)

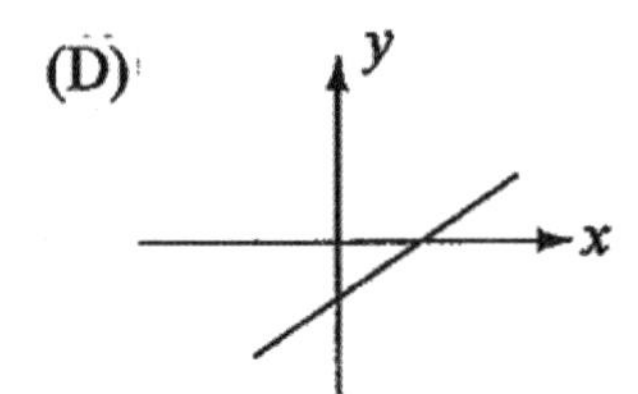

(E)

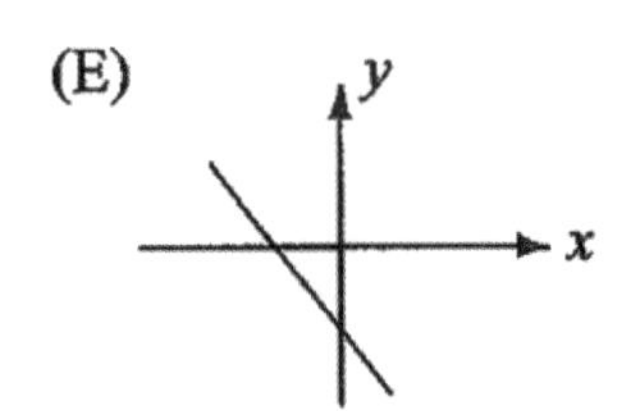

5. How many arrangements can be made from the letters *a, b, c, d,* and *e* in a row if letter *a* is not at either end ?

(A) 60
(B) 72
(C) 96
(D) 100
(E) 120

6. If $3^x + 3^x + 3^x = 27^x$, what is the value of x ?

(A) 4
(B) 2
(C) $\frac{1}{2}$
(D) $\frac{1}{3}$
(E) $\frac{1}{4}$

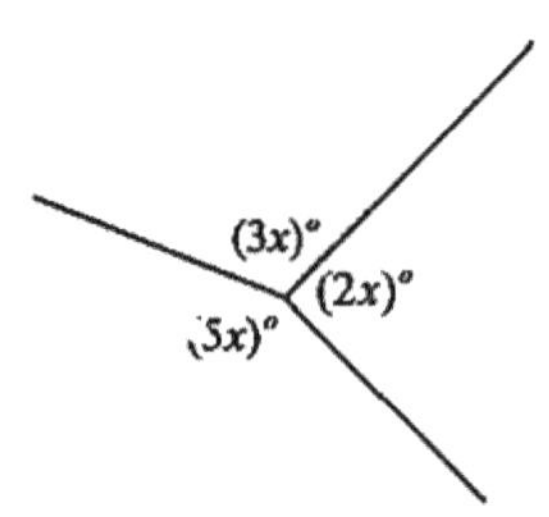

Note: Figure not drawn to scale.

7. In the figure above, what is the value of x ?

(A) 40
(B) 36
(C) 25
(D) 20
(E) 18

GO ON TO THE NEXT PAGE

8. A club consists of 30 boys and 25 girls. They want to select a girl as president, a boy as vice president, and a treasurer who may be either sex. How many choices are possible ?

(A) 10,800
(B) 21,350
(C) 25,480
(D) 39,750
(E) 40,100

9. If y is varies inversely as x, and if $y = 12$ when $x = \frac{2}{3}$, find y when $x = \frac{4}{5}$.

(A) 15
(B) 13
(C) 12
(D) 10
(E) 8

Note: Figure not drawn to scale.

10. In the figure above, what part of the circle is shaded ?

(A) $\frac{1}{14}$
(B) $\frac{2}{13}$
(C) $\frac{1}{12}$
(D) $\frac{5}{56}$
(E) $\frac{3}{29}$

11. What percent is 12 of 32 ?

(A) 35%
(B) 37.5%
(C) 39.5%
(D) 40%
(E) 41.3%

12. The ratio of sugar to water in a 230-gram sugar-water solution is 3 : 20. How many grams of sugar are there in the solution ?

(A) 34.5
(B) 30
(C) 26
(D) 24
(E) 20

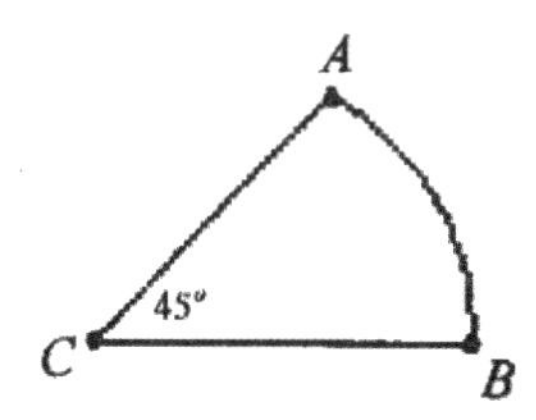

13. In the figure above, AB is the arc of a circle. If the length of the arc AB is 3π, what is the area of the sector ABC ?

(A) 24π
(B) 20π
(C) 18π
(D) 15π
(E) 12π

GO ON TO THE NEXT PAGE

14. If $a > b$ and $c < 0$, which of the following must be true ?

I. $ac > bc$
II. $ac < bc$
III. $a - c < b - c$

(A) I only
(B) II only
(C) I and II only
(D) II and III only
(E) I, II, and III

15. If $x^{\oplus}$ be defined such that $x^{\oplus} = x^3 - 3x$, what is the value of $4^{\oplus} + 2^{\oplus}$?

(A) 20
(B) 30
(C) 40
(D) 54
(E) 60

16. What is the 10th term of the following sequence ?
5, 11, 18, 26, 35, ……

(A) 97
(B) 95
(C) 90
(D) 84
(E) 72

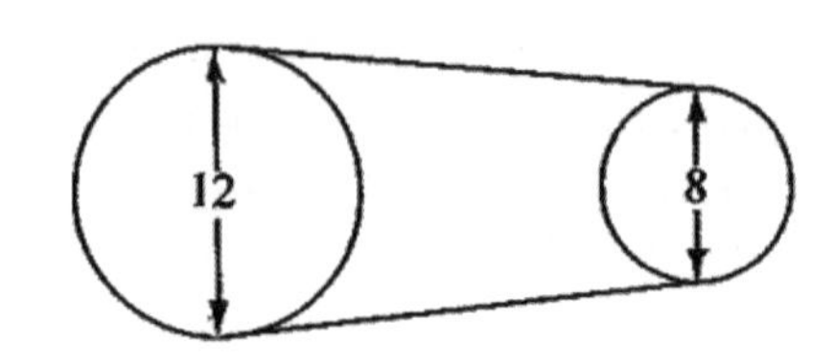

17. In the figure above, a large pulley is belted to a small pulley. If the large pulley runs 80 revolutions per minute, what is the speed of the small pulley, in revolutions per minute ?

(A) 120
(B) 140
(C) 168
(D) 174
(E) 192

18. $3\sqrt{18} + 2\sqrt{8} - 6\sqrt{50} =$

(A) $15\sqrt{3}$
(B) $12\sqrt{2}$
(C) $-17\sqrt{2}$
(D) $-20\sqrt{5}$
(E) $-25\sqrt{3}$

GO ON TO THE NEXT PAGE

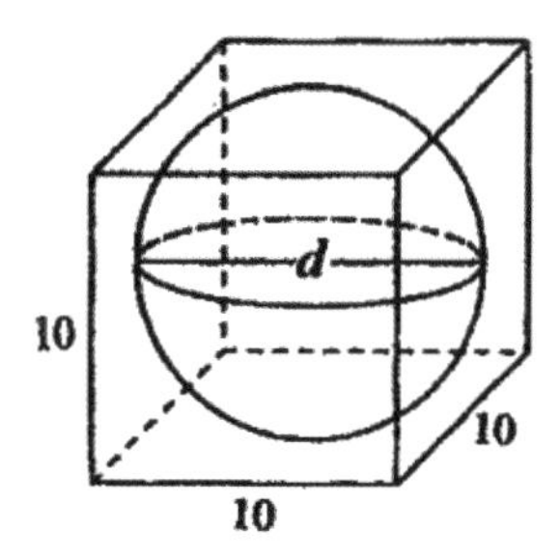

19. In the figure above, a sphere is inscribed in a cube. What is the ratio of the volume of the sphere to the volume of the cube ? (Volume of a sphere = $\frac{4}{3}\pi r^3$, $\pi = 3.14$)

(A) 0.314
(B) 0.523
(C) 0.705
(D) 0.782
(E) 0.815

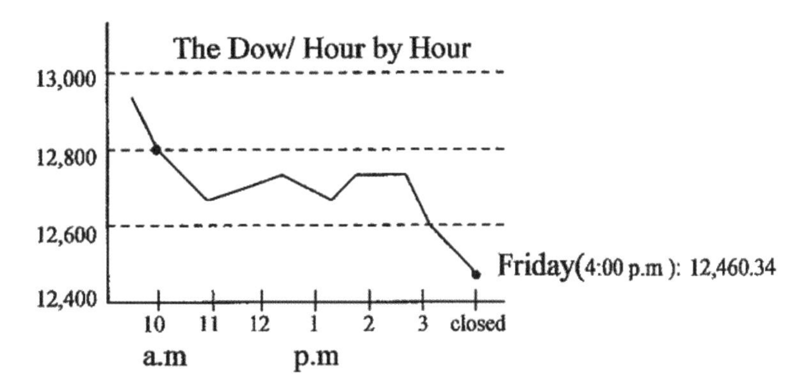

20. In the line graph above shows the Dow Jones industrial index for stock over one day, hour by hour. What percent did the Dow Johns industrial index go down from 10:00 a.m to 4 p.m ?

(A) 5.13 %
(B) 4.82 %
(C) 3.31 %
(D) 2.73 %
(E) 2.65 %

STOP

If you finish before time is called, you may check your work on this section only.
Do not turn to any other section in this test.

SECTION 2

Time — 25 minutes
18 Questions

Direction: This section contains two types of questions. You have 25 minutes to complete both types. For questions 1-8, solve each problem and decide which is the best of the choices given. Fill in the corresponding circle on the answer sheet. You may use any available space for scratch work.

Notes:

1. The use of a calculator is permitted.
2. All numbers used are real numbers.
3. Figures that accompany problems in this test are intended to provide information useful in solving the problems. They are drawn as accurately as possible EXCEPT when it is stated in a specific problem that the figure is not drawn to scale. All figures lie in a plane unless otherwise indicated.
4. Unless otherwise specified, the domain of any function f is assumed to be the set of all real number x for which $f(x)$ is a real number.

Reference Information

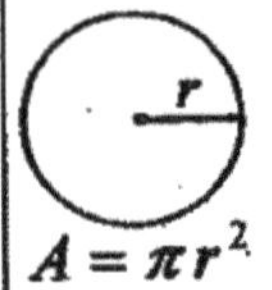

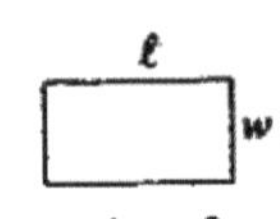

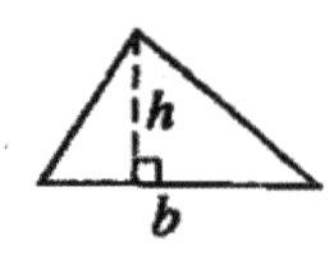

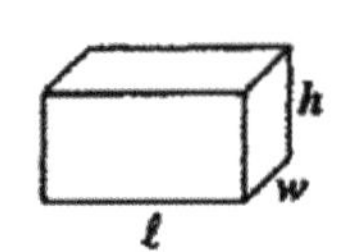

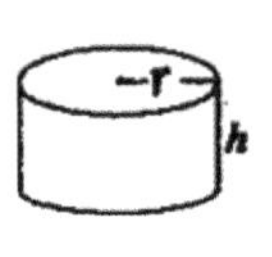

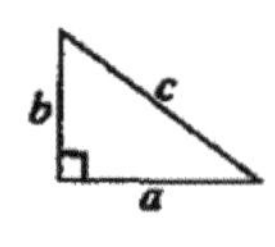

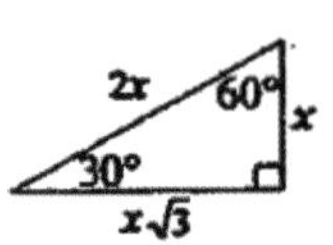

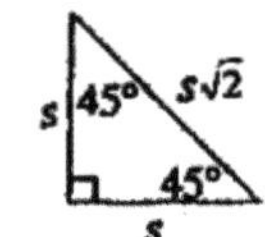

$A = \pi r^2$ $\quad A = \ell w$

$C = 2\pi r$ $\quad A = \frac{1}{2}bh$ $\quad V = \ell wh$ $\quad V = \pi r^2 h$ $\quad c^2 = a^2 + b^2$ $\quad$ **Special Right Triangles**

The number of degrees of arc in a circle is 360.
The sum of the measures in degrees of the angles of a triangle is 180.

1. If $x(y+2) = 3$ and x is a real number, which of the following CANNOT be the value of y ?

(A) 2
(B) 1
(C) 0
(D) −1
(E) −2

2. If $x^2 = y + 2$ and x is a real number, which of the following CANNOT be the value of y ?

(A) 1
(B) 0
(C) −1
(D) −2
(E) −3

3. The diameter of the solar system is about 118,000,000,000 miles. Which of the following is the scientific notation of this number ?

(A) 1.18×10^{12}
(B) 11.8×10^{10}
(C) 118×10^{9}
(D) 1.18×10^{11}
(E) 1.18×10^{10}

GO ON TO THE NEXT PAGE

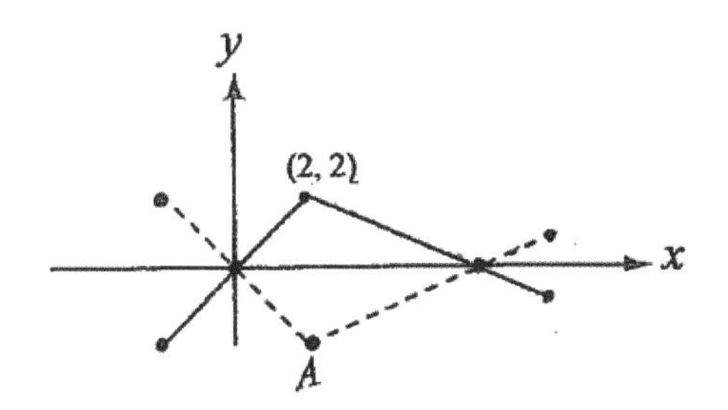

4. In the figure above, the dashed line is the reflection of the solid line about the x-axis. What is the coordinates of A ?

(A) (2, −2)
(B) (−2, 2)
(C) (−2, −1)
(D) (2, −1)
(E) (5, 0)

5. If a, b, and c are positive integers greater than 1, $ab = 45$ and $bc = 27$. Which of the following must be true ?

(A) $c > a > b$
(B) $b > c > a$
(C) $b > a > c$
(D) $a > c > b$
(E) $a > b > c$

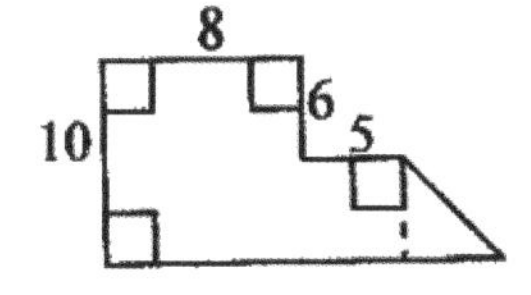

Note: Figure not drawn to scale

6. In the figure above, what is the perimeter ?

(A) 42
(B) 44
(C) 46
(D) 48
(E) 50

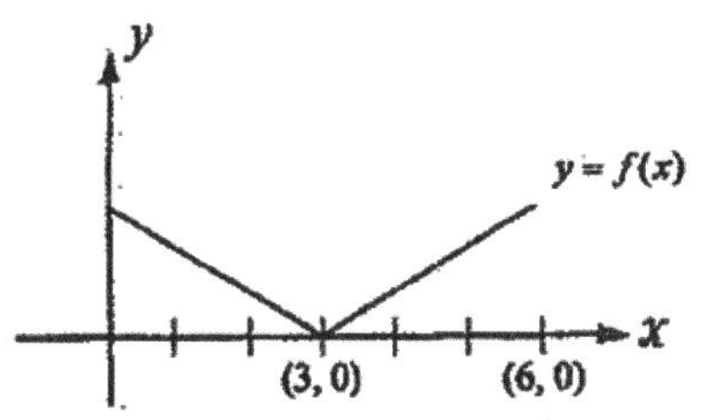

7. In the figure above, which of the following are x-coordinates of two points on the graph of the function whose y-coordinates are equal ?

(A) 1 and 4
(B) 1 and 6
(C) 2 and 4
(D) 2 and 5
(E) 2 and 6

8. The stretched length beyond its nature length of the wire spring is directly proportional to the mass of the weight attached. A 50 pounds weight causes a wire spring stretched 6 inches, what will be the stretched length of the wire spring if a 95 pounds weight is attached ?

(A) 7.2 inches
(B) 8.5 inches
(C) 10.6 inches
(D) 11.4 inches
(E) 12.3 inches

GO ON TO THE NEXT PAGE

Direction: For Student-Produced Response questions 9-18, use the grids at the bottom of the answer sheet page on which you have answered questions 1-8.

Each of the remaining 10 questions requires you to solve the problem and enter your answer by marking the circle in the special grid, as shown in the examples below. You may use any available space for scratch work.

Answer: 7/12

Write answer in box → | 7 | / | 1 | 2 | ←Fraction line

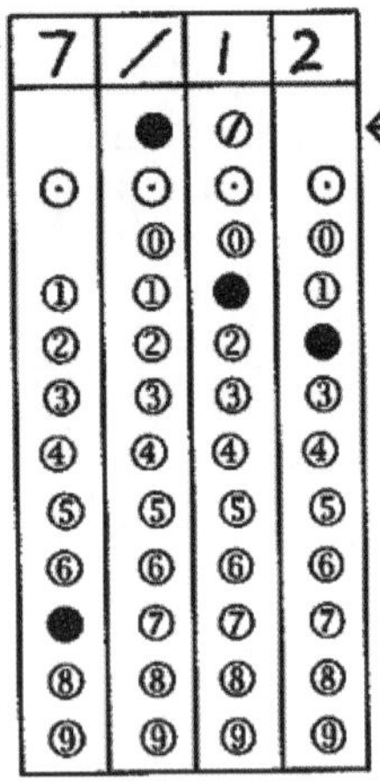

Grid in→ result

Answer: 2.5

| | 2 | . | 5 | ←Decimal point

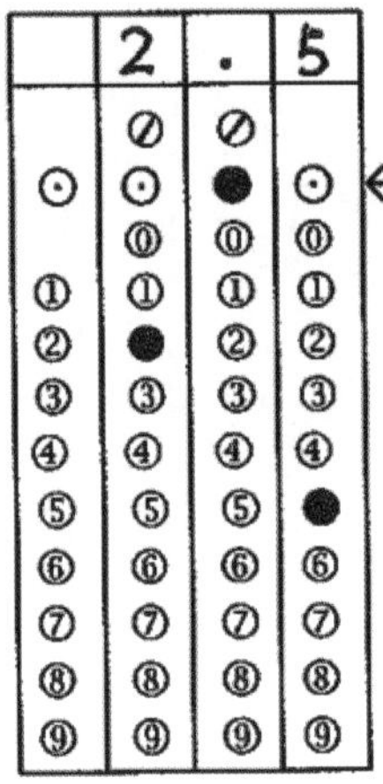

Answer: 201
Either position is correct.

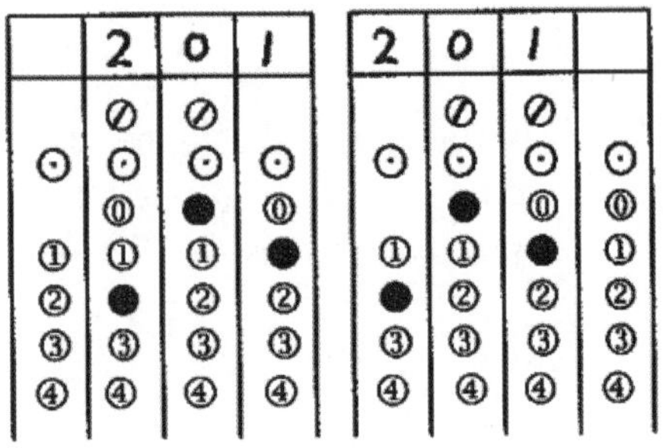

| | 2 | 0 | 1 | | 2 | 0 | 1 | |

Note: You may start your answers in any column, space permitting. Columns not needed should be left blank.

- Mark no more than one circle in any column.
- Because the answer sheet will be machine-scored, **you will receive credit only if the circles are filled in correctly.**
- Although not required, it is suggested that you write your answers in the boxes at the top of the columns to help you fill in the circles accurately.
- Some problems may have more than one correct answer. In such cases, grid only one answer.
- No question has a negative answer.
- **Mixed numbers such as** $3\frac{1}{2}$ must be gridded as 3.5 or 7/2. (If | 3 | 1 | / | 2 | is gridded, it will be interpreted as $\frac{31}{2}$, not $3\frac{1}{2}$.
- **Decimal Answers:** If you obtain a decimal answer with more digits than the grid can accommodate, it may be either rounded or truncated, but it must fill the entire grid. For example, if you obtain an answer such as 0.6666····, you should record you result as 0.666 or 0.667. **A less accurate value such as 0.66 or 0.67 will be scored as incorrect.**
  Acceptable ways to grid 2/3 are:

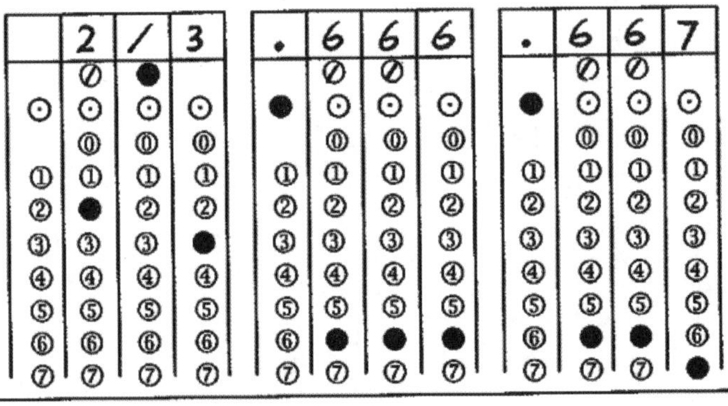

| | 2 | / | 3 | | . | 6 | 6 | 6 | | . | 6 | 6 | 7 |

9. If $5(a-1)=3a+2$, then $a=$

10. If $a-b=-2$ and $2a-b=4$, then $a+b=$

GO ON TO THE NEXT PAGE

11. What is the smallest number which is divided by 6, divided by 8, or divided by 11. All divisions have the same remainder of 5 ?

12. What is the value of n for $(n^{1/2})^{2/3} = 4$?

13. The vertices of a triangle ABC in the coordinate plane are $A(3, 4)$, $B(3, -1)$, and $C(-2, -1)$. What is the area of the triangle ?

14. What is the distance between two points, $A(1, -6)$ and $B(7, 2)$?

(–2, 4) (3, 4)

(–6, 1)

(–6, –2) (3, –2)

15. In the figure above, the coordinates of a polygon are shown. It has right angles at three corners. What is the area of the polygon ?

16. The height (h) in feet of an object reaches in t seconds after it is thrown upward with an initial speed of v feet per second is given by the formula $h = vt - 16t^2$. A rocket is fired upward with an initial speed of 224 feet per second. What is the height in feet of the rocket reaches in 4 seconds after being fired upward ?

17. In the school theater, there are 372 students at present and is $6/17$ full. What is the total number of students that the theater can hold when it is full ?

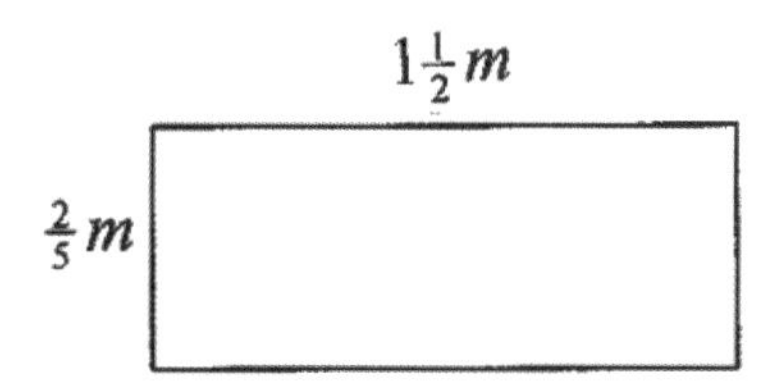

18. The drawing above shows the length and the width of a rectangular field on a map. What is the actual width in meters if the actual length is 300 meters ?

STOP

If you finish before time is called, you may check your work on this section only. Do not turn to any other section in the test.

SECTION 3

Time — 20 minutes
16 Questions

Direction: For this section, solve each problem and decide which is the best of the choices given. Fill in the corresponding circle on the answer sheet. You may use any available space for scratch work.

Notes:
1. The use of a calculator is permitted.
2. All numbers used are real numbers.
3. Figures that accompany problems in this test are intended to provide information useful in solving the problems. They are drawn as accurately as possible EXCEPT when it is stated in a specific problem that the figure is not drawn to scale. All figures lie in a plane unless otherwise indicated.
4. Unless otherwise specified, the domain of any function f is to be the set of all real number x for which $f(x)$ is a real number.

Reference Information

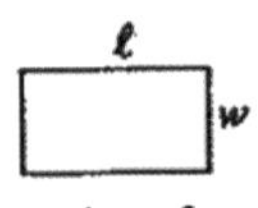

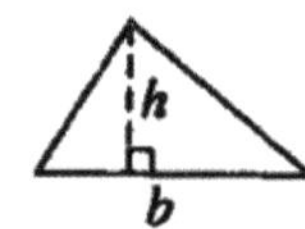

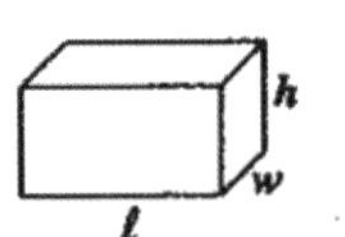

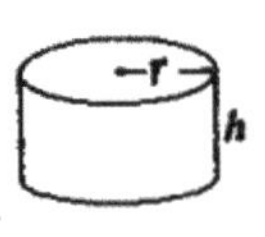

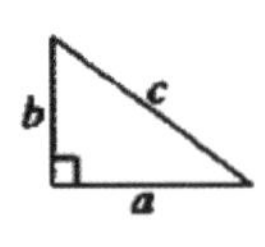

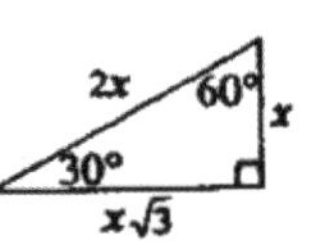

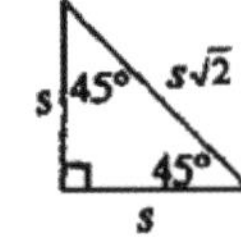

$A = \pi r^2$ $A = \ell w$
$C = 2\pi r$ $A = \frac{1}{2}bh$ $V = \ell wh$ $V = \pi r^2 h$ $c^2 = a^2 + b^2$ Special Right Triangles

The number of degrees of arc in a circle is 360.
The sum of the measures in degrees of the angles of a triangle is 180.

1. If $x = 3$ and $y = -2$, then $\frac{1}{2}x^2 - 3y^3 =$

(A) 28.5
(B) 27
(C) 24.5
(D) 20
(E) 19.5

2. 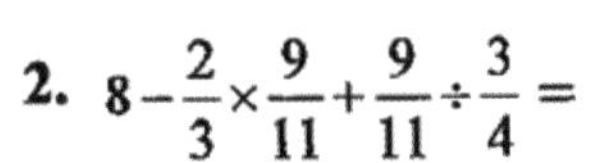
$8 - \frac{2}{3} \times \frac{9}{11} + \frac{9}{11} \div \frac{3}{4} =$

(A) $9\frac{1}{8}$
(B) $9\frac{5}{8}$
(C) $8\frac{6}{11}$
(D) $7\frac{5}{11}$
(E) $7\frac{9}{11}$

Note: Figure not drawn to scale.

3. In the figure above, $AB = BC$. If $x = 30$, what is the length of side AC ?

(A) $4\sqrt{2}$
(B) 4
(C) $2\sqrt{2}$
(D) 2
(E) 5

GO ON TO THE NEXT PAGE

4. If three points $(-1, 3)$, $(-3, 2)$, and $(-4,\ a-5)$ are on a straight line, what is the value of a ?

(A) 9
(B) 8.5
(C) 7.5
(D) 7
(E) 6.5

5. The average of Roger's scores (arithmetic mean) for 8 math tests is 76. What does the average score of the next 2 tests have to be if the average of the entire 10 tests equals 80 ?

(A) 88
(B) 90
(C) 92
(D) 96
(E) 98

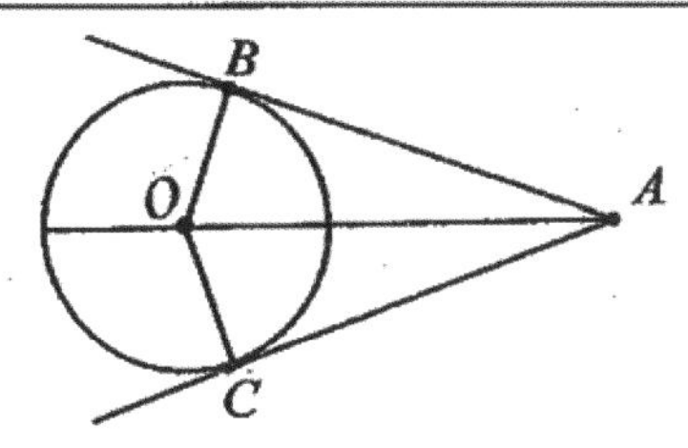

Note: Figure not drawn to scale.

6. In the figure above, $\overline{AB}$ and $\overline{AC}$ are tangent to the circle O at B and C. If $OA = 10$ and $OB = 6$, what is the value of $AB + AC$?

(A) 24
(B) 22
(C) 20
(D) 18
(E) 16

7. If $g(x) = \sqrt{x-4}$, what is the range of the function $g(x)$?

(A) All real numbers
(B) All real numbers greater than or equal to 4
(C) All real numbers greater than 0
(D) All real number greater than 4
(E) All real number except 4

8. A plane intersects a rectangular solid. Which of the following could be the intersection ?

I. A triangle
II. A rectangle
III. A parallelogram

(A) I only
(B) II only
(C) I and III only
(D) II and III only
(E) I, II, and III

9. The sum of two numbers is 42. The larger number is 18 more than one-half of the smaller number. What is the value of the larger number ?

(A) 26
(B) 28
(C) 30
(D) 32
(E) 34

GO ON TO THE NEXT PAGE

10. If $n!$ be defined such that

$$n! = n(n-1)(n-2)\cdots 3\cdot 2\cdot 1,$$

What is the value of $5! \div 4!$?

(A) 7
(B) 6
(C) 5
(D) 4
(E) 3

11. If $|2a-8| = 6$, then $a =$

(A) 1, 7
(B) 2, 6
(C) −1, 7
(D) −2, −6
(E) 0, 7

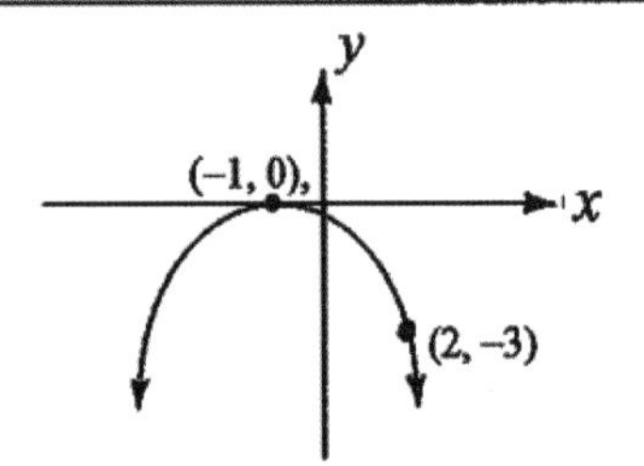

12. Which of the following is the equation of the graph of parabola above ?

(A) $y = 3(x-1)^2$
(B) $y = -3(x-1)^2$
(C) $y = \frac{1}{3}(x-1)^2$
(D) $y = \frac{1}{3}(x+1)^2$
(E) $y = -\frac{1}{3}(x+1)^2$

13. How many of the integers between 1 and 250 are multiples of 4 or 5 ?

(A) 50
(B) 62
(C) 80
(D) 94
(E) 100

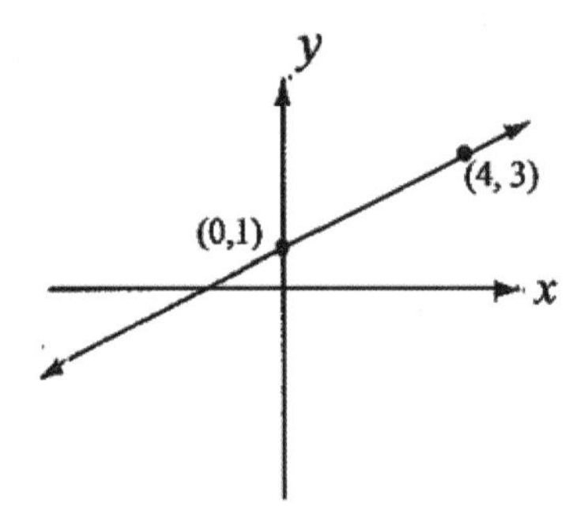

14. In the figure above shows the graph of a straight line. If each point on the line is translated horizontally 3 units to the left, what is the value of y-intercept of the translating line ?

(A) $\frac{3}{2}$
(B) $\frac{5}{2}$
(C) 3
(D) $\frac{7}{2}$
(E) 4

GO ON TO THE NEXT PAGE

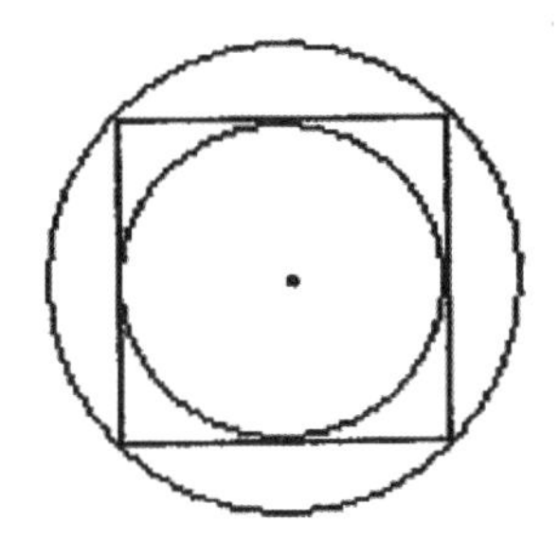

15. In the figure above, a large circle is circumscribed about a square and a small circle is inscribed in the same square. What is the ratio of the area of the small circle to the large circle ?

(A) 1 : 2
(B) 1 : 4
(C) 2 : 3
(D) 2 : 5
(E) 3 : 5

A-Plus Company
Annual Sale of Books

Year	Number Sold
2003	4,000
2004	4,600
2005	5,200
2006	6,000
2007	6,900
2008	7,900

16. In the table above, which yearly period had the largest percent increase in sale ?

(A) 2003~2004
(B) 2004~2005
(C) 2005~2006
(D) 2006~2007
(E) 2007~2008

STOP

If you finish before time is called, you may check your work on this section only.
Do not turn to any other section in the test.

PRACTICE TEST 4
(SAT Math)

Page 345 ~ 357

Answer Key

Section 1: **1.** B **2.** A **3.** D **4.** E **5.** B **6.** C **7.** B **8.** D **9.** D **10.** D

11. B **12.** B **13.** C **14.** B **15.** D **16.** B **17.** A **18.** C **19.** B **20.** E

Section 2: **1.** E **2.** E **3.** D **4.** A **5.** C **6.** E **7.** C **8.** D

9. $7/2$ or 3.5 **10.** 14 **11.** 269 **12.** 64 **13.** $25/2$ or 12.5 **14.** 10 **15.** 48 **16.** 640

17. 1054 **18.** 80

Section 3: **1.** A **2.** C **3.** B **4.** E **5.** D **6.** E **7.** B **8.** E **9.** A **10.** C

11. A **12.** E **13.** E **14.** B **15.** A **16.** C

CALCULATE YOUR SCORE

Questions that you omit are not counted as correct answers or incorrect answers

	Correct	Incorrect
Section 1		
Questions 1 ~ 20	( )	( )
Section 2	+	+
Questions 1 ~ 8	( )	( )
	+	
Questions 9 ~ 18	( )	
Section 3	+	+
Questions 1 ~ 16	( )	( )

Total Unrounded Raw Score ( ) – ( ) × 0.25 = ( )
Total Rounded Raw Score ( )
Your Math Scaled Score (See Table Below) ( )

SAT MATH SCORE CONVERSION TABLE

Raw Score	Scaled Score	Raw Score	Scaled Score	Raw Score	Scaled Score	Raw Score	Scaled Score
–11 to 0	200	16	400-460	32	520-600	48	670-730
1	220-310	17	410-470	33	520-610	49	680-750
2	240-330	18	420-470	34	530-610	50	690-770
3	250-350	19	430-480	35	540-620	51	730-780
4	270-370	20	430-490	36	550-620	52	760-800
5	280-370	21	440-510	37	560-630	53	780-800
6	290-400	22	450-510	38	560-640	54	800
7	300-410	23	460-520	39	570-650		
8	310-410	24	460-520	40	580-650		
9	320-420	25	470-540	41	600-660		
10	340-430	26	480-540	42	610-670		
11	340-430	27	490-550	43	620-670		
12	360-440	28	490-560	44	620-680		
13	380-440	29	500-560	45	630-690		
14	390-450	30	510-570	46	640-700		
15	400-460	31	510-590	47	660-710		

PRACTICE TEST 5
(SAT MATH)

SECTION 1
Time — 25 minutes
20 Questions

Direction: For this section, solve each problem and decide which is the best of the choices given. Fill in the corresponding circle on the answer sheet. You may use any available space for scratch work.

Notes:
1. The use of a calculator is permitted.
2. All numbers used are real numbers.
3. Figures that accompany problems in this test are intended to provide information useful in solving the problems. They are drawn as accurately as possible EXCEPT when it is stated in a specific problem that the figure is not drawn to scale. All figures lie in a plane unless otherwise indicated.
4. Unless otherwise specified, the domain of any function f is to be the set of all real number x for which $f(x)$ is a real number.

Reference Information

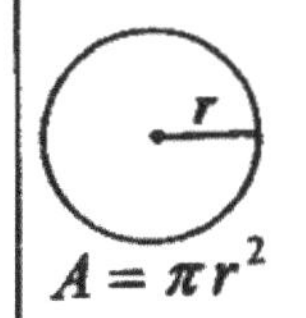

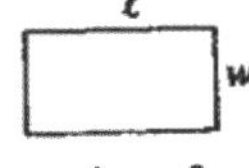

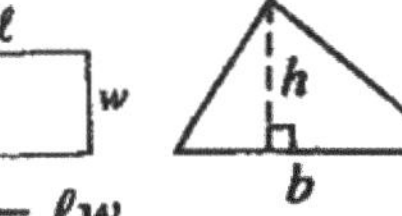

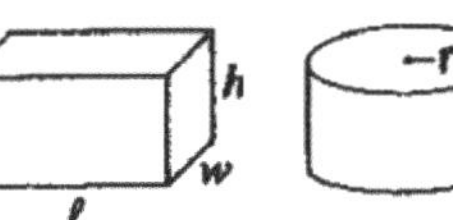

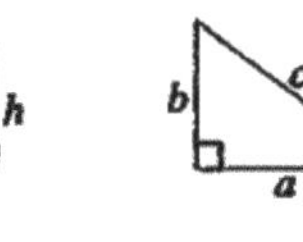

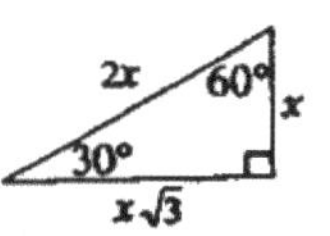

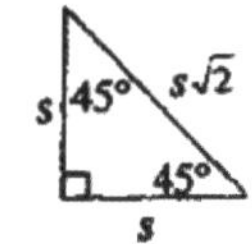

$A = \pi r^2$ $\quad C = 2\pi r$ $\quad A = \ell w$ $\quad A = \frac{1}{2}bh$ $\quad V = \ell wh$ $\quad V = \pi r^2 h$ $\quad c^2 = a^2 + b^2$ $\quad$ Special Right Triangles

The number of degrees of arc in a circle is 360.
The sum of the measures in degrees of the angles of a triangle is 180.

1. If $(a-4)^2 = 36$ and $a > 0$, what is the value of a^3 ?

(A) 8
(B) 64
(C) 96
(D) 1000
(E) 1080

$$\underset{A}{\bullet}\overset{2x}{\text{———}}\underset{B}{\bullet}\overset{5x}{\text{——————}}\underset{C}{\bullet}$$

2. In the figure above, the length of the line segment AC is 42, what is the length of the line segment BC ?

(A) 30
(B) 28
(C) 26
(D) 24
(E) 22

GO ON TO THE NEXT PAGE

3. If $|x-4|<12$, which of the following cannot be the value of x?

(A) −9
(B) −7
(C) 0
(D) 4
(E) 15

4. What are the domain of $y=5+\sqrt{x-2}$?

(A) All real numbers
(B) All nonnegative real numbers
(C) All real numbers greater than or equal to 1
(D) All real numbers greater than or equal to 2
(E) All real numbers greater than or equal to 5

5. What are the range of $y=5+\sqrt{x-2}$?

(A) All real numbers
(B) All nonnegative real numbers
(C) All real numbers greater than or equal to 1
(D) All real numbers greater than or equal to 2
(E) All real numbers greater than or equal to 5

6. If $2^{5x}=64$, then $x=$

(A) 6
(B) 4
(C) $\frac{3}{2}$
(D) $\frac{5}{2}$
(E) $\frac{6}{5}$

7. The quadratic function is given by $h(x)=-\frac{1}{2}(x-1)^2+3$. Which of the following could be the graph of h?

(A)
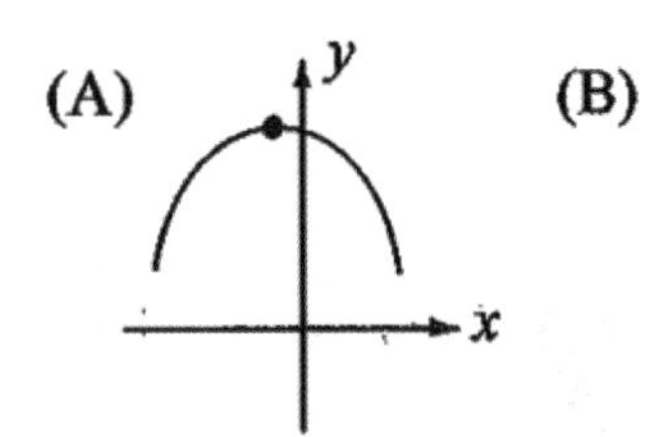

(B)
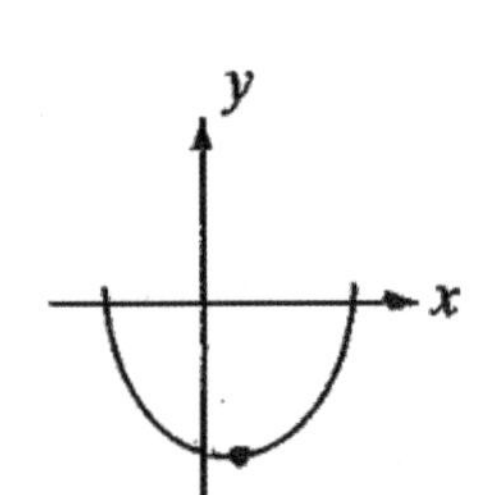

(C)
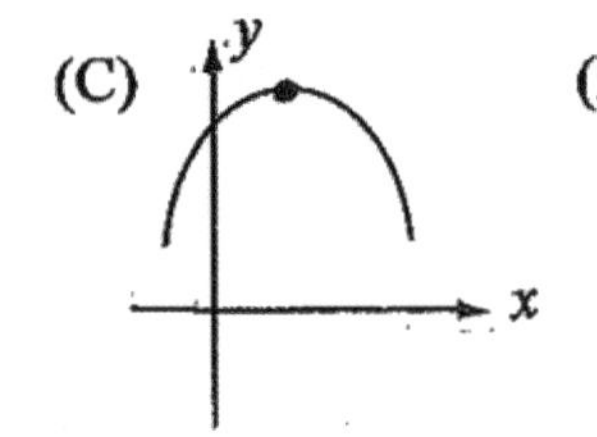

(D)
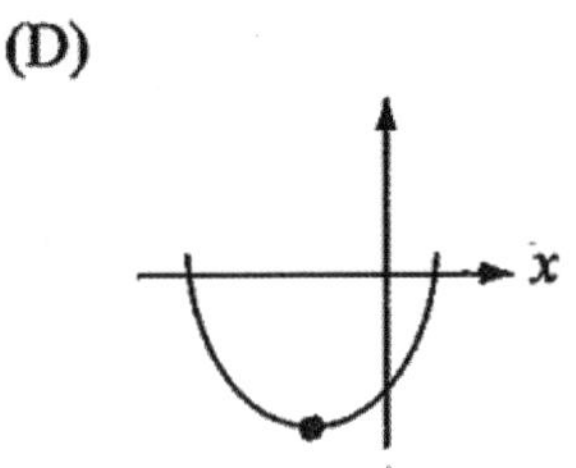

(E)
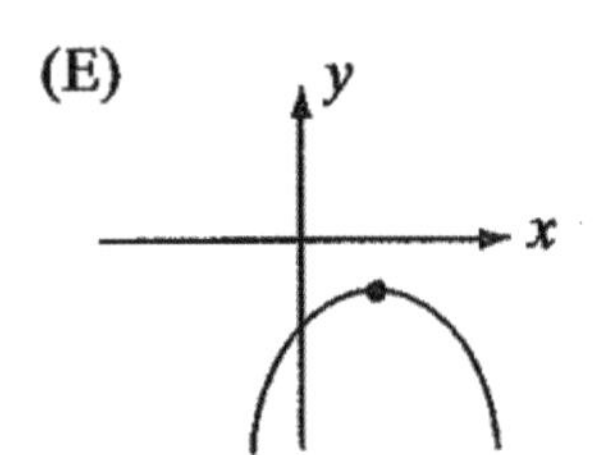

8. If $a=2+\dfrac{1}{b}$ and $b>1$, which of the following could be the value of a?

(A) 1.5
(B) 1.9
(C) 2.8
(D) 3.2
(E) 3.6

GO ON TO THE NEXT PAGE

9. A number multiplied by 8 is equal to the number increased by 8. What is the value of the number ?

(A) $\frac{4}{3}$
(B) $\frac{8}{7}$
(C) $\frac{7}{8}$
(D) $\frac{3}{4}$
(E) 16

10. If $r + 3s - 4t = 5$, what is the value of r in terms of s and t ?

(A) $3s - 4t - 5$
(B) $4t - 3s - 5$
(C) $-3s - 4t - 5$
(D) $5 - 4t + 3s$
(E) $5 + 4t - 3s$

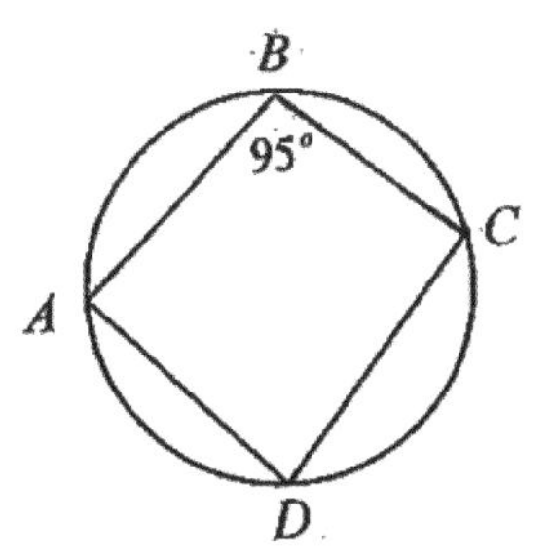

Note: Figure not drawn to scale.

11. In the figure above, quadrilateral $ABCD$ is inscribed in the circle, $\angle ABC = 95°$. What is the measure of $\angle ADC$?

(A) 70°
(B) 75°
(C) 80°
(D) 85°
(E) 90°

12. If x is divided by 4, the remainder is 1. If x is divided by 5, the remainder is 2. If x is divided by 6, the remainder is 3. What is the smallest value of x ?

(A) 25
(B) 37
(C) 45
(D) 57
(E) 60

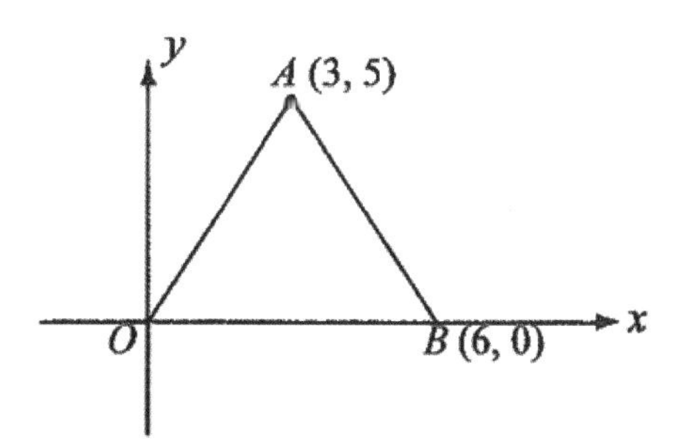

13. In the figure above, $OA = AB$, what is area of the triangle OAB ?

(A) 10
(B) 12
(C) 13
(D) 15
(E) 20

GO ON TO THE NEXT PAGE

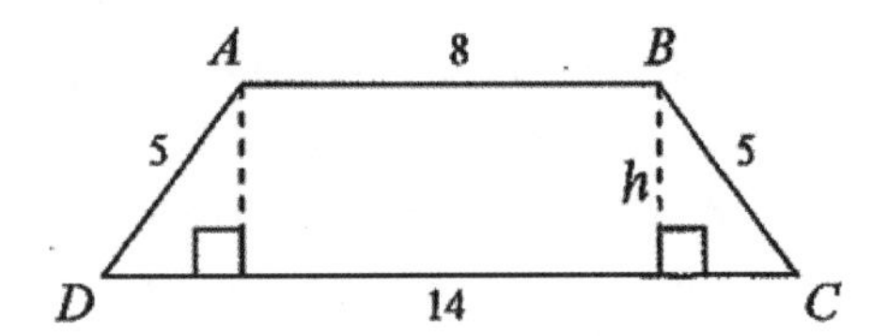

Note: Figure not drawn to scale.

14. In the figure above, what is the area of the quadrilateral *ABCD* ?

(A) 44
(B) 88
(C) 96
(D) 112
(E) 120

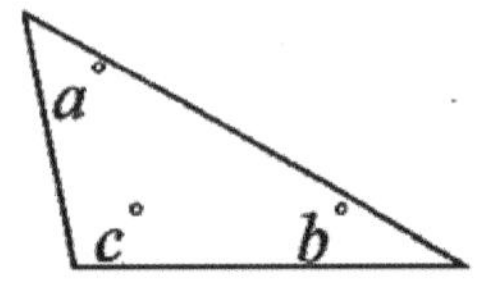

Note: Figure not drawn to scale.

15. In the triangle shown above, a, b, and c are all integers, $c > 90$ and $a = b + 12$. What is the greatest possible value of a ?

(A) 47
(B) 48
(C) 49
(D) 50
(E) 51

16. Three-fifths of the books were sold at $4 each, one-half of the rest of the book were sold at $2 each, and the remaining 65 books were sold at $1 each. What was the total amount of dollars earned from the books sold ?

(A) $1250
(B) $1040
(C) $ 975
(D) $ 520
(E) $ 395

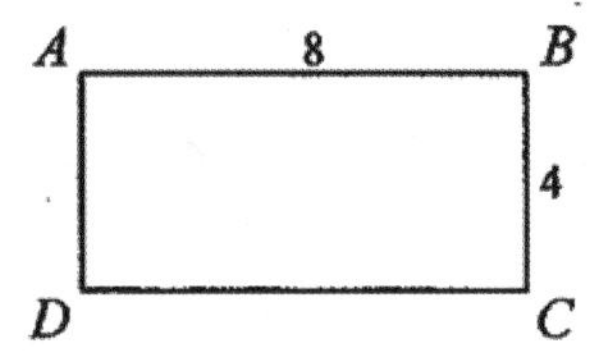

Note: Figure not drawn to scale.

17. In the figure about, a solid is made by rotating the rectangle *ABCD* 360° about the side *AD*. What is the volume of the solid ?

(A) 278π
(B) 256π
(C) 218π
(D) 187π
(E) 124π

18. The ratio of the lengths of three sides of a triangle is 3 : 4 : 5. What is the length of the longest side if the perimeter is 84 ?

(A) 21
(B) 28
(C) 35
(D) 42
(E) 48

GO ON TO THE NEXT PAGE

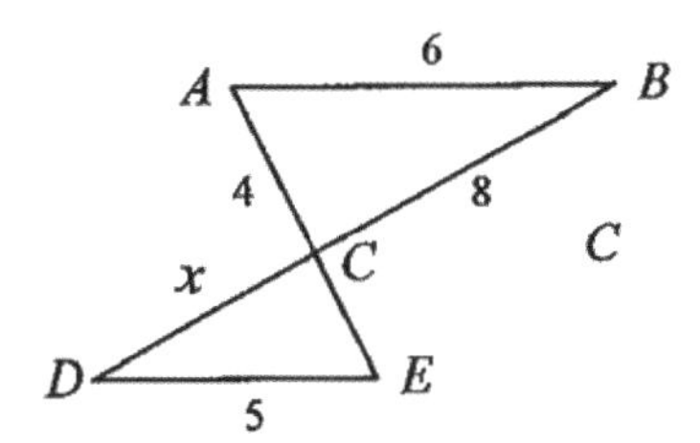

Note: Figure not drawn to scale.

19. In the figure above, $\Delta ABC \sim \Delta CDE$, what is the value of x ?

(A) 9.60
(B) 6.67
(C) 5.82
(D) 5.43
(E) 3.33

The Dow/ Day by Day

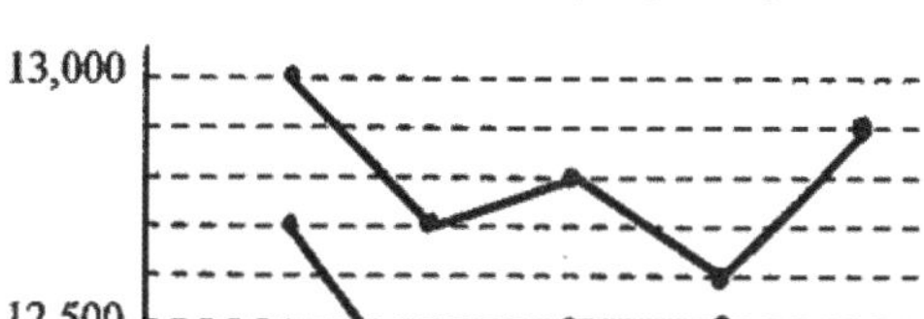
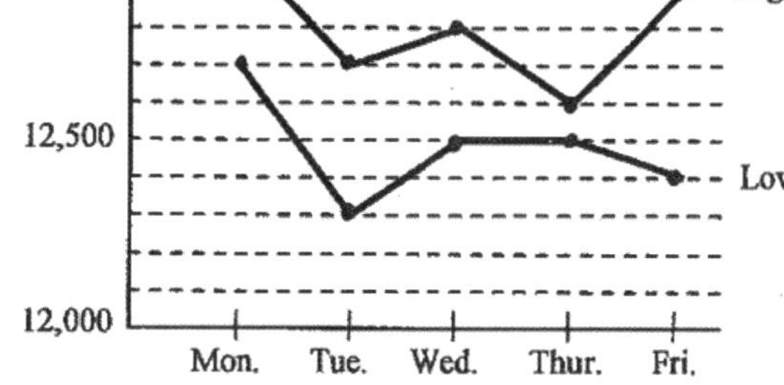

20. The line graph above shows the Dow Jones industrial index for stock from Monday to Friday. Which day has the biggest difference between the high and the low.

(A) Monday
(B) Tuesday
(C) Wednesday
(D) Thursday
(E) Friday

STOP

If you finish before time is called, you may check your work on this section only. Do not turn to any other section in this test.

SECTION 2

Time — 25 minutes
18 Questions

Direction: This section contains two types of questions. You have 25 minutes to complete both types. For questions 1-8, solve each problem and decide which is the best of the choices given. Fill in the corresponding circle on the answer sheet. You may use any available space for scratch work.

Notes:
1. The use of a calculator is permitted.
2. All numbers used are real numbers.
3. Figures that accompany problems in this test are intended to provide information useful in solving the problems. They are drawn as accurately as possible EXCEPT when it is stated in a specific problem that the figure is not drawn to scale. All figures lie in a plane unless otherwise indicated.
4. Unless otherwise specified, the domain of any function f is assumed to be the set of all real number x for which $f(x)$ is a real number.

Reference Information

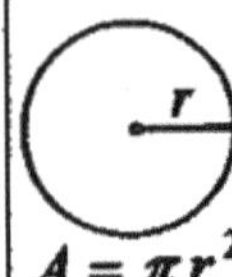

$A = \pi r^2$
$C = 2\pi r$

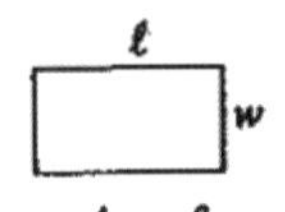

$A = \ell w$

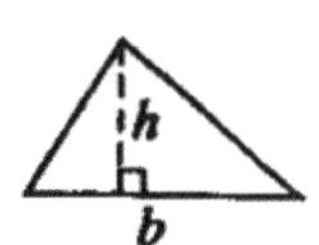

$A = \frac{1}{2}bh$

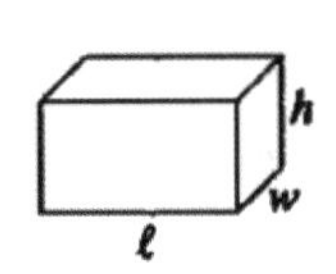

$V = \ell wh$

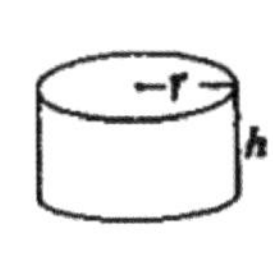

$V = \pi r^2 h$

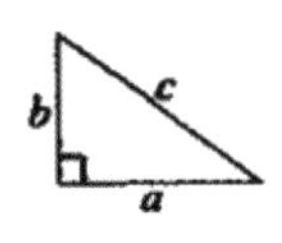

$c^2 = a^2 + b^2$

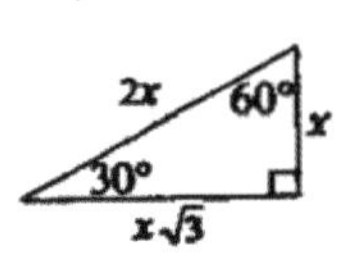

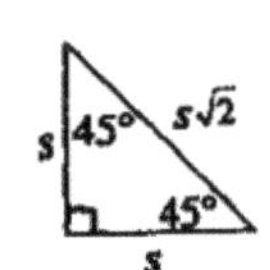

Special Right Triangles

The number of degrees of arc in a circle is 360.
The sum of the measures in degrees of the angles of a triangle is 180.

1. Which of the following numbers has the digit 5 in the thousandths place ?

(A) 5000.500
(B) 500.050
(C) 0.005
(D) 0.050
(E) 0.500

2. If $2(x-3)=8$, then $x =$

(A) 3
(B) 4
(C) 5
(D) 6
(E) 7

3. Which of the following is the sum of $2\frac{1}{3}$ and its reciprocal ?

(A) $2\frac{16}{21}$
(B) $2\frac{13}{18}$
(C) $3\frac{15}{19}$
(D) $3\frac{19}{21}$
(E) 6

GO ON TO THE NEXT PAGE

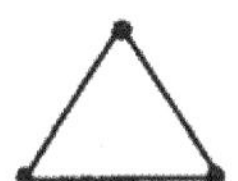

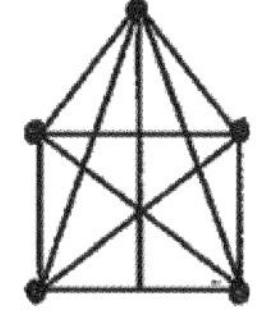

4. The figures above shown the line segments can be drawn between 2, 3, 4, and 5 points, no three of which lie on the same line. How many line segments can be drawn between 7 points ?

(A) 20
(B) 21
(C) 22
(D) 23
(E) 28

5. If $x^2 - 12x + 32 = 0$, then $x =$

(A) 2, 16
(B) 4, 8
(C) 5, 4
(D) 6, 2
(E) 7, 8

6. Steve ate one-third of the pizza. Maria ate three-fifths of the remaining part. What fractional part of the whole pizza did Maria eat ?

(A) $\frac{4}{5}$
(B) $\frac{3}{5}$
(C) $\frac{2}{5}$
(D) $\frac{3}{4}$
(E) $\frac{4}{7}$

Questions 7-8 refer to the following figures.

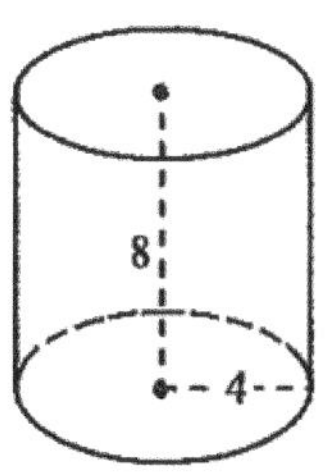

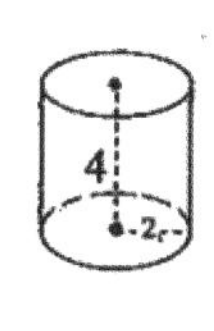

7. In the figure above, what is the ratio of the total surface area of the large cylinder to the total surface area of the small cylinder ?

(A) 3 : 1
(B) 4 : 1
(C) 6 : 1
(D) 8 : 1
(E) 9 : 1

8. In the figure above, what is the ratio of the volume of the large cylinder to the volume of the small cylinder ?

(A) 3 : 1
(B) 4 : 1
(C) 6 : 1
(D) 8 : 1
(E) 9 : 1

GO ON TO THE NEXT PAGE

Direction: For Student-Produced Response questions 9-18, use the grids at the bottom of the answer sheet page on which you have answered questions 1-8.

Each of the remaining 10 questions requires you to solve the problem and enter your answer by marking the circle in the special grid, as shown in the examples below. You may use any available space for scratch work.

Answer: $7/12$

Write answer in box → 7 / 1 2 ← Fraction line

Grid in → result

Answer: 2.5

2 . 5 ← Decimal point

Answer: 201

Either position is correct.

2 0 1 | 2 0 1

Note: You may start your answers in any column, space permitting. Columns not needed should be left blank.

- Mark no more than one circle in any column.
- Because the answer sheet will be machine-scored, **you will receive credit only if the circles are filled in correctly.**
- Although not required, it is suggested that you write your answers in the boxes at the top of the columns to help you fill in the circles accurately.
- Some problems may have more than one correct answer. In such cases, grid only one answer.
- No question has a negative answer.
- **Mixed numbers such as** $3\frac{1}{2}$ must be gridded as 3.5 or 7/2. (If 3 1 / 2 is gridded, it will be interpreted as $\frac{31}{2}$, not $3\frac{1}{2}$.

- **Decimal Answers:** If you obtain a decimal answer with more digits than the grid can accommodate, it may be either rounded or truncated, but it must fill the entire grid. For example, if you obtain an answer such as 0.6666····, you should record you result as 0.666 or 0.667. **A less accurate value such as 0.66 or 0.67 will be scored as incorrect.**

Acceptable ways to grid 2/3 are:

2 / 3 | . 6 6 6 | . 6 6 7

9. If $2x-7=11$, what is the value of $6x-21$?

10. If $\frac{a}{5}=\frac{6}{b}$, what is the value of $(ab)^2$?

GO ON TO THE NEXT PAGE

11. How many different arrangements can be made for 6 students seated in a row ?

12. How many different arrangements can be made for 6 students seated in a row if one of them does not want to sit in the first seat ?

13. In a high school class, there are 23 students who took Algebra, 17 students took Biology. These figure include 5 students took both courses. How many students are in this class ?

14. If $a+b=10$, $b+c=14$, and $a+c=18$, what is the value of $a+b+c$?

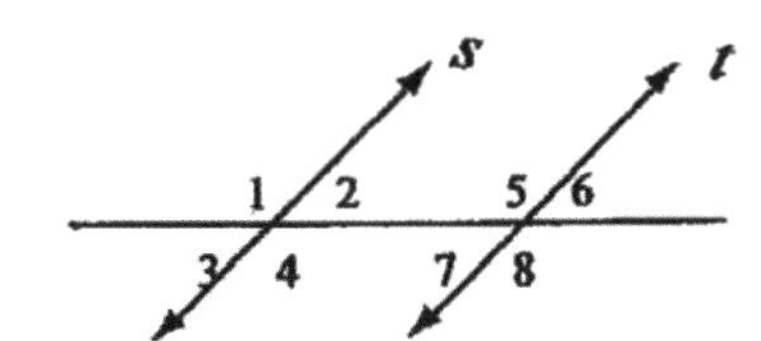

15. In the figure above, line *s* and line *t* are parallel, $<2=48°$. What is the degree measure of <5 ? (Do not grid the degree symbol.)

16. There are 50 students in a class. 24 of them took Algebra, 32 of them took Biology, and 2 of them took neither courses. How many students took both courses ?

17. The average height of 12 students is 172 *cm*. The average height of 4 other students is 164 *cm*. What is the average height in *cm* of all 16 students ?

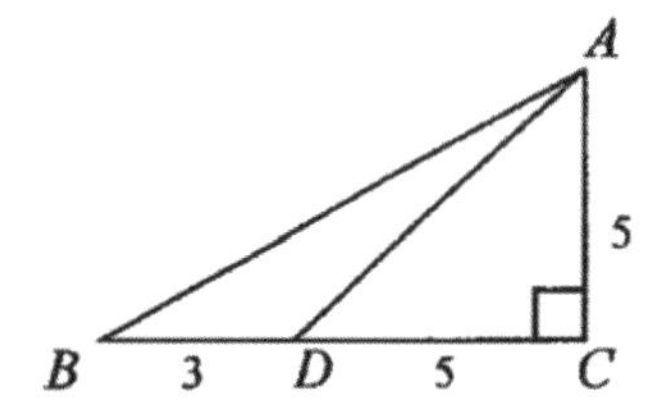

Note: Figure not drawn to scale.

18. In the figure above, what is the area of the triangle *ABD* ?

STOP

If you finish before time is called, you may check your work on this section only. Do not turn to any other section in the test.

SECTION 3

Time — 20 minutes
16 Questions

Direction: For this section, solve each problem and decide which is the best of the choices given. Fill in the corresponding circle on the answer sheet. You may use any available space for scratch work.

Notes:

1. The use of a calculator is permitted.
2. All numbers used are real numbers.
3. Figures that accompany problems in this test are intended to provide information useful in solving the problems. They are drawn as accurately as possible EXCEPT when it is stated in a specific problem that the figure is not drawn to scale. All figures lie in a plane unless otherwise indicated.
4. Unless otherwise specified, the domain of any function f is to be the set of all real number x for which $f(x)$ is a real number.

Reference Information

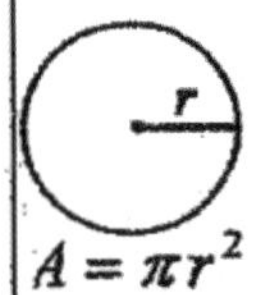

$A = \pi r^2$ $A = \ell w$
$C = 2\pi r$

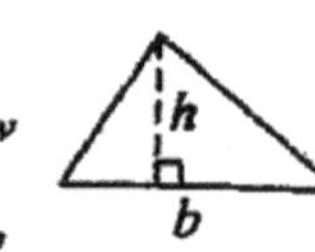

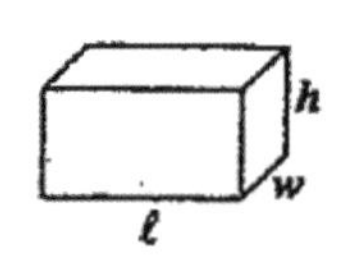

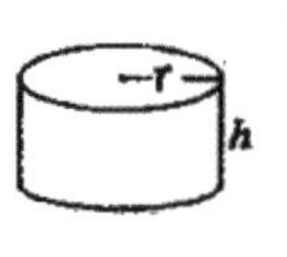

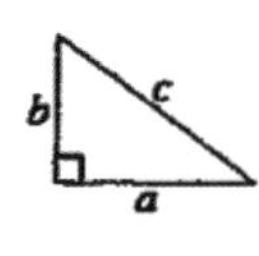

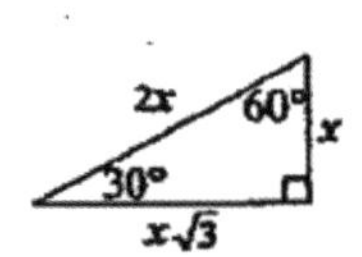

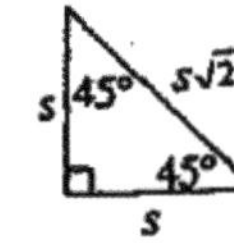

$A = \frac{1}{2}bh$ $V = \ell wh$ $V = \pi r^2 h$ $c^2 = a^2 + b^2$ Special Right Triangles

The number of degrees of arc in a circle is 360.
The sum of the measures in degrees of the angles of a triangle is 180.

1. If $2^x + 2^x = \sqrt{8}$, then $x =$

(A) $\frac{1}{2}$
(B) $\frac{3}{4}$
(C) 1
(D) 2
(E) $\frac{5}{2}$

2. If $2^x \times 2^x = \sqrt{8}$, then $x =$

(A) $\frac{1}{2}$
(B) $\frac{3}{4}$
(C) 1
(D) 2
(E) $\frac{5}{2}$

A 14 B
5
D E C

Note: Figure not drawn to scale.

3. In the figure above, $\overline{DE}$ is five-sevenths the length of $\overline{DC}$. What is the area of quadrilateral $ABCE$?

(A) 41
(B) 42
(C) 43
(D) 44
(E) 45

GO ON TO THE NEXT PAGE

4. The average price of gasoline was $3.70. This year, the average price of gasoline is $4.29. By what percent has the price of gasoline increased ?

(A) 13.75 %
(B) 15.95 %
(C) 17.25 %
(D) 18.37 %
(E) 19.24 %

5. A car traveled at an average of 60 miles per hour for a trip in 8 hours. If it had traveled at an average of 50 miles per hour, how many minutes longer the trip would have taken ?

(A) 80
(B) 87
(C) 90
(D) 96
(E) 99

6. If $f(x)=1+|x-2|$, what is the domain of $f(x)$?

(A) All real numbers
(B) All nonnegative real numbers
(C) All real numbers greater than 1
(D) All real numbers greater than or equal to 1
(E) All real numbers greater than 2

7. If $f(x)=1+|x-2|$, what is the range of $f(x)$?

(A) All real numbers
(B) All nonnegative real numbers
(C) All real numbers greater than 1
(D) All real numbers greater than or equal to 1
(E) All real numbers greater than 2

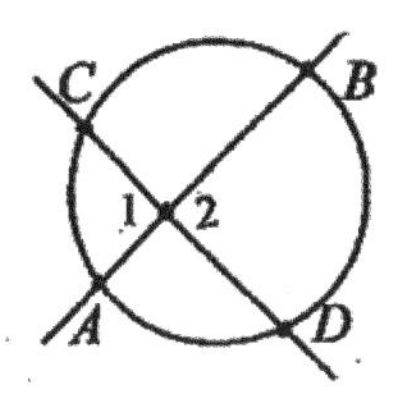

Note: Figure not drawn to scale.

8. In the figure above, two secant lines $\overline{AB}$ and $\overline{CD}$ intersect in the interior of a circle, arc $AC = 40^o$ and arc $BD = 100^o$. What is the degree measure of <1 ?

(A) 40°
(B) 50°
(C) 60°
(D) 70°
(E) 80^0

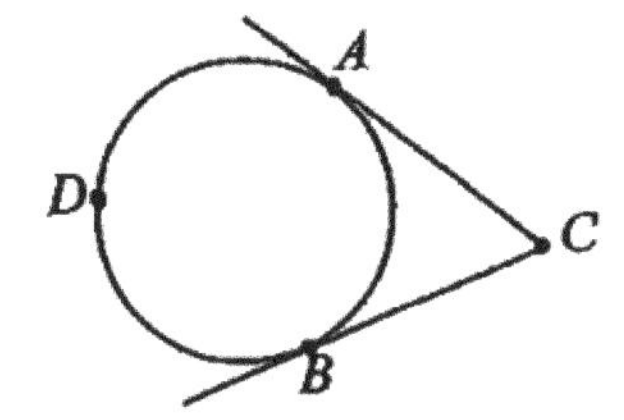

Note: Figure not drawn to scale.

9. In the figure above, two tangent lines $\overline{AC}$ and $\overline{BC}$ intersect in the exterior of a circle, arc $AB = 140^o$ and arc $ADB = 220^o$. What is the degree measure of $<C$?

(A) 40°
(B) 50°
(C) 60°
(D) 70°
(E) 80°

GO ON TO THE NEXT PAGE

10. Given $h(x) = 3x^2 - 2$, what is the value of $2h(-2)$?

(A) –28
(B) –20
(C) 20
(D) 28
(E) 68

11. How many multiples of 3 are between 3 and 250 ?

(A) 81
(B) 82
(C) 83
(D) 84
(E) 85

12. Two sides of a triangle have lengths 5 and 8. Which of the following could be the length of the third side ?

I. 14
II. 13
III. 12

(A) I only
(B) II only
(C) III only
(E) I and III only
(D) I, II, and III

13. If $(x+2)(x-h)$ is equivalent to $x^2 - kx - 8$, h and k are constants. What is the value of k ?

(A) 2
(B) 3
(C) 5
(D) 9
(E) 10

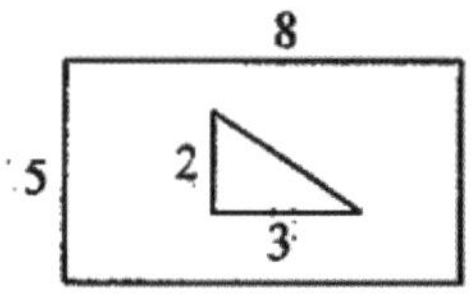

8
5
2
3

14. In the figure above, a small triangle is placed in the center of a large rectangle. The small triangle has a base 3 and height 2. The large rectangle has a length 8 and width 5. If you throw a dart at random to hit the large rectangle, what is the probability that the dart will hit the small triangle ?

(A) $\frac{3}{20}$
(B) $\frac{1}{20}$
(C) $\frac{7}{40}$
(D) $\frac{3}{40}$
(E) $\frac{1}{40}$

GO ON TO THE NEXT PAGE

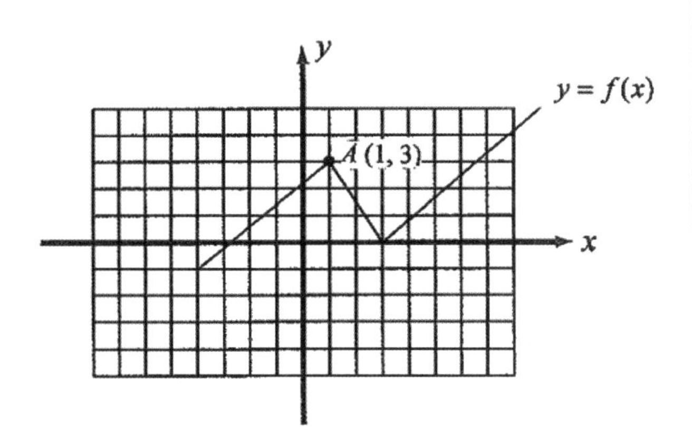

15. In the figure above, which of the following are the new coordinates of point A on the graph if the function $y = f(x)$ is translated by $y = f(x+3) - 4$?

(A) $(-2, -1)$
(B) $(4, 8)$
(C) $(-3, -4)$
(D) $(2, -1)$
(E) $(3, -4)$

ABC Manufacturing Company
Sales of Telephones in U.S.A
☎ = 10,000

Zones	Number Sold
Pacific	☎☎☎☎☎
Mountain	☎
Central	☎☎☎☎
Eastern	☎☎☎☎☎☎
Alaska	☎☎
Hawaii	☎☎

16. In the table above, the total of which two zones have exactly 50% of all telephone sales ?

(A) Pacific and Central
(B) Central and Hawaii
(C) Pacific and Hawaii
(D) Mountain and Alaska
(E) Central and Eastern

STOP

If you finish before time is called, you may check your work on this section only.
Do not turn to any other section in the test.

PRACTICE TEST 5
(SAT Math)

Page 359 ~ 371

Answer Key

Section 1: **1.** D **2.** A **3.** A **4.** D **5.** E **6.** E **7.** C **8.** C **9.** B **10.** E

11. D **12.** D **13.** D **14.** A **15.** E **16.** C **17.** B **18.** C **19.** B **20.** E

Section 2: **1.** C **2.** E **3.** A **4.** B **5.** B **6.** C **7.** B **8.** D

9. 33 **10.** 900 **11.** 720 **12.** 600 **13.** 35 **14.** 21 **15.** 132 **16.** 8

17. 170 **18.** 7.5

Section 3: **1.** A **2.** B **3.** E **4.** B **5.** D **6.** A **7.** D **8.** D **9.** A **10.** C

11. C **12.** C **13.** A **14.** D **15.** A **16.** E

CALCULATE YOUR SCORE

Questions that you omit are not counted as correct answers or incorrect answers

	Correct	Incorrect
Section 1		
Questions 1 ~ 20	( )	( )
Section 2	+	+
Questions 1 ~ 8	( )	( )
	+	
Questions 9 ~ 18	( )	
Section 3	+	+
Questions 1 ~ 16	( )	( )

Total Unrounded Raw Score ( ) – ( ) × 0.25 = ( )
Total Rounded Raw Score ( )
Your Math Scaled Score (See Table Below) ( )

SAT MATH SCORE CONVERSION TABLE

Raw Score	Scaled Score	Raw Score	Scaled Score	Raw Score	Scaled Score	Raw Score	Scaled Score
–11 to 0	200	16	400-460	32	520-600	48	670-730
1	220-310	17	410-470	33	520-610	49	680-750
2	240-330	18	420-470	34	530-610	50	690-770
3	250-350	19	430-480	35	540-620	51	730-780
4	270-370	20	430-490	36	550-620	52	760-800
5	280-370	21	440-510	37	560-630	53	780-800
6	290-400	22	450-510	38	560-640	54	800
7	300-410	23	460-520	39	570-650		
8	310-410	24	460-520	40	580-650		
9	320-420	25	470-540	41	600-660		
10	340-430	26	480-540	42	610-670		
11	340-430	27	490-550	43	620-670		
12	360-440	28	490-560	44	620-680		
13	380-440	29	500-560	45	630-690		
14	390-450	30	510-570	46	640-700		
15	400-460	31	510-590	47	660-710		

PRACTICE TEST 6
(SAT MATH)

SECTION 1
Time — 25 minutes
20 Questions

Direction: For this section, solve each problem and decide which is the best of the choices given. Fill in the corresponding circle on the answer sheet. You may use any available space for scratch work.

Notes:
1. The use of a calculator is permitted.
2. All numbers used are real numbers.
3. Figures that accompany problems in this test are intended to provide information useful in solving the problems. They are drawn as accurately as possible EXCEPT when it is stated in a specific problem that the figure is not drawn to scale. All figures lie in a plane unless otherwise indicated.
4. Unless otherwise specified, the domain of any function f is to be the set of all real number x for which $f(x)$ is a real number.

Reference Information

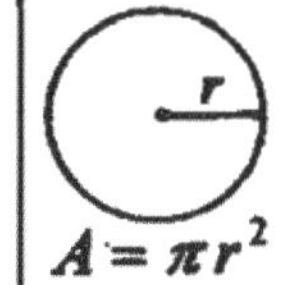

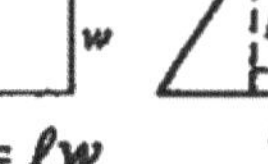
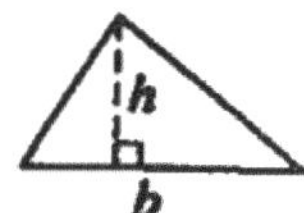
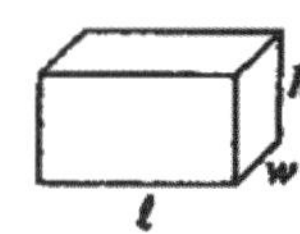

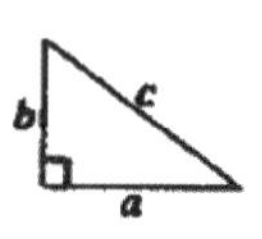
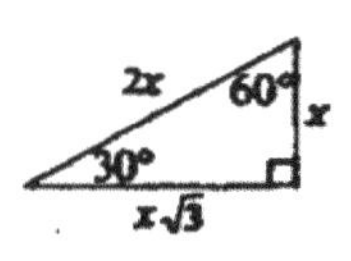
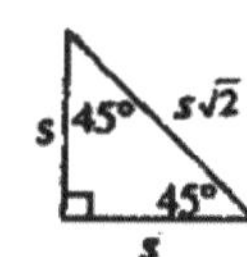

$A = \pi r^2$ $\quad$ $C = 2\pi r$ $\quad$ $A = \ell w$ $\quad$ $A = \frac{1}{2}bh$ $\quad$ $V = \ell wh$ $\quad$ $V = \pi r^2 h$ $\quad$ $c^2 = a^2 + b^2$ $\quad$ Special Right Triangles

The number of degrees of arc in a circle is 360.
The sum of the measures in degrees of the angles of a triangle is 180.

1. In the equation $y = kx$, k is a constant. If $y = 6$ when $x = 2$, what is the value of x when $y = 15$?

(A) -5
(B) -3
(C) 1
(D) 3
(E) 5

2. In a computer store, you want to buy one of the four monitors, one of the three keyboards, and one of the six computers. If all of the choices are compatible, how many possible choices can you have ?

(A) 11
(B) 22
(C) 36
(D) 72
(E) 90

GO ON TO THE NEXT PAGE

3. A school with 3,500 students has 70 students who are left-handed this year. What is the probability of a student chosen at random will be left-handed ?

(A) $\frac{1}{35}$
(B) $\frac{1}{40}$
(C) $\frac{1}{50}$
(D) $\frac{1}{65}$
(E) $\frac{1}{70}$

4. Scientists tag 100 salmons in a river. Later, they caught 150 salmons. Of these, 5 have tags. Estimate how many salmons are in the river ?

(A) 2,000
(B) 3,000
(C) 4,000
(D) 5,000
(E) 6,000

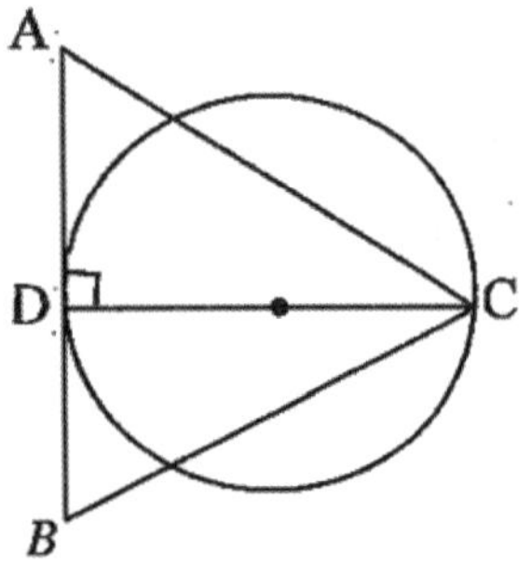

5. In the figure above, ΔABC is equilateral with side length 4. CD is the diameter of the circle. What is the area of the circle ?

(A) 3π
(B) 4π
(C) 5π
(D) 6π
(E) 7π

6. If $3^{2x} = 27^{x-2}$, what is the value of x ?

(A) 2
(B) 3
(C) 5
(D) 6
(E) 9

7. The average (arithmetic mean) of a, b, and c is 9. The average of a, b, c, and d is 12. What is the value of d ?

(A) 20
(B) 21
(C) 22
(D) 23
(E) 24

8. How many lines can be determined by nine noncollinear points ?

(A) 21
(B) 24
(C) 25
(D) 28
(E) 36

GO ON TO THE NEXT PAGE

9. What could be their intersection if two planes intersect ?

I. A plane
II. A line
III. A point

(A) I only
(B) II only
(C) I and II only
(D) II and III only
(E) I, II, and III

10. If $f(x) = ax + b$ is a linear function, $f(0) = k$, $f(2) = n$, and $f(1) = 8$, what is the value of $k + n$?

(A) 12
(B) 13
(C) 14
(D) 15
(E) 16

3x° 57°

11. In the figure above, what is the value of x ?

(A) 40
(B) 41
(C) 42
(D) 43
(E) 45

12. If $f(x) = 2x^3 - x^2 - 2$, what is the value of $f(-2)$?

(A) 2
(B) 6
(C) -8
(D) -12
(E) -22

13. A rocket is fired upward with an initial speed of 160 feet per second. The height (h) of the rocket is given by the formula $h = 160t - 16t^2$, where t is the time in seconds. How many seconds will the rocket reach the ground again ?

(A) 8 seconds
(B) 9 seconds
(C) 10 seconds
(D) 11 seconds
(E) 12 seconds

14. If the symbol $a \boxtimes b$ be defined as the product of positive integers larger than a and less than b. What is the value of $4 \boxtimes 8$?

(A) 294
(B) 282
(C) 275
(D) 266
(E) 210

GO ON TO THE NEXT PAGE

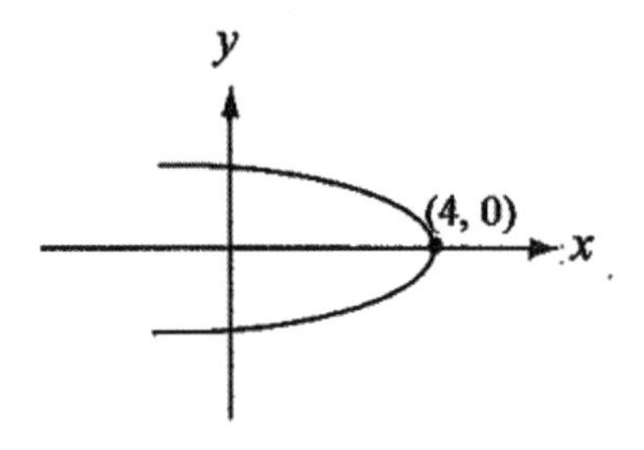

Note: Figure not drawn to scale.

15. The graph above is a parabola that is symmetric about the x-axis. Which of the following could be the equation of the parabola ?

(A) $x = y^2 + 4$
(B) $x = y^2 - 4$
(C) $x = -y^2 - 4$
(D) $x = -y^2 + 4$
(E) $x = -(y-2)^2$

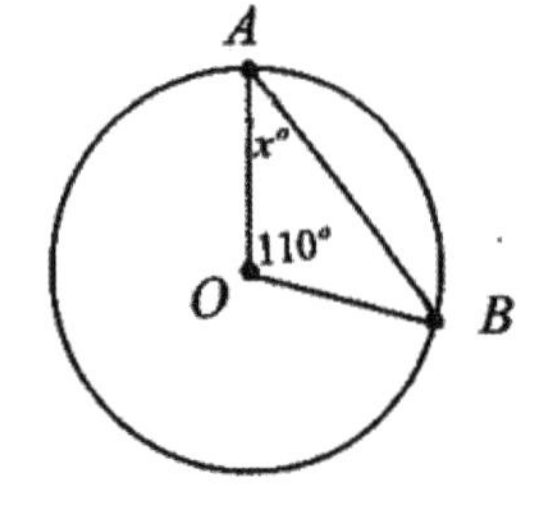

Note: Figure not drawn to scale.

16. In the figure above, O is the center of the circle. What is the value of x ?

(A) 20
(B) 25
(C) 30
(D) 35
(E) 40

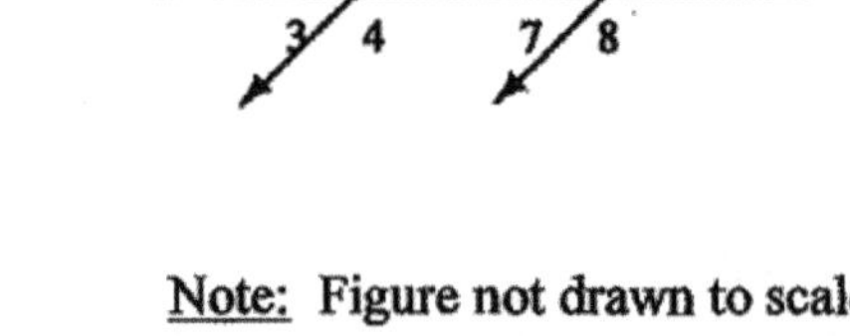

Note: Figure not drawn to scale.

17. In the figure above, line s is parallel to line t, and $<2 = 32^o$. What is the degree measure of <8 ?

(A) 148°
(B) 146°
(C) 130°
(D) 125°
(E) 120°

18. The slope of the line that passes through the points (2, –3) and (4, a) is 4. What is the value of a ?

(A) 4
(B) 5
(C) 6
(D) 7
(E) 8

GO ON TO THE NEXT PAGE

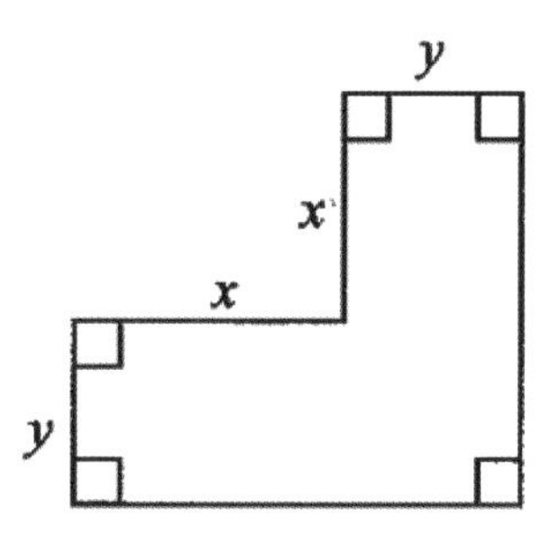

19. In the figure above, which of the following is the expression for the perimeter ?

(A) $4x+3y$
(B) $3x+4y$
(C) $4(x+y)$
(D) $4(x-y)$
(E) $3(x+y)$

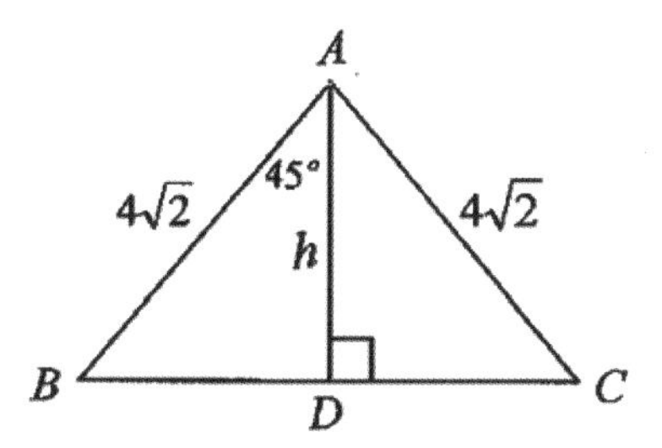

Note: Figure not drawn to scale.

20. In the figure above, what is the area of ΔABC ?

(A) 14
(B) 15
(C) 16
(D) 17
(E) 18

STOP

If you finish before time is called, you may check your work on this section only. Do not turn to any other section in this test.

SECTION 2

Time — 25 minutes
18 Questions

Direction: This section contains two types of questions. You have 25 minutes to complete both types. For questions 1-8, solve each problem and decide which is the best of the choices given. Fill in the corresponding circle on the answer sheet. You may use any available space for scratch work.

Notes:

1. The use of a calculator is permitted.
2. All numbers used are real numbers.
3. Figures that accompany problems in this test are intended to provide information useful in solving the problems. They are drawn as accurately as possible EXCEPT when it is stated in a specific problem that the figure is not drawn to scale. All figures lie in a plane unless otherwise indicated.
4. Unless otherwise specified, the domain of any function f is assumed to be the set of all real number x for which $f(x)$ is a real number.

Reference Information

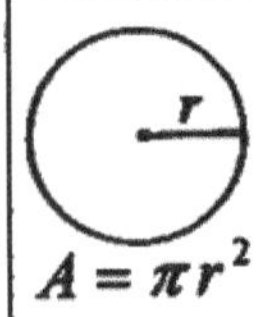

$A = \pi r^2$
$C = 2\pi r$

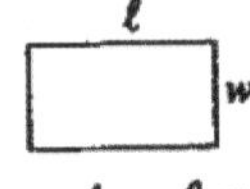

$A = \ell w$

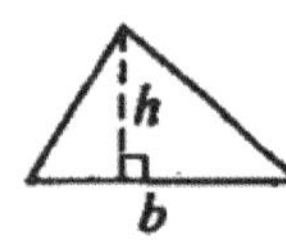

$A = \frac{1}{2}bh$

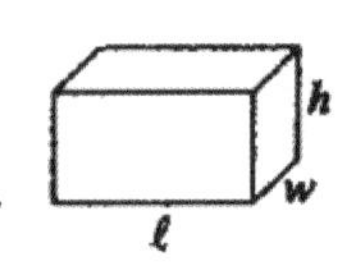

$V = \ell wh$

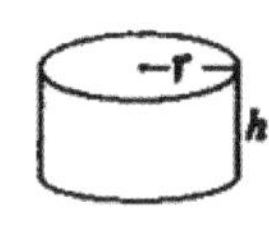

$V = \pi r^2 h$

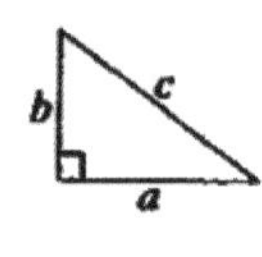

$c^2 = a^2 + b^2$

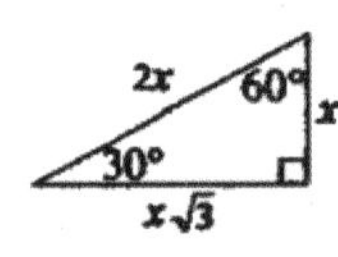

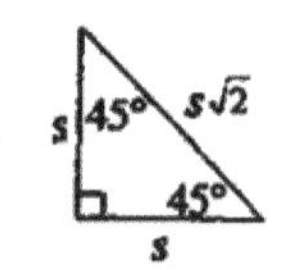

Special Right Triangles

The number of degrees of arc in a circle is 360.
The sum of the measures in degrees of the angles of a triangle is 180.

1. If $x = 4$ and $y = 2$, then $\frac{1}{2}x^2 - 3y^2 + 1 =$

(A) 3
(B) 1
(C) 0
(D) −1
(E) −3

2. If $2x^2 + 5x - 3 = 0$ and $x > 0$, what is the value of x?

(A) 3
(B) 2
(C) 1
(D) $\frac{1}{2}$
(E) 0

3. How many multiples of 4 are there between 4 and 350 ?

(A) 85
(B) 87
(C) 89
(D) 90
(E) 92

GO ON TO THE NEXT PAGE

4. If $p = 2r + 5$ and $p + 3r = 20$, what is the value of $p + r$?

(A) 14
(B) 15
(C) 16
(D) 17
(E) 18

5. What number is $5\frac{4}{5}\%$ of 55 ?

(A) 2.58
(B) 2.96
(C) 3.02
(D) 3.19
(E) 4.24

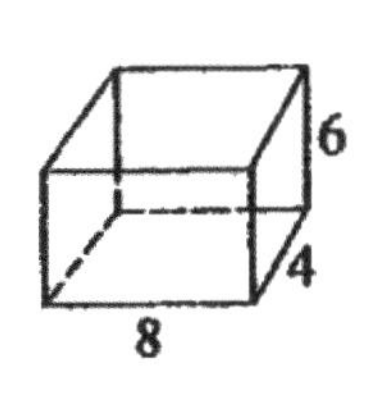

6. The above two rectangular solids are similar. What is the ratio of their surface areas ?

(A) $\frac{8}{3}$
(B) $\frac{8}{5}$
(C) $\frac{9}{4}$
(D) $\frac{9}{5}$
(E) $\frac{11}{7}$

7. A 15-gallon salt-water solution contains 20% pure salt. How much water should be added to produce the solution to 15% salt ?

(A) 1 gallon
(B) 2 gallons
(C) 3 gallons
(D) 4 gallons
(E) 5 gallons

8. A 15-gallon salt-water solution contains 20% pure salt. How much pure salt should be added to produce the solution 25% salt ?

(A) 1 gallon
(B) 2 gallons
(C) 3 gallons
(D) 4 gallons
(E) 5 gallons

GO ON TO THE NEXT PAGE

Direction: For Student-Produced Response questions 9-18, use the grids at the bottom of the answer sheet page on which you have answered questions 1-8.

Each of the remaining 10 questions requires you to solve the problem and enter your answer by marking the circle in the special grid, as shown in the examples below. You may use any available space for scratch work.

Answer: $7/12$ — Write answer in box → | 7 | / | 1 | 2 | — Grid in result → (←Fraction line)

Answer: 2.5 — | | 2 | . | 5 | (←Decimal point)

Answer: 201

Either position is correct.

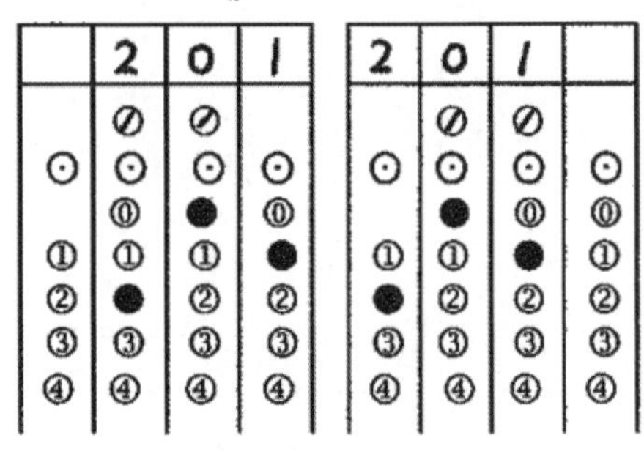

Note: You may start your answers in any column, space permitting. Columns not needed should be left blank.

- Mark no more than one circle in any column.
- Because the answer sheet will be machine-scored, **you will receive credit only if the circles are filled in correctly.**
- Although not required, it is suggested that you write your answers in the boxes at the top of the columns to help you fill in the circles accurately.
- Some problems may have more than one correct answer. In such cases, grid only one answer.
- No question has a negative answer.
- **Mixed numbers such as $3\frac{1}{2}$** must be gridded as 3.5 or 7/2. (If | 3 | 1 | / | 2 | is gridded, it will be interpreted as $\frac{31}{2}$, not $3\frac{1}{2}$.)

- **Decimal Answers:** If you obtain a decimal answer with more digits than the grid can accommodate, it may be either rounded or truncated, but it must fill the entire grid. For example, if you obtain an answer such as 0.6666⋯, you should record you result as 0.666 or 0.667. **A less accurate value such as 0.66 or 0.67 will be scored as incorrect.** Acceptable ways to grid 2/3 are:

	2	/	3
.	6	6	6
.	6	6	7

9. If $5x-3(x+1)=x+10$, then $x=$

10. Twice the sum of a number and 3 is 22. What is the value of the number ?

GO ON TO THE NEXT PAGE

11. Line k is parallel to the line $3x-4y=5$. What is the slope of line k?

12. Line p is perpendicular to the line $7x+8y=9$. What is the slope of line p?

13. Roger picked up two-fifths of the books in the shelf. Maria picked up one-half of the remaining books. Jack picked up 6 books that were left. How many books were in the shelf to begin with ?

14. If n is divided by 4, the remainder is 2.
If n is divided by 5, the remainder is 3.
If n is divided by 6, the remainder is 4.
What is the smallest value of n?

15. If $a+3b=11$, $b+3c=15$, and $c+3a=10$, what is the value of $a+b+c$?

Questions 16-17 refer to the following figure.

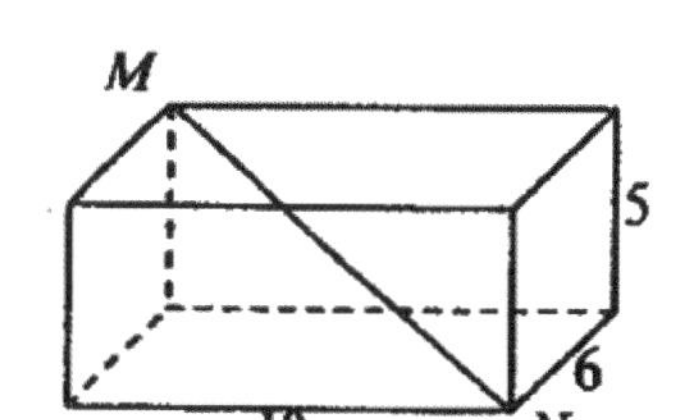

Note: Figure not drawn to scale.

16. In the rectangular solid above, what is the value of total surface area ?

---

17. In the rectangular solid above, what is the length of the diagonal MN?

18. If $x^2=36$, $y^2=9$, $x>0$ and $y>0$, what is the value of $(x-y)^2$?

STOP

If you finish before time is called, you may check your work on this section only.
Do not turn to any other section in the test.

SECTION 3

Time — 20 minutes
16 Questions

Direction: For this section, solve each problem and decide which is the best of the choices given. Fill in the corresponding circle on the answer sheet. You may use any available space for scratch work.

Notes:
1. The use of a calculator is permitted.
2. All numbers used are real numbers.
3. Figures that accompany problems in this test are intended to provide information useful in solving the problems. They are drawn as accurately as possible EXCEPT when it is stated in a specific problem that the figure is not drawn to scale. All figures lie in a plane unless otherwise indicated.
4. Unless otherwise specified, the domain of any function f is to be the set of all real number x for which $f(x)$ is a real number.

Reference Information

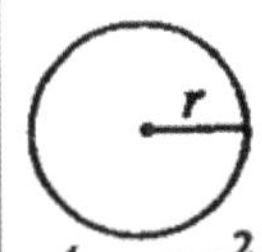

$A = \pi r^2$

$C = 2\pi r$

$A = \ell w$

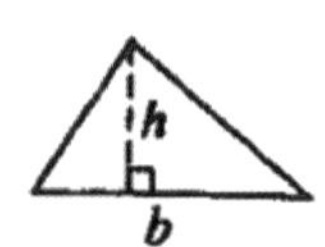

$A = \frac{1}{2}bh$

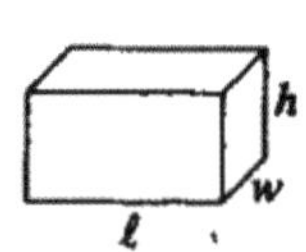

$V = \ell wh$

$V = \pi r^2 h$

$c^2 = a^2 + b^2$

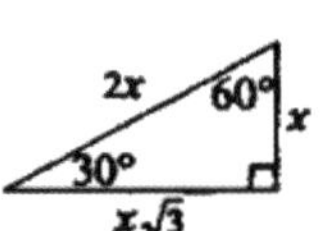

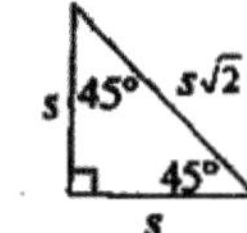

Special Right Triangles

The number of degrees of arc in a circle is 360.
The sum of the measures in degrees of the angles of a triangle is 180.

1. John completed two-fifths of the questions in 28 minutes. How many minutes does he need to complete all of the questions ?

(A) 52 minutes
(B) 64 minutes
(C) 70 minutes
(D) 73 minutes
(E) 78 minutes

2. If $\frac{3}{8}$ of n is 33, what is $\frac{5}{8}$ of n ?

(A) 44
(B) 55
(C) 66
(D) 77
(E) 88

3. If 5 students sit in a row, how many different arrangements are possible if two of them want to sit next to each other ?

(A) 24
(B) 48
(C) 60
(D) 60
(E) 72

GO ON TO THE NEXT PAGE

4. Which of the following could be used to show that the statement " $3^x > 2^x$ " is not true ?

I. $x = -1$
II. $x = 0$
III. $x = 1$

(A) I only
(B) II only
(C) III only
(D) I and II only
(E) I, II, and III

5. If $(1, k)$ is a solution to the system of inequalities:

$$\begin{cases} y > x^2 - 4 \\ y < x + 3 \end{cases}$$

What is the range (interval) of the values of k ?

(A) $-3 < k < 4$
(B) $-4 < k < -3$
(C) $-5 < k < 3$
(D) $3 < k < 4$
(E) $3 < k < 5$

6. If $\frac{x}{3} + \frac{x}{4} = 14$, then $x =$

(A) 15
(B) 18
(C) 20
(D) 24
(E) 28

7. If a line passes through the origin and is perpendicular to the line $y = -\frac{1}{2}x + 4$. The point of intersection of these two lines is $(a, 6)$. What is the value of a ?

(A) 1
(B) 2
(C) 3
(D) 4
(E) 5

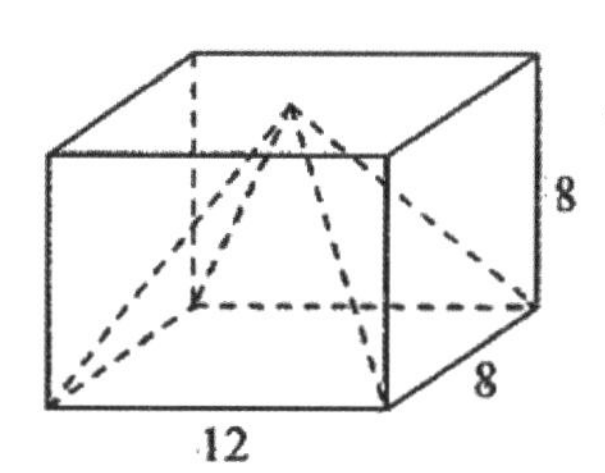

Note: Figure not drawn to scale.

8. In the figure above, a pyramid is cut out of a rectangular prism. What is the volume of the pyramid ?

(A) 256
(B) 312
(C) 384
(D) 524
(E) 750

GO ON TO THE NEXT PAGE

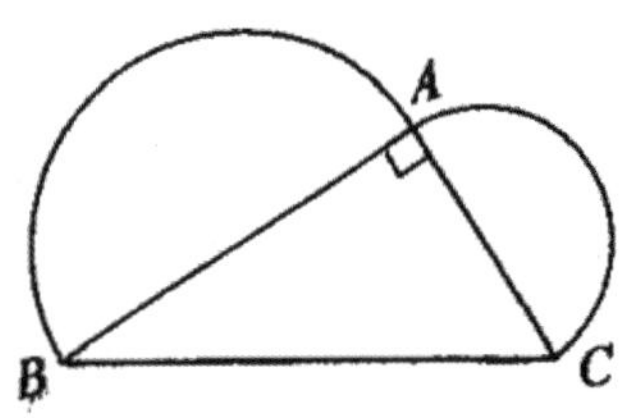

9. In the figure above, the arc AB of the semicircle has length 4π and the arc AC of the semicircle has length 3π. What is the area of of ΔABC ?

(A) 24
(B) 20
(C) 15
(D) 9
(E) 6

---

10. If $f(x) = \dfrac{x+1}{x+2}$, what is the domain of $f(x)$?

(A) All real numbers except 0
(B) All real numbers greater than -1
(C) All real numbers greater than 2
(D) All real numbers except -2
(E) All real numbers except 1

---

11. If $f(x) = \dfrac{x+1}{x+2}$, what is the range of $f(x)$?

(A) All real numbers except 0
(B) All real numbers greater than -1
(C) All real numbers greater than 2
(D) All real numbers except -2
(E) All real numbers except 1

Questions 12~13 refer to the following figure.

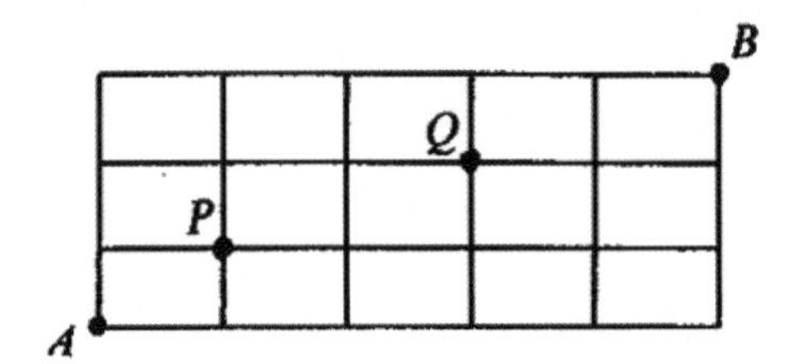

12. In the figure above, how many short-cut paths (moving upward or to the right along the grid lines) can be selected from A to B ?

(A) 20
(B) 34
(C) 56
(D) 64
(E) 72

---

13. In the figure above, how many short-cut paths (moving upward or to the right along the grid lines) can be selected from A to B that do not include P or Q ?

(A) 16
(B) 14
(C) 12
(D) 10
(E) 8

GO ON TO THE NEXT PAGE

A-Plus Company
Monthly Sale of Books

Month	Number Sold
January	300 copies
February	340
March	520
April	560
May	680

14. In the table above, what percent of the expected annual sale was reached during the first five-month period if the annual sale of 8,000 copies of books is expected ?

(A) 50%
(B) 45%
(C) 40%
(D) 30%
(E) 25%

15. A sequence is defined by $a_n = \frac{1}{n+1} - \frac{1}{n+2}$, where $n = 1, 2, 3, \cdots, n$. What is the sum of 30 terms of this sequence ?

(A) $\frac{1}{2}$

(B) $\frac{15}{32}$

(C) $\frac{11}{32}$

(D) $\frac{5}{31}$

(E) $\frac{1}{6}$

Party	Candidate 1	Candidate 2	Undecided
Republican	28	16	20
Democrat	24	32	18
Independent	18	24	20

16. The above table shows the result of the poll of 200 people to determine their voting preferences by political party for each candidate. What percent of the people had decided their voting preferrences on the candidates ?

(A) 42%
(B) 48%
(C) 52%
(D) 68%
(E) 71%

STOP

If you finish before time is called, you may check your work on this section only. Do not turn to any other section in the test.

PRACTICE TEST 6
(SAT Math)

Page 373 ~ 385

Answer Key

Section 1: **1.** E **2.** D **3.** C **4.** B **5.** A **6.** D **7.** B **8.** E **9.** C **10.** E

11. B **12.** E **13.** C **14.** E **15.** D **16.** D **17.** A **18.** B **19.** C **20.** C

Section 2: **1.** E **2.** D **3.** B **4.** A **5.** D **6.** C **7.** E **8.** A

9. 13 **10.** 8 **11.** $\frac{3}{4}$ or .75 **12.** $\frac{8}{7}$ or 1.14 **13.** 20 **14.** 58 **15.** 9 **16.** 280

17. 12.7 **18.** 9

Section 3: **1.** C **2.** B **3.** B **4.** D **5.** A **6.** D **7.** C **8.** A **9.** A **10.** D

11. E **12.** C **13.** B **14.** D **15.** B **16.** E

CALCULATE YOUR SCORE

Questions that you omit are not counted as correct answers or incorrect answers

	Correct	Incorrect
Section 1		
Questions 1 ~ 20	( )	( )
	+	+
Section 2		
Questions 1 ~ 8	( )	( )
	+	
Questions 9 ~ 18	( )	
	+	+
Section 3		
Questions 1 ~ 16	( )	( )

Total Unrounded Raw Score ( ) – ( ) × 0.25 = ( )

Total Rounded Raw Score ( )

Your Math Scaled Score (See Table Below) ( )

SAT MATH SCORE CONVERSION TABLE

Raw Score	Scaled Score	Raw Score	Scaled Score	Raw Score	Scaled Score	Raw Score	Scaled Score
–11 to 0	200	16	400-460	32	520-600	48	670-730
1	220-310	17	410-470	33	520-610	49	680-750
2	240-330	18	420-470	34	530-610	50	690-770
3	250-350	19	430-480	35	540-620	51	730-780
4	270-370	20	430-490	36	550-620	52	760-800
5	280-370	21	440-510	37	560-630	53	780-800
6	290-400	22	450-510	38	560-640	54	800
7	300-410	23	460-520	39	570-650		
8	310-410	24	460-520	40	580-650		
9	320-420	25	470-540	41	600-660		
10	340-430	26	480-540	42	610-670		
11	340-430	27	490-550	43	620-670		
12	360-440	28	490-560	44	620-680		
13	380-440	29	500-560	45	630-690		
14	390-450	30	510-570	46	640-700		
15	400-460	31	510-590	47	660-710		

PRACTICE TEST 7
(SAT SUBJECT TEST MATH LEVEL I)

Time — 60 minutes
50 questions

Direction: For each of the following problems, decide which is the BEST of the choices given. If the exact numerical value is not one of the choices, select the choices that best approximates this value. Then fill in the corresponding circle on the answer sheet.

Notes: (1) A scientific or graphing calculator will be necessary for answering some (but not all) of the questions in this test. For each question you will have to decide whether or not you should use a calculator.
(2) The only angle measure used on this test is degree measure. Make sure your calculator is in the degree mode.
(3) Figure that accompany problems in this test are intended to provide information useful in solving the problems. They are drawn as accurately as possible EXCEPT when it is stated in a specific problem that its figure is not drawn to scale. All figures lie in a plane unless otherwise indicated.
(4) Unless otherwise specified, the domain of any function f is assumed to be the set of all real numbers x for which $f(x)$ is a real number. The range of f is assumed to be the set of all real numbers $f(x)$, where x is in the domain of f.
(5) Reference information that may be useful in answering the questions in the test can be found below.

REFERENCE INFORMATION

THE FOLLOWING INFORMATION IS FOR YOUR REFERENCE IN ANSWERING SOME OF THE QUESTIONS IN THE TEST.

Volume of a right circular cone with radius r and height h: $V = \frac{1}{3}\pi r^2 h$

Lateral Area of a right circular cone with circumference of the base c and slant height ℓ: $S = \frac{1}{2}c\ell$

Volume of a sphere with radius r: $V = \frac{4}{3}\pi r^3$

Surface Area of a sphere with radius r: $S = 4\pi r^2$

Volume of a pyramid with base B and height h: $V = \frac{1}{3}Bh$

USE THIS SPACE FOR SCRATCHWORK

1. Which of the following could be the function of the graph below ?

(A) $y = 2^x$
(B) $y = 2^x - 3$
(C) $y = 2^x + 3$
(D) $y = 2^{-x} - 3$
(E) $y = 2^{-x} + 3$

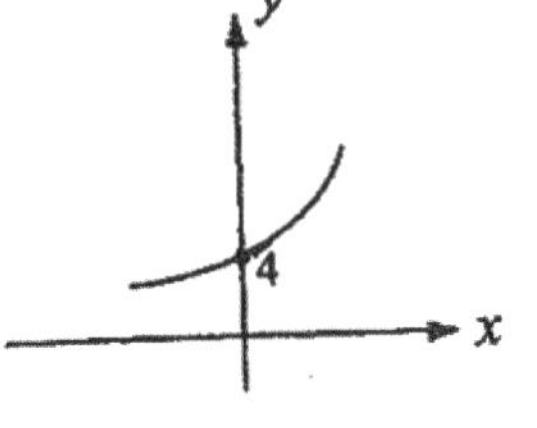

GO ON TO THE NEXT PAGE

MATHEMATICS LEVEL 1 TEST — Continued

USE THIS SPACE FOR SCRATCHWORK

2. $\log_2 64 =$

(A) 4
(B) 5
(C) 6
(D) 8
(E) 32

3. A 5,000 cubic inches of water is poured into a cylindrical container with base radius 10 inches. What is the height in inches of the water ? ($\pi = 3.14$)

(A) 15.92
(B) 21.34
(C) 18.95
(D) 16.24
(E) 15.18

4. Which of the following is a contraposition to the true statement " If two angles are vertical angles, then they are congruent. " ?

(A) If two angles are congruent, then they are vertical angles.
(B) If two angles are not congruent, then they are not vertical angles.
(C) If two angles are not vertical angles, then they are not congruent.
(D) If two angles are congruent, then they are vertical angles.
(E) If two angles are congruent, then they are not vertical angles.

5. Which of the following is a counterexample to the false statement " If two angles are congruent, then they are vertical angles. " ?

(A) If two angles are congruent, then they are not vertical angles.
(B) If two angles are not congruent, then they are vertical angles.
(C) If two angles are vertical angles, then they are not congruent.
(D) If two angles are congruent, then they could be two of the three angles in a triangle.
(E) If two angles are not vertical angles, then they are not congruent.

GO ON TO THE NEXT PAGE

MATHEMATICS LEVEL 1 TEST — Continued

USE THIS SPACE FOR SCRATCHWORK

6. What is the domain of $f(x)=\frac{|x|}{x}$?

(A) All real numbers
(B) All real numbers of $x>0$
(C) All real numbers of $x<0$
(D) All real numbers except 0
(E) $x=1$ or $x=-1$

7. What is the range of $f(x)=\frac{|x|}{x}$?

(A) All real numbers
(B) All real numbers of $y>0$
(C) All real numbers of $y<0$
(D) All real numbers except 1
(E) $y=1$ or $y=-1$

8. If $a^{3x+2}=a^{4x-1}$ for all values of a, what is the value of x ?

(A) 7 (B) $\frac{3}{4}$ (C) $\frac{4}{3}$ (D) 5 (E) 3

9. An alarm system uses a four-digit code from 0 to 9, no 0 can be used for the first digit. How many possible codes are there ?

(A) 9,000
(B) 6,561
(C) 5,040
(D) 5,000
(E) 3,042

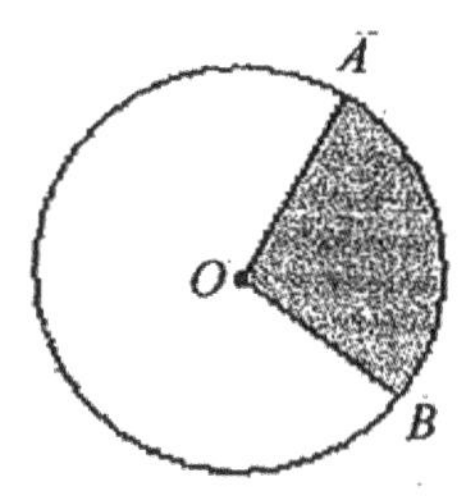

10. In the figure above, $<AOB=100°$, what fraction of the circle is shaded ?

(A) $\frac{2}{5}$ (B) $\frac{2}{3}$ (C) $\frac{3}{5}$ (D) $\frac{4}{9}$ (E) $\frac{5}{18}$

GO ON TO THE NEXT PAGE

MATHEMATICS LEVEL 1 TEST — Continued

USE THIS SPACE FOR SCRATCHWORK

11. For what value of x is $\dfrac{3x}{2x-5}$ undefined ?

(A) $-\frac{2}{5}$ (B) 0 (C) $\frac{5}{2}$ (D) 3 (E) 5

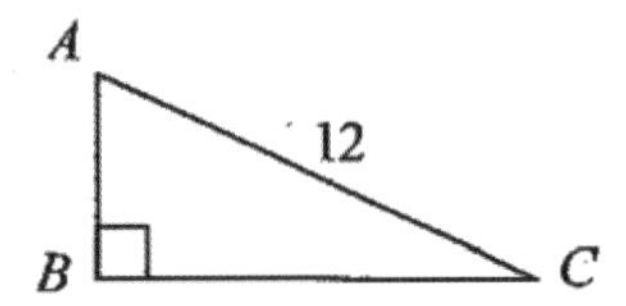

12. In the figure above, ΔABC is a right triangle, $<C = 25^{o}$, what is the length of side AB ?

(A) 5.07 (B) 6.28 (C) 7 (D) 8.36 (E) 9

13. Where does the graph of $f(x)=\dfrac{x^3+4x^2-5x}{x-1}$ intersect the x-axis ?

(A) −5, 0 (B) −1, 0 (C) 0, 1 (D) 0, 5 (E) 2, 5

14. The function $C(n)=200+20n$ is used to represent the total cost C in dollars to produce calculators, n is the number of calculators produced. Base on this function, if the total cost was $5,000, how many calculators were produced ?

(A) 200 (B) 210 (C) 220 (D) 230 (E) 240

15. If $\sin x = \frac{5}{8}$, then $1+\csc x =$

(A) $1\frac{3}{5}$ (B) 2 (C) $2\frac{3}{5}$ (D) 3 (E) $3\frac{1}{2}$

GO ON TO THE NEXT PAGE

MATHEMATICS LEVEL 1 TEST — Continued

USE THIS SPACE FOR SCRATCHWORK

16. In a triangle ΔABC, $<A:<B:<C=2:4:6$, what is the measure in degrees of $<B$?

(A) 30 (B) 40 (C) 60 (D) 90 (E) 120

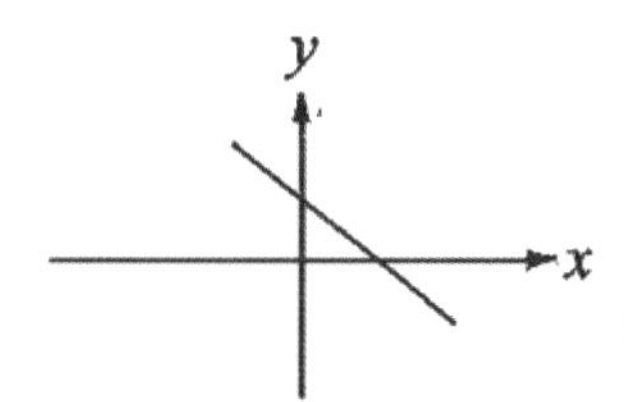

17. The graph of a linear function is shown above. Which of the following could represent its slope and y-intercept ?

(A) positive slope and positive y-intercept
(B) positive slope and negative y-intercept
(C) negative slope and positive y-intercept
(D) negative slope and negative y-intercept
(E) negative slope and no y-intercept

18. Which of the following numbers is a counterexample to the statement "$|n+5|=|n|+5$"?

(A) -1 (B) 0 (C) 1 (D) 2 (E) 3

19. How many seat arrangements can be made for 6 students in a row if one of them does not want to take the first seat ?

(A) 720
(B) 700
(C) 650
(D) 600
(E) 550

GO ON TO THE NEXT PAGE

MATHEMATICS LEVEL 1 TEST — Continued

USE THIS SPACE FOR SCRATCHWORK

20. What is the nth of the following sequence ?
$1, 2, 4, 8, 16, 32, \cdots\cdots$.

(A) 2^n (B) 2^{n-1} (C) 2^{n-2} (D) n^2 (E) $2n$

21. The volume of a sphere is 288π. What is the surface area of the sphere ?

(A) 122π (B) 136π (C) 144π (D) 169π (E) 189π

22. In the figure below, a sphere is inscribed in a right cylinder, $r_1 = r_2 = 3$. What is the ratio of the volume of the sphere to the volume of cylinder ? (Hint: volume of a sphere $= \frac{4}{3}\pi r^3$)

(A) $\frac{2}{3}$
(B) $\frac{4}{3}$
(C) $\frac{2}{5}$
(D) $\frac{4}{5}$
(E) $\frac{3}{7}$

Note: Figure not drawn to scale

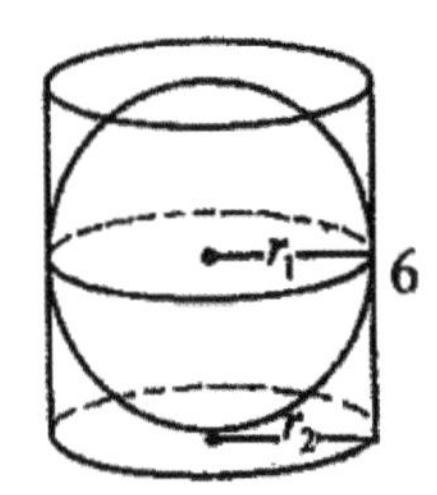

23. A box contains 4 red balls, 8 white balls, and 6 green balls. What is the probability of drawing a red ball and then drawing another green ball from remaining balls ? (without replacement)

(A) $\frac{5}{9}$ (B) $\frac{8}{9}$ (C) $\frac{4}{51}$ (D) $\frac{5}{8}$ (E) $\frac{2}{27}$

24. A box contains 4 red balls, 8 white balls, and 6 green balls. What is the probability of drawing a red ball and then a green ball if the ball is put back after the first draw ? (with replacement)

(A) $\frac{5}{9}$ (B) $\frac{8}{9}$ (C) $\frac{4}{51}$ (D) $\frac{5}{8}$ (E) $\frac{2}{27}$

GO ON TO THE NEXT PAGE

MATHEMATICS LEVEL 1 TEST — Continued

USE THIS SPACE FOR SCRATCHWORK

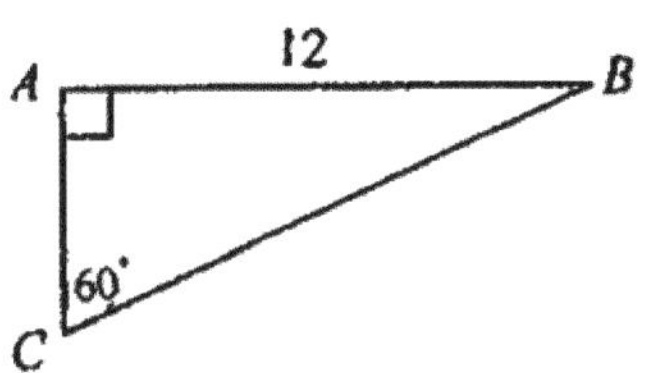

25. In the figure above, what is the length of side $\overline{BC}$?

(A) $8\sqrt{3}$ (B) 8 (C) $6\sqrt{3}$ (D) 6 (E) $4\sqrt{3}$

26. On a map, 1.2 inches represents 3 miles. If the actual distance between city A and city B is 12 miles. What would be the distance between these two cities on the map ?

(A) 3.5 in. (B) 4.2 in. (C) 4.8 in. (D) 5.4 in. (E) 5.7 in.

27. In the figure below, a square is cut out from a rectangle of length 30 and width 20. What is the area of the dashed region ?

(A) 350
(B) 400
(C) 450
(D) 500
(E) 550

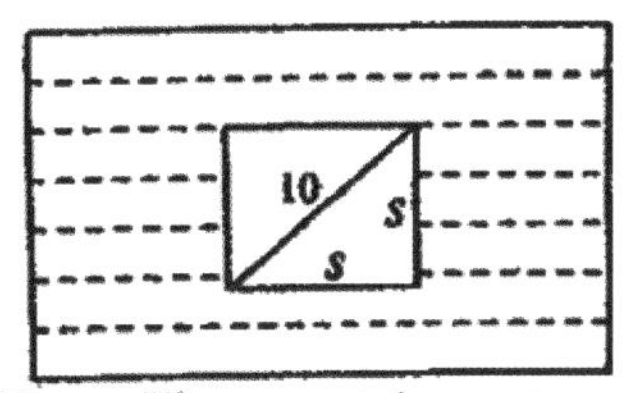

Note: Figure not drawn to scale.

28. If $f(x) = x^2 + 24$ and n is a positive number such that $f(2n) = 2f(n)$. What is the value of n?

(A) 2 (B) $2\sqrt{3}$ (C) $3\sqrt{2}$ (D) 5 (E) $4\sqrt{3}$

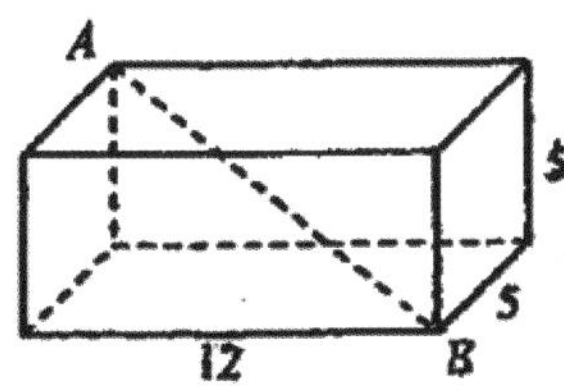

Note: Figure not drawn to scale.

29. In the figure above, what is the length of the diagonal $\overline{AB}$ in the rectangular solid ?

(A) 13.93 (B) 17.32 (C) 17.98 (D) 18.23 (E) 18.56

GO ON TO THE NEXT PAGE

MATHEMATICS LEVEL 1 TEST — Continued

USE THIS SPACE FOR SCRATCHWORK

30. $\csc x - \cos x \cot x =$

(A) $\sin x$ (B) $\cos x$ (C) $\tan x$ (D) $\sec x$ (E) $\csc x$

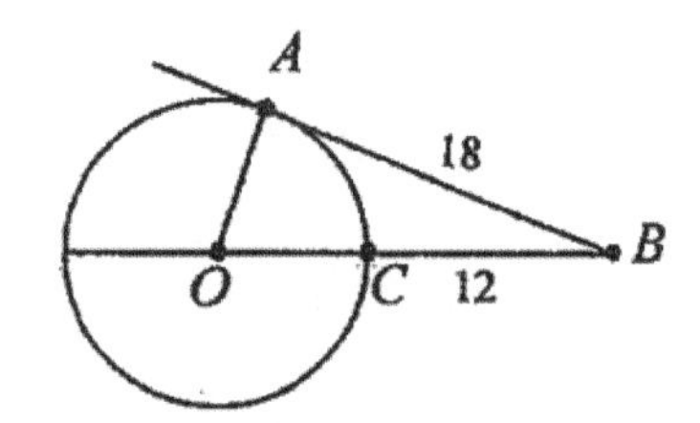

Note: Figure not drawn to scale.

31. In the figure above, $BC = 12$ and $AB = 18$. $\overline{AB}$ is a tangent line to the circle O at A. What is the length of the diameter of the circle ?

(A) 12 (B) 13 (C) 14 (D) 15 (E) 16

32. In the figure below, what is the length of the small arc AB in the circle O ? ($\pi \approx 3.14$)

(A) 9.42
(B) 6.28
(C) 6.49
(D) 5.64
(E) 4.19

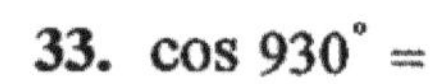

33. $\cos 930^\circ =$

(A) $\frac{\sqrt{3}}{2}$ (B) $-\frac{\sqrt{3}}{2}$ (C) $\frac{\sqrt{2}}{2}$ (D) $-\frac{\sqrt{2}}{2}$ (E) $-\frac{1}{2}$

34. What is the product of the roots of the equation $2x^2 - 6x + 7 = 0$?

(A) 4 (B) 3.5 (C) 3 (D) 2.5 (E) 2

GO ON TO THE NEXT PAGE

MATHEMATICS LEVEL 1 TEST — Continued

USE THIS SPACE FOR SCRATCHWORK

35. What is the last digit written of 2^{40} ?

(A) 0 (B) 2 (C) 4 (D) 6 (E) 8

36. If $f(x) = x^3 - 4$ and $f^{-1}(x)$ is the inverse function of $f(x)$, what is the value of $f^{-1}(10)$?

(A) 1.22 (B) 2.41 (C) 2.79 (D) 3.12 (E) 3.87

37. $(\sin x + \cos x)^2 + (\sin x - \cos x)^2 =$

(A) 5 (B) 4 (C) 3 (D) 2 (E) 1

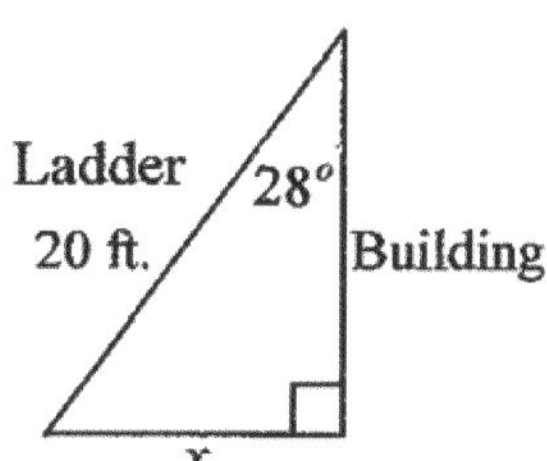

38. In the figure above, a ladder of length 20 feet leans against the side of a building by an angle of 28° with the building. What is the distance in feet from the base of the ladder to the bottom of the building ?

(A) 9.24 (B) 9.39 (C) 9.58 (D) 9.76 (E) 9.99

39. What is the domain of $f(x) = \frac{x}{x+2} - \frac{1}{x}$?

(A) All real numbers
(B) All real numbers greater than 0
(C) All real numbers less than 2
(D) All real numbers except 0 and –2
(E) All real numbers except 0 and 2

GO ON TO THE NEXT PAGE

MATHEMATICS LEVEL 1 TEST — Continued

USE THIS SPACE FOR SCRATCHWORK

40. If $y = \log_x 2^x$, then $x =$

(A) $\sqrt[y]{2^x}$ (B) $\sqrt[x]{2^y}$ (C) x^{2y} (D) y^{2x} (E) xy^x

41. What are all values of x for which $|x^3 - 32| \geq 5$?

(A) $x \leq 3$ or $x \geq 3.33$
(B) $3 \leq x \leq 3.33$
(C) $-3.17 \leq x \leq 3$
(D) $x \leq -3$ or $x \geq 3.17$
(E) $0 \leq x \leq 3.33$

42. What is the value of k such that the equation $9x^2 + 30x + k = 0$ has only one real-number root ?

(A) 21 (B) 22 (C) 23 (D) 24 (E) 25

43. Which of the following has the greatest value ?

(A) 2^{400} (B) 3^{300} (C) 5^{200} (D) $\left(\sqrt{2}\right)^{840}$ (E) $\left(\sqrt{3}\right)^{620}$

44. The quartiles of the test scores of the students in a class are $Q_1 = 75$, $Q_2 = 83.5$, and $Q_3 = 88$. What are the interquartile range of the test scores ?

(A) 4.5 (B) 6.5 (C) 13 (D) 19 (E) 83.5

45. A two-digit number is picked at random. What is the probability that it is a multiple of 3 ?

(A) $\frac{1}{5}$ (B) $\frac{2}{9}$ (C) $\frac{1}{3}$ (D) $\frac{1}{4}$ (E) $\frac{4}{9}$

GO ON TO THE NEXT PAGE

MATHEMATICS LEVEL 1 TEST — Continued

USE THIS SPACE FOR SCRATCHWORK

46. What is the distance between the two centers of the circles $(x+2)^2+(y-4)^2=16$ and $(x-2)^2+(y+4)^2=9$?

(A) 6 (B) $6\sqrt{2}$ (C) $4\sqrt{5}$ (D) 9 (E) $5\sqrt{8}$

47. If $\sin x+\cos x=\sqrt{2}\sin x$, then $\tan x+\cot x=$

(A) 2 (B) $2\sqrt{2}$ (C) 3 (D) $3\sqrt{2}$ (E) 4

48. If $\log_a 24-\log_a x=\log_a 2$, then $x=$

(A) 9 (B) 10 (C) 11 (D) 12 (E) 13

49. In the figure below, the quadrilateral *ABCD* is reflected across the line *m*. What is the coordinates of the reflection of point *A* ?

(A) (7, 0)
(B) (8, 0)
(C) (9, 0)
(D) (10, 0)
(E) (11, 0)

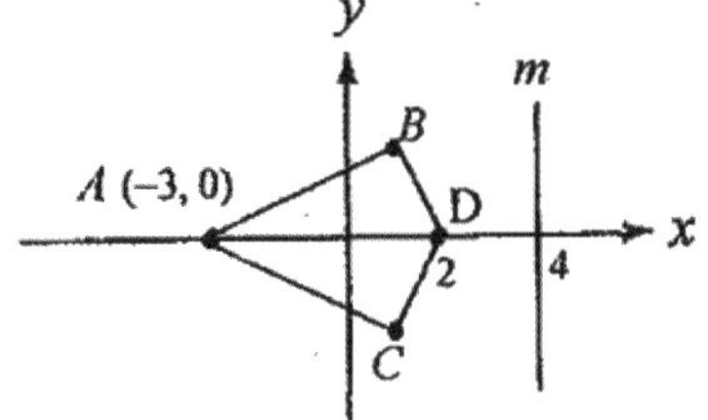

50. What is the least-square regression line for the following data ?

x	1	2	3	4	5
y	3.35	3.58	3.81	4.04	4.27

(A) $y=6.47x-3.12$
(B) $y=0.23x+3.12$
(C) $y=0.26x+3.03$
(D) $y=1.435x-0.36$
(E) $y=2.14x-6.43$

STOP

IF YOU FINISH BEFORE TIME IS CALLED, YOU MAY CHECK YOUR WORK ON THIS PAGE ONLY. DO NOT TURN TO ANY OTHER TEST IN THIS BOOK.

PRACTICE TEST 7
(SAT SUBJECT TEST MATH LEVEL 1)

Page 387 ~ 397

Answer Key

1. C	2. C	3. A	4. B	5. D	6. D	7. E	8. E	9. A	10. E
11. C	12. A	13. A	14. E	15. C	16. C	17. C	18. A	19. D	20. B
21. C	22. A	23. C	24. E	25. A	26. C	27. E	28. B	29. A	30. A
31. D	32. E	33. B	34. B	35. D	36. B	37. D	38. B	39. D	40. A
41. A	42. E	43. E	44. C	45. C	46. C	47. B	48. D	49. E	50. B

CALCULATE YOUR SCORE

Questions that you omit are not counted as correct answers or incorrect answers.

	Correct	Incorrect
Questions 1 ~ 50	( )	( )

Total Unrounded Raw Score ( ) – ( ) × 0.25 = ( )

Total Rounded Raw Score ( )

Your Math Scaled Score (See Table Below) ( )

SAT SUBJECT TEST MATH CONVERSION TABLE

Raw Score	Scaled Score	Raw Score	Scaled Score	Raw Score	Scaled Score	Raw Score	Scaled Score
–12 to 0	200~340	16	470	32	620	48	780
1	350	17	480	33	630	49	790
2	360	18	480	34	640	50	800
3	370	19	490	35	650		
4	370	20	500	36	660		
5	380	21	510	37	670		
6	380	22	520	38	680		
7	390	23	530	39	690		
8	400	24	540	40	700		
9	410	25	550	41	710		
10	420	26	560	42	720		
11	430	27	570	43	720		
12	430	28	570	44	730		
13	440	29	580	45	740		
14	450	30	600	46	750		
15	460	31	610	47	770		

PRACTICE TEST 8
(SAT SUBJECT TEST MATH LEVEL I)

Time — 60 minutes
50 questions

Direction: For each of the following problems, decide which is the BEST of the choices given. If the exact numerical value is not one of the choices, select the choices that best approximates this value. Then fill in the corresponding circle on the answer sheet.

Notes: (1) A scientific or graphing calculator will be necessary for answering some (but not all) of the questions in this test. For each question you will have to decide whether or not you should use a calculator.
(2) The only angle measure used on this test is degree measure. Make sure your calculator is in the degree mode.
(3) Figure that accompany problems in this test are intended to provide information useful in solving the problems. They are drawn as accurately as possible EXCEPT when it is stated in a specific problem that its figure is not drawn to scale. All figures lie in a plane unless otherwise indicated.
(4) Unless otherwise specified, the domain of any function f is assumed to be the set of all real numbers x for which $f(x)$ is a real number. The range of f is assumed to be the set of all real numbers $f(x)$, where x is in the domain of f.
(5) Reference information that may be useful in answering the questions in the test can be found below.

REFERENCE INFORMATION

THE FOLLOWING INFORMATION IS FOR YOUR REFERENCE IN ANSWERING SOME OF THE QUESTIONS IN THE TEST.
Volume of a right circular cone with radius r and height h: $V = \frac{1}{3}\pi r^2 h$
Lateral Area of a right circular cone with circumference of the base c and slant height ℓ: $S = \frac{1}{2}c\ell$
Volume of a sphere with radius r: $V = \frac{4}{3}\pi r^3$
Surface Area of a sphere with radius r: $S = 4\pi r^2$
Volume of a pyramid with base B and height h: $V = \frac{1}{3}Bh$

USE THIS SPACE FOR SCRATCHWORK

1. What is the domain of $y = 5 + \sqrt{2-x}$?

(A) All real numbers (B) $x \le 2$ (C) $x \ge 0$
(D) $x \ge 2$ (E) $x \ge -2$

2. If $x \ne 0$, then $\dfrac{x^3}{2x^{-2}} =$

(A) $\frac{1}{2}x$ (B) $\frac{1}{4}x$ (C) $\frac{1}{2}x^5$ (D) $\frac{1}{4}x^5$ (E) $2x$

GO ON TO THE NEXT PAGE

MATHEMATICS LEVEL 1 TEST — Continued

USE THIS SPACE FOR SCRATCHWORK

3. $\sqrt{4\sin^2 2\theta + 4\cos^2 2\theta} =$

(A) $2\sin\theta + 2\cos\theta$
(B) $2\sin 2\theta + 2\cos 2\theta$
(C) 1
(D) 2
(E) 4

4. If $\left(\sqrt[3]{n}\right)^{\frac{1}{2}} = 2$, the $n =$

(A) 64 (B) 36 (C) 27 (D) 8 (E) 4

Question 5 ~ 6 refer to the following figure.

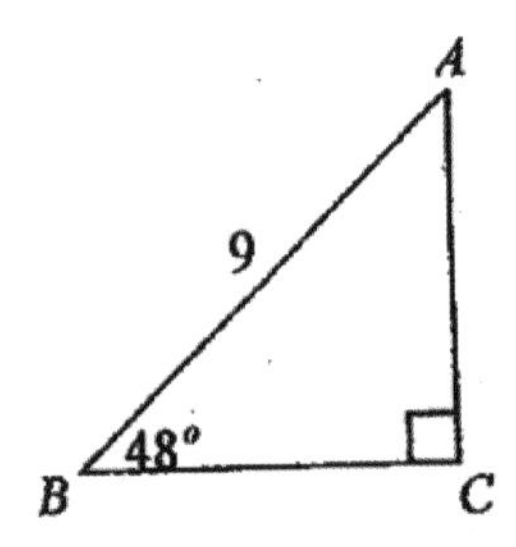

5. In the triangle above, what is the length of side BC ?

(A) 6.02 (B) 6.69 (C) 7 (D) 7.92 (E) 8.89

6. In the triangle above, what is the length of side AC ?

(A) 6.02 (B) 6.69 (C) 7 (D) 7.92 (E) 8.89

GO ON TO THE NEXT PAGE

MATHEMATICS LEVEL 1 TEST — Continued

USE THIS SPACE FOR SCRATCHWORK

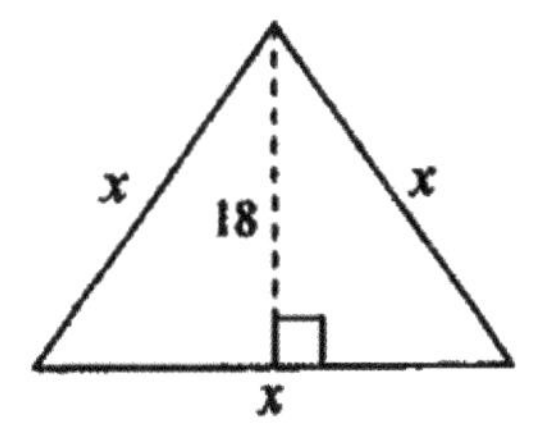

Note: Figure not drawn to scale.

7. In the figure above, what is the measure of x ?

   (A) $18\sqrt{3}$ (B) $12\sqrt{3}$ (C) $10\sqrt{3}$ (D) $9\sqrt{3}$ (E) $6\sqrt{3}$

8. If the ratio of the radii of two similar cylinders is $\frac{3}{2}$, What is the ratio of their volume ?

   (A) $\frac{3}{2}$ (B) $\frac{9}{4}$ (C) $\frac{9}{2}$ (D) $\frac{21}{4}$ (E) $\frac{27}{8}$

9. How many degrees of arc are there in $5/12$ of a circle ?

   (A) 70° B) 90° (C) 120° (D) 150° (E) 270°

10. What is the sum of $1+2+3+\cdots\cdots+60$?

    (A) 1750 (B) 1830 (C) 1928 (D) 2154 (E) 2415

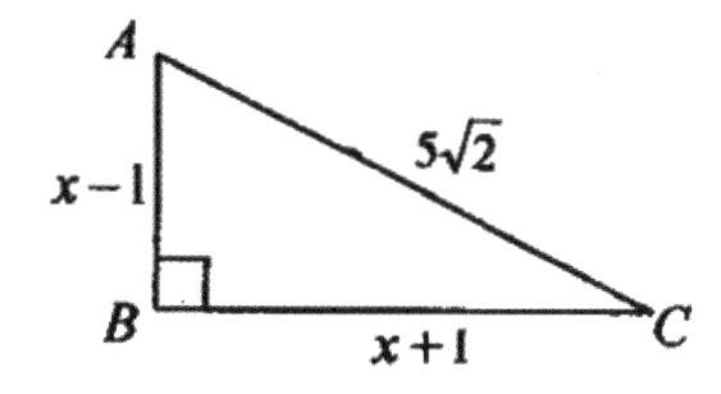

Note: Figure not drawn to scale.

11. In the right triangle above, what is the length of side BC ?

    (A) 3.25 (B) 4.32 (C) 5.24 (D) 5.90 (E) 6.42

GO ON TO THE NEXT PAGE

MATHEMATICS LEVEL 1 TEST — Continued

USE THIS SPACE FOR SCRATCHWORK

12. What is the range of $y = 8 + |2 - x|$?

(A) All real numbers
(B) $y \geq 8$
(C) $y \leq 2$
(D) $y \leq 8$
(E) $y \geq 0$

13. If $(2.52)^a = (4.26)^b$, what is the value of $\frac{b}{a}$?

(A) 0.23 (B) 0.59 (C) 0.64 (D) 1.56 (E) 1.69

14. If $f(x) = 4x - 12$, what is the solution of the equation $f(3x) = -48$?

(A) −8 (B) −3 (C) 5 (D) 10 (E) 18

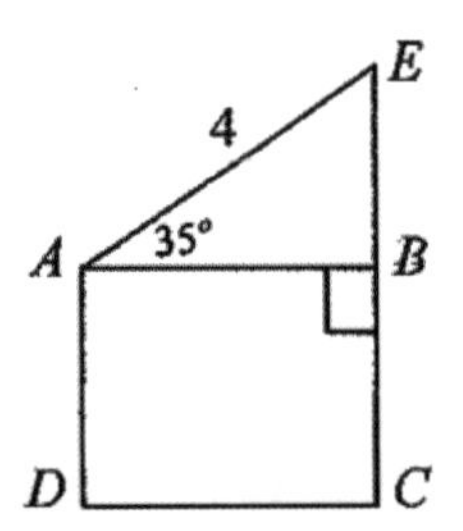

15. In the figure above, *ABCD* is a square. The length of *AE* is 4. What is the area of the square ?

(A) 4.24 (B) 4.86 (C) 5.26 (D) 8.12 (E) 10.74

16. If the measure of each exterior angle of a regular polygon is 12°, how many sides does it have ?

(A) 15 (B) 22 (C) 25 (D) 30 (E) 34

GO ON TO THE NEXT PAGE

MATHEMATICS LEVEL 1 TEST — Continued

USE THIS SPACE FOR SCRATCHWORK

17. If $f(x) = \ln x$, the $f^{-1}(-2) =$

(A) 0.14 (B) 0.28 (C) 0.69 (D) 0.89 (E) 7.32

18. $10x^2 - 35x - 20 =$

(A) $5(2x-1)(x-4)$
(B) $5(2x+1)(x-4)$
(C) $5(2x-1)(x+4)$
(D) $(10x+2)(x-10)$
(E) $(5x-2)(2x-10)$

19. If $x \neq 0$, then $\dfrac{x^2-4x-5}{x} \div \dfrac{x-5}{x^2+x} =$

(A) $x+1$
(B) $(x+1)^2$
(C) $(x-5)^2/x^2$
(D) $(x-5)^2$
(E) $(x+1)(x-5)$

20. If $\dfrac{x+y}{2} = \dfrac{x-y}{3}$, what is the ratio of x to y ?

(A) −5 (B) −2 (C) 1 (D) 2 (E) 5

21. One card is drawn at random from a 52-card bridge deck. What is the probability that it is an ace or a heart ?

(A) $\frac{17}{52}$ (B) $\frac{1}{52}$ (C) $\frac{4}{13}$ (D) $\frac{2}{13}$ (E) $\frac{1}{4}$

GO ON TO THE NEXT PAGE

MATHEMATICS LEVEL 1 TEST — Continued

USE THIS SPACE FOR SCRATCHWORK

22. Which of the following could be the quadratic equation if the sum of its two roots is $\frac{2}{3}$ and the product of its two roots is $-\frac{4}{3}$?

(A) $x^2 + 2x - 4 = 0$
(B) $x^2 - 2x + 4 = 0$
(C) $3x^2 - 2x + 4 = 0$
(D) $3x^2 + 2x + 4 = 0$
(E) $3x^2 - 2x - 4 = 0$

23. $\frac{(a^7)^2 \times a^8}{a^2} =$

(A) a^{11} (B) a^{15} (C) a^{20} (D) a^{36} (E) a^{40}

24. Which of the following could be the functions of the graph below ?

(A) $y = x^2 + 1$
(B) $y = x^2 - 1$
(C) $y = x^2 + 2x + 2$
(D) $y = x^2 + 2x$
(E) $y = -x^2 - 2x$

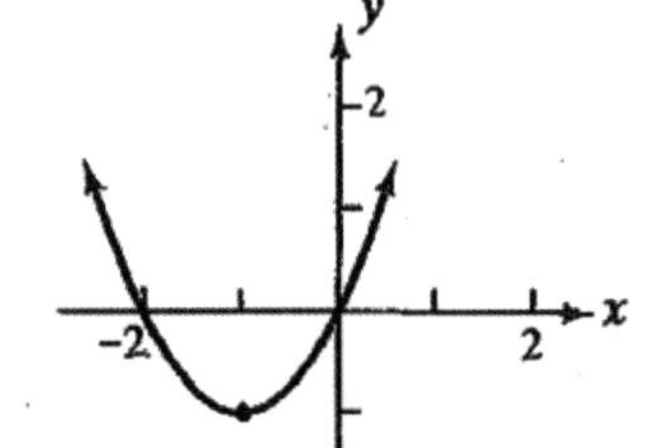

25. If $f(x) = 2x^3 - 1$ and $f^{-1}(x)$ is the inverse function of $f(x)$, then $f^{-1}(3) =$

(A) 0.02 (B) 1.26 (C) 7.28 (D) 8.24 (E) −53

26. What is the minimum integer of k such that the equation $9x^2 + 30x + k = 0$ has no real-number root ?

(A) 23 (B) 24 (C) 25 (D) 26 (E) 27

GO ON TO THE NEXT PAGE

MATHEMATICS LEVEL 1 TEST — Continued

USE THIS SPACE FOR SCRATCHWORK

27. $\log(\sin x) + \log(\csc x) =$

(A) 4 (B) 3 (C) 2 (D) 1 (E) 0

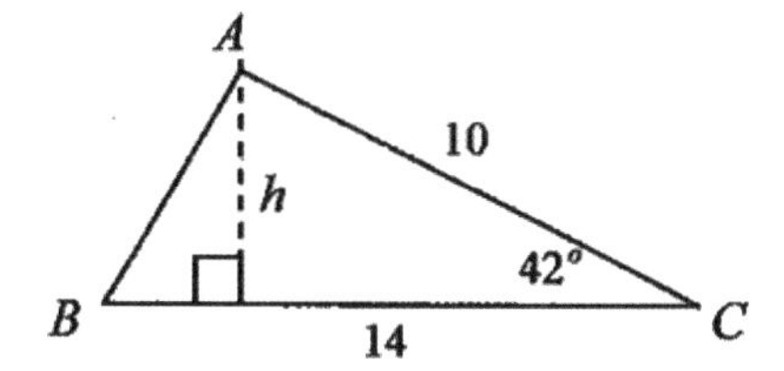

Note: Figure not drawn to scale.

28. In the figure above, what is the area of the triangle ABC?

(A) 46.84 (B) 52.02 (C) 63.03 (D) 68.14 (E) 72.15

29. If y varies inversely as x, and if $y = 10$ when $x = 2$, what is the value of y when $x = 12$?

(A) $\frac{3}{5}$ (B) $\frac{5}{3}$ (C) 3 (D) 5 (E) 60

30. If z varies directly as x and inversely as y, and if $z = 9$ when $x = 15$ and $y = 3$. What is the value of z when $x=45$ and $y = 2$?

(A) 25 (B) 25.5 (C) 29 (D) 29.5 (E) 40.5

31. $1 + \dfrac{1}{1+\frac{1}{2}} =$

(A) $1\frac{1}{2}$ (B) $1\frac{1}{3}$ (C) $1\frac{2}{3}$ (D) $3\frac{1}{2}$ (E) 4

GO ON TO THE NEXT PAGE

MATHEMATICS LEVEL 1 TEST — Continued

USE THIS SPACE FOR SCRATCHWORK

32. The height of a free-falling object is a function of time that the object travels. The height (h) in meters that a rock falls from a height of 50 meters on earth after x seconds is approximately by the function

$$h(x) = 50 - 4.9x^2.$$

When does the rock strike the ground ?

(A) 3.19 seconds
(B) 4.28 seconds
(C) 4.98 seconds
(D) 5.24 seconds
(E) 5.87 seconds

33. $\frac{\tan\theta}{\sec\theta}(\cot\theta) =$

(A) $\sin^2\theta$ (B) $\cos^2\theta$ (C) $\sin\theta$ (D) $\cos\theta$ (E) $\csc\theta$

34. A triangle has the length of sides 4, 12, and x. What is the possible value of x ?

(A) $4 < x < 12$ (B) $8 < x < 16$ (C) $10 < x < 15$
(D) $12 < x < 14$ (E) $0 < x < 48$

35. What is the domain of $f(x) = \frac{2}{x\sqrt{x+5}}$?

(A) All real numbers
(B) All real numbers between 0 and 5
(C) All real numbers less than 0 or greater than 2
(D) All real numbers greater than −5 but except 0
(E) All real numbers less than 0 or greater than 5

36. $\log_3 9^5 =$

(A) 3 (B) 5 (C) 8 (D) 9 (E) 10

GO ON TO THE NEXT PAGE

MATHEMATICS LEVEL 1 TEST — Continued

USE THIS SPACE FOR SCRATCHWORK

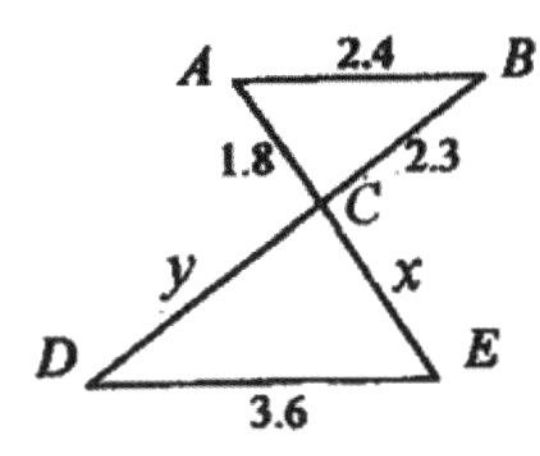

Note: Figure not drawn to scale.

37. In the figure above, $\Delta ABC \sim \Delta CDE$, what is the value of $\frac{x}{y}$?

(A) 0.78 (B) 1.28 (C) 1.5 (D) 2 (E) 3

38. If $x > 0$, then $\dfrac{x^{2/3}}{x^{-1}} =$

(A) $x^{1/3}$ (B) $x^{3/5}$ (C) $x^{5/3}$ (D) x (E) x^2

39. If $i^2 = -1$, then $(3+4i)(3-4i) =$

(A) $9-16i$ (B) -7 (C) 7 (D) 25 (E) $9+16i$

40. How many different ways to arrange 10 flags of different colors taken 6 flags at a time ?

(A) 210
(B) 25200
(C) 50400
(D) 151200
(E) 302400

41. If $\sin a^o = 0.573$, then $\sin(a^o - 13^o) =$

(A) 0.175 (B) 0.196 (C) 0.257 (D) 0.312 (E) 0.374

GO ON TO THE NEXT PAGE

MATHEMATICS LEVEL 1 TEST — Continued

USE THIS SPACE FOR SCRATCHWORK

42. If $m = \dfrac{1+\tan x}{\sec x}$ and $n = \dfrac{1-\tan x}{\sec x}$, then $m^2 + n^2 =$

(A) 5 (B) 4 (C) 3 (D) 2 (E) 1

43. The true statement " If you have measles, then you have a rash. " is given. In logical reasoning, which of the following cannot be true ?

(A) You have no measles, but you have a rash.
(B) You have no rash, then you have measles.
(C) You have no rash, then you have no measles.
(D) You have a rash, and you have no measles.
(E) You have no measles, and you have no rash.

44. What is the range of the function $f(x) = \sqrt{9x^2 - 4}$?

(A) All real numbers
(B) All real numbers of $y \geq 0$
(C) All real numbers of $y > 0$
(D) All real numbers of $y \leq 0$
(E) $y = 0$

45. If $g(x) = x - 2$ and $h(g(x)) = x - 1$, then $h(x) =$

(A) $2x-1$ (B) $2x+1$ (C) $x+1$ (D) $x-1$ (E) $2x+2$

46. What is the formula for the nth term of the arithmetic sequence 3, 8, 13, 18, 23, · · · · · ?

(A) $n^2 - 1$
(B) $2n+2$
(C) $2n-5$
(D) $3n-1$
(E) $5n-2$

GO ON TO THE NEXT PAGE

MATHEMATICS LEVEL 1 TEST — Continued

USE THIS SPACE FOR SCRATCHWORK

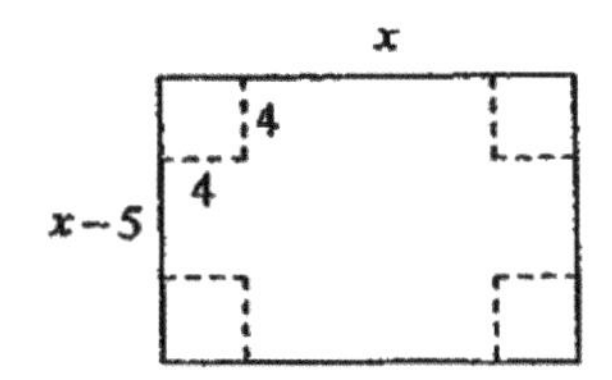

47. In the graph above, an open box is to be made from a cardboard by cutting 4-inch squares from each corner and folding up the sides. The width of the cardboard is 5 inches less than the length. If the volume of the box is 336 cubic inches, what are the dimensions of the original cardboard ?

(A) 22 by 15 inches
(B) 21 by 16 inches
(C) 20 by 15 inches
(D) 19 by 14 inches
(E) 18 by 13 inches

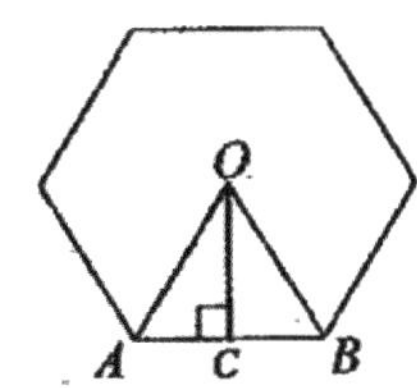

48. In the graph above, what is the perimeter of the regular hexagon (6 sides) if its apothem $OC = 12\sqrt{3}$?

(A) 100 (B) 121 (C) 144 (D) 169 (E) 187

49. $\log_7 12 =$

(A) 1.28 (B) 1.58 (C) 2.10 (D) 2.54 (E) 2.62

50. What is the least-squares regression line for the points (1, 4), (2, 5), (4, 6), (5, 7) ?

(A) $y = 0.4x + 3.8$
(B) $y = 0.5x + 3.6$
(C) $y = 0.6x + 3.5$
(D) $y = 0.7x + 3.4$
(E) $y = 0.9x + 3.2$

STOP

IF YOU FINISH BEFORE TIME IS CALLED, YOU MAY CHECK YOUR WORK ON THIS PAGE ONLY. DO NOT TURN TO ANY OTHER TEST IN THIS BOOK.

PRACTICE TEST 8
(SAT SUBJECT TEST MATH LEVEL 1)

Page 399 ~ 409

Answer Key

1. B	**2.** C	**3.** D	**4.** A	**5.** A	**6.** B	**7.** B	**8.** E	**9.** D	**10.** B
11. D	**12.** B	**13.** C	**14.** B	**15.** E	**16.** D	**17.** A	**18.** B	**19.** B	**20.** A
21. C	**22.** E	**23.** C	**24.** D	**25.** B	**26.** D	**27.** E	**28.** A	**29.** B	**30.** E
31. C	**32.** A	**33.** D	**34.** B	**35.** D	**36.** E	**37.** A	**38.** C	**39.** D	**40.** D
41. E	**42.** D	**43.** B	**44.** B	**45.** C	**46.** E	**47.** C	**48.** C	**49.** A	**50.** D

CALCULATE YOUR SCORE

Questions that you omit are not counted as correct answers or incorrect answers.

	Correct	Incorrect
Questions 1 ~ 50	( )	( )

Total Unrounded Raw Score (   ) − (   ) × 0.25 = (   )

Total Rounded Raw Score (   )

Your Math Scaled Score (See Table Below) (   )

SAT SUBJECT TEST MATH CONVERSION TABLE

Raw Score	Scaled Score	Raw Score	Scaled Score	Raw Score	Scaled Score	Raw Score	Scaled Score
−12 to 0	200~340	16	470	32	620	48	780
1	350	17	480	33	630	49	790
2	360	18	480	34	640	50	800
3	370	19	490	35	650		
4	370	20	500	36	660		
5	380	21	510	37	670		
6	380	22	520	38	680		
7	390	23	530	39	690		
8	400	24	540	40	700		
9	410	25	550	41	710		
10	420	26	560	42	720		
11	430	27	570	43	720		
12	430	28	570	44	730		
13	440	29	580	45	740		
14	450	30	600	46	750		
15	460	31	610	47	770		

PRACTICE TEST 9
(SAT SUBJECT TEST MATH LEVEL II)

Time — 60 minutes
50 questions

Direction: For each of the following problems, decide which is the BEST of the choices given. If the exact numerical value is not one of the choices, select the choices that best approximates this value. Then fill in the corresponding circle on the answer sheet.

Notes: (1) A scientific or graphing calculator will be necessary for answering some (but not all) of the questions in this test. For each question you will have to decide whether or not you should use a calculator.
(2) For some questions in this test you may have to decide whether your calculator should be in radian mode or the degree mode.
(3) Figure that accompany problems in this test are intended to provide information useful in solving the problems. They are drawn as accurately as possible EXCEPT when it is stated in a specific problem that its figure is not drawn to scale. All figures lie in a plane unless otherwise indicated.
(4) Unless otherwise specified, the domain of any function f is assumed to be the set of all real numbers x for which $f(x)$ is a real number. The range of f is assumed to be the set of all real numbers $f(x)$, where x is in the domain of f.
(5) Reference information that may be useful in answering the questions in the test can be found below.

REFERENCE INFORMATION

THE FOLLOWING INFORMATION IS FOR YOUR REFERENCE IN ANSWERING SOME OF THE QUESTIONS IN THE TEST.
Volume of a right circular cone with radius r and height h: $V = \frac{1}{3}\pi r^2 h$
Lateral Area of a right circular cone with circumference of the base c and slant height ℓ: $S = \frac{1}{2}c\ell$
Volume of a sphere with radius r: $V = \frac{4}{3}\pi r^3$
Surface Area of a sphere with radius r: $S = 4\pi r^2$
Volume of a pyramid with base B and height h: $V = \frac{1}{3}Bh$

USE THIS SPACE FOR SCRATCHWORK

1. If $\sqrt[4]{\sqrt[3]{\sqrt{n}}} = 2$, what is the value of n?

(A) 2^9 (B) 2^{12} (C) 2^{13} (D) 2^{24} (E) 2^{30}

2. If $4^x + 4^x + 4^x + 4^x = 64^x$, what is the value of x?

(A) 4 (B) 2 (C) 1 (D) $\frac{1}{2}$ (E) $\frac{1}{4}$

GO ON TO THE NEXT PAGE

MATHEMATICS LEVEL 2 TEST — Continued

USE THIS SPACE FOR SCRATCHWORK

3. How many arrangements can be made from the letters $a, b, c, d, e,$ and f in a row if letter a is not at either side ?

(A) 120 (B) 240 (C) 320 (D) 360 (E) 720

4. Which of the following has the lowest value ?

(A) 10^{100} (B) 100^{50} (C) $(10^{50})^2$ (D) $(\sqrt{1000})^{60}$ (E) $(\sqrt[3]{100})^{120}$

5. If $i = \sqrt{-1}$, then $i^5 + i^7 - i^8 =$

(A) 3 (B) 1 (C) -1 (D) i (E) $2i$

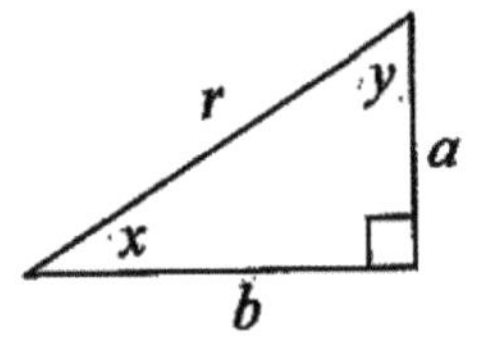

6. In the triangle above, which of the following must be true ?

I. $\sin x = \cos y$

II. $\tan x = \cot y$

III. $\sec x = \frac{r}{a}$

(A) I only
(B) II only
(C) I and II only
(D) II and III only
(E) I, II, III

7. What is the smallest number which is divided by 6, divided by 9, or divided by 13. All divisions have the same reminder of 4 ?

(A) 239 (B) 238 (C) 237 (D) 236 (E) 235

GO ON TO THE NEXT PAGE

MATHEMATICS LEVEL 2 TEST — Continued

USE THIS SPACE FOR SCRATCHWORK

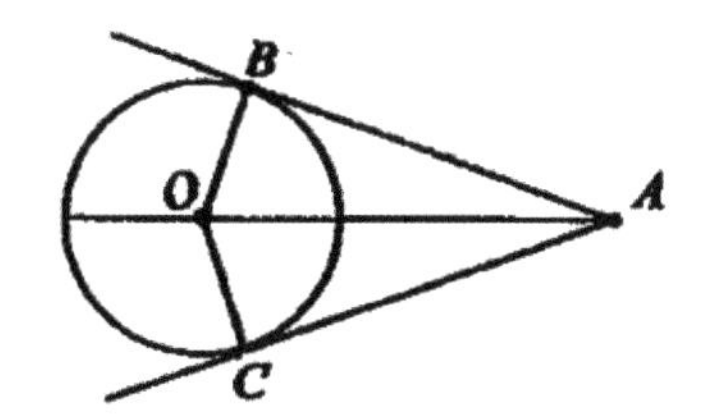

8. In the figure above, $\overline{AB}$ and $\overline{AC}$ are tangent to the circle O at B and C. If $OA = 13$ and $OB = 4$, what is the value of $\overline{AB} + \overline{AC}$?

   (A) 24.74 (B) 26.24 (C) 27.81 (D) 28.12 (E) 29.57

9. How many of the integers between 1 and 500 are multiples of 4 or 5 ?

   (A) 100 (B) 125 (C) 175 (D) 200 (E) 225

10. If each point of the straight line $y = \frac{2}{3}x + 4$ is translated horizontal 6 units to the right, what is the value of the y–intercept of the translating line ?

    (A) 10 (B) 8 (C) 6 (D) 4 (E) 0

11. What is the 20th term of the following sequence ?
    $9, 27, 81, 243, \cdots\cdots.$

    (A) 3^{18} (B) 3^{19} (C) 3^{20} (D) 3^{21} (E) 3^{22}

12. What is the sum of the following infinite geometric series ?
    $1 + \frac{1}{2} + \frac{1}{4} + \frac{1}{8} + \frac{1}{16} + \cdots\cdots.$

    (A) 1 (B) 2 (C) 3 (D) 4 (E) 5

GO ON TO THE NEXT PAGE

MATHEMATICS LEVEL 2 TEST — Continued

USE THIS SPACE FOR SCRATCHWORK

13. A plane intersects a right circular cone. Which of the following could be the intersection ?

I. A circle
II. An ellipse
III. A parabola

(A) I only
(B) II only
(C) I and II only
(D) II and III only
(E) I, II, and III

14. What is the distance between two points ($2\sqrt{5}$, $-\sqrt{3}$) and ($-\sqrt{5}$, $2\sqrt{3}$) ?

(A) $4\sqrt{2}$ (B) $6\sqrt{2}$ (C) $9\sqrt{2}$ (D) $10\sqrt{2}$ (E) 12

15. What is the equation of a circle if the endpoint of the diameter are at (5, 4) and (–1, –2) ?

(A) $(x+2)^2+(y+1)^2=18$
(B) $(x-2)^2+(y+1)^2=18$
(C) $(x-2)^2+(y-1)^2=18$
(D) $(x+2)^2+(y+1)^2=12$
(E) $(x-2)^2+(y-1)^2=12$

16. If $f(g(x))=2x-1$ and $f(x)=x-4$, then $g(x)=$

(A) $2x-3$ (B) $2x+3$ (C) $x-5$ (D) $x+5$ (E) $2x-4$

17. If $f(x)=\sqrt{x}$ and $g(x)=9x$, then $f(g(x))=$

(A) $3\sqrt{x}$ (B) $9\sqrt{x}$ (C) $3x^2$ (D) $3x$ (E) $9x$

GO ON TO THE NEXT PAGE

MATHEMATICS LEVEL 2 TEST — Continued

USE THIS SPACE FOR SCRATCHWORK

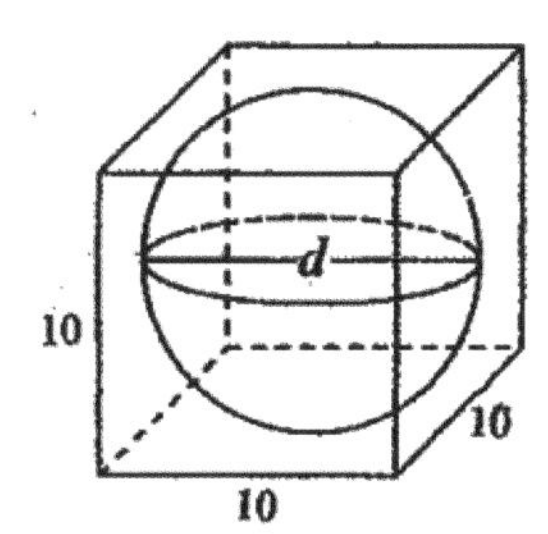

18. In the figure above, a sphere is inscribed in a cube. What is the ratio of the surface area of the sphere to the surface area of the cube ? ($\pi = 3.14$)

    (A) 0.815 (B) 0.782 (C) 0.705 (D) 0.523 (E) 0.314

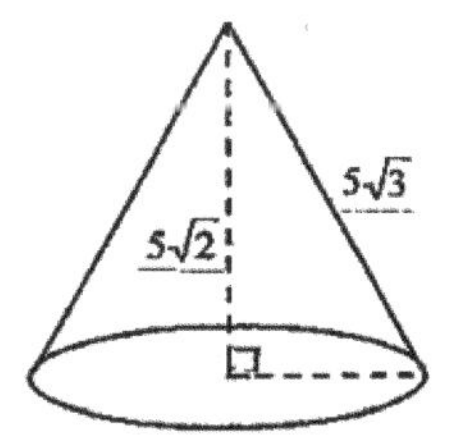

Note: Figure not drawn to scale.

19. The figure above is a right circular cone. What is the lateral surface area ?

    (A) $75\sqrt{3}\,\pi$ (B) $50\sqrt{2}\,\pi$ (C) $25\sqrt{3}\,\pi$ (D) 50π (E) 75π

20. What is the length of the major axis of the ellipse whose equation is $45x^2 + 15y^2 = 135$?

    (A) 1.73 (B) 3 (C) 3.46 (D) 6 (E) 18

21. What is the length of the transverse axis of the hyperbola whose equation is $45y^2 - 15x^2 = 135$?

    (A) 1.73 (B) 3 (C) 3.46 (D) 6 (E) 18

22. If $f(x) = \log_5 x$, then $f^{-1}(-2) =$

    (A) -25 (B) $\frac{1}{25}$ (C) $\frac{2}{5}$ (D) 25 (E) 32

GO ON TO THE NEXT PAGE

MATHEMATICS LEVEL 2 TEST — Continued

USE THIS SPACE FOR SCRATCHWORK

23. There are 13 students in a class. How many different basketball teams of 5 students can be formed ?

(A) 220 (B) 728 (C) 792 (D) 924 (E) 1287

24. 13 students are divided into two sport teams. How many different ways can be formed, each basketball team with 5 students and each volleyball team with 8 students ?

(A) 220 (B) 728 (C) 792 (D) 924 (E) 1287

25. In the logical reasoning, what is the assumption of an indirect proof used to verify a valid conclusion for the statement " If $x=5$, then $\sqrt{x}$ is not an integer. " ?

(A) $x \neq 5$
(B) $\sqrt{x}$ is an integer
(C) $\sqrt{x}$ is not an integer
(D) $x=\sqrt{5}$
(E) x is an integer

26. In the logical reasoning, which of the following could be a counterexample used to verify the statement

" $3^{x+1} > 2^x$ " ?

(A) $x=-3$ (B) $x=-2$ (C) $x=-1$ (D) $x=0$ (E) $x=2$

27. $\dfrac{(n+1)!}{(n-2)!(n-1)} =$

(A) n (B) $n+1$ (C) $n(n+1)$ (D) $n(n-1)$ (E) $(n+1)(n-1)$

GO ON TO THE NEXT PAGE

MATHEMATICS LEVEL 2 TEST — Continued

USE THIS SPACE FOR SCRATCHWORK

28. If ${}_nC_2 = 4 \cdot {}_nC_1$, then $n =$

(A) 6 (B) 7 (C) 8 (D) 9 (E) 10

29. How many different ways to arrange 10 flags of different colors taken 6 flags at a time around a track field ?

(A) 210
(B) 25200
(C) 50400
(D) 151200
(E) 302400

30. What is the arc length of a circle of radius 4 meters and subtended by a central angle of 0.54 radian ?

(A) 4.32 meters
(B) 3.24 meters
(C) 2.16 meters
(D) 1.65 meters
(E) 1.08 meters

31. If $\sin^2 x + 2\sin x - 1 = 0$, what is the radian value of x ?

(A) 0.41 (B) 0.43 (C) 0.85 (D) 0.92 (E) 1.28

32. If $0 < x < \pi/2$ and $\sec x = \frac{\sqrt{5}}{2}$, what is the value of $\tan x$?

(A) 2.24 (B) 2 (C) 0.89 (D) 0.5 (E) 0.45

33. $\lim_{x \to 1} \frac{x^2 + x - 2}{x - 1} =$

(A) 0 (B) 1 (C) 2 (D) 3 (E) undefined

GO ON TO THE NEXT PAGE

MATHEMATICS LEVEL 2 TEST — Continued

USE THIS SPACE FOR SCRATCHWORK

34. Which of the following are the asymptotes of the function $y = \dfrac{2x+1}{x-1}$?

I. $x = 1$
II. $y = 1$
III. $y = 2$

(A) I only
(B) II only
(C) I and II only
(D) I and III only
(E) I, II, and III

35. If $\cos(\sin^{-1} x) = \frac{4}{5}$ and $0 \le x \le \frac{\pi}{2}$, what is the value of x ?

(A) 0.60 (B) 0.64 (C) 0.75 (D) 0.80 (E) 1.33

36. If $(3.5)^a = (5.3)^b$, then $\dfrac{a}{b} =$

(A) 0.18 (B) 0.66 (C) 0.75 (D) 1.33 (E) 1.51

37. $\ln_{0.1} 3 =$

(A) −0.477 (B) 0.001 (C) 0.477 (D) 1.116 (D) 3.401

38. The population (p) that a certain type of bacteria increases at time t (in hours) follows the equation $p(t) = ae^{0.05t}$, where a is the initial population. How many hours will it take for an initial population of 800 of this bacteria to reach 2,000 ?

(A) 7.96 (B) 10.31 (C) 14.32 (D) 15.64 (E) 18.33

GO ON TO THE NEXT PAGE

MATHEMATICS LEVEL 2 TEST — Continued

USE THIS SPACE FOR SCRATCHWORK

Questions 39 ~ 40 refer to the following function.

$$f(x)=\begin{cases} x^2-2\ , & x\le 2 \\ \frac{1}{2}x+2\ , & x>2 \end{cases}$$

39. What is the domain of the function above ?

(A) All real numbers
(B) $x \le 2$
(C) $x = 2$
(D) $x > 2$
(E) $x \ge 2$

40. What is the range of the function above ?

(A) All real numbers
(B) $y \ge -2$
(C) $y > 3$
(D) $2 < y \le 3$
(E) $y \le 2$

41. If $f(g(x)) = 4x^2 - 4x$ and $f(x) = x^2 - 1$, then $g(x) =$

(A) $4x+1$
(B) $4x-4$
(C) $2x-1$
(D) $2x-2$
(E) $2x+1$

42. If $f(g(x)) = 2x-1$ and $g(x) = 2x+3$, then $f(x) =$

(A) $x-1$
(B) $x-2$
(C) $x-3$
(D) $x-4$
(E) $x-5$

GO ON TO THE NEXT PAGE

MATHEMATICS LEVEL 2 TEST — Continued

USE THIS SPACE FOR SCRATCHWORK

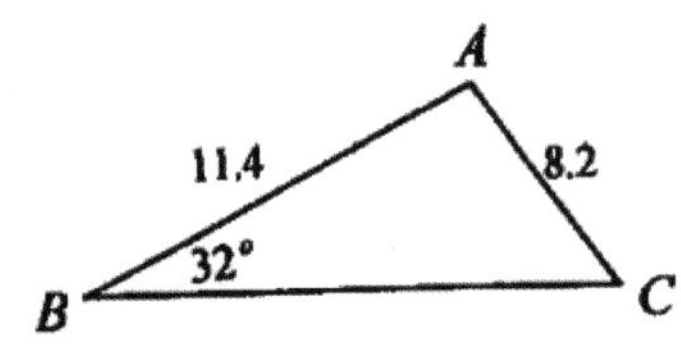

Note: Figure not drawn to scale.

43. In the figure above, what is the measure of $<C$?

(A) 41.12° (B) 45.47° (C) 47.45° (D) 49.12° (E) 50°

Questions 44 ~ 45 refer to the following figure.

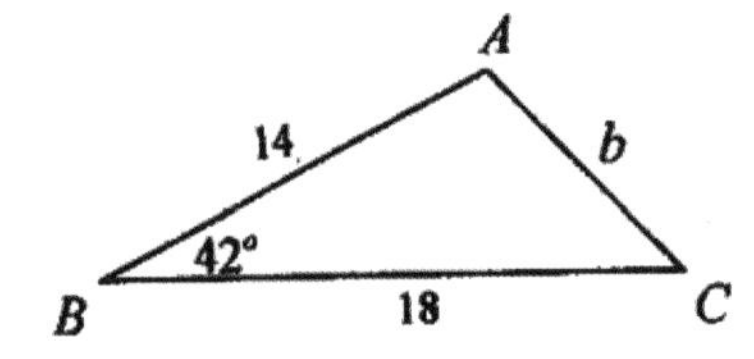

Note: Figure not drawn to scale.

44. In the triangle above, what is the area ?

(A) 80.2 (B) 84.3 (C) 86.2 (D) 88.7 (E) 92.4

45. In the triangle above, what is the length of b ?

(A) 9.89 (B) 10.54 (C) 12.06 (D) 13.96 (E) 15

46. What is the graph of the parametric equation represented by $x = 2\cos\theta$ and $y = 4\sin\theta$?

(A) a line
(B) a parabola
(C) an ellipse
(D) a circle
(E) a hyperbola

GO ON TO THE NEXT PAGE

MATHEMATICS LEVEL 2 TEST — Continued

USE THIS SPACE FOR SCRATCHWORK

47. What is the rectangular form of the polar equation $r = 4\sin\theta$?

(A) $x = 4$
(B) $y = 4$
(C) $x^2 - y^2 - 4y = 0$
(D) $x^2 + y^2 - 4x = 0$
(E) $x^2 + y^2 - 4y = 0$

48. What is the value of the determinant $\begin{vmatrix} \log 5 & \log 2 \\ \log 2 & \log 5 \end{vmatrix}$?

(A) $\log\frac{5}{2}$ (B) $\log\frac{2}{5}$ (C) $\log\frac{25}{4}$ (D) $\log\frac{4}{25}$ (E) $\log 21$

49. If $2\sin^2 x - \sin x = 1$ and $0 \le x \le 90^\circ$, then $x =$

(A) 30° (B) 45° (C) 60° (D) 75° (E) 90°

50. What is the least-squares quadratic regression for the following data ?

x	1	2	3	4	5
y	3.7	0.8	−1.7	−3.8	−5.5

(A) $y = 0.3x^2 - 5.3x + 8.7$
(B) $y = 0.2x^2 - 3.5x + 7$
(C) $y = 0.4x^2 + 2.6x + 0.7$
(D) $y = 2.1x^2 + 4.6x - 3$
(E) $y = 3.2x^2 - 1.5x + 2$

STOP

IF YOU FINISH BEFORE TIME IS CALLED, YOU MAY CHECK YOUR WORK ON THIS PAGE ONLY. DO NOT TURN TO ANY OTHER TEST IN THIS BOOK.

PRACTICE TEST 9
(SAT SUBJECT TEST MATH LEVEL II)

Page 411 ~ 421

Answer Key

1. B	**2.** D	**3.** E	**4.** E	**5.** C	**6.** C	**7.** B	**8.** A	**9.** D	**10.** E
11. D	**12.** B	**13.** E	**14.** B	**15.** C	**16.** B	**17.** A	**18.** D	**19.** C	**20.** D
21. C	**22.** B	**23.** E	**24.** E	**25.** B	**26.** A	**27.** C	**28.** D	**29.** B	**30.** C
31. B	**32.** D	**33.** D	**34.** D	**35.** A	**36.** D	**37.** A	**38.** E	**39.** A	**40.** B
41. C	**42.** D	**43.** C	**44.** B	**45.** C	**46.** C	**47.** E	**48.** A	**49.** E	**50.** B

CALCULATE YOUR SCORE

Questions that you omit are not counted as correct answers or incorrect answers.

	Correct	Incorrect
Questions 1 ~ 50	( )	( )

Total Unrounded Raw Score ( ) – ( ) × 0.25 = ( )

Total Rounded Raw Score ( )

Your Math Scaled Score (See Table Below) ( )

SAT SUBJECT TEST MATH CONVERSION TABLE

Raw Score	Scaled Score	Raw Score	Scaled Score	Raw Score	Scaled Score	Raw Score	Scaled Score
–12 to 0	200~340	16	470	32	620	48	780
1	350	17	480	33	630	49	790
2	360	18	480	34	640	50	800
3	370	19	490	35	650		
4	370	20	500	36	660		
5	380	21	510	37	670		
6	380	22	520	38	680		
7	390	23	530	39	690		
8	400	24	540	40	700		
9	410	25	550	41	710		
10	420	26	560	42	720		
11	430	27	570	43	720		
12	430	28	570	44	730		
13	440	29	580	45	740		
14	450	30	600	46	750		
15	460	31	610	47	770		

PRACTICE TEST 10
(SAT SUBJECT TEST MATH LEVEL II)

Time — 60 minutes
50 questions

Direction: For each of the following problems, decide which is the BEST of the choices given. If the exact numerical value is not one of the choices, select the choices that best approximates this value. Then fill in the corresponding circle on the answer sheet.

Notes: (1) A scientific or graphing calculator will be necessary for answering some (but not all) of the questions in this test. For each question you will have to decide whether or not you should use a calculator.

(2) For some questions in this test you may have to decide whether your calculator should be in radian mode or the degree mode.

(3) Figure that accompany problems in this test are intended to provide information useful in solving the problems. They are drawn as accurately as possible EXCEPT when it is stated in a specific problem that its figure is not drawn to scale. All figures lie in a plane unless otherwise indicated.

(4) Unless otherwise specified, the domain of any function f is assumed to be the set of all real numbers x for which $f(x)$ is a real number. The range of f is assumed to be the set of all real numbers $f(x)$, where x is in the domain of f.

(5) Reference information that may be useful in answering the questions in the test can be found below.

REFERENCE INFORMATION

THE FOLLOWING INFORMATION IS FOR YOUR REFERENCE IN ANSWERING SOME OF THE QUESTIONS IN THE TEST.

Volume of a right circular cone with radius r and height h: $V = \frac{1}{3}\pi r^2 h$

Lateral Area of a right circular cone with circumference of the base c and slant height ℓ: $S = \frac{1}{2}c\ell$

Volume of a sphere with radius r: $V = \frac{4}{3}\pi r^3$

Surface Area of a sphere with radius r: $S = 4\pi r^2$

Volume of a pyramid with base B and height h: $V = \frac{1}{3}Bh$

USE THIS SPACE FOR SCRATCHWORK

1. If $|x-4| < 12$, which of the following cannot be the value of x ?

(A) −7 (B) −3 (C) 0 (D) 15 (E) 16

2. If $4^x + 4^x = 30$, what is the value of x ?

(A) 1.95 (B) 1.04 (C) 0.57 (D) 0.42 (E) 0.34

GO ON TO THE NEXT PAGE

MATHEMATICS LEVEL 2 TEST — Continued

USE THIS SPACE FOR SCRATCHWORK

3. What is the range of $2-\sqrt{2-x}$?

(A) All real numbers
(B) $y \ge 0$
(C) $y \le 0$
(D) $y \ge 2$
(E) $y \le 2$

4. A parabola is given by the function $f(x)=2x^2-8x+4$.
Which of the following is the vertex of the parabola ?

(A) (2, 4) (B) (4, 2) (C) (2, –4) (D) (–2, 4) (E) (–4, –2)

5. A circle is given by the equation $x^2+y^2+4x-8y+11=0$.
Which of the following is the center of the circle ?

(A) (2, –4) (B) (4, –2) (C) (–4, 2) (D) (–2, 4) (E) (2, 4)

6. What is the radian measure of $82°$?

(A) 0.68 (B) 0.84 (C) 0.96 (D) 1.25 (E) 1.43

7. If $f(x)=\sin x$, then $f^{-1}(0.74)=$

(A) 0.47 radian
(B) 0.54 radian
(C) 0.83 radian
(D) 0.89 radian
(E) 0.95 radian

8. If $\tan x = 100$, then $\log \sin x - \log \cos x =$

(A) 4 (B) 3 (C) 2 (D) 1 (E) 0

GO ON TO THE NEXT PAGE

MATHEMATICS LEVEL 2 TEST — Continued

USE THIS SPACE FOR SCRATCHWORK

9. How many different ways to arrange 10 flags of different colors if 6 are red, 3 are white, and 1 is black ?

(A) 840 (B) 920 (C) 980 (D) 1020 (E) 1245

10. How many different arrangements are possible for 6 different keys placed on a key ring ?

(A) 40 (B) 60 (C) 80 (D) 100 (E) 120

11. If $n^3 - 3n^2 + 2n - 210 = 0$ and $n \in R$, then $n =$

(A) 3 (B) 4 (C) 5 (D) 6 (E) 7

12. What is the 8th term of the following recursive sequence ?

$$a_1 = 5, \ a_{n+1} = a_n - 2$$

(A) −7 (B) −8 (C) −9 (D) −10 (E) −11

13. $\sum_{k=1}^{9} \log \frac{k+1}{k} =$

(A) 10 (B) 6 (C) 4 (D) 2 (E) 1

14. What is the sum of the following infinite geometric series ?

$$3 - 1 + \frac{1}{3} - \frac{1}{9} + \frac{1}{27} + \cdots\cdots$$

(A) 1 (B) 1.15 (C) 1.75 (D) 2.25 (E) 2.5

GO ON TO THE NEXT PAGE

MATHEMATICS LEVEL 2 TEST — Continued

USE THIS SPACE FOR SCRATCHWORK

15. If two roots of a cubic equation with real coefficients are 1 and $1-2i$, what is the equation ?

(A) $x^3-2x^2+4x-3=0$
(B) $x^3-3x^2+7x-5=0$
(C) $x^3+2x^2-4x+1=0$
(D) $x^3+3x-7x+3=0$
(E) $x^3-5x+2x+2=0$

16. A ball is thrown upward from the top of a tower 96 feet high with an initial upward speed of 80 feet per second. The height of the ball is given by the formula $h=96+vt-16t^2$, where v is the initial speed and t is the time in seconds. What is the maximum height of the ball reaches ?

(A) 120 feet
(B) 156 feet
(C) 196 feet
(D) 228 feet
(E) 243 feet

17. $\log_{12} 7 =$

(A) 0.23 (B) 0.78 (C) 1.28 (D) 1.58 (E) 2.54

18. What is the 7th term of $(x-y^2)^{15}$?

(A) $6435\ x^9y^{12}$
(B) $5005\ x^9y^{12}$
(C) $3003\ x^9y^{12}$
(D) $2002\ x^9y^{12}$
(E) $8008\ x^9y^{12}$

GO ON TO THE NEXT PAGE

MATHEMATICS LEVEL 2 TEST — Continued

USE THIS SPACE FOR SCRATCHWOR

Questions 19 ~ 21 refer to the following statement.

L(x) is the line segment joining the points (–5, 4) and (3, 6). What is the corresponding line segment after each transformation is applied ?

19. The line segment above is reflected about the line $x = 1$.

(A) (7, –4) and (1, 6)
(B) (–7, 4) and (1, 6)
(C) (7, 4) and (1, 6)
(D) (7, 4) and (–1, 6)
(E) (7, –4) and (1, –6)

20. The line segment above is translated to the right 2 units and reflected across the x-axis.

(A) (–3, –4) and (5, –6)
(B) (–3, –4) and (5, 6)
(C) (–3, 4) and (–5, 6)
(D) (4, –3) and (5, –6)
(E) (3, –4) and (–5, –6)

21. The line segment above is rotated by 90° (counterclockwise) with the origin as the center of rotation.

(A) (4, –5) and (3, –6)
(B) (5, –4) and (3, –6)
(C) (5, 4) and (–3, 6)
(D) (4, 5) and (6, 3)
(E) (–4, –5) and (–6, 3)

22. Your eye doctor performs an eye test and dilates the radius of your iris by 30%. Assuming your iris is a circle, by what percent has the area been increased ?

(A) 27% (B) 60% (C) 69% (D) 90% (E) 120%

GO ON TO THE NEXT PAGE

MATHEMATICS LEVEL 2 TEST — Continued

USE THIS SPACE FOR SCRATCHWORK

23. If $\boldsymbol{u}$ and $\boldsymbol{v}$ are vectors, $\boldsymbol{u}=(4,3)$ and $\boldsymbol{v}=(6,-1)$, what is the resultant $\boldsymbol{z}=2\boldsymbol{u}-3\boldsymbol{v}$?

(A) $(3,-8)$
(B) $(-8,\ 9)$
(C) $(-8,\ 3)$
(D) $(-10,9)$
(E) $(10,-3)$

24. If $\boldsymbol{u}$ and $\boldsymbol{v}$ are vectors, $\boldsymbol{u}=4\mathbf{i}+3\mathbf{j}$ and $\boldsymbol{v}=6\mathbf{i}-\mathbf{j}$, what is the degree measure of the angle between $\boldsymbol{u}$ and $\boldsymbol{v}$?

(A) 46.3 (B) 52.4 (C) 55.7 (D) 59.2 (E) 62.6

25. The expression $z=4(\cos 120^o+i\sin 120^o)$ is the polar form of a complex number. What is the rectangular form $z=a+b\,i$?

(A) $z=2+\sqrt{2}\,i$
(B) $z=2-\sqrt{2}\,i$
(C) $z=2-\sqrt{3}\,i$
(D) $z=2+\sqrt{3}\,i$
(E) $z=-2+2\sqrt{3}\,i$

26. The expression $z=-2+2\sqrt{3}\,i$ is the rectangular form of a complex number. What is the value (modulus) of $|z|$?

(A) $\sqrt{3}$ (B) 3 (C) $2\sqrt{5}$ (D) 4 (E) $3\sqrt{5}$

27. A rectangular solid of length 4, width 3, and height $\sqrt{11}$ is inscribed in a sphere. What is the volume of the sphere ?

(A) 42π (B) 36π (C) 30π (D) 18π (E) 12π

GO ON TO THE NEXT PAGE

MATHEMATICS LEVEL 2 TEST — Continued

USE THIS SPACE FOR SCRATCHWOR

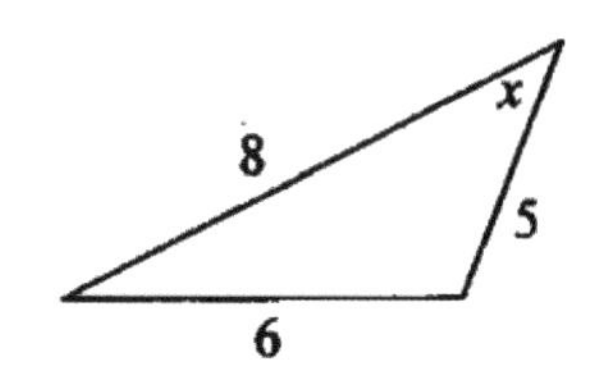

Note: Figure not drawn to scale.

28. In the triangle above, what is the degree measure of x ?

(A) 48.5 (B) 46.7 (C) 44.2 (D) 40.1 (E) 36.4

29. If $\sin x = a$ and $\cos x = b$, then $\sin(2x) =$

(A) ab (B) $2ab$ (C) $3ab$ (D) $5ab$ (E) $8ab$

30. What is the rectangular form of the parametric equation represented by $x = 1 - \cot^2 \theta$ and $y = 2 + \csc^2 \theta$?

(A) $x - y = 4$
(B) $x + y = 3$
(C) $x + y = 4$
(D) $x - y = 3$
(E) $x^2 + y^2 = 2$

31. Which of the following is the center of the hyperbola whose equation is $x^2 - y^2 - 6x - 4y + 1 = 0$?

(A) (4, 3) (B) (3, 2) (C) (–4, 3) (D) (–3, 2) (E) (3, –2)

32. $\lim_{x \to \infty} \frac{x^2 + 1}{2x^4 - 1} =$

(A) 0 (B) $\frac{1}{2}$ (C) 1 (D) 4 (E) ∞

GO ON TO THE NEXT PAGE

MATHEMATICS LEVEL 2 TEST — Continued

USE THIS SPACE FOR SCRATCHWORK

33. $\lim_{x \to -2} \dfrac{x^2 - x - 6}{x^2 - 4} =$

(A) -2 (B) $-\frac{3}{2}$ (C) $\frac{5}{4}$ (D) 6 (E) undefined

34. If $\begin{vmatrix} a & d & g \\ b & e & h \\ c & f & k \end{vmatrix} = p$, then $\begin{vmatrix} 10a & 2d & 2g \\ 5b & e & h \\ 5c & f & k \end{vmatrix} =$

(A) $10p$ (B) $20p$ (C) $50p$ (D) $70p$ (E) $100p$

35. What is the graph of the equation represented by the following determinant ?

$$\begin{vmatrix} x^2 & 3y^2 & x \\ 1 & 3 & 1 \\ 4 & 3 & 2 \end{vmatrix} = 0$$

(A) a straight line
(B) an ellipse
(C) a circle
(D) a hyperbola
(E) a parabola

36. If $\sin 2x + \cos 2x = 1$ and $0^o \le x \le 90^o$, then $x =$

(A) 22.5^o
(B) 90°
(C) 50° or 75°
(D) 40°
(E) 0° or 45°

GO ON TO THE NEXT PAGE

MATHEMATICS LEVEL 2 TEST — Continued

USE THIS SPACE FOR SCRATCHWOR

37. which of the following is the rectangular form of the polar equation $r = \sin\theta$?

(A) $x + y = 1$
(B) $x^2 + y = 1$
(C) $x^2 - y^2 + y = 0$
(D) $x^2 + y^2 - y = 0$
(E) $x^2 + y^2 - x = 0$

38. What is the radian measure of $\sin^{-1}(\cos 70^\circ)$?

(A) −0.35 (B) −0.17 (C) 0.17 (D) 1.4 (E) 0.35

39. $\csc\left(\cos^{-1}\dfrac{\sqrt{3}}{4}\right) =$

(A) 1.11 (B) 1.17 (C) 2.46 (D) 3.12 (E) 3.89

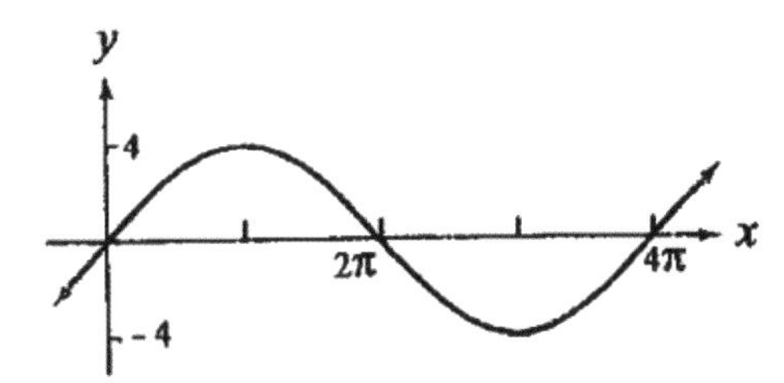

40. Which of the following periodic functions could be represented by the above graph ?

(A) $y = 2\sin\dfrac{\pi}{2}$
(B) $y = 2\sin\pi$
(C) $y = 4\sin 2\pi$
(D) $y = 4\sin\dfrac{\pi}{2}$
(E) $y = 4\sin 4\pi$

GO ON TO THE NEXT PAGE

MATHEMATICS LEVEL 2 TEST — Continued

USE THIS SPACE FOR SCRATCHWORK

41. An ellipse has foci at $(\pm 4, 0)$ and its major axis has length 10. Which of the following could be the equation ?

(A) $9y^2 + 25x^2 = 225$
(B) $25x^2 + 16y^2 = 225$
(C) $9x^2 + 25y^2 = 225$
(D) $25x^2 + 16y^2 = 225$
(E) $25y^2 + 16x^2 = 225$

42. A hyperbola has foci at $(0, \pm 3)$ and eccentricity $\frac{3}{2}$. Which of the following could be the equation ?

(A) $5x^2 - 4y^2 = 20$
(B) $5y^2 - 4x^2 = 20$
(C) $9y^2 - 5x^2 = 45$
(D) $x^2 - 2y^2 = 6$
(E) $y^2 - 2x^2 = 6$

43. In three dimensional coordinate system, what is the distance between the points $(1, 3, -5)$ and $(2, -4, 6)$.

(A) 18.2 (B) 17.4 (C) 16.3 (D) 15.9 (E) 13.1

44. If $i = \sqrt{-1}$, then $i^{100} - i^{99} =$

(A) i (B) -1 (C) 1 (D) $1 + i$ (E) $1 - i$

45. If ${}_nP_2 = 8 \cdot {}_nP_1$, then $n =$

(A) 5 (B) 6 (C) 7 (D) 8 (E) 9

GO ON TO THE NEXT PAGE

MATHEMATICS LEVEL 2 TEST — Continued

USE THIS SPACE FOR SCRATCHWOR

46. Which of the following is the center of the sphere $x^2+y^2+z^2-4x+10y-6z+29=0$?

(A) (2, 5, 3)
(B) (2, –5, 3)
(C) (3, 2, –5)
(D) (–5, 2, 3)
(E) (5, 2. –3)

47. What is the standard deviation of the following data ?
8, 15, 15, 31, 41

(A) 11.21 (B) 12.13 (C) 13.42 (D) 14.53 (E) 16.12

48. What is the amplitude of the graph of $y = 3\sin x + 4\cos x$?

(A) 5 (B) 6 (C) 7 (D) 8 (E) 9

49. What is the slope of the tangent line at the point (2, 5) on the graph of $y = 3x^2 + 2x - 4$?

(A) 10 (B) 11 (C) 12 (D) 13 (E) 14

50. What is the least-squares exponential regression for the following data ?

x	1	2	3	4	5	6
y	9.60	7.68	6.14	4.92	3.93	3.15

(A) $y = 19.2 \times 0.5^x$
(B) $y = 10.7 \times 0.9^x$
(C) $y = 12 \times 0.8^x$
(D) $y = 13.7 \times 0.7^x$
(E) $y = 14 \times 0.6^x$

STOP

IF YOU FINISH BEFORE TIME IS CALLED, YOU MAY CHECK YOUR WORK ON THIS PAGE ONLY.
DO NOT TURN TO ANY OTHER TEST IN THIS BOOK.

PRACTICE TEST 10
(SAT SUBJECT TEST MATH LEVEL II)

Page 423 ~ 433

Answer Key

1. E	**2.** A	**3.** E	**4.** C	**5.** D	**6.** E	**7.** C	**8.** C	**9.** A	**10.** B
11. E	**12.** C	**13.** E	**14.** D	**15.** B	**16.** C	**17.** B	**18.** B	**19.** D	**20.** A
21. E	**22.** C	**23.** D	**24.** A	**25.** E	**26.** D	**27.** B	**28.** A	**29.** B	**30.** C
31. E	**32.** B	**33.** C	**34.** A	**35.** B	**36.** E	**37.** D	**38.** E	**39.** A	**40.** D
41. C	**42.** B	**43.** E	**44.** D	**45.** E	**46.** B	**47.** B	**48.** A	**49.** E	**50.** C

CALCULATE YOUR SCORE

Questions that you omit are not counted as correct answers or incorrect answers.

	Correct	Incorrect
Questions 1 ~ 50	( )	( )

Total Unrounded Raw Score ( ) – ( ) × 0.25 = ( )

Total Rounded Raw Score ( )

Your Math Scaled Score (See Table Below) ( )

SAT SUBJECT TEST MATH CONVERSION TABLE

Raw Score	Scaled Score	Raw Score	Scaled Score	Raw Score	Scaled Score	Raw Score	Scaled Score
–12 to 0	200~340	16	470	32	620	48	780
1	350	17	480	33	630	49	790
2	360	18	480	34	640	50	800
3	370	19	490	35	650		
4	370	20	500	36	660		
5	380	21	510	37	670		
6	380	22	520	38	680		
7	390	23	530	39	690		
8	400	24	540	40	700		
9	410	25	550	41	710		
10	420	26	560	42	720		
11	430	27	570	43	720		
12	430	28	570	44	730		
13	440	29	580	45	740		
14	450	30	600	46	750		
15	460	31	610	47	770		

Answer Explanations
(PRACTICE TEST 1)
PSAT / NMSQT
Page 311 ~ 319

Section 1: 1. D Given $2x+3y=12$, we have : $8x+12y-3=4(2x+3y)-3=4(12)-3=48-3=45$.

2. B (Read page 30) Given $3^{2x}=81$, we have $3^{2x}=3^4$, $2x=4$, $x=2$.

3. D (Read page 155) Two angles forming a linear pair are supplementary.
We have : $4x+40=180$, $4x=140$, $x=35$.

4. C (Read page 197) Volume = base area × height
We have : $5000=10\times 25\times h$, $h=\frac{5000}{250}=20$ inches.

5. B Cost $=\frac{\$10}{2}\times 20=\$100$, sales = \$1.50 × (16×8) = \$192, profit = \$192 − \$100 = \$92.

6. A (Read page 159) The degree measure of the two base angles of the small triangle $=180-150=30$. We have : $x=180-(50+40+30)=60$.

7. E (Read page 34)

Teams	1	2	3	4	5	6	7
Games	0	1	3	6	10	15	21
		+1	+2	+3	+4	+5	+6

8. D (Read page 181) Given $P(4, 3)$, $M(0, 0)$, $Q(a, b)$
$\frac{4+a}{2}=0$, $4+a=0$, $a=-4$. ; $\frac{3+b}{2}=0$, $3+b=0$, $b=-3$.
We have: $Q(a, b)=(-4, -3)$.

9. B Draw a number line:

A —16— B —16— C — p —16— D

$\overline{AP}=16+16+8=40$.

10. B Let n = the smaller number , $n+1$ = the larger number
We have the equation: $n(n+1)=600$
$n^2+n-600=0$, $(n+25)(n-24)=0$, $n=24$.
The larger number: $n+1=24+1=25$.

11. E (Read page 21) $280\times\frac{5}{9+5}=280\times\frac{5}{14}=100$ red balls.

12. C (Read page 292) $26\times 25\times 24=15{,}600$.

13. D (Read page 291) Area $=\pi r^2$, we have : $p=\frac{\pi\cdot 2^2}{\pi\cdot 6^2}=\frac{4}{36}=\frac{1}{9}$.

14. B (Read page 25) $\frac{K-J}{J}=\frac{20-10}{10}=\frac{10}{10}=1=100\%$.

15. A (Read page 21) $\angle C=180\times\frac{5}{3+4+5}=180\times\frac{5}{12}=75$.

16. E (Read page 281) Given $\frac{a+b}{2}=21$, we have $a+b=42$.
Given $\frac{a+b+c}{3}=26$, we have $\frac{42+c}{3}=26$, $42+c=78$, $c=36$.

Answer Explanations
(PRACTICE TEST 1)
PSAT / NMSQT
Page 311 ~ 319

Section 1: 17. B (Read page 21) $200 \times \frac{100}{4} = 200 \times 25 = 5{,}000$ salmons.

18. D (Read page 155)
I. $b = d$. It is true. (Vertical angles are equal.)
II. $a + c$ and $b + d$ could not be equal. It is not true.
III. $a + b = c + d = 180$. It is true.

19. D (Read page 281)
65, 66, 69, 69, 70, 71, 75, 75, 80, 82, 82, 86
Median = (71 + 75) ÷ 2 = 73.

20. A (Read page 25) $\frac{40}{30+20+40+35+50+45} = \frac{40}{220} \approx 0.18 \approx 18\%$.

Section 2: 21. B $60 \times \frac{1}{3} = 20$ miles.

22. C $2(a-4+3a) = 12$, $2(4a-4) = 12$, $4a-4 = 6$, $4a = 10$, $a = \frac{10}{4} = 2\frac{1}{2}$.

23. C (Read page 163) Apply Pythagorean Theorem.
$x^2 = (100\sqrt{2})^2 + (100\sqrt{2})^2 = 20000 + 20000 = 40000$, $x = 200$ feet.

24. D (Read page 72)
$5x + 6y = 15$, $6y = -5x + 15$, $y = -\frac{5}{6}x + \frac{5}{2}$, we have: $m = -\frac{5}{6}$.

25. E (Read page 169) Perimeter $= \ell + 2(k+b) + a + (\ell - a) = 2(k + \ell + b)$.

26. E $\frac{x+y}{x-y} = \frac{\frac{1}{2}+\frac{1}{3}}{\frac{1}{2}-\frac{1}{3}} = \frac{\frac{3}{6}+\frac{2}{6}}{\frac{3}{6}-\frac{2}{6}} = \frac{5/6}{1/6} = \frac{5}{\not{6}} \times \frac{\not{6}}{1} = 5$.

27. B (Read pages 41 and 292) There are $5 \times 4 \times 3 \times 2 \times 1 = 120$ seat arrangements.
$4 \times 3 \times 2 \times 1 = 24$ seat arrangements for a student who takes the first seat.
$120 - 24 = 96$ seat arrangements if a student doesn't take the first seat.

28. A (Read page 159) $(13-9) < x < (9+13)$, $4 < x < 22$.

29. 30 Given $4x + 5y = 10$, we have : $12x + 15y = 3(4x + 5y) = 3(10) = 30$.

30. 316 Let n = 1st. integer, we have : $n + (n+1) + (n+2) + (n+3) = 1285$
$4n + 6 = 1258$, $4n = 1252$, $n = 313$.
The greatest integer is: $n + 3 = 313 + 3 = 316$.

31. 4.5 Let n = the number
We have : $\frac{n}{3} = n - 3$, $n = 3n - 9$, $2n = 9$, $n = 4.5$.

32. 37.5 The original solution has $25 \times 40\% = 10$ gallons of pure salt.
Let x = gallons of water added
We have the equation:
$\frac{10}{25+x} = 0.16$, $10 = 4 + 0.16x$, $6 = 0.16x$, $x = 37.5$ gallons of water added.

33. 3 $(\sqrt{5}+\sqrt{2})(\sqrt{5}-\sqrt{2}) = (\sqrt{5})^2 - (\sqrt{2})^2 = 5 - 2 = 3$.

Answer Explanations
(PRACTICE TEST 1)
PSAT / NMSQT
Page 311 ~ 319

Section 2: 34. 3.75 Working together, they can complete $\frac{1}{10}+\frac{1}{6}=\frac{6+10}{60}=\frac{16}{60}=\frac{4}{15}$ of the work in one day. It will take $1\div\frac{4}{15}=1\times\frac{15}{4}=3.75$ hours to paint the warehouse.

35. 800 (Read page 163) Apply Pythagorean Theorem.

$$s^2+s^2=20^2,\ 2s^2=400,\ s^2=200,\ s=\sqrt{200}.$$

The area of the dashed region $=(40\times25)-(\sqrt{200})^2=1000-200=800$.

36. 6 (Read page 197 and 205)

The ratio of the area of the sphere to the area of the cube is:

$$\frac{4\pi(r^2)}{6(2r)^2}=\frac{4\pi(r^2)}{24(r^2)}=\frac{\pi}{6},\ \text{Given } \frac{\pi}{6}=\frac{\pi}{n},\ \text{we have } n=6.$$

37. 3 (Read page 69) Given $f(x)=x^2+18$, $f(2n)=2f(n)$, n is a positive number.

We have : $f(2n)=(2n)^2+18=4n^2+18$

$2f(n)=2(n^2+18)=2n^2+36$

We have : $4n^2+18=2n^2+36,\ 2n^2=18,\ n^2=9,\ n=3$.

38. 3 (Read page 291) $4500\times\frac{1}{30}\times\frac{1}{50}=3$.

> In the court room, three defendants are waiting for the judge to sentence them for the time they need to stay in jail for the crimes they have committed.
>
> Judge: Do you have any good reason why you think I should sentence you lesser time in jail ?
> Defendant A: I was your high school classmate.
> Judge: 1 week.
> Defendant B: I built your house.
> Judge: 1 month.
> Defendant C: I introduced a girl friend to you 20 years ago. She is your wife now.
> Judge: 20 years.

Answer Explanations
(PRACTICE TEST 2)
PSAT / NMSQT
Page 321 ~ 329

Section 1:

1. D (Read page 30) $4^{3x} = 16^{x+2}$, $4^{3x} = (4^2)^{x+2}$, $4^{3x} = 4^{2x+4}$, $3x = 2x+4$, $x = 4$.

2. D (Read page 281) $\frac{4+8+12+20+a}{5} = 18$, $\frac{44+a}{5} = 18$, $44+a = 90$, $a = 46$.

3. A (Read page 159) ΔABC is an isosceles triangle. $\angle B = \angle C = 50^\circ$.
$x = 180 - 50 - 50 = 80$.

4. B Given $a^2 + b^2 = 32$ and $ab = 14$,
$(a-b)^2 = a^2 - 2ab + b^2 = (a^2 + b^2) - 2ab = 32 - 2(14) = 32 - 28 = 4$.

5. E There are 25 pairs of (50+1). The sum = $(50+1) \times 25 = 1275$.

6. C (Read page 110) $\frac{10}{x-2} + \frac{3}{2-x} = \frac{10}{x-2} - \frac{3}{x-2} = \frac{7}{x-2}$.

7. B (Read page 25) $\frac{74400 - 60000}{60000} = \frac{14400}{60000} = 24\%$.

8. E $\frac{3}{8_2} \times \frac{4^1}{5} = \frac{x}{10} \times \frac{1}{7}$, $\frac{3}{10} = \frac{x}{70}$, $10x = 210$, $x = 21$.

9. A The original solution has $30 \times 20\% = 6$ gallons of pure alcohol.
Let x = the pure alcohol added
The equation is: $\frac{6+x}{30+x} = 0.25$, $6 + x = 7.5 + 0.25x$, $0.75x = 1.5$
$x = \frac{1.5}{0.75} = 2$ gallons of pure alcohol added.

10. E (Read page 163) Apply Pythagorean Theorem.
$(x-1)^2 + (x+1)^2 = 10^2$
$x^2 - 2x + 1 + x^2 + 2x + 1 = 100$, $2x^2 + 2 = 100$, $2x^2 = 98$, $x^2 = 49$, $x = 7$.
$AC = x - 1 = 7 - 1 = 6$.

11. E $\frac{a-6}{a} = \frac{2}{5}$, $5(a-6) = 2a$, $5a - 30 = 2a$, $3a = 30$, $a = 10$.

12. E $5x + y = 2x - 1$, $y = -5x + 2x - 1$, $y = -3x - 1$.

13. C (Read page 84) $|x-4| < 7$, $-7 < x - 4 < 7$, $-3 < x < 11$.

14. D (Read page 30) $4^8 \times 16^{20} = 4^{2n}$, $4^8 \times (4^2)^{20} = 4^{2n}$, $4^8 \times 4^{40} = 4^{2n}$
$4^{48} = 4^{2n}$, $48 = 2n$, $n = 24$.

15. A $\$2400 \times 25\% = \$2400 \times 0.25 = \$600$.

16. C $\$2400 \times 3\frac{2}{3}\% = \$2400 \times \frac{11}{3}\% = \$2400 \times \frac{11}{3} \cdot \frac{1}{100} = \$2400 \times \frac{11}{300} = \$88$.

17. B (Read page 72) $y = mx + b$, $m = -7$ and $b = 5$, $y = -7x + 5$, $7x + y = 5$.

18. D (Read page 195) $V = \pi r^2 \cdot h = \pi(4^2) \cdot 6 = 96\pi$.

19. E (Read page 75)
The slope of the line $y = 4x - 8$ is 4. The slope of its perpendicular line is $-\frac{1}{4}$.
The equation is $y = -\frac{1}{4}x + b$, where b can be any number.

Answer Explanations
(PRACTICE TEST 2)
PSAT / NMSQT
Page 321 ~ 329

Section 1: 20. C (Read page 97) Vertical shift 2 units upward.

Section 2: 21. C $10+\sqrt{a}=19$, $\sqrt{a}=9$, $a=81$.

22. A $a=30\times 20\%=30\times 0.20=6$, $b=48\times 50\%=48\times 0.50=24$,
$a\times b=6\times 24=144$.

23. B (Read page 69) Given $f(x)=3x+12$ and $f(2x)=-18$
$f(2x)=3(2x)+12=-18$, $6x=-30$, $x=-5$.

24. A (Read page 159) $3x+x+2x=180$, $6x=180$, $x=30$.
$<A=3x^{\circ}=3(30^{\circ})=90^{\circ}$.

25. D Let x = miles he drove
The equation is: $80+0.35x=157$, $0.35x=77$, $x=\frac{77}{0.35}=220$ miles.

26. E $360^{\circ}\times\frac{3}{4}=270^{\circ}$.

27. E (Read page 69) $y=f(a)=2$. The graph shows $y=2$ when $x=a\leq -2$.

28. E (Read page 165) ΔABE is a 30-60-90 triangle.
$EB=2$, $AB=2\sqrt{3}$, the area = $(AB)^2=(2\sqrt{3})^2=12$.

29. 4 (Read page 70) Find the x-intercept of $4x+5y=16$,
let $y=0$, $4x+5(0)=16$, $4x=16$, $x=4$.

30. 10 $45\times\frac{2}{7+2}=45\times\frac{2}{9}=10$ liters of pure alcohol.

31. 45 (Read page 165) ΔACD is a 30-60-90 triangle.
$AD=\frac{1}{2}(AC)=\frac{1}{2}(30\sqrt{3})=15\sqrt{3}$. $CD=\sqrt{3}(AD)=\sqrt{3}(15\sqrt{3})=45$.

32. $\frac{2}{15}$ or 0.133 (Read page 292)
$$p=\frac{4}{10}\times\frac{3}{9}=\frac{12}{90}=\frac{2}{15}\text{ or }0.133.$$

33. 200 (Read page 145)
$t=5$, $h=120t-16t^2=120(5)-16(5^2)=600-400=200$ feet.

34. 13 $\frac{x-6}{x}=\frac{7}{13}$, $13(x-6)=7x$, $13x-78=7x$, $6x=78$, $x=13$.

35. $\frac{4}{25}$ or 0.16 (Read page 295)
There are $5\times 5=25$ outcomes.
A sum of 5 includes (1, 4), (2,3), (3,2), and (4, 1). $p=\frac{4}{5\times 5}=\frac{4}{25}$ or 0.16.

36. 3,600 (Read page 170) $180^{\circ}\times(n-2)=180^{\circ}\times(22-2)=180^{\circ}\times 20=3600^{\circ}$.

Answer Explanations
(PRACTICE TEST 2)
PSAT / NMSQT
Page 321 ~ 329

Section 2: 37. 75.4 (Read page 195) Surface area $SA = 2\pi r^2 + 2\pi r \cdot h$
$= 2\pi(2^2) + 2\pi(2)(4) = 8\pi + 16\pi = 24\pi = 24(3.14) = 75.4$.

38. 90 (Read page 70) $(4 \otimes 8) - (3 \otimes 7) = (5 \times 6 \times 7) - (4 \times 5 \times 6) = 210 - 120 = 90$.

A young girl walks around in a shopping mall. Her hair and eyes are painted with multiple colors. An old man looks at her frequently.

Girl: What is wrong with you, old man ? You never got wild when you were young ?

Man: I am wondering either your father or mother is a peacock. They got wild when they were young.

Answer Explanations
(PRACTICE TEST 3)
SAT MATH
Page 331 ~ 343

Section 1:

1\. D (Read page 69) Given $3x - y = 4$, $y = 3x - 4$. A point $(x, y) = (x, 3x - 4)$.

2\. A Solve the equation : $6(a+3) = 24$, $6a + 18 = 24$, $6a = 6$, $a = 1$.

3\. B (Read page 17) The numbers divisible by 3 and 8 are 24, 48, 72, ... They are all divisible by 6.

4\. A

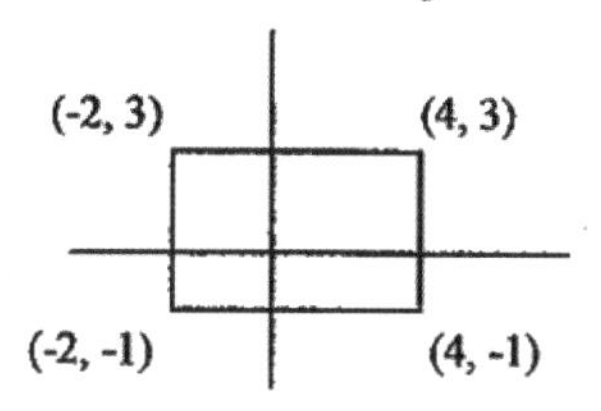

Length $= 4 - (-2) = 6$, Width $= 3 - (-1) = 4$.

Area $= 6 \times 4 = 24$

5\. D (Read page 83) Test each choice.

I. $|0+0| = |0| + |0|$ It is true.

II. $|2+(-1)| \neq |2| + |-1|$ It is not true.

III. $|2+1| = |2| + |1|$ It is true.

6\. E (Read page 34)

2, 3, 5, 8, 12, 17, 23, 30, 38, 47

+1 +2 +3 +4 +5 +6 +7 +8 +9

7\. C (Read page 157)

$< DCE = < ACB = 180^\circ - 90^\circ - 30^\circ = 60^\circ$.

$< D = 180^\circ - 80^\circ - 60^\circ = 40^\circ$.

8\. D (Read page 197) $SA = 2(\frac{1}{2} \cdot 3 \cdot 4) + 3 \times 4 + 4 \times 5 + 4 \times 4 = 12 + 12 + 20 + 16 = 60$.

9\. C Average speed $= \dfrac{60 \times 2 + 65 \times 3}{2+3} = \dfrac{120 + 195}{5} = \dfrac{315}{5} = 63$ miles per hour.

10\. E The number of the product sold on February is 50. None of the months has 100 of the product sold.

11\. A (Read page 110) $\dfrac{5}{a-b} + \dfrac{2}{b-a} = \dfrac{5}{a-b} - \dfrac{2}{a-b} = \dfrac{3}{a-b}$.

12\. C (Read page 178)

The area of the circle is: $A = \pi r^2 = \pi(9^2) = 81\pi$.

The area of the shaded region

$= 81\pi - (81\pi)(\frac{40}{360}) = 81\pi - (81\pi)(\frac{1}{9}) = 81\pi - 9\pi = 72\pi$.

13\. B (Read page 15) $1.6 \times \dfrac{312}{1.2} = 1.6 \times 260 = 416$ meters.

14\. D (Read page 26) $\$7,500 \times (1 + 0.04) = \$7,500 \times 1.04 = \$7,800$.

15\. C (Read page 34) 18, 14, 16, 15, **15.5**, ...

16\. A (Read page 69) $y = f(x)$, $y = f(a) = 0$ at $x = a$ where a equals 2 or -5.

17\. C $3a + b + c = 10$, $3a + c = b$

$(3a + c) + b = 10$, $b + b = 10$, $2b = 10$, $b = 5$.

Answer Explanations
(PRACTICE TEST 3)
SAT MATH
Page 331 ~ 343

Section 1: 18. A (Read page 157) $a+b+90+40+30=360$.
$a+b+160=360$, $a+b=200$, $a=200-b$.

19. B (Read page 197) Apply Pythagorean Theorem in three-dimensions.
$\overline{MN}=\sqrt{13^2+4^2+5^2}=\sqrt{169+16+25}=\sqrt{210}\approx 14.49$.

20. E $x^2+ax+12=(x-3)(x+b)$
$x^2+ax+12=x^2-3x+bx-3b$
$x^2+ax+12=x^2-(3-b)x-3b$
Comparing the coefficients: We have $a=-(3-b)$ and $12=-3b$.
$12=-3b$ gives $b=-4$
$a=-(3-b)=-[3-(-4)]=-(3+4)=-7$.

Section 2: 1. D (Page 30) $(n^2)^3=64$, $n^6=2^6$. $n=2$.

2. B $\frac{110}{5}=22$ miles per gallon. $\frac{242}{22}=11$ gallons to travel 242 miles.

3. D Let $n=$ The number of books to begin with.
The equation is: $(1-\frac{2}{3})n=15$, $\frac{1}{3}n=15$, $n=45$.

4. C Let x = Carol's money, $3x$ = Roger's money.
The equation is: $3x+x=480$, $4x=480$, $x=120$,
$3x=3(120)=360$ (Roger's money), $\frac{360}{480}=\frac{3}{4}$.

5. A (Read page 77) $-8\le 2a-4\le 14$
$-4\le 2a\le 18$, $-2\le a\le 9$.

6. E (Read page 69) $f(x)=3x-2$ and $\frac{1}{2}f(\sqrt{k})=3$
$\frac{1}{2}(3\sqrt{k}-2)=3$, $3\sqrt{k}-2=6$, $3\sqrt{k}=8$, $\sqrt{k}=\frac{8}{3}$, $k=\frac{64}{9}$.

7. B Let $\frac{M}{F}=$ the ratio of males to females at the beginning of the year.
The ratio of males to females at the end of the year is:
$$\frac{1.12M}{0.8F}=\frac{112}{80}\left(\frac{M}{F}\right)=\frac{7}{5}\left(\frac{M}{F}\right).$$

8. E Let $x=$ the radius of the smaller semicircle
$y=$ the radius of the larger semicircle
The sum of the area of the two semicircles
Areas $=\frac{1}{2}\pi(\frac{x}{2})^2+\frac{1}{2}\pi(\frac{y}{2})^2=\frac{1}{8}\pi(x^2+y^2)=\frac{1}{8}\pi(64)=8\pi$.

9. 3 $(\sqrt{5}+\sqrt{2})(\sqrt{5}-\sqrt{2})=(\sqrt{5})^2-(\sqrt{2})^2=5-2=3$

10. 6.25 Let $n=$ the number
The equation is: $\frac{n}{5}=n-5$, $n=5n-25$, $4n=25$, $n=6.25$.

Answer Explanations
(PRACTICE TEST 3)
SAT MATH
Page 331 ~ 343

Section 2: 11. 11 $7(a-1)=5(a+3)$, $7a-7=5a+15$, $2a=22$, $a=11$.

12. 3 (Read page 30) $3^{n+2}=243$, $3^{n+2}=3^5$, $n+2=5$, $n=3$.

13. 38.4 (Read page 76) $a=kb$, $24=k(5)$, $k=\frac{24}{5}=4.8$. The equation is : $a=4.8k$

We have : $a=4.8b=4.8(8)=38.4$ when $b=8$.

14. 3.74 (Read page 143) Apply Pythagorean Theorem.

$$(x+6)^2+(x-6)^2=10^2$$
$$x^2+\cancel{12x}+36+x^2-\cancel{12x}+36=100$$
$$2x^2+72=100$$
$$2x^2=28,\ x^2=14,\ x=\sqrt{14}\approx 3.74.$$

15. 15 (Read page 30) $a^n\cdot a^5=a^{20}$ and $a\neq 0$.

$a^{n+5}=a^{20}$, $n+5=20$, $n=15$.

16. 55 (Read page 155) The degree measure of each of the vertical angles is 80°.

$80+45+x=180$, $125+x=180$, $x=55$.

17. 2 (Read page 71) Given $(-2,-7)$, $(0,-2)$, $(k,3)$

Slope: $m_1=\dfrac{-2-(-7)}{0-(-2)}=\dfrac{-2+7}{2}=\dfrac{5}{2}$, $m_2=\dfrac{3-(-2)}{k-0}=\dfrac{3+2}{k}=\dfrac{5}{k}$,

$m_1=m_2$, $\dfrac{5}{2}=\dfrac{5}{k}$, $k=2$.

18. 40 (Read page 281)

$$\text{Mean}=\frac{60+40+30+50+20+40}{6}=\frac{240}{6}=40.$$

Section 3: 1. D (Read page 23) $8p=14$, $p=\dfrac{14}{8}=1.75=175\%$.

2. A (Read page 21) $21\times\dfrac{2}{2+5}=21\times\dfrac{2}{7}=6$ liters.

3. C (Read page 122) Let b = cost of one pound of beef
p = cost of one pound of pork.

We have the system of equations:

$$\begin{cases}3b+2p=8.20\\2b+p=5.40\end{cases}$$

Solve the above system of equations, we have $b=\$2.60$.

4. C (Read page 163) Apply Pythagorean Theorem, we can find the height of each triangle is 4.

Surface Area: $SA=2(5\times 8)+(8\times 6)+2(\frac{1}{2}\times 6\times 4)=80+48+24=152$.

5. B (Read page 122) $\begin{cases}x+2y=21\\3x-y=7\end{cases}$

Solve the above system of equations, we have $x=5$, $y=8$, $x+y=5+8=13$.

Answer Explanations
(PRACTICE TEST 3)
SAT MATH
Page 331 ~ 343

Section 3: 6. D Let x = the average weight of females
$2x$ = the average weight of males

$$\frac{2x(1)}{x(7)+2x(1)}=\frac{2x}{9x}=\frac{2}{9}.$$

7. E (Read page 197) The base area $= 3\times 4+\frac{1}{2}\cdot 2\cdot 3=12+3=15$.
The volume = base area × height = 15 × 5 = 75.

8. D (Read page 39) $9\times 10\times 10=900$.

9. B (Read page 86) $y=a(x-h)+k$, Vertex (h, k), a is any positive number.
The equation could be $y=(x-7)^2-2$ with vertex (7, −2).

10. A The length of each piece $= 12\times 3+2=38$ inches.
The length of 8 pieces: $x = 38\times 8=304$ inches.

$$x=\frac{304}{12}=25\tfrac{1}{3} \text{ feet}.$$

11. C $x+y=14$ and $6<x<12$, we have $y=14-x$.
x = 7, 8, 9, 10, or 11
y = 7, 6, 5, 4, or 3
The greatest possible value of $x-y=11-3=8$.

12. C (Read page 281)
Let a = 1 st. integer, we have: $a, a+2, a+4, a+6, a+8, a+10, a+11$.

$$\text{The mean} = \frac{a+a+2+a+4+a+6+a+8+a+11}{7}=\frac{7a+42}{7}=a+6.$$

We have : $a+6=n$, $a=n-6$. The median $= a+6=(n-6)+6=n$.

13. C (Read page 201) It is a right circular cone.

$$\text{Volume} = \tfrac{1}{3}(\pi\cdot r^2\cdot h)=\tfrac{1}{3}\pi\cdot 10^2\cdot 6=200\pi.$$

14. E (Read page 291)

$$p=\frac{1}{2}\times\frac{1}{2}\times\frac{1}{2}\times\frac{1}{2}\times\frac{1}{2}=\left(\frac{1}{2}\right)^5=\frac{1}{32}.$$

15. D (Read page 281)
40, 40, 50, **60**, 70, 70, 90. The median is 60.

16. B (Read page 97)
$g(x)=f(x)-1$ moves the graph of $y=f(x)$, 1 unit downward.
The y-intercept moves from −1 to −2.

Answer Explanations
(PRACTICE TEST 4)
SAT MATH
Page 345 ~ 357

Section 1: 1. B (read page 281) Given $\frac{a+b+c+d}{4}=12$ and $\frac{a+b}{2}=14$.
We have : $a+b+c+d=48$ and $a+b=28$. Therefore, $28+c+d=48$,
$c+d=20$, the average of c and d = $\frac{c+d}{2}=\frac{20}{2}=10$.

2. A (Read page 89) $a^2-b^2=(a+b)(a-b)=4\times 2=8$.

3. D Each hour he can finite $\frac{5}{7}\div 15=\frac{5}{7}\times\frac{1}{15}=\frac{1}{21}$ of the job. He can finite the whole Job in 21 hours. Or : $15\div\frac{5}{7}=15\times\frac{7}{5}=21$ hours.

4. E (Read page 71) Both C and E having a line with negative slope.
E has a line with a negative y-intercept.

5. B (read pages 41and 292)
For 5 letters, there are $5\times 4\times 3\times 2\times 1=120$ arrangements.
For letter a at the first, there are $4\times 3\times 2\times 1=24$ arrangements.
For letter a at the last, there are $4\times 3\times 2\times 1=24$ arrangement.
If a is not at either end, there are $120-24-24=72$ arrangements.

6. C (Read page 30) $3^x+3^x+3^x=27^x$,
$3(3^x)=27^x$, $3^{1+x}=3^{3x}$, $1+x=3x$, $1=2x$, $x=\frac{1}{2}$.

7. B (Read page 157) $2x+3x+5x=360$, $10x=360$, $x=36$.

8. D (Read pages 41 and 292) $30\times 25\times 53=39{,}750$ choices.

9. D (Read page 118) $y=\frac{k}{x}$, $12=\frac{k}{2/3}$, $k=12\times\frac{2}{3}=8$. $y=\frac{8}{x}=\frac{8}{4/5}=8\times\frac{5}{4}=10$.

10. D $1-\frac{1}{2}-\frac{1}{8}-\frac{2}{7}=\frac{56}{56}-\frac{28}{56}-\frac{7}{56}-\frac{16}{56}=\frac{5}{56}$.

11. B (Read page 23) $p=\frac{12}{32}=0.375=37.5\%$.

12. B (Read page 21) $230\times\frac{3}{20+3}=230\times\frac{3}{23}=30$ grams of sugar.

13. C (Read page 178)
Circumference of the circle: $c=3\pi\times\frac{360}{45}=24\pi$.
Find the radius of the circle: $2\pi r=24\pi$, $r=12$.
The area of the sector: $A=\pi\cdot 12^2\cdot\frac{45}{360}=\pi\cdot 144\cdot\frac{1}{8}=18\pi$.

14. B Given $a>b$ and $c<0$, c is negative.
Since c is a negative number, we have $ac<bc$. I is not true. II is true.
Since $a>b$, we have $a-c>b-c$. III is not true.

15. D (Read page 70) Given $x^{\oplus}=x^3-3x$, we have
$4^{\oplus}+2^{\oplus}=(4^3-3\times 4)+(2^3-3\times 2)=64-12+8-6=54$.

16. B (Read page 34) 5, 11, 18, 26, 35, 45, 56, 68, 81, **95**
+6 +7 +8 +9 +10 +11 +12 +13 +14

Answer Explanations
(PRACTICE TEST 4)
SAT MATH
Page 345 ~ 357

Section 1: 17. A (Read page 173)

The distance (circumference) of each revolution of the pulley run is $\pi \times D$.

The distance of the large pulley runs 80 revolutions is: $\pi \times 12 \times 80 = 960\pi$.

The speed of the small pulley is: $\frac{960\pi}{8\pi} = 120$ revolutions per minute.

18. C $3\sqrt{18}+2\sqrt{8}-6\sqrt{50} = 3\sqrt{9\cdot 2}+2\sqrt{4\cdot 2}-6\sqrt{25\cdot 2} = 9\sqrt{2}+4\sqrt{2}-30\sqrt{2} = -17\sqrt{2}$.

19. B (Read page 197 and 205) Ratio $= \frac{\frac{4}{3}\pi \cdot r^3}{(2r)^3} = \frac{\frac{4}{3}\pi \cdot r^3}{8\cdot r^3} = \frac{4\pi}{3}\cdot\frac{1}{8} = \frac{\pi}{6} \approx 0.523$.

20. E (Read page 25) $p = \frac{12800-12460.34}{12800} = \frac{339.66}{12800} = 0.0265 \approx 2.65\%$

Section 2: 1. E Given $x(y+2)=3$, we have $x = \frac{3}{y+2}$, $y+2 \neq 0$, $y \neq -2$.

A fraction with zero denominator is undefined. It means that the denominator cannot be 0.

2. E Given $x^2 = y+2$, we have $x = \pm\sqrt{y+2}$.

Given x is a real number, we have $y+2 \geq 0$(positive), $y \geq -2$.

3. D (Read page 29) $118{,}000{,}000{,}000 = 1.18 \times 10^{11}$.

4. A (Read page 191) The coordinates of A is (2, −2).

5. C Given $ab=45$ and $bc=27$, we have $ab > bc$, $a > c$.

$5\times 9 = 45$	$3\times 9 = 27$
or	or
$9\times 5 = 45$	$9 \times 3 = 27$

Therefore, $ab = 5\times 9$ and $bc = 9\times 3$.

We have: $a=5, b=9, c=3$. It gives: $b > a > c$.

6. E The hypotenuse of the right triangle with lengths of two legs, 4 and 3.

Apply Pythagorean Theorem, the length of the hypotenuse is 5.

The perimeter $= 10+8+6+5+5+16 = 50$.

7. C The possible x-coordinates whose y-coordinates are equal are:

1 and 5, 2 and 4, 3 and 6, ····· .

8. D (Read page 76)

Let x = The mass of the weight attached, y = the stretched length.

We have the equation : $y = kx$, $6 = k(50)$, $k = \frac{6}{50} = 0.12$.

We have the equation: $y = 0.12x$. If $x = 95$, $y = 0.12(95) = 11.4$ inches.

9. $\frac{7}{2}$ or 3.5 $5(a-1) = 3a+2$, $5a-5 = 3a+2$, $2a = 7$, $a = \frac{7}{2} = 3.5$.

10. 14 (Read page 122) Solve the system of equations: $\begin{cases} a-b=-2 \\ 2a-b=4 \end{cases}$,

we have $a=6$ and $b=8$. Therefore : $a+b = 6+8 = 14$.

Answer Explanations
(PRACTICE TEST 4)
SAT MATH
Page 345 ~ 357

Section 2: 11. 269 (Read page 20) $2 \underline{| 6, 8, 11}$ $3, 4, 11$

LCM $= 2\times3\times4\times11 = 264$, the number is $264 + 5 = 269$.

12. 64 (Read page 30) $(n^{1/2})^{2/3} = 4$, $n^{1/3} = 4$, $(n^{1/3})^3 = 4^3$, $n = 64$.

13. 25/2 or 12.5 Draw a coordinate plane and locate the given points, we can find the triangle with base 5 and height 5.
Area $= \frac{1}{2}bh = \frac{1}{2}\times5\times5 = \frac{25}{2} = 12.5$.

14. 10 (Read page 181) Given two points $A(1,-6)$ and $B(7,2)$
$d = \sqrt{(7-1)^2 + [2-(-6)]^2} = \sqrt{6^2+8^2} = \sqrt{36+64} = \sqrt{100} = 10$.

15. 48 The area of the rectangle with length 9 and width 6 $= 9\times6 = 54$.
The area of the triangle on the left-top with base 4 and height 3 $= \frac{1}{2}\times4\times3 = 6$.
The area of the figure $= 54 - 6 = 48$.

16. 640 (Read page 145) $h = vt - 16t^2 = 224(4) - 16(4)^2 = 896 - 256 = 640$ feet.

17. 1054 Let n = total number of students that the theater can hold.
$n\times\frac{6}{17} = 372$, $n = 372\times\frac{17}{6} = 1054$ students.
Or: $n = 372\div\frac{6}{17} = 372\times\frac{17}{6} = 1054$ students.

18. 80 (Read page 15) The scale factor $= 300\div1\frac{1}{2} = 300\div\frac{3}{2} = 300\times\frac{2}{3} = 200$ times of the drawing. The actual width $= \frac{2}{5}\times200 = 80$ meters.

Section 3: 1. A $x = 3$ and $y = -2$, $\frac{1}{2}x^2 - 3y^3 = \frac{1}{2}(3)^2 - 3(-2)^3 = \frac{1}{2}(9) - 3(-8) = 4.5 + 24 = 28.5$.

2. C $8 - \frac{2}{3}\times\frac{9}{11} + \frac{9}{11}\div\frac{3}{4} = 8 - \frac{6}{11} + \frac{9}{11}\times\frac{4}{3} = 8 - \frac{6}{11} + \frac{12}{11} = 8 + \frac{6}{11} = 8\frac{6}{11}$.

3. B Given $x = 30$, we have $\angle B = 2x^\circ = 2(30^\circ) = 60^\circ$.
$\angle A = \angle C = \frac{180^\circ - 60^\circ}{2} = \frac{120^\circ}{2} = 60^\circ$. The triangle is equilateral (equiangular).
Therefore: $AB = AC = BC = 4$.

4. E (Read page 71) Given three points $(-1, 3)$, $(-3, 2)$, $(-4, a-5)$.
The slopes between all points are equal.
Slope $m = \frac{2-3}{-3-(-1)} = \frac{-1}{-2} = \frac{1}{2}$. We have: $m = \frac{a-5-2}{-4-(-3)} = \frac{1}{2}$, $\frac{a-7}{-1} = \frac{1}{2}$,
$2a - 14 = -1$, $2a = 13$, $a = 6.5$.

5. D (Read page 281) Let x = the average score of the next 2 tests.
We have the equation: $\frac{76\times8 + (x\times2)}{10} = 80$, $608 + 2x = 800$, $2x = 192$, $x = 96$.

Answer Explanations
(PRACTICE TEST 4)
SAT MATH
Page 345 ~ 357

Section 3: 6. E (Read page 173) ΔOAB is a right (90°) triangle. Apply Pythagorean Theorem:
$\overline{OB}^2 + \overline{AB}^2 = \overline{OA}^2$, $6^2 + \overline{AB}^2 = 10^2$, $\overline{AB}^2 = 100 - 36 = 64$, $AB = 8$.
$AC = AB = 8$, $AB + AC = 8 + 8 = 16$.

7. B (Read page 148) $g(x) = \sqrt{x-4}$, $x - 4 \geq 0$, $x \geq 4$.

8. E A plane slices through three adjacent faces at a corner of the solid, the intersection is a triangle.

9. A Let $n =$ the larger number, $42 - n =$ the smaller number.
We have the equation :
$n = 18 + \frac{1}{2}(42 - n)$, $n = 18 + 21 - \frac{1}{2}n$, $\frac{3}{2}n = 39$, $n = \frac{39\times2}{3} = \frac{78}{3} = 26$.

10. C (Read page 41) $\frac{5!}{4!} = \frac{5\cdot4\cdot3\cdot2\cdot1}{4\cdot3\cdot2\cdot1} = 5$.

11. A (Read page 84) Given $|2a - 8| = 6$
We have : $2a - 8 = 6$ or $2a - 8 = -6$
$2a = 14$ | $2a = 2$
$a = 7$ | $a = 1$

12. E (Read page 86) $y - k = a(x - h)^2$, The vertex is $(h, k) = (-1, 0)$
The equation is $y = a(x+1)^2$.
The given graph opens downward, a must be negative.

13. E (Read page 52) Between 1 and 250,
$1\times4, 2\times4, \cdots, 62\times4 = 248$, there are 62 multiples of 4.
$1\times5, 2\times5, \cdots, 50\times5 = 250$, there are 50 multiples of 5.
$1\times20, 2\times20, \cdots, 12\times20 = 240$, there are 12 multiples of 20.
There are 12 multiples of 20 are overlap within the multiples of 4 and 5.
Therefore, there are $62 + 50 - 12 = 100$ of these integers are multiples of 4 or 5, or both 4 and 5.

14. B (Read page 97) We have two points (0, 1) and (4, 3) on the given line.
Slope of the line: $m = \frac{3-1}{4-0} = \frac{1}{2}$. Equation of the line: $y = \frac{1}{2}x + 1$.
The line is translated 3 units to the left. We have the new equation:
$y = \frac{1}{2}(x+3) + 1 = \frac{1}{2}x + \frac{3}{2} + 1 = \frac{1}{2}x + \frac{5}{2}$, y-intercept $= \frac{5}{2}$.

15. A The radius of the larger circle is the hypotenuse of a 45-45-90 triangle.
$$\text{Ratio} = \frac{\pi\cdot r_1^2}{\pi\cdot r_2^2} = \frac{r_1^2}{r_2^2} = \frac{1^2}{(\sqrt{2})^2} = \frac{1}{2}$$

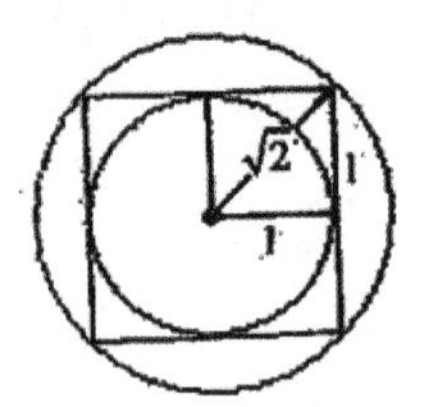

16. C (Read page 25)
2003 ~ 2004 : $\frac{600}{4000} = 15\%$ 2004 ~ 2005 : $\frac{600}{4600} \approx 13.04\%$
2005 ~ 2006 : $\frac{800}{5200} \approx 15.38\%$ √ 2006 ~ 2007 : $\frac{900}{6000} = 15\%$
2007 ~ 2008 : $\frac{1000}{6900} \approx 14.49\%$

Answer Explanations
(PRACTICE TEST 5)
SAT MATH
Page 359 ~ 371

Section 1:

1. D (Read page 85) Given $(a-4)^2 = 36$ and $a > 0$.
$a - 4 = \pm 6$, $a = 10$, $a^3 = 10^3 = 1{,}000$.

2. A (Read page 13) $BC = 42 \times \frac{5x}{2x+5x} = 42 \times \frac{5x}{7x} = 42 \times \frac{5}{7} = 30$.

3. A (Read page 84) $|x-4| < 12$
$-12 < x - 4 < 12$
$-8 < x < 16$. You may test each choice for the answer.

4. D (Read page 148) $y = 5 + \sqrt{x-2}$, where $x - 2 \geq 0$, $x \geq 2$.

5. E (Read page 148) $y = 5 + \sqrt{x-2}$, where $x - 2 \geq 0$, $y \geq 5$.

6. E (Read page 30) $2^{5x} = 64$, $2^{5x} = 2^6$, $5x = 6$, $x = \frac{6}{5}$.

7. C (Read page 86) $h(x) = -\frac{1}{2}(x-1)^2 + 3$ has vertex (1, 3).
$a = -\frac{1}{2} < 0$, the parabola $h(x)$ opens downward.

8. C Given $a = 2 + \frac{1}{b}$ and $b > 1$, we have $0 < \frac{1}{b} < 1$ and $2 < a < 3$.

9. B Let n = the number
We have the equation: $8n = n + 8$, $7n = 8$, $n = \frac{8}{7}$.

10. E Given $r + 3s - 4t = 5$, solve for r, $r = 5 - 4t + 3s$.

11. D (Read page 174) $\angle ADC = 180^\circ - \angle ABC = 180^\circ - 95^\circ = 85^\circ$.
In a circle, the opposite angles of an inscribed quadrilateral are supplementary.

12. D (Read page 20) The LCM of 4, 5, and 6 is 60. $x = 60 - 3 = 57$.

13. D (Read page 177)
The base length = 6, the height = 5. Area = $\frac{1}{2}bh = \frac{1}{2}(6)(5) = 15$.

14. A (Read page 177) The length of one leg of the triangle is $(14-8) \div 2 = 3$.
Apply Pythagorean Theorem. We have the height of the Trapezoid : $h = 4$.
The area of the trapezoid (quadrilateral) $= \frac{1}{2}(8+14)(4) = 44$.

15. E Given $a + b + c = 180$ and $c > 90$, we have $a + b < 90$, $a < 90 - b$
Given $a = b + 12$, therefore $a < (90 - b)$, $a < [90 - (a - 12)]$, $a < (90 - a + 12)$,
$2a < 102$, $a < 51$.

16. C We have: 65 books are $1 - \frac{3}{5} - (\frac{2}{5} \cdot \frac{1}{2}) = 1 - \frac{3}{5} - \frac{1}{5} = \frac{1}{5}$ of the total books.
The total number of books $= 65 \div \frac{1}{5} = 65 \times 5 = 325$ books.
There are $325 \times \frac{3}{5} = 195$ books were sold for \$4 each.
$(325 - 195) \times \frac{1}{2} = 65$ books were sold for \$2 each.
$325 - 195 - 65 = 65$ books were sold for \$1 each.
The total amount of dollars earned is :
$\$4 \times 195 + \$2 \times 65 + \$1 \times 65 = \$975$.

Answer Explanations
(PRACTICE TEST 5)
SAT MATH
Page 359 ~ 371

Section 1: 17. B (Read page 195) The solid is a right cylinder.
Volume $= \pi r^2 \cdot h = \pi(8^2) \cdot 4 = 256\pi$.

18. C $84 \times \frac{5}{3+4+5} = 84 \times \frac{5}{12} = 35$.

19. B (Read page 167) If two triangles are similar, the ratio of the lengths of their corresponding sides is a constant (equal). $\frac{6}{5} = \frac{x}{8}$, $x = \frac{48}{5} = 9.6$.

20. E Monday: $13{,}000 - 12{,}700 = 300$. Tuesday: $12{,}700 - 12{,}300 = 400$.
Wednesday: $12{,}800 - 12{,}500 = 300$. Thursday: $12{,}600 - 12{,}500 = 100$.
Friday: $12{,}900 - 12{,}400 = 500$. √

Section 2: 1. C (Read page 10 of the book "A-Plus Notes for Beginning Algebra")

2. E $2(x-3) = 8$, $2x - 6 = 8$, $2x = 14$, $x = 7$.

3. A $2\frac{1}{3} = \frac{7}{3}$, the reciprocal of $2\frac{1}{3}$ is $\frac{3}{7}$. $\frac{7}{3} + \frac{3}{7} = \frac{49}{21} + \frac{9}{21} = \frac{58}{21} = 2\frac{16}{21}$.

4. B (Read page 34)

Points	2	3	4	5	6	7
Lines	1	3	6	10	15	21

+2 +3 +4 +5 +6

5. B (Read page 85) Solve $x^2 - 12x + 32 = 0$, $(x-4)(x-8) = 0$, $x = 4$ or 8.

6. C $(1 - \frac{1}{3}) \times \frac{3}{5} = \frac{2}{3} \times \frac{3}{5} = \frac{2}{5}$.

7. B (Read page 15) $\frac{A_1}{A_2} = \frac{2(\pi \cdot 4^2) + 2\pi \cdot 4 \cdot 8}{2(\pi \cdot 2^2) + 2\pi \cdot 2 \cdot 4} = \frac{32\pi + 64\pi}{8\pi + 16\pi} = \frac{96\pi}{24\pi} = \frac{4}{1}$.

Or: $\frac{A_1}{A_2} = \left(\frac{r_1}{r_2}\right)^2 = \left(\frac{4}{2}\right)^2 = \frac{16}{4} = \frac{4}{1}$.

8. D (Read page 15) $\frac{V_1}{V_2} = \frac{\pi \cdot 4^2 \cdot 8}{\pi \cdot 2^2 \cdot 4} = \frac{128\pi}{16\pi} = \frac{8}{1}$.

Or: $\frac{V_1}{V_2} = \left(\frac{r_1}{r_2}\right)^3 = \left(\frac{4}{2}\right)^3 = \frac{64}{8} = \frac{8}{1}$.

9. 33 Given $2x - 7 = 11$, we have $6x - 21 = 3(2x-7) = 3(11) = 33$.

10. 900 Given $\frac{a}{5} = \frac{6}{b}$, we have $ab = 30$. $(ab)^2 = 30^2 = 900$.

11. 720 (Read page 41) $6 \times 5 \times 4 \times 3 \times 2 \times 1 = 720$ seat arrangements.

12. 600 (Read page 41) $6 \times 5 \times 4 \times 3 \times 2 \times 1 = 720$ seat arrangements.
There are $5 \times 4 \times 3 \times 2 \times 1 = 120$ arrangements for one student sitting in the first seat. $720 - 120 = 600$ seat arrangements.

Answer Explanations
(PRACTICE TEST 5)
SAT MATH
Page 359 ~ 371

Section 2: 13. 35 (Read page 36) Draw a Venn Diagram.
$23+17-5=35$ students.

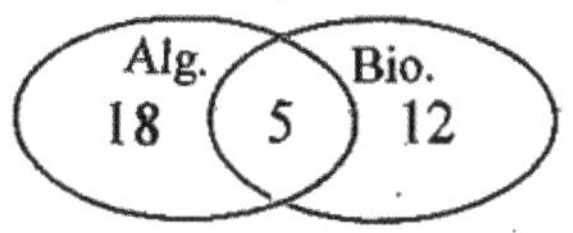

14. 21 Given $a+b=10,\ b+c=14,\ a+c=18,$
We have $(a+b)+(b+c)+(a+c)=10+14+18=42$.
$2(a+b+c)=42,\ a+b+c=21$.

15. 132 (Read page 157) $<5=<1=180^\circ-<2=180^\circ-48^\circ=132^\circ$.

16. 8 (Read page 36) Draw a Venn Diagram.
$50-2=48$ students who took one or both courses.
$24+32-48=8$ students took both courses.

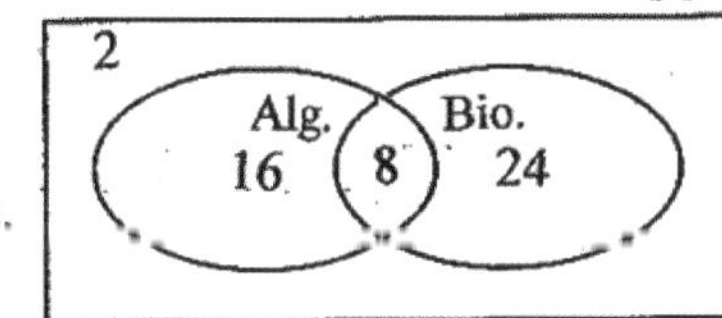

17. 170 Average height $=\dfrac{172\times12+164\times4}{12+4}=\dfrac{2064+656}{16}=\dfrac{2720}{16}=170\ cm.$

18. 7.5 (Read page 177) Area of $\Delta ABD=$ Area of $\Delta ABC-$ Area of ΔADC
$=\frac{1}{2}\times8\times5-\frac{1}{2}\times5\times5=20-12.5=7.5$.

Section 3: 1. A (Read page 120) $2^x+2^x=\sqrt{8},\ 2(2^x)=\sqrt{2^3},$
$2^{1+x}=2^{3/2},\ 1+x=\frac{3}{2},\ x=\frac{1}{2}$.

2. B (Read page 120) $2^x\cdot2^x=\sqrt{8},\ 2^{x+x}=\sqrt{2^3},\ 2^{2x}=2^{3/2},\ 2x=\frac{3}{2},\ x=\frac{3}{4}$.

3. E (Read page 177) $\overline{EC}=14\times\frac{2}{7}=4$.
Area of the trapezoid $ABCE=\frac{1}{2}(14+4)\times5=45$.

4. B (Read page 25) $p=\dfrac{4.29-3.70}{3.70}=\dfrac{0.59}{3.70}\approx0.15945\approx15.95\%$.

5. D If the car traveled 50 miles per hour, it needed $\frac{60\times8}{50}=9.6$ hours for the same trip. It would take $9.6-8=1.6$ hours $=96$ minutes longer.

6. A (Read page 148) x can be any real number.

7. D (Read page 148) Since $|x-2|\ge0$, we have $y=1+|x-2|\ge1$.

8. D (Read page 174) $<1=\dfrac{100^\circ+40^\circ}{2}=70^\circ$.

9. A (Read page 174) $<C=\dfrac{220^\circ-140^\circ}{2}=40^\circ$.

10. C Given $h(x)=3x^2-2$, we have $2h(-2)=2[3\cdot(-2)^2-2]=2(12-2)=20$.

11. C (Read page 52) $1\times3=3,\ 2\times3=6,\ 3\times3=9,\ \cdots,\ \mathbf{83}\times3=249$.

12. C (Read page 159, Third-Side Rule) $5+8>12$.

Answer Explanations
(PRACTICE TEST 5)
SAT MATH
Page 359 ~ 371

Section 3: 13. A Given $(x+2)(x-h)=x^2-kx-8$

$$x^2-xh+2x-2h=x^2-kx-8$$
$$x^2+(2-h)x-2h=x^2-kx-8$$

Comparing the coefficients, we have : $2h=8,\ h=4$

$$2-h=-k\ ,2-4=-k,\ -2=-k,\ k=2\ .$$

14. D (Read page 291) $p=\dfrac{\frac{1}{2}\times3\times2}{5\times8}=\dfrac{3}{40}$.

15. A (Read page 97)
The graph of $y=f(x+3)-4$ moves the graph of $y=f(x)$, 3 units to the left and 4 units downward.

16. E There are 20,000 telephones sold.
We have : $\dfrac{4000+6000}{20000}=\dfrac{10000}{20000}=\dfrac{1}{2}=50\%$. They are Central and Eastern.

A lady said to her friend.
Lady: I always keep quiet whenever my husband argues with me.
Friend: How do you control your anger ?
Lady: I clean the toilet.
Friend: How does that help ?
Lady: I use his toothbrush.

Answer Explanations
(PRACTICE TEST 6)
SAT MATH
Page 373 ~ 385

Section 1:

1. E (Read page 76) Given $y = kx$, $y = 6$ and $x = 2$, we have $k = \frac{y}{x} = \frac{6}{2} = 3$.
The equation is: $y = 3x$. Given $y = 15$, $15 = 3x$, we have $x = 5$.

2. D (Read page 39) $4 \times 3 \times 6 = 72$ choices.

3. C (Read page 291) $p = \frac{70}{3500} = \frac{1}{50}$.

4. B It estimates $\frac{150}{5} = 30$ salmons for each tagged salmon.
There are $100 \times 30 = 3{,}000$ salmons in the river.

5. A (Read page 165) Given $AC = 4$, $AD = 2$, ΔACD is a 30-60-90 triangle.
$CD = \sqrt{3}(AD) = \sqrt{3}(2) = 2\sqrt{3}$, r (radius of the circle) $= \sqrt{3}$.
The area of the circle : $A = \pi r^2 = \pi(\sqrt{3})^2 = 3\pi$.

6. D (Read page 30) $3^{2x} = 27^{x-2}$, $3^{2x} = (3^3)^{x-2}$, $3^{2x} = 3^{3x-6}$, $2x = 3x - 6$, $x = 6$.

7. B (Read page 281) Given $a + b + c = 3 \times 9 = 27$,
we have $\frac{a+b+c+d}{4} = 12$, $\frac{27+d}{4} = 12$, $27 + d = 48$, $d = 21$.

8. E (Read page 34)

Points	2	3	4	5	6	7	8	9
Lines	1	3	6	10	15	21	28	36
		+2	+3	+4	+5	+6	+7	+8

9. C The intersection of two planes is a line, or a plane.

10. E (Read page 70) Given $f(x) = ax + b$,
we have $f(0) = k$, then $f(0) = a(0) + b = k$, $b = k$.
$f(2) = n$, then $f(2) = a(2) + b = n$, $2a + b = n$.
$f(1) = 8$, then $f(1) = a(1) + b = 8$, $a + b = 8$.
Therefore, $k + n = b + 2a + b = 2(a + b) = 2(8) = 16$.

11. B $3x + 57 = 180$, $3x = 123$, $x = 41$.

12. E (Read page 70) $f(x) = 2x^3 - x^2 - 2$, $f(-2) = 2(-2)^3 - (-2)^2 - 2 = -16 - 4 - 2 = -22$.

13. C (Read page 145) $h = 160t - 16t^2$, $h = 0$ (reach the ground).
We have : $160t - 16t^2 = 0$, $16t(10 - t) = 0$, $t = 10$ seconds.

14. E (Read page 70) $5 \times 6 \times 7 = 210$.

15. D (Read page 86) The equation of the parabola is $x - h = a(y - k)^2$,
The parabola opens to the left. a is a negative number.
Given vertex (4, 0), we have : $x - 4 = a(y - 0)^2$, $x = ay^2 + 4$, where $a = -1$.

16. D OA and OB are the radius of the circle. $OA = OB$.
ΔOAB is isosceles. We have : $\angle A = \angle B = x^\circ$
Therefore, $x + x + 110 = 180$, $2x + 110 = 180$, $2x = 70$, $x = 35$.

Answer Explanations
(PRACTICE TEST 6)
SAT MATH
Page 373 ~ 385

Section 1:

17. A (Read page 157) $\angle 6 = \angle 2 = 32^\circ$, $\angle 8 = 180^\circ - \angle 6 = 180^\circ - 32^\circ = 148^\circ$.

18. B (Read page 71) Given points (2, −3), (4, a), and $m = 4$.

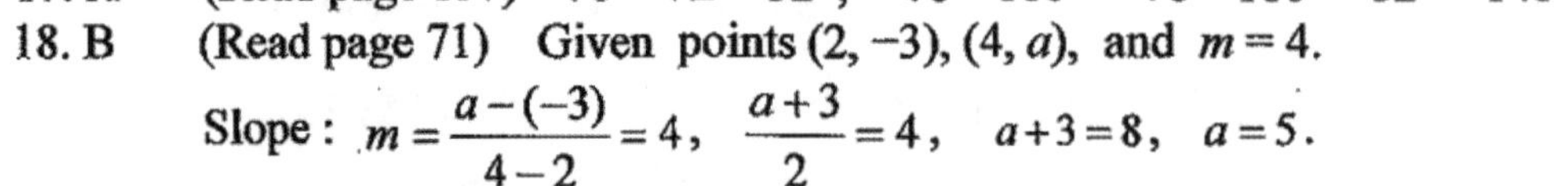

Slope : $m = \dfrac{a-(-3)}{4-2} = 4$, $\dfrac{a+3}{2} = 4$, $a+3=8$, $a=5$.

19. C (Read page 177)
$4x + 4y = 4(x+y)$.

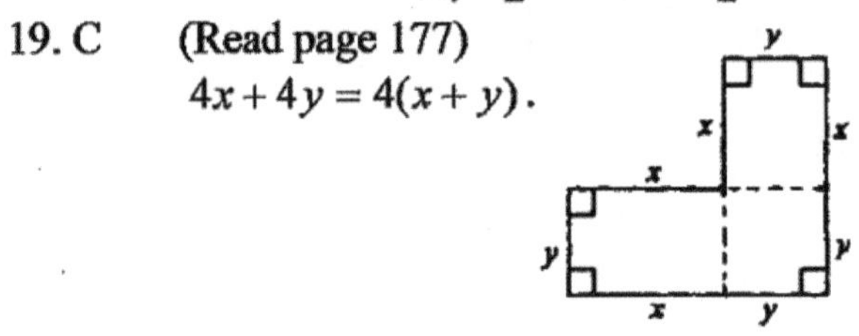

20. C (Read page 177) $\angle B = 90^\circ - 45^\circ = 45^\circ$, $\angle C = \angle B = 45^\circ$, $\angle A = 90^\circ$.
ΔABC and ΔABD are right triangles. Apply Pythagorean Theorem :
$(BC)^2 = (4\sqrt{2})^2 + (4\sqrt{2})^2 = 32 + 32 = 64$, $BC = 8$, $BD = 4$.
$h^2 + 4^2 = (4\sqrt{2})^2$, $h^2 + 16 = 32$, $h^2 = 16$, $h = 4$.
Area $= \frac{1}{2}(8)(4) = 16$.

Section 2:

1. E Given $x = 4$, $y = 2$. $\frac{1}{2}x^2 - 3y^2 + 1 = \frac{1}{2}(4)^2 - 3(2)^2 + 1 = 8 - 12 + 1 = -3$.

2. D (Read page 85). $2x^2 + 5x - 3 = 0$, $x > 0$

$$(2x-1)(x+3) = 0$$
$$2x - 1 = 0 \text{ or } x + 3 = 0$$
$$2x = 1 \quad | \quad x = -3$$
$$x = \tfrac{1}{2} \checkmark$$

3. B (Read page 52) $1 \times 4 = 4$, $2 \times 4 = 8$, $3 \times 4 = 12$, ..., $87 \times 4 = 348$.

4. A Substitute $p = 2r + 5$ into $p + 3r = 20$,
we have : $2r + 5 + 3r = 20$, $5r = 15$, $r = 3$,
$p = 2r + 5 = 2(3) + 5 = 11$. $p + r = 11 + 3 = 14$.

5. D (Read page 23) $55 \times 5\frac{4}{5}\% = 55 \times 5.8\% = 55 \times 0.058 = 3.19$.

6. C (Read page 15)
$\dfrac{A_1}{A_2} = \dfrac{2(12\times6 + 12\times9 + 9\times6)}{2(8\times4 + 8\times6 + 6\times4)} = \dfrac{468}{208} = \dfrac{9}{4}$. Or: $\dfrac{A_1}{A_2} = \left(\dfrac{S_1}{S_2}\right)^2 = \left(\dfrac{12}{8}\right)^2 = \left(\dfrac{3}{2}\right)^2 = \dfrac{9}{4}$.

7. E There are $15 \times 20\% = 15 \times 0.2 = 3$ gallons of pure salt in the original solution.
Let x = water added, we have the equation :
$\dfrac{3}{15+x} = 0.15$, $2.25 + 0.15x = 3$, $0.15x = 0.75$, $x = 5$ gallons of water.

8. A There are $15 \times 20\% = 15 \times 0.2 = 3$ gallons of pure salt in the original solution.
Let x = salt added, we have the equation :
$\dfrac{3+x}{15+x} = 0.25$, $3 + x = 3.75 + 0.25x$, $0.75x = 0.75$, $x = 1$ gallons of salt.

Answer Explanations
(PRACTICE TEST 6)
SAT MATH
Page 373 ~ 385

Section 2: 9. 13 Solve $5x-3(x+1)=x+10$, $5x-3x-3=x+10$, $2x-3=x+10$, $x=13$.

10. 8 Let $n=$ the number, we have the equation :

$$2(n+3)=22,\ 2n+6=22,\ 2n=16,\ n=8.$$

11. $\frac{3}{4}$ or 0.75 (Read page 72) Two lines have the same slope.

$3x-4y=5$, $4y=3x-5$, $y=\frac{3}{4}x-\frac{5}{4}$, slope $m=\frac{3}{4}$.

12. $\frac{8}{7}$ or 1.14 (Read page 75)

The slopes of two lines are negative reciprocals of each other.

$7x+8y=9$, $8y=-7x+9$, $y=-\frac{7}{8}x+\frac{9}{8}$, slope $m=-\frac{7}{8}$.

The slope of the perpendicular line : $m'=-\left(-\frac{8}{7}\right)=\frac{8}{7}$.

13. 20 Let $n=$ the number of books to begin with.

We have the equation :

$$n-\left(\frac{2}{5}n+\frac{1}{2}\cdot\frac{3}{5}n\right)=6,\ n-\frac{2}{5}n-\frac{3}{10}n=6,\ \frac{10n-4n-3n}{10}=6,$$

$\frac{3n}{10}=6$, $3n=60$, $n=20$ books.

14. 58 (Read page 20) Let $n=$ the smallest number

$$\begin{array}{r|l} 2 & 4,\ 5,\ 6 \\ \hline & 2,\ 5,\ 3 \end{array}$$

LCM $=2\times2\times5\times3=60$.

$4-2=2$, $5-3=2$, $6-4=2$. $n+2=60$ is divisible by 4, 5, and 6.

Therefore, the smallest value of n is : $n=60-2=58$.

15. 9 Given $a+3b=11$, $b+3c=15$, $c+3a=10$.

$(a+3b)+(b+3c)+(c+3a)=11+15+10$

$4a+4b+4c=36$, $4(a+b+c)=36$, $a+b+c=9$.

16. 280 (Read page 197)

Surface area (SA) $=2(6\times10)+2(5\times10)+2(5\times6)=120+100+60=280$.

17. 12.7 (Read page 197) Apply Pythagorean Theorem in three-dimension.

$\overline{MN}=\sqrt{10^2+6^2+5^2}=\sqrt{100+36+25}=\sqrt{161}\approx12.7$.

18. 9 Given $x^2=36$, $y^2=9$, $x>0$, $y>0$.

We have : $x=6$, $y=9$. $(x-y)^2=(6-9)^2=(-3)^2=9$.

Section 3: 1. C $28\div\frac{2}{5}=28\times\frac{5}{2}=70$ minutes.

2. B $\frac{3}{8}n=33$, $3n=264$, $n=88$. $\frac{5}{8}n=\frac{5}{8}(88)=55$.

3. B (Read page 39)

2	×	×	×

$4\times3\times2\times1=24$, $24\times2=48$ seat arrangements.

Answer Explanations
(PRACTICE TEST 6)
SAT MATH
Page 373 ~ 385

Section 3: 4. D (Read page 30) Given $3^x > 2^x$

I. $x = -1$, $3^{-1} > 2^{-1}$ is not true because $3^{-1} = \frac{1}{3}$ and $2^{-1} = \frac{1}{2}$, $\frac{1}{3} < \frac{1}{2}$.

II. $x = 0$, $3^0 > 2^0$ is not true because $3^0 = 1$ and $2^0 = 1$, $3^0 = 2^0$.

III. $x = 1$, $3^1 > 2^1$ is true.

5. A (Read page 122) Given the system of equation $y > x^2 - 4$ and $y < x + 3$, $(1, k)$ is a solution.

Therefore, $k > 1^2 - 4$, $k > -3$.

$k < 1 + 3$, $k < 4$. We have : $-3 < k < 4$.

6. D $\frac{x}{3} + \frac{x}{4} = 14$, $\frac{4x + 3x}{12} = 14$, $7x = 168$, $x = 24$.

7. C (Read page 75) Given $y = -\frac{1}{2}x + 4$, we have the slope $m = -\frac{1}{2}$.

The slope of the perpendicular line is : $m' = 2$.

The equation of the perpendicular line passing the origin : $y = 2x$.

The line passes the point $(a, 6)$. We have : $6 = 2a$, $a = 3$.

8. A (Read page 203) $V = \frac{1}{3}bh = \frac{1}{3}\cdot(12 \times 8)\cdot 8 = \frac{1}{3}(96)(8) = 256$.

9. A (Read page 173) The circumference of a circle is : $C = \pi \cdot D$

We have : $2(4\pi) = \pi(AB)$, $AB = 8$.

$2(3\pi) = \pi(AC)$, $AC = 6$.

The area of $\Delta ABC = \frac{1}{2}(8)(6) = 24$.

10. D (Read page 148) $x + 2 \neq 0$, $x \neq -2$.

11. E (Read page 149) Divide the denominator into the numerator, we have :

$y = \frac{x+1}{x+2} = 1 - \frac{1}{x+2}$, $y = 1$ is a horizontal asymptote.

12. C (Read page 40 and page 394 in the book "A-Plus Notes for Algebra")

A → B : → → → → → ↑ ↑ ↑ $\frac{8!}{5!3!} = \frac{8\cdot 7\cdot 6}{3\cdot 2\cdot 1} = 56$ paths.

13. B (Read page 40)

14. D $p = \frac{300 + 340 + 520 + 560 + 680}{8000} = \frac{2400}{8000} = 0.3 = 30\%$.

15. B (Read page 45)

$$\sum_{n=1}^{30}\left(\frac{1}{n+1} - \frac{1}{n+2}\right) = \sum_{n=1}^{30}\left(\frac{1}{n+1}\right) - \sum_{n=1}^{30}\left(\frac{1}{n+2}\right)$$

$$= \left(\frac{1}{2} + \frac{1}{3} + \frac{1}{4} + \cdots + \frac{1}{30} + \frac{1}{31}\right) - \left(\frac{1}{3} + \frac{1}{4} + \cdots + \frac{1}{30} + \frac{1}{31} + \frac{1}{32}\right)$$

$$= \frac{1}{2} - \frac{1}{32} = \frac{16}{32} - \frac{1}{32} = \frac{15}{32}.$$

16. E $\frac{20 + 18 + 20}{200} = \frac{58}{200} \approx 0.29 \approx 29\%$, $1 - 29\% = 71\%$.

Answer Explanations
(PRACTICE TEST 7)
SAT Subject Test Math Level I
Page 387 ~ 397

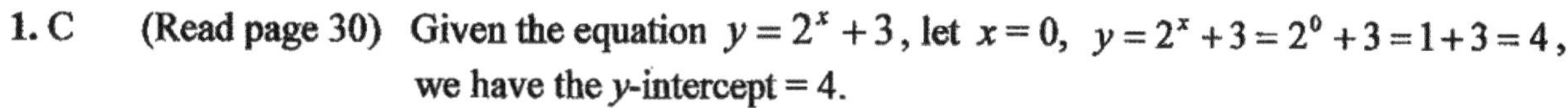

1. C (Read page 30) Given the equation $y = 2^x + 3$, let $x = 0$, $y = 2^x + 3 = 2^0 + 3 = 1 + 3 = 4$, we have the y-intercept = 4.

2. C (Read page 33) $\log_2 64 = \log_2 2^6 = 6\log_2 2 = 6(1) = 6$.

3. A (Read page 195) Volume $V = \pi \cdot r^2 \cdot h$, $h = \frac{V}{\pi \cdot r^2} = \frac{5000}{(3.14)\cdot 10^2} \approx 15.92$ inches.

4. B (Read page 37)

5. D (Read page 38)

6. D (Read page 148)

7. E (Read page 148)

8. E Given $a^{3x+2} = a^{4x-1}$, we have $3x + 2 = 4x - 1$, $x = 3$.

9. A (Read page 39) $9 \times 10 \times 10 \times 10 = 9{,}000$

10. E $\frac{100}{360} = \frac{10}{36} = \frac{5}{18}$.

11. C Given $f(x) = \frac{3x}{2x-5}$. The function is undefined if $2x - 5 = 0$. We have $x = \frac{5}{2}$.

12. A (Read page 239) $\sin < C = \frac{AB}{AC}$, $AB = AC \sin < C = 12 \sin 25^\circ \approx 5.07$.

13. A Let $f(x) = 0$, we have $\frac{x^3 + 4x^2 - 5x}{x-1} = \frac{x(x^2 + 4x - 5)}{x-1} = \frac{x(x+5)\cancel{(x-1)}}{\cancel{x-1}} = 0$.
Therefore : $x(x+5) = 0$, $x = 0$ or -5.

14. E Given $C(n) = 200 + 20n$, we have $5000 = 200 + 20n$, $4800 = 20n$, $n = 240$ calculators.

15. C (Read page 237) Given $\sin x = \frac{5}{8}$, we have $1 + \csc x = 1 + \frac{1}{\sin x} = 1 + \frac{8}{5} = 2\frac{3}{5}$.

16. C (Read page 159) $< B = 180^\circ \times \frac{4}{2+4+6} = 180^\circ \times \frac{4}{12} = 60^\circ$.

17. C (Read page 71) The line slants down from left to right. The slope is negative. y-intercept is positive.

18. A (Read page 38) If $n = -1$, then $|n + 5| \neq |n| + 5$.

19. D (Read page 39) For 6 students, there are $6 \times 5 \times 4 \times 3 \times 2 \times 1 = 720$ seat arrangements.
There are $5 \times 4 \times 3 \times 2 \times 1 = 120$ seat arrangements for one student taking the first seat.
There are $720 - 120 = 600$ seat arrangements if one of them does not take the first seat.

20. B (Read page 45) 1, 2, 4, 8, 16, 32, ···· is a geometric series.
The common ratio $r = \frac{2}{1} = 2$. The nth term $a_n = a_1 r^{n-1} = (1)(2^{n-1}) = 2^{n-1}$.

21. C (Read page 205) Volume $V = \frac{4}{3}\pi \cdot r^3$, $\frac{4}{3}\cancel{\pi} \cdot r^3 = 288\cancel{\pi}$, $4r^3 = 864$, $r^3 = 216$, $r = 6$.
The surface area $SA = 4\pi \cdot r^2 = 4\pi \cdot 6^2 = 144\pi$.

Answer Explanations
(PRACTICE TEST 7)
SAT Subject Test Math Level I
Page 387 ~ 397

22. A (Read page 195 and 205) $\frac{V_1}{V_2} = \frac{\frac{4}{3}\pi \cdot r^3}{\pi \cdot r^2 \cdot h} = \frac{\frac{4}{3}r}{h} = \frac{\frac{4}{3}\cdot 3}{6} = \frac{4}{6} = \frac{2}{3}$.

23. C (Read page 292) $p = \frac{4}{18} \times \frac{6}{17} = \frac{4}{51}$.

24. E (Read page 292) $p = \frac{4}{18} \times \frac{6}{18} = \frac{2}{27}$.

25. A (Read page 239) We have $\sin 60^\circ = \frac{12}{BC}$, $BC = \frac{12}{\sin 60^\circ} = \frac{12}{\sqrt{3}/2} = \frac{24}{\sqrt{3}} = 8\sqrt{3}$.

26. C Let $x =$ the distance between these two cities on the map.
We have the equation : $\frac{1.2}{3} = \frac{x}{12}$, $3x = 14.4$, $x = 4.8$ inches.

27. E (Read page 163) Apply Pythagorean Theorem.
$s^2 + s^2 = 10^2$, $2s^2 = 100$, $s^2 = 50$, $s = \sqrt{50}$.
The area of the dashed region : $A = 30 \times 20 - \sqrt{50} \times \sqrt{50} = 600 - 50 = 550$.

28. B Given $f(x) = x^2 + 24$ and $n > 0$.
we have $f(2n) = (2n)^2 + 24 = 4n^2 + 24$
$2f(n) = 2(n^2 + 24) = 2n^2 + 48$
$f(2n) = 2f(n)$, we have $4n^2 + 24 = 2n^2 + 48$, $2n^2 = 24$, $n^2 = 12$, $n = \sqrt{12} = 2\sqrt{3}$.

29. A (Read page 197) Apply Pythagorean Theorem in three-dimensions.
$\overline{AB} = \sqrt{12^2 + 5^2 + 5^2} = \sqrt{144 + 25 + 25} = \sqrt{194} \approx 13.93$.

30. A (Read page 241)
$$\csc x - \cos x \cot x = \frac{1}{\sin x} - \cos x \cdot \frac{\cos x}{\sin x} = \frac{1}{\sin x} - \frac{\cos^2 x}{\sin x} = \frac{1 - \cos^2}{\sin x} = \frac{\sin^2 x}{\sin x} = \sin x.$$

31. D (Read page 173) We have $\angle OAB = 90^\circ$.
Let $x =$ the radius of the circle, $x = OA = OC$.
Apply Pythagorean Theorem : $x^2 + 18^2 = (x+12)^2$, $\cancel{x^2} + 324 = \cancel{x^2} + 24x + 144$
$180 = 24x$, $x = \frac{180}{24} = 7.5$, diameter $D = 2(7.5) = 15$.

32. E (Read page 173) Circumference $C = 2\pi \cdot r = 2\pi \cdot 6 = 12\pi$.
The length of the minor arc $AB = 12\pi \times \frac{40}{360} = 12\pi \times \frac{1}{9} = \frac{4}{3}\pi \approx 4.19.$.

33. B (Read page 247)
$\cos 930^\circ = \cos(2 \times 360^\circ + 210^\circ) = \cos 210^\circ = \cos(180^\circ + 30^\circ) = -\cos 30^\circ = -\frac{\sqrt{3}}{2}$.

34. B (Read page 95) Given $2x^2 - 6x + 7 = 0$, $a = 2$, $b = -6$, $c = 7$.
We have $r_1 \cdot r_2 = \frac{c}{a} = \frac{7}{2} = 3.5$.

Answer Explanations
(PRACTICE TEST 7)
SAT Subject Test Math Level I
Page 387 ~ 397

35. D $2^1 = 2$, $2^2 = 4$, $2^3 = 8$, $2^4 = 16$
$2^5 = 32$, $2^6 = 64$, $2^7 = 128$, $2^8 = 256$
.
$2^{40} =$
The last digits repeat by 2, 4, 8, 6. The last digit of 2^{40} is 6.

36. B (Read page 105)
Given the function $f(x) = x^3 - 4$, we have $y = x^3 - 4$
Interchange x and y: $x = y^3 - 4$, $y^3 = x + 4$, $y = \sqrt[3]{x+4}$.
We have the inverse function $f^{-1}(x) = \sqrt[3]{x+4}$
$f^{-1}(10) = \sqrt[3]{10+4} = \sqrt[3]{14} \approx 2.41$.

37. D (Read page 241)
$(\sin x + \cos x)^2 + (\sin x - \cos x)^2 = \sin^2 x + 2\sin x\cos x + \cos^2 x + \sin^2 x - 2\sin x\cos x + \cos^2 x$
$= 2(\sin^2 x + \cos^2 x) = 2(1) = 2$.

38. B (Read page 239)
Let $x =$ the distance from the base of the ladders to the bottom of the building.
We have : $\sin 28^\circ = \frac{x}{20}$, $x = 20\sin 28^\circ \approx 9.39$.

39. D (Read page 148) Given $f(x) = \frac{x}{x+2} - \frac{1}{x}$, we have $x + 2 \neq 0$ and $x \neq 0$.
Therefore, $x \neq -2$ and 0.

40. A (Read page 33) $y = \log_x 2^x$, $y = x\log_x 2$, $\frac{y}{x} = \log_x 2$, $x^{y/x} = 2$, $x = 2^{x/y} = \sqrt[y]{2^x}$.

41. A (Read page 104) $|x^3 - 32| \geq 5$
$x^3 - 32 \leq -5$ or $x^3 - 32 \geq 5$
$x^3 \leq 27$ | $x^3 \geq 37$
$x \leq 3$ | $x \geq 3.33$

42. E (Read page 86) Given $9x^2 + 30x + k = 0$, $a = 9$, $b = 30$, $c = k$.
$b^2 - 4ac = 30^2 - 4(9)(k) = 0$, $36k = 900$, $k = 25$.

43. E $2^{400} = (2^4)^{100} = (16)^{100}$, $3^{300} = (3^3)^{100} = (27)^{100}$, $5^{200} = (5^2)^{100} = (25)^{100}$
$(\sqrt{2})^{840} = (2^{1/2})^{840} = 2^{420} = (2^{4.2})^{100} \approx (18.38)^{100}$
$(\sqrt{3})^{620} = (3^{1/2})^{620} = 3^{310} = (3^{3.1})^{100} \approx (30.14)^{100}$ √

44. C (Read page 283) Interquartile $= Q_3 - Q_1 = 88 - 75 = 13$.

45. C (Read page 52) $3\times1 = 3$, $3\times2 = 6$, $3\times3 = 9$, $3\times4 = 12$, ⋯⋯, $3\times33 = 99$.
There are $33 - 3 = 30$ multiples of 3 in $99 - 9 = 90$ two-digit numbers.
Probability $p = \frac{30}{90} = \frac{1}{3}$.

Answer Explanations
(PRACTICE TEST 7)
SAT Subject Test Math Level I
Page 387 ~ 397

46. C (Read page 181 & 187) The centers are $(-2, 4)$ and $(2, -4)$.
Distance $d = \sqrt{(-4-4)^2 + (2+2)^2} = \sqrt{64+16} = \sqrt{80} = 4\sqrt{5}$.

47. B (Read page 241)
Given $\sin x + \cos x = \sqrt{2}\sin x$, $1 + \dfrac{\cos x}{\sin x} = \sqrt{2}$, $1 + \cot x = \sqrt{2}$, $\cot x = \sqrt{2} - 1$.

$$\tan x + \cot x = \frac{1}{\sqrt{2}-1} + \sqrt{2} - 1 = \frac{\sqrt{2}+1}{(\sqrt{2}-1)(\sqrt{2}+1)} + \sqrt{2} - 1 = \frac{\sqrt{2}+1}{2-1} + \sqrt{2} - 1 = \sqrt{2} + 1 + \sqrt{2} - 1 = 2\sqrt{2}.$$

48. D (Read page 33)
$$\log_a 24 - \log_a x = \log_a 2$$
$$\log_a \frac{24}{x} = \log_a 2, \quad \frac{24}{x} = 2, \quad x = 12.$$

49. E (Read page 191) The distance from the point $A(-3, 0)$ to the line of reflection $x = 4$ is 7. The reflection of point A has the horizontal distance 7 units to the right of the line m, and $4 + 7 = 11$ units from the origin.

50. B (Read page 298)
Method 1 : Using the formula on page 297 and example on page 308.
Method 2 : Using a graphing calculator.

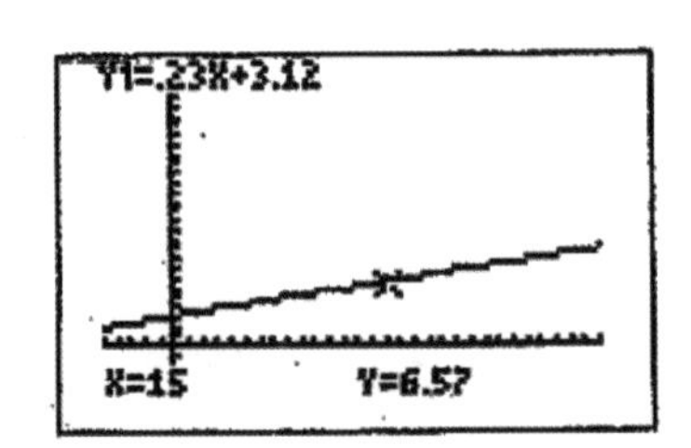

In the mall, a young girl asks her boy-friend to buy a lipstick for her.
Boy: You look better without using lipstick.
Girl: Fortunately, I did not ask you to buy a pair of pants for me.

Answer Explanations
(PRACTICE TEST 8)
SAT Subject Test Math Level I
Page 399 ~ 409

1. B (Read page 148) Given $y = 5 + \sqrt{2-x}$, where $2 - x \geq 0$, $-x \geq -2$, $x \leq 2$.

2. C (Read page 30) $\dfrac{x^3}{2x^{-2}} = \dfrac{1}{2}x^{3-(-2)} = \dfrac{1}{2}x^{3+2} = \dfrac{1}{2}x^5$.

3. D (Read page 241) $\sqrt{4\sin^2 2\theta + 4\cos^2 2\theta} = \sqrt{4(\sin^2 2\theta + \cos^2 2\theta)} = \sqrt{4(1)} = 2$.

4. A (Read page 30) $\left(\sqrt[3]{n}\right)^{1/2} = 2$, we have $\left(n^{1/3}\right)^{1/2} = 2$, $n^{1/6} = 2$, $\left(n^{1/6}\right)^6 = 2^6$, $n = 64$.

5. A (Read page 237) $\cos 48^\circ = \dfrac{BC}{AB}$, $BC = AB\cos 48^\circ = 9\cos 48^\circ \approx 6.02$.

6. B (Read page 237) $\sin 48^\circ = \dfrac{AC}{AB}$, $AC = AB\sin 48^\circ \approx 6.69$.

7. B (Read page 163) Apply Pythagorean Theorem.
$18^2 + (\frac{1}{2}x)^2 = x^2$, $324 + \frac{1}{4}x^2 = x^2$, $324 = \frac{3}{4}x^2$, $3x^2 = 1296$, $x^2 = 432$, $x = 12\sqrt{3}$.

8. E (Read page 15) $\dfrac{V_1}{V_2} = \left(\dfrac{r_1}{r_2}\right)^3 = \left(\dfrac{3}{2}\right)^3 = \dfrac{27}{8}$.

9. D (Read page 173) $360^\circ \times \dfrac{5}{12} = 150^\circ$.

10. B There are 30 pairs of (60+1). $(60+1) \times 30 = 61 \times 30 = 1830$.

11. D (Read page 163) Apply Pythagorean Theorem.

$$(x-1)^2 + (x+1)^2 = (5\sqrt{2})^2$$
$$x^2 - \cancel{2x} + 1 + x^2 + \cancel{2x} + 1 = 50$$
$$2x^2 + 2 = 50$$
$$2x^2 = 48$$
$$x^2 = 24,\quad x = \sqrt{24} = 2\sqrt{6}.\quad BC = x + 1 = 2\sqrt{6} + 1 \approx 5.90.$$

12. B (Read page 148) Given $y = 8 + |2 - x|$, we have $2 - x \geq 0$.
Therefore, $y \geq 8$.

13. C (Read page 33) $(2.52)^a = (4.26)^b$
$\log(2.52)^a = \log(4.26)^b$
$a\log 2.52 = b\log 4.26$, we have : $\dfrac{b}{a} = \dfrac{\log 2.52}{\log 4.26} \approx 0.64$.

14. B (Read page 69) Given $f(x) = 4x - 12$ and $f(3x) = -48$.
We have : $f(3x) = 4(3x) - 12 = -48$
$12x - 12 = -48$, $12x = -36$, $x = -3$.

15. E (Read page 237) $\cos 35^\circ = \dfrac{AB}{AE} = \dfrac{AB}{4}$, $AB = 4\cos 35^\circ$.
Area of the square : $A = (AB)^2 = (4\cos 35^\circ)^2 \approx 10.74$.

Answer Explanations
(PRACTICE TEST 8)
SAT Subject Test Math Level I
Page 399 ~ 409

16. D (Read page 170) The sum of the degree measures of the exterior angles of a polygon with n sides is 360 .

The sides of a regular polygon with the degree measure of each exterior angle of 12 are :

$$n = \frac{360^\circ}{12^\circ} = 30 \text{ sides.}$$

17. A (Read page 105) Given $f(x) = \ln x$, we have $y = \ln x$.

Interchange x and y, we have $x = \ln y$, $y = e^x$ is the inverse function.

$f^{-1}(x) = e^x$, $f^{-1}(-2) = e^{-2} \approx 0.14$.

18. B (Read page 85) $10x^2 - 35x - 20 = 5(2x^2 - 7x - 4) = 5(2x+1)(x-4)$.

19. B (Read page 110) $\frac{x^2-4x-5}{x} \div \frac{x-5}{x^2+x} = \frac{x^2-4x-5}{x} \cdot \frac{x^2+x}{x-5} = \frac{(x-5)(x+1)}{x} \cdot \frac{x(x+1)}{x-5} = (x+1)^2$.

20. A $\frac{x+y}{2} = \frac{x-y}{3}$, $3x+3y = 2x-2y$, $x = -5y$, $\frac{x}{y} = -5$.

21. C (Read page 42 and 293) $p(A) = \frac{4}{52}$, $p(B) = \frac{13}{52}$, $p(A \cap B) = \frac{1}{52}$.

$p(A \cup B) = p(A) + p(B) - p(A \cap B) = \frac{4}{52} + \frac{13}{52} - \frac{1}{52} = \frac{16}{52} = \frac{4}{13}$.

22. E (Read page 95) If $ax^2 + bx + c = 0$,

then the roots $r_1 + r_2 = -\frac{b}{a} = \frac{2}{3}$ and $r_1 \cdot r_2 = \frac{c}{a} = -\frac{4}{3}$.

We have : $a = 3$, $b = -2$, $c = -4$, the equation is $3x^2 - 2x - 4 = 0$.

23. C (Read page 30) $\frac{(a^7)^2 \times a^8}{a^2} = \frac{a^{14} \times a^8}{a^2} = \frac{a^{22}}{a^2} = a^{20}$.

24. D (Read page 86)
The graph is a parabola with vertex $(-1, -1)$, x-intercepts 0 and -2, and open upward.
Find the x-intercepts on each equation by letting $y = 0$.
$y = x^2 + 2x$, $x^2 + 2x = 0$, $x(x+2) = 0$, $x = 0, -2$ (intercepts), open upward.
Hint: (E) $y = -x^2 - 2x$ has x-intercepts 0 and -2, open downward.)

25. B (Read page 105) Given $f(x) = 2x^3 - 1$, we have $y = 2x^3 - 1$.

Interchange x and y, we have : $x = 2y^3 - 1$, $2y^3 = x + 1$, $y^3 = \frac{x+1}{2}$,

$y = \sqrt[3]{\frac{x+1}{2}}$ is the inverse function. $f^{-1}(x) = \sqrt[3]{\frac{x+1}{2}}$, $f^{-1}(3) = \sqrt[3]{\frac{3+1}{2}} = \sqrt[3]{2} \approx 1.26$.

26. D (Read page 86) Given the equation $9x^2 + 30x + k = 0$, $a = 9$, $b = 30$, $c = k$.
If $b^2 - 4ac < 0$, the equation has no real-number solution.
$b^2 - 4ac = 30^2 - 4 \cdot 9 \cdot k < 0$, $900 < 36k$, $k > \frac{900}{36}$, $k > 25$. The minimum integer of k is 26.

Answer Explanations
(PRACTICE TEST 8)
SAT Subject Test Math Level I
Page 399 ~ 409

27. E (Read page 33) $\log(\sin x)+\log(\csc x)=\log(\sin x\cdot\csc x)=\log\left(\sin x\cdot\frac{1}{\sin x}\right)=\log(1)=0$.

28. A (Read page 239) The height : $h=10\sin 42^\circ$.
The area of the triangle : $A=\frac{1}{2}(base)(height)=\frac{1}{2}(14)(10\sin 42^\circ)=70\sin 42^\circ\approx 46.84$.

29. B (Read page 118) Given y varies inversely as x, we have $y=\frac{k}{x}$, k is a constant.

Given $y=10$ when $x=2$, we have $10=\frac{k}{2}$, $k=20$. $y=\frac{20}{x}$ is the equation.

Therefore, when $x=12$, we have $y=\frac{20}{x}=\frac{20}{12}=\frac{5}{3}$.

30. E (Read page 118)
Given z varies directly as x and inversely as y, we have $z=k\frac{x}{y}$, k is a constant.

Given $z=9$ when $x=15$ and $y=3$, we have $9=k\frac{15}{3}$, $k=\frac{9}{5}$, $z=\frac{9}{5}\left(\frac{x}{y}\right)$ is the equation.

Therefore, when $x=45$ and $y=2$, we have $y=\frac{9}{5}\cdot\frac{45}{2}=\frac{81}{2}=40.5$.

31. C $1+\frac{1}{1+\frac{1}{2}}=1+\frac{1}{3/2}=1+\frac{2}{3}=1\frac{2}{3}$.

32. A (Read page 145) $h=50-4.9x^2$. When the rock strike the ground, the height $h=0$.
$0=50-4.9x^2$, $4.9x^2=50$, $x=\sqrt{\frac{50}{4.9}}\approx 3.19$ seconds.

33. D (Read page 241) $\frac{\tan\theta}{\sec\theta}(\cot\theta)=\frac{\tan\theta}{\sec\theta}\cdot\frac{1}{\tan\theta}=\frac{1}{\sec\theta}=\cos\theta$.

34. B (Read page 159, Third-Side Rule) $12-4<x<12+4$, $8<x<16$.

35. D (Read page 148) $f(x)=\frac{2}{x\sqrt{x+5}}$, we have $x\neq 0$, and $x+5>0$, $x>-5$.

36. E (Read page 33) $\log_3 9^5=\log_3(3^2)^5=\log_3 3^{10}=10\log_3 3=10(1)=10$.

37. A (Read page 167) If ΔABC and ΔCDE are similar, the ratio of the lengths of their corresponding sides is a constant (the ratios of their corresponding sides are proportional).

We have : $\frac{x}{1.8}=\frac{y}{2.3}$, $\frac{x}{y}=\frac{1.8}{2.3}\approx 0.78$.

38. C (Read page 30) $\frac{x^{2/3}}{x^{-1}}=x^{2/3-(-1)}=x^{2/3+1}=x^{5/3}$.

39. D $(3+4i)(3-4i)=3^2-(4i)^2=9-16i^2=9-16(-1)=9+16=25$.

40. D (Read page 41) ${}_{10}P_6=\frac{10!}{(10-6)!}=\frac{10!}{4!}=\frac{10\cdot 9\cdot 8\cdot 7\cdot 6\cdot 5\cdot 4!}{4!}=151{,}200$.

Answer Explanations
(PRACTICE TEST 8)
SAT Subject Test Math Level I
Page 399 ~ 409

41. E (Read page 273) Given $\sin a^\circ = 0.573$, we have $a^\circ = \sin^{-1}(0.573) \approx 34.960^\circ$.
Therefore, $\sin(a^\circ - 13^\circ) = \sin(34.960^\circ - 13^\circ) \approx 0.374$.

42. D (Read page 241) $m^2 + n^2 = \left(\frac{1+\tan x}{\sec x}\right)^2 + \left(\frac{1-\tan x}{\sec x}\right)^2 = \frac{(1+\tan x)^2 + (1-\tan x)^2}{\sec^2 x}$
$= \frac{1 + 2\tan x + \tan^2 x + 1 - 2\tan x + \tan^2 x}{\sec^2 x} = \frac{2 + 2\tan^2 x}{\sec^2 x} = \frac{2(1+\tan^2 x)}{\sec^2 x} = \frac{2\sec^2 x}{\sec^2 x} = 2.$

43. B (Read page 37) A, D, and E could be true because they are not If-then statements.
B and C are If-then statements. B contradicts the rule of contraposition.

44. B (Read page 148) Given $f(x) = \sqrt{9x^2 - 4}$, we have $9x^2 - 4 \ge 0$. Therefore $f(x) \ge 0$.

45. C (Read page 107) Given $g(x) = x - 2$ and $h(g(x)) = x - 1$, we have
$h(g(x)) = x - 1 = (x-2) + 1 = g(x) + 1$. Therefore $h(x) = x + 1$.
Or: Test each choice. (C) $h(x) = x + 1$, $h(g(x)) = (x-2) + 1 = x - 1$.

46. E (Read page 35) $a_n = a_1 + (n-1) \cdot d = 3 + (n-1) \cdot 5 = 3 + 5n - 5 = 5n - 2$.

47. C The box has the length $(x - 8)$, the width $(x - 5 - 8)$, and the height 4. Volume = 336.
We have the equation : $336 = (x-8)(x-13)(4)$
$84 = (x-8)(x-13)$
$84 = x^2 - 21x + 104$
$x^2 - 21x + 20 = 0$, $(x-20)(x-1) = 0$, $x = 20$.
Therefore, the original cardboard has the length 20 and width 15.

48. C (Read page 171) $< AOB = \frac{360^\circ}{6} = 60^\circ$. ΔAOC is a 30-60-90 triangle.
$OC = \sqrt{3}(AC)$, $AC = \frac{OC}{\sqrt{3}} = \frac{12\sqrt{3}}{\sqrt{3}} = 12$, $AB = 24$. The perimeter $p = 24 \times 6 = 144$.

49. A (Read page 33) $\log_7 12 = \frac{\log 12}{\log 7} \approx 1.28$.

50. D (Read page 298) Find the least-squares linear regression by a graphing calculator.
Method 1 : Using the formula on page 297 and example on page 308.
Method 2 : Using a graphing calculator.

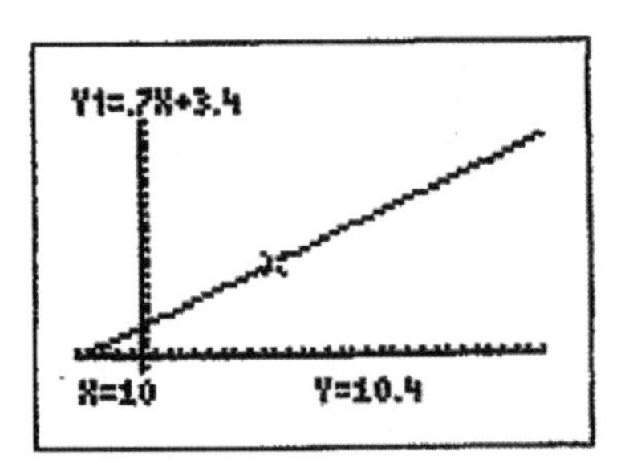

INDEX

A

B

C

D

E

INDEX

F

G

H

I

J

L

M

INDEX

N

O

P

Q

R

INDEX

S

T

U

V

W

Z

Notes

The A-Plus Notes Series:

1. **A-Plus Notes for Beginning Algebra (Pre-Algebra and Algebra 1)**
2. **A-Plus Notes for Algebra (Algebra 2 and Pre-Calculus)**
   (A Graphing Calculator Approach)
3. **A-Plus Notes for SAT Math (SAT/PSAT and Subject Tests)**
   (Covers Geometry and Trigonometry)
4. **A-Plus Notes for GRE/GMAT Math (General and Subject Tests)**
5. **A-Plus Notes for ACT Math (Coming Soon)**
6. **A-Plus Notes for Calculus (Coming Soon)**

If you are an educator and want to write your own book (all subjects) to improve the education in schools, please contact us.

" A person's life has an end.
Honor or dishonor does not last long.
Only creative writing is eternal. "

Confucius

" 生命有其所終，
榮辱止於其身，
未若文章之無窮。"

孔夫子

Simply complete and mail the order form (or a copy of the form).

A-PLUS NOTES FOR SAT MATH

ORDER FORM

Qty	Paperback	Hardcover	Quantity	$ Amount
Regular	$24.50	$34.50		
10 to 50	23.50	33.50		
51 to 100	22.50	32.50		
101 and up	21.50	31.50		

Prices are subject to change without notice.
Please call (310) 542-0029.

Subtotal ________

$2.50 per copy for S/H & Tax ________

Total ________

Ship to:(PLEASE PRINT)
Name ______________________________
School name ______________________________
Address ______________________________
City/State/Zip ______________________________
Daytime Phone ______________________________

Make check payable to **A-Plus Notes Learning Center**.
Mail with form to:

A-Plus Notes Learning Center
2211 Graham Ave.
Redondo Beach, CA 90278

Allow 2 to 3 weeks for delivery. Thank you.

Notes